Fostering Self-Regulated Learning through ICT

Giulana Dettori
Institute for Educational Technology - National Research Council (CNR), Italy

Donatella Persico
Institute for Educational Technology - National Research Council (CNR), Italy

Information Science REFERENCE

INFORMATION SCIENCE REFERENCE

Hershey · New York

Director of Editorial Content: Kristin Klinger
Director of Book Publications: Julia Mosemann
Acquisitions Editor: Lindsay Johnston
Development Editor: Michael Killian
Publishing Assistant: Milan Vracarich, Jr.
Typesetter: Milan Vracarich, Jr.
Production Editor: Jamie Snavely
Cover Design: Lisa Tosheff

Published in the United States of America by
Information Science Reference (an imprint of IGI Global)
701 E. Chocolate Avenue
Hershey PA 17033
Tel: 717-533-8845
Fax: 717-533-8661
E-mail: cust@igi-global.com
Web site: http://www.igi-global.com

Library of Congress Cataloging-in-Publication Data

Fostering self-regulated learning through ICT / Giuliana Dettori and Donatella Persico, editors.
p. cm.
Includes bibliographical references and index.
Summary: "This book presents the relationship between SRL and ICT from several standpoints, addressing both theoretical and applicative issues, providing examples from a range of disciplinary fields and educational settings"--Provided by publisher.
ISBN 978-1-61692-901-5 (hardcover) -- ISBN 978-1-61692-903-9 (ebook) 1. Educational technology. 2. Mobile communication systems in education. 3. Instructional systems--Design. I. Dettori, Giuliana, 1955- II. Persico, Donatella, 1957-
LB1028.3.F675 2011
371.33--dc22

2010016309

British Cataloguing in Publication Data
A Cataloguing in Publication record for this book is available from the British Library.

List of Reviewers

Rylan Egan, *Simon Fraser University, Canada*
Cath Ellis, *University of Huddersfield, UK*
Filomena Faiella, *University of Salerno, Italy*
Sue Folley, *University of Huddersfield, UK*
Paola Forcheri, *IMATI-CNR, Italy*
Elizabeth Guerin, *University of Florence, Italy*
M. Carmen González Torres, *University of Navarra, Spain*
Brice R Harris, *Western Illinois University, USA*
Maria Grazia Ierardi, *IMATI-CNR, Italy*
Bracha Kramarski, *Bar-Ilan University, Israel*
David Kumrow, *California State University, USA*
M. Alessandra Mariotti, *University of Siena, Italy*
Mark McMahon, *Edith Cowan University, Australia*
Mariachiara Pettenati, *University of Florence, Italy*
Antje Proske, *TU Dresden, Germany*
Pedro Rosário, *University of Minho, Portugal*
M. Luisa Sanz de Acedo Lizarraga, *Public University of Navarre, Spain*
Karl Steffens, *University of Cologne, Germany*
Vighnarajah, *Universiti Putra Malaysia, Malaysia*

Table of Contents

Detailed Table of Contents

Recent research suggests that technology-enhanced learning environments (TELEs) represent an opportunity for students to build their ability to self-regulate. This chapter reviews 55 empirical studies and interprets their findings to answers the following questions: (1) What is the theoretical basis for understanding the possible relations among SRL and TELEs? (2) What types of TELE have been used to study these relations? (3) When participants engage in self-regulatory behaviours in a well-designed TELE, do they show greater learning than their peers who engage in fewer self-regulatory behaviours? (4) How have TELEs been shown to promote self-regulatory tendencies in learners? and (5) How do pre-existing self-regulated learning tendencies influence the ways in which learners interact with technology enhanced learning environments? This review suggests that TELEs can promote SRL and are best used by those who can self-regulate learning. SRL training should occur before the task, or be embedded in the TELE.

This chapter examines current literature concerning the measurement of online SRL behaviours and the application of this online SRL measurement with regard to profiling SRL behaviours in TELEs. The

This chapter challenges the traditional differentiation between metacognitive monitoring and control in text-based self-regulated learning (SRL). Building on Pieshl (2009), the authors present a case for conceptualizing and measuring calibration as the interaction between metacognitive monitoring and control under the assumption that learners adjust metacognitive judgments as they monitor and control their learning both within and between trials. To this end, three separate but related measures of calibration are described – assessment, internal, and strategic calibration – to address questions such as what kind of test will be given; how will the user perform on such a test; and what can he/she do to improve performance. Each type of calibration is mutually exclusive; overall calibration accuracy, however, relies on the hierarchical interplay among all three types. Finally, examples are provided of how trace data for each type of calibration may be collected in a multimedia-learning environment.

Chapter 6
Using Student Assessment Choice and e-Assessment to Achieve Self-Regulated Learning 89
Cath Ellis, University of Huddersfield, UK
Sue Folley, University of Huddersfield, UK

This chapter explores how we can harness technology to foster self regulated learning in assessment practices. Innovation in assessment traditionally lags behind that in other areas of teaching and learning. It is important, however, to make sure that assessment methods and practices are aligned with the learning objectives. For assessment to be a beneficial learning experience for students it is important that they are afforded more autonomy and agency over what, when and how they are assessed. This chapter reflects on the 'problem' that assessment and feedback presents and on what the research is showing academics need to concentrate on. Secondly it considers how eAssessment tools can provide the way forward to achieving these objectives and helping students to develop more self-regulated learning strategies. Finally we will explore how the use of these tools can allow students greater autonomy over the whole assessment process, and the essential role that technology played in achieving this.

Chapter 7
The Role of SRL and TELEs in Distance Education: Narrowing the Gap 105
Maureen Snow Andrade, Utah Valley University, USA
Ellen L. Bunker, Brigham Young University Hawaii, USA

Self-regulated learning (SRL), defined as learners taking responsibility for their own learning is a critical component for success in distance education. Distance education contexts, typically TELEs (Technology Enhanced Learning Environments), also have the potential to foster SRL. This chapter focuses on the importance of SRL in distance education, specifically in higher education and lifelong learning contexts, and how SRL can mediate the gap between the learner and instructor and decrease the distance that may be created by Information and Communication Technology (ICT). The chapter reviews the use of ICT in distance education, explicates key terms related to SRL, presents a model for course design, and illustrates how behaviours of key stakeholders can support development of SRL.

The primary purpose of this chapter is to present techniques and strategies that can be incorporated in online courses to promote students' use of self-regulated learning strategies. In addition, the authors discuss why self-regulated learning skills are particularly critical in online learning environments, present a model of self-regulated learning, discuss issues related to measuring self-regulated learning, address the issue of whether or not self-regulated learning can be taught, and discuss why online learning environments are ideal to scaffold self-regulation. The authors present several strategies and techniques they have found successful for promoting self-regulated learning that can be readily incorporated and implemented in online courses. The chapter concludes with a scenario that represents an idealized model of how to promote self-regulated learning in an online learning environment by employing an intelligent tutoring component as a tool to support students' use and development of self-regulated learning tactics and strategies.

This chapter analyses SRL in a virtual community interacting through asynchronous textual communication, consising of the trainee teachers of a post-graduate blended course in Educational Technology. The study aims to compare SRL practice in different types of collaborative activities carried out online. The investigation method is based on interaction analysis, an approach allowing a systematic study of the content of the messages exchanged by the community members. The outcomes of such analysis consist of quantitative data on SRL-related events that took place during the learning process, which allows the comparison of activities according to the degree and type of self-regulation displayed by the learners. The results of the study suggest that the nature of the task influences the way students self-regulate. The difference, however, does not lie in the total amount of detected SRL indicators but in their type, therefore suggesting that different types tasks might induce different kinds of SRL actions.

The aim of this chapter is to provide an outline of the main theoretical issues in the field of Self-Regulated Learning which have inspired the design and implementation of a blended learning course of English as a Foreign Language (EFL) at the University of Salerno (Italy). In particular, the first part of

Due to the specific nature of the subject matter and the learners, the traditional teaching methods and pedagogical constructs may not be suitable. Based on a blended learning model, E-Health teaching at the Centre for Online Health University of Queensland, Australia has shown its capacity to provide a unique learning experience to students. While designing e-Health curriculum, a particular attention has been paid to aspects such as flexibility of learning processes, students' control in learning, self observation and self evaluation. These are, in fact, core principles of self regulated learning (SRL) that have been incorporated in the teaching and learning process of e-Health. This chapter examines in details the elements of SRL embedded in e-Health teaching and the role of SRL in maximizing the learning outcomes.

Chapter 22
Informal Self-Regulated Learning in Corporate Organizations.. 364
Wim Veen, Delft University of Technology, The Netherlands
Jan-Paul van Staalduinen, Delft University of Technology, The Netherlands
Thieme Hennis, Delft University of Technology, The Netherlands

Sharing knowledge is one of the most challenging tasks modern companies have to deal with. A vast amount of knowledge exists within organizations; however it is often difficult to find and to judge its value. As a consequence, learning and knowledge building seem to be a lonely activity, separated from everyday work. Transfer of knowledge acquired in formal courses has little impact and effect on day-to-day work. Knowledge management systems have also proven to be ineffective as they fail in presenting the knowledge employees are looking for. So how can we improve learning in organizations using ICT? To find an answer to this question we might learn from the generation that has grown up with modern communication technologies. This Homo Zappiens shows us that we can increasingly rely on technology to connect and get organized as a group. In a networked society, innovation and knowledge reside in a network, rather than in each separate individual. This chapter describes self-regulated learning within a network (Networked Learning) and presents a model for it.

Chapter 23
Face-to-Face and Web-Forum Interventions Promoting SRL Skills at University............................ 380
Barbara De Marco, University of Milan Bicocca, Italy
Nicoletta Businaro, University of Milan Bicocca, Italy
Eleonora Farina, University of Milan Bicocca, Italy,
Ottavia Albanese, University of Milan Bicocca, Italy

Based on recent findings about SRL, this chapter outlines three educational interventions aimed at fostering students' learning competence. Its particular focus is on the interaction between collaborative learning in Technology Enhanced Learning contexts and the development of SRL competencies. Two interventions are described, involving collaborative activities conducted face-to-face and in web-based learning environments, aimed at promoting the SRL skills of first year university students. Based on the outcomes of these two projects, a further project for different departments was undertaken. This last intervention was designed to facilitate collaborative reflection on the components and processes of SRL through e-tivities and discussion forums. This study suggests that collaboration in analyzing and

Preface

In order to understand the importance of investigating the relationship between Self-Regulated Learning (SRL) and Information and Communication Technology (ICT), it is worth looking at the picture of pedagogical developments in the past century. The need to move away from a vision of learning as a transmissive process, where the teacher plays the role of "sage on the stage", has been widely acknowledged. Both theoretical and applied research indicate educational paradigms like cognitivism, constructivism, and their social versions as stimulating and effective approaches, not only as concerns learners' development of content-related competence, but also regarding their overall cognitive and personal growth. According to these views, the role of teachers is to promote student-centered learning by designing learning environments which encourage motivation, self-efficacy and metacognitive awareness. On the other hand, learners are expected to become more active, reflective and responsible for their own learning, to different degrees and in different ways, according to their needs and potential, as well as the nature of learning objectives and content. The role of technology, in this process, can be of primary importance, because computers empower students by acting like amplifiers of their cognitive, social and creative abilities. As Marc Prensky[i] straightforwardly puts it, "technology's role – and its only role – should be to support students teaching themselves (with, of course, their teachers' guidance)".

Despite many years of investigation and field experiments, including the implementation of governments' polices addressing the introduction of ICT in schools, an effective combination of the above mentioned learning theories and the integration of ICT in schools is still far from being widely practiced, probably because it entails a compound and significant effort to create "student-centered", "problem-based", "personalisable" learning environments and to integrate their use in educational settings. In addition, to take advantage of such environments students need to become active and responsible in their approach to learning. Fostering learners' self-regulation is therefore a necessary step to actually implement promising pedagogical approaches and improve learning.

The pervasiveness of technology in all aspects of our life makes SRL even more necessary in view of a world where learning can no longer take place once and for all, and lifelong learning appears to be the only way to cope with a fast evolving society. Technology, however, also makes SRL skills more difficult to achieve, in that it adds extra variables that must be controlled at the cognitive, metacognitive, motivational and emotional levels. For example, having at one's disposal a variety of expressive media and information sources entails the ability to make choices that require a high degree of metacognitive awareness, while handling cognitive overload or dealing with the sense of isolation in distance learning entail a good deal of control on the motivational and emotional level. The need for the development of SRL skills is not limited to formal learning contexts, but is a real, compelling necessity also to exploit the opportunities, and cope with the challenges, of informal learning contexts, as well as of work and life in the knowledge society. At the same time, technology-rich environments put their users quite naturally in an active position, therefore providing a suitable environment for practicing these skills. In conclusion, technology-rich environments both require and foster SRL, which makes the relationship between SRL and ICT quite complex.

This book deals with this relationship from several standpoints, addressing both theoretical and applicative issues, providing examples from a range of disciplinary fields and educational settings. It includes 24 chapters by a total of 57 authors from 28 different institutions in 13 countries. All together, they give a wide-angle view on the relationship between SRL and ICT, well representing the current state of the art. Moreover, they help to deepen an understanding of the nature of SRL, spotting a variety of relevant aspects, as well as of possible approaches to its study.

The book starts with a review that analyses the theoretical basis for understanding the possible relations among SRL and Technology-Enhanced Learning Environments (TELEs). Bernacki, Aguilar and Byrnes (Chapter 1) examine 55 empirical studies and interpret their findings to draw a picture of the current support to SRL in TELEs from the point of view of opportunities offered and learners' propensity to take advantage of them. These authors focus on several aspects, such as types of TELEs considered in the literature, learning entailed and influence of personal SRL tendencies. The picture that arises from this review substantiates the claim that TELEs can promote SRL but are also best used by self-regulated learners.

Among the aspects highlighted in this review, the importance of developing approaches to analyze learners' SRL attitude, as well as to evaluate or measure their SRL competence stands out as a critical one, because achievements in this field are a necessary condition to develop and test any SRL-improving approach. This is the focus of the following four chapters.

Barnard-Brak, Lan and Paton (Chapter 2) tackle the issue of measuring SRL behaviors in online learning environments. They analyze the problems entailed by several ways of measuring SRL currently in use, and propose to overcome them by profiling SRL behaviors. The five distinct SRL profiles that arise from their analysis are discussed in a social cognitive perspective and are related to metacognitive factors and academic achievement.

Assessing SRL behavior is the focus of De la Fuente and Lozano (Chapter 3), who describe and discuss an ICT-based assessment tool addressed to young children, designed according to a psychological model of SRL as well as a model for the design of Computer Assisted Assessment. They also point out the need for any kind of assessment tool to comply with ethical standards; these encompass elements such as competency, interpretation and use of computer-produced reports, characteristics of the person to be evaluated, confidentiality, as well as equivalence of paper-and-pencil and computer-supported versions of a same test. The assessment tool they propose appears to satisfy all these requirements.

Colombo and Antonietti (Chapter 4) relate SRL strategies and cognitive styles in multimedia learning. They designed an experiment to investigate the interplay between a number of variables: topic's perceived complexity, students' learning strategies in multimedia and related metacognitive awareness and students' cognitive styles. Learning outcomes were assessed based on mere retention and on two types of problem solving tasks. The study suggests that participants tended to self-regulate their strategies according to topic complexity and that cognitive styles play a minor role in self-regulation but seem to influence metacognitive awareness of the strategies applied.

Egan and Zhou (Chapter 5) deal with learners' ability to correctly predict their performance in assessment tests. They propose the Trichotomous Calibration Model, which includes three forms of calibration: assessment calibration (i.e. learner's ability to predict the main features of an upcoming assessment), strategic calibration (i.e. the match between perceived task difficulties and strategies chosen to tackle it) and internal calibration (i.e. the accuracy of learners' judgment of their achievement on a future assessment). Trichotomous Calibration can be measured by means of nStudy, a software tool that keeps track of students' choices while learning, without interfering with the learning process. This model sheds light on how students can improve their assessment predictions, so as to adjust their learning strategies accordingly. This is a nice example of how measuring some regulation aspects can influence SRL improvement.

Enhancing learners' self-regulation in TELEs is the unifying aim of all the other chapters, which show a wide variety of approaches and points of view: some analyze issues that are of concern in a variety of learning contexts, others focus on particular aspects in disciplinary fields; some make use of general-purpose technological tools and propose methods and strategies to create SRL-supportive learning environments, others make use of specific software tools which present favorable features or were expressly designed to foster SRL; some analyze experiences carried out within single courses, others tackle SRL improvement by acting on the organization of a whole school, institution or enterprise.

Let us start with four studies tackling relevant issues of broad applicability.

Pertinent to any academic level and subject, and regarding many kinds of TELEs, is the contribution of Ellis and Folley (Chapter 6), who point out that a focus on SRL in education should lead to a radical transformation of learning assessment approaches. They observe that learners are not fully in the position to take advantage of teachers' encouragements and support to self-regulate if learning assessment is not in line with learning objectives and strategies. In order to actually foster SRL, assessment should be redesigned, giving space to learners' choice as concerns a number of aspects: format, subject, criteria, timing and results. Technology offers a wide range of tools to support the realization of all these choices. Their proposal indeed appears as the ultimate step to take in order to make learners take responsibility for their own learning and growth.

Improving SRL in distance education, in particular in higher education and lifelong learning contexts, is the focus of Andrade and Bunker (Chapter 7), who discuss how SRL can help to narrow the gap between learner and instructor in online learning. Their aim is to identify how distance education contexts can support the development of SRL through course design, instructor feedback and institutional support. After reviewing the use of ICT in distance education, they present a model for course design aiming to help develop SRL in distance learners, and illustrate it with applications for key stakeholders.

Harris, Lindner and Piña (Chapter 8) concentrate on techniques that can be incorporated in online courses to promote student's use of self-regulated learning strategies. After discussing related issues, they present a number of strategies and techniques that appear successful for promoting SRL and can easily be incorporated in online courses. Their proposal is exemplified by means of a scenario where an intelligent tutoring component is used to support students' development of SRL strategies.

Delfino, Dettori and Persico (Chapter 9) discuss the possible influence of task nature on learners' self-regulation, analyzing the case of an online course based on a socio-constructivist approach. The aim of the study is to inform the design of online collaborative learning activities supportive of self-regulation. To this end, they apply interaction analysis to learners' messages in the discussion forums of four different tasks, looking for indicators of self-regulated learning behaviors. The outcomes of their study show that task influence does not concern the total amount of SRL indicators found in the messages, but rather their type. In addition, the way tasks are proposed and scaffolded appears to have an influence on how students self-regulate.

The following six chapters concentrate on specific disciplinary fields where self-regulation competence appears useful to help learners overcome what are usually considered intrinsic difficulties of the subjects. All of them use wide-application technology and give rise to SRL-supportive learning environments thanks to suitable set ups and methodologies.

Calabrese and Faiella (Chapter 10) deal with issues related to the design of online activities that effectively support SRL in language learning. Based on the most recent theoretical and methodological approaches to language learning, they argue that the very nature of this subject requires learners to be self-regulated in order to obtain any result. Hence, it is particularly important for language courses to be designed and structured so as to favor SRL. This becomes even more important for online courses, as they require a good degree of autonomy of the students. They point out the main aspects to be taken care of in online language course design, and illustrate their proposal with the example of on online English course at university level.

The online module of a blended English course for higher education is also the focus of Hirata (Chapter 11), who seeks to improve students' autonomy so as to help them exploit the wealth of language-related resources available on the Internet. This issue is particularly critical in Japan because traditional cultural factors discourage learners to act in a self-directed or self-regulated way. She experimented two different approaches to web-based, self-directed language learning (i.e. data-driven language activities and website critical evaluation). The outcomes of both experiences were positive, and showed that students' own planning, monitoring and evaluation helped them to take advantage of, and appreciate, the two proposed activities, as well as to develop a positive and responsible attitude with regards to learning.

Mathematics problem solving is instead the concern of Tung and Chin (Chapter 12), who use video as a feedback tool to stimulate learners' self-regulation, within an Activity System Theory perspective. Students are asked to reason aloud and are videotaped while solving mathematical problems. The video becomes an opportunity for self-observation, helping the students to reflect on their own problem solving behaviors. This helps them to understand the reasons for failures and figure out possible ways to overcome them, thanks also to the assistance of a tutor who prompts them with suitable questions when necessary. The application of this approach with secondary school students produced positive outcomes and led the authors to draw a set of guidelines to assist educators in the development of similar learning activities.

Helping learners to overcome failure in mathematics problem solving by improving their self-regulation is also the aim of Mariotti and Maffei (Chapter 13), who worked out an approach to effectively use the feedback provided by a computer-algebra system in remedial activities. Their detailed analysis of the case of some high school students highlights the effectiveness of their proposal and how improvement in algebraic problem solving took place in parallel with the acquisition of SRL competence.

Yet another way to use feedback as a support to self-regulation in mathematics problem solving is proposed by Kramarski (Chapter 14). Here students work collaboratively in virtual communities, sharing problems and solutions and explaining their thinking and solution approaches. Critically examining each other's reasoning, with the support of metacognitive prompts, leads the learners to become aware of and monitor their own thinking, which has a positive influence on mathematical reasoning. The author describes an experience in which two different metacognitive teaching methods were applied, showing that these had different influences on the outcomes of the learning process.

SRL prompts are used by Olakanmi, Blake and Scanlon (Chapter 15) to improve the academic performance of high school students learning science in a computer-based simulation environment. The outcomes show that such prompts effectively supported the learning process, helping the students to take control of their activity through critical thinking, and to obtain better academic achievements than a control group working in the same environment without SRL-prompts. This study aims to provide a platform to help understand how a teaching approach based on SRL-prompts is applicable and effective in different kinds of TELEs on different topics.

While the previous group of studies rely on methodological approaches to create supportive environments with wide-application technological tools, the following four chapters take advantage of specific software tools including features which appear particularly supportive of SRL.

Vighnarajah, Wong and Abu Bakar (Chapter 16) describe how a group of high school students exercised self-regulation in a blended course in physics, thanks to the use of an online environment especially designed to enhance the practice of SRL. Measurements of students' SRL competence with the MSLQ questionnaire and statistical analysis of the gathered data testify that this online environment actually proved to be a valid support to SRL.

An environment developed to promote self-monitoring and regulation is used by McMahon (Chapter 17) to help university students improve their capacity to read critically, apply reading concepts to solve problems and develop higher order conceptual understanding. This environment provides a range of tools, such as annotations and discussion facilities, to assist learners in monitoring their reading comprehension. Application of this environment in an undergraduate class highlighted its actual usefulness to support individuals who lack effective strategies of reading comprehension.

Two online tools developed specifically to stimulate student's self-regulation are described and analyzed by Sanz de Acedo Lizarraga, Ardaiz and Sanz de Acedo Baquedano (Chapter 18). One of these tools is a wiki designed to support group regulation during the collaborative generation, analysis and assessment of ideas, while the other tool supports the creation of small working groups within large classes by helping to detect affinities among ideas individuals have proposed, by supporting goal setting as well as generation and evaluation of ideas. A pilot study carried out with university students working on tasks requiring creativity and innovation encouraged the application of these tools to support group self-regulation in creative tasks.

Proske, Narciss and Körndle (Chapter 19) focus on a platform designed to scaffold students' self regulation by means of a number of tools addressing different aspects and phases of SRL. An interesting feature of this platform is that teachers who use it to set up a learning environment can choose which tools to include, hence tailoring the learning environment on the SRL aspects they wish to support most. In the study described in this chapter, the focus is on the use of a learning plan tool by the students of a university course, which positively affected students' actively working on the achievement of their learning goals.

Is it practical that a large organization, like a university, or a specialization school, or an enterprise, leave the development of learners' self-regulation to the initiative of single individuals or single courses? A synergy within the whole organization would likely result more economical and give rise to better results in that, even though some self-regulation skills are context-dependent, many others can be advantageously applied across fields, such as the development of a self-reflective attitude, or the habit to make goal-driven plans and monitor their development. This motivates the need to foster a SRL culture at organizational level. This is the focus of the following three chapters.

Bergamin, Bettoni, Ziska and Eggs (Chapter 20) look at the relationship between SRL and ICT from the point of view of a university which aims to embed ICT in educational activities in a collaborative way, supporting and improving the quality of teaching and learning. The solution proposed is based on the concept of a "university-wide media culture", that the authors try to develop by means of a collaborative "Reference Course Model". This approach specifies principles, structures and procedures to be applied in course organization and provides a collaborative framework to encourage both the individual learners and the whole community to gradually take control of their own learning.

Edirippulige and Marasinghe (Chapter 21) focus on SRL in the context of eHealth, that is, the delivery of healthcare through ICT communication tools. Becoming active in this field requires physicians not only to learn new knowledge and skills, but also to transform their attitudes and behaviors so as to start a new way of practicing. This entails designing eHealth educational programs fostering learners' self-regulation, by promoting meta-cognition, supporting strategic action, developing ability to monitor one's own thinking and actions and sustaining motivation. In order to meet these needs, a global SRL approach has been applied so that courses contribute to encourage individual goal setting and self-monitoring, provide opportunities for self-reflection, motivate learners by means of suitable feedback, stimulate them to make important decisions about their learning process and release them from highly structured, traditional assessment tasks.

Veen, van Staalduinen and Hennis (Chapter 22) propose the Networked Learning Model for developing and managing knowledge in a self-regulated way within complex working settings. The networked system Yuno, which is based on this model, was successfully tested in the real context of a large, knowledge-intensive company. The model and system incorporate various principles of SRL and are inspired by the way young generations share and develop new knowledge through digital, social networks. According to these authors, Networked Learners are self-motivated, strategic thinkers with a high degree of self-awareness. Their strength lies in their natural attitude towards networked activity, which will hopefully allow them to solve complex problems by reducing their complexity, thanks to competence sharing within the communities of practice they belong to.

The book is concluded by two chapters that reflect on the comparison of several exploratory studies, which allows their authors to draw suggestions to support different aspects of self-regulation in technology-rich learning environments.

De Marco, Businaro, Farina and Albanese (Chapter 23) explore three different ways to foster learning competence of first year university students. Their focus is on the interaction between collaborative learning in Technology Enhanced Learning contexts and the development of SRL competencies.

Pilling-Cormick (Chapter 24) reflects on the central role that control plays when learning takes place in technology-rich environments and on how self-regulated and self-directed learning are integrally related to it. Without some form of learner control, it is extremely difficult for both learners and educators to be truly successful in learning with technology. Recognizing that technology does not always allow full learner control, it becomes vital to discover ways to operate within the constraints of the environment to improve such situation. The analysis of three pilot studies developed in different contexts and with different technological tools is her starting point to share recommendations for designing instruction in TELEs with a focus on learner control.

All these chapters, together, provide a rich and compound picture of the state of the art on SRL in TELEs and the several directions in which the field is developing. Many software tools, technology-rich environments, subjects, educational levels and approaches are dealt with. Among them, online learning environments are, by far, the most often considered. This is not surprising, since this kind of TELE is increasingly used to develop the most diverse activities, not only in formal learning, but also to support non-traditional or informal learning situations. Online learning environments, however, do not constitute the only focus of this book, as a variety of other settings are also considered and most of the SRL-supportive methodological approaches proposed throughout the chapters are not strictly dependent on a particular kind of learning environment. In all cases, the studies presented witness a strong possible synergy between SRL and ICT, provided a sound methodology (e.g. metacognitive support, teaching strategies, prompts of different kinds, task structuring, scaffolding, etc.) is applied to exploit the environments' potential. University appears to be the educational context most frequently addressed. This is not surprising, again, since university represents the last formative step before entering into productive life and starting to cope autonomously with life-long-learning. Hence, at this level, it is mandatory that students consolidate their ability to self-regulate their learning. Many other age ranges, however, are also considered, from young children in Chapter 3 up to the working context in Chapter 22. Several chapters are strictly focused on improving learning in some specific subject where SRL skills appear particularly critical. Most insight that can be gained from the studies in this book, however, is not strictly dependent on a particular age range or disciplinary field and can be applied to broader contexts. All this makes this collection of scholarly papers a precious source of educational reflections of large applicability.

The logical sequence adopted to organize the book's Table of contents is not the only one possible, as a wide variety of aspects intertwine in SRL and many connections among the chapters emerge when reading their content. For instance, the radical and stimulating suggestions of Chapter 6 concerning student choice in assessment are echoed in Chapter 21, which mentions release from highly structured assessment tasks among the measures undertaken to foster participants' SRL, hence showing in a practical case that student-centered assessment is actually viable and profitable. We will leave to the readers the pleasure of discovering so many connections, in the hope that this book may help them to appreciate the considered topic and discover ever new and interesting facets of fostering SRL with ICT.

Giuliana Dettori
Institute for Educational Technology - National Research Council (CNR), Italy

Donatella Persico
Institute for Educational Technology - National Research Council (CNR), Italy

ENDNOTE

i Prensky, M. (2008). The Role of Technology in teaching and the classroom. Educational Technology, 48(6), 64.

Acknowledgment

We heartily wish to thank all the people who supported our labor along this journey of book's creation: the authors, who contributed their knowledge and wisdom, and patiently accepted to satisfy the two rounds of revisions requested; the members of the editorial advisory board and the reviewers, who generously shared their competence to improve the book's quality; and the editorial staff at IGI Global, who believed in our proposal and helpfully assisted us during its realization.

Giuliana Dettori
Institute for Educational Technology of CNR, Italy

Donatella Persico
Institute for Educational Technology of CNR, Italy

Chapter 1
Self-Regulated Learning and Technology-Enhanced Learning Environments:
An Opportunity-Propensity Analysis

Matthew L. Bernacki
Temple University, USA

Anita C. Aguilar
Temple University, USA

James P. Byrnes
Temple University, USA

ABSTRACT

Recent research suggests that technologically enhanced learning environments (TELEs) represent an opportunity for students to build their ability to self-regulate, and for some, leverage their ability to apply self-regulated learning (SRL) to acquire knowledge. This chapter reviews 55 empirical studies and interprets their findings to answer the following questions: (1) What is the theoretical basis for understanding the possible relations among SRL and TELEs? (2) What types of TELE have been used to study these relations? (3) When participants engage in SRL behaviors in a well-designed TELE, do they show greater learning than their peers who engage in fewer SRL behaviors? (4) How have TELEs been shown to promote SRL tendencies in learners? and (5) How do pre-existing SRL tendencies influence the ways in which learners interact with TELEs? Our review suggests that TELEs can promote SRL and are best used by those who can self-regulate learning. SRL training should occur before the task, or be embedded in the TELE. Knowledge acquisition in TELEs is supported by learner self-regulation and by design features that include immediate and adaptive feedback and tools which support SRL behaviors.

DOI: 10.4018/978-1-61692-901-5.ch001

INTRODUCTION

Technology-enhanced learning environments (TELEs) have become increasingly prevalent over the past 25 years. Although the growth in TELEs is due to a number of factors, the most influential include the widespread availability of relevant technologies (e.g., personal computers, wireless communication, teleconferencing, etc.; One Laptop Per Child, 2009; Pea, Wulf, Elliot & Darling, 2003), the need to serve large numbers of students who reside in locations that are far removed from brick and mortar institutions, (Sloan Consortium, 2006) increased appreciation for the fact that technology can present information and capture performance in ways that traditional instruction cannot (Mayer, 2005; Winne & Perry, 2000) and generational shifts in comfort levels with technology. In addition, the increased popularity of distance and online learning options have created opportunities for new courses that are motivating even the most reluctant faculty to offer at least some of their programs online (Waits & Lewis, 2003).

Each time a new form of TELEs emerges, it is usually promoted as holding considerable promise (Winne, 2005). However, the gap between such predictions and reality have forced many to acknowledge the role that students play in getting the most out of a TELE. Even a well-designed technology will only have its desired effects if teachers and students take advantage of what it has to offer. In what follows, we expand on this basic premise and examine the empirical evidence related to TELEs as we provide answers to the following five questions: (1) What is the theoretical basis for understanding the possible relations among SRL and TELEs? (2) What types of TELE have been used to study these relations? (3) When participants engage in SRL behaviors in a well-designed TELE, do they show greater learning than their peers who engage in fewer self-regulatory behaviors? (4) How have TELEs been shown to promote SRL tendencies in learners? and (5) How do pre-existing SRL tendencies influence the ways in which learners interact with TELEs? After providing answers to these questions in turn, we draw conclusions.

QUESTION 1: WHAT IS THE THEORETICAL BASIS FOR UNDERSTANDING THE POSSIBLE RELATIONS AMONG SRL AND TELES?

As will become evident in subsequent sections of this chapter, researchers who have examined the linkage between SRL and TELEs have implicitly or explicitly adopted a particular theoretical stance to predict and explain the behavior of their participants. Some authors have also evaluated TELEs in terms of how well these environments support SRL as defined by particular theories (e.g., Zimmerman & Tsikalas, 2005). As such, it is useful to begin our review by engaging in a brief theoretical and meta-theoretical analysis of the literature on SRL in TELEs prior to describing the findings of empirical studies. At the core of our analysis are three issues. The first is *relevance*—the (reasonable) presumption that SRL may be particularly germane to TELEs. The second is *parsimony*—the problem of multiple, partially overlapping theories in the literature, which generally explain the same phenomenon with slightly different terminology. The third is *utility* – specifically, the utility of an Opportunity-Propensity framework for understanding the relations between SRL and TELEs.

Relevance

It is important to note that environments differ in the extent to which students need to be self-regulated in order to be successful. If a learning environment is highly structured, engaging, and focused on the acquisition of a simple (non-demanding) skill or task, students need not be self-regulated in order to be successful in that environment. In contrast, self-regulation is particularly required when: (a) the environment is focused on complex, multi-step tasks in which possible solution strategies and outcomes are not known in advance (so the learner must plan and monitor performance), (b) it is easy for the learner to become distracted, lose interest, or forget the main goals of the task, (c) the task requires the use of strategies (e.g., note-taking) to overcome the processing limitations of the mind, and (d) learners must engage in helpful behaviors (e.g., planning, monitoring, strategy use, etc.) on their own, without guidance, pressure, or prompting from others. In such environments, learners who engage in SRL behaviors are far more likely to be successful than learners who do not engage in SRL behaviors. Given that many (though conceivably not all) TELEs focus on complex problems, require strategies to overcome processing limitations, and so on, it is reasonable to expect that a self-regulated learner would show higher rates of learning in a TELE than their less self-regulated peers.

Parsimony

The foregoing discussion, however, makes clear the fact that one can only know whether a TELE requires SRL (or whether it is helpful to be self-regulated in that environment) only if one has in mind a definition of what it means to be self-regulated. Unfortunately, the literature contains a number of distinct and partially overlapping theoretical models (and definitions) of self-regulation. As a result, one must answer the question, "Does this TELE require SRL?" with the answer, "It depends on the SRL theorist you have in mind." To illustrate some of the differences among specific approaches, consider the contrasting models of Zimmerman and his colleagues, on the one hand, and Boekaerts and her colleagues on the other.

Zimmerman and Tsikalas (2005) argue that self-regulation emerges in learning contexts in three cyclical phases: forethought, performance and self-reflection. During the *forethought* phase, self-regulated students engage in metacognitive processes (i.e., *task analysis, goal setting, strategic planning*) and self-motivational processes (*task interest, values, intrinsic interest, self-efficacy*). During the *performance phase,* they engage in metacognitive strategies (e.g., *self-instruction, attention focusing*), behavioral strategies, metacognitive monitoring and behavioral recording. Finally, during the *self-reflection* phase, they reflect upon and react to their performance (e.g., *causal attributions* for success; feelings of satisfaction).

Boekaerts and Niemivirta (2000), in contrast, suggest that the key theoretical construct is the notion of *appraisal.* Depending on how a task is appraised, a learner decides whether or not to even attempt it (a metacognitive belief), and also decides which way to proceed to successfully complete the task. Metacognitive beliefs, in turn, are moderated by motivational factors. If learners develop positive appraisals, they advance to a goal process which involves goal setting and action. If the appraisal is negative, learners choose instead to not complete a task and protect their ego, resources, and well-being. Depending upon the learning context, two different action patterns are generated. If it is deemed that this context is similar to one previously encountered, an automatic action pattern is followed, and the learner proceeds immediately to goal setting and carrying out an action plan. A learning context which has not been previously encountered requires additional consideration, where learners must complete the appraisal process and determine if (a) the task is within their ability to complete, (b) it represents any threat to their well-being to attempt and (c) the task is worth completing.

Although it is possible to work out points of overlap in these two accounts, it should be clear that two researchers could make very different predictions about the degree of SRL that would take place in the very same TELE, depending on which of these theories each person advocated. Because this ultimately is an untenable obstacle that limits scientific progress and the development of truly effective forms of intervention, we propose that a useful solution to this problem is to propose a consensus definition of SRL that distills and integrates the key constructs evident in various approaches. In particular, we define SRL as having the following attributes:

- **SRL is Metacognitive**, in the sense that the learner engages in effective forms of planning, organizing, task analysis, goal-setting and monitoring of progress.
- **SRL is Strategic,** in the sense that the learner utilizes effective domain-general (e.g., help-seeking, note-taking) and domain-specific strategies (e.g., reading strategies) that help them overcome processing limitations, overcome emotional distress and/or promote better comprehension and retention of material.
- **SRL is Adaptive,** in the sense that the learner adjusts appropriately to changes in circumstances and demonstrates an emotional and motivational profile that is associated with achievement (e.g., a calibrated sense of ability, self-efficacy, being concerned about the right kind of things)
- **SRL is Engaged,** in the sense that the learner is focused and remains focused on learning the material and is able to avoid being distracted.
- **SRL is Self-initiating,** in the sense that they do not need others to urge them to begin tasks, remain focused, organize themselves, use strategies and so on. They engage in self-regulatory behaviors on their own because they want to be successful and understand how these behaviors help them be more successful.

Utility

Anyone who has experience as an educator would probably view the aforementioned list of attributes of a self-regulated learner (i.e., metacognitive, strategic, adaptive, engaged and self-initiating) as representing an ideal that few students demonstrate. Researchers who study SRL within TELEs would also probably agree. Many students fail to take full advantage of even a well-designed TELE. A useful way to understand this phenomenon is to cast it within an Opportunity-Propensity (O-P) framework that has been used to successfully explain the acquisition of knowledge in other kinds of settings (e.g., Byrnes & Miller, 2007; Byrnes & Wasik, 2009). The basic premise of the O-P framework is that learners are more likely to attain high levels of achievement within a particular domain (e.g., mathematics) if two necessary conditions are met: (a) they are given genuine opportunities to enhance their skills in the domain (the opportunity condition) and (b) they are willing and able to take advantage of these opportunities (the propensity condition). When individual or group differences are observed in achievement, advocates of the O-P framework would account for this outcome by determining the extent to which the opportunity and propensity conditions had been fulfilled in individuals who performed poorly. In particular, the account suggests the utility of testing the following three hypotheses: (1) low performers were presented with fewer opportunities to learn than high performers, (2) low performers were presented with as many opportunities as high performers but the former were unable to benefit from these opportunities (e.g., due to lack of preparation) and (3) low performers were presented with as many opportunities as high performers but the former were unwilling to engage fully and benefit.

Starting with this central premise, one then considers how each of the factors proposed in the literature might relate either to opportunities to learn or to the propensity to take advantage of opportunities to learn. Byrnes and Miller (2007) define opportunities to learn as *culturally defined contexts in which an individual is presented with content to learn* (e.g., by a teacher or parent, an author, etc.) or *given opportunities to practice skills*. Thus, opportunities can occur both within school and outside of school. We argue that any variables related to exposure (e.g., coursework, content coverage, a teacher's emphasis, homework, amount of repetition, etc.) or teaching quality (e.g., use of proven techniques, communication skills, classroom management, equitable treatment of students) would fall into the domain of an opportunity factor (Opdenakker, Van Damme, De Fraine, Van Langeghem & Onghena, 2002; Pressley, Wharton-McDonald & Raphael, 2002; Tate, 1995). That is, children would be expected to show higher achievement if they are taught by a skilled teacher who treats all children fairly and equitably and if they are adequately and systematically exposed to the content required on end-of-year assessments. When applied to the current theme of the present book, learners are presented with an opportunity to learn when they find themselves in a well-designed TELE.

In contrast, propensity factors are *any factors that relate to the ability or willingness to learn content once it has been exposed or presented in particular contexts* (Byrnes & Miller, 2007). As such, factors such as domain-specific aptitude, pre-existing knowledge, motivation and self-regulation all pertain to the propensity component. That is, children would be more willing and able to take advantage of learning opportunities if they bring to these learning opportunities prerequisite skills, aptitudes, the desire to learn the content, and the spontaneous tendency to utilize effective strategies where appropriate (Byrnes, 2003; Byrnes & Miller, 2007; Carroll, 1989; Corno et al., 2002; Jones & Byrnes, 2006; Pintrich, 2000; Reynolds & Walberg, 1991; Wigfield, Byrnes, & Eccles, 2005). Thus, self-regulation is an important aspect of propensity.

Viewed in this light, learning successes and failures within TELEs can be diagnosed in a retrospective manner. For example, when students evince relatively low levels of self-regulation or low levels of learning, these disappointing results could either be due to the fact that the TELE did not represent a genuine opportunity to learn (e.g., because it was poorly designed or confusing) or due to the fact that students failed to enter the TELE with a sufficient level of prior knowledge, aptitude, motivation and self-regulation.

QUESTION 2: WHAT TYPES OF TELE HAVE BEEN USED TO EXAMINE THE RELATIONS BETWEEN SRL AND TELES?

Until this point, we have been focused primarily on the assumptions made by various theorists as they define SRL, and the ways in which we can collectively discuss them using like terms. We do so in order to explore the potential interaction between aspects of SRL and learning in TELEs.

We now turn our attention to the remaining questions posed in the introduction, and attempt to answer such inquiries by relying on the body of empirical evidence collected to date. To complete this task, we reviewed 75 empirical studies. Because many studies attempted to answer multiple questions, we first summarize common methodologies, and then treat findings as they relate to each question separately. Twenty of the studies had too many shortcomings to draw firm conclusions, so we limit our discussion to the 55 listed in Table 1.

When reviewing empirical studies, it became clear that three main types of TELEs were employed in these studies. The first TELE is a didactic learning environment that was designed to teach students how to be self-regulated, either through pre-task training, or through prompting and scaffolding of specific SRL activities. These TELEs often tutor or prompt students to self-regulate learning by encouraging planning, monitoring, strategy use or reflection (e.g. Cognitive Tutor; Aleven, McLaren, Roll & Koedinger, 2006). A second class of TELEs are less instructional, but instead are designed to facilitate students' naturally occurring self-regulation of learning (e.g. gStudy; Nesbit, Winne, Jamieson-

Table 1. Classification of learning environment structures

Didactic TELEs	Facilitative TELEs	Unenhanced Environments
Aleven et al. (2006)[1,2]	Azevedo & Cromley (2004)* [1,3]	Balcytiene (1999)[1,3]
Azevedo, Seibert, Guthrie, Cromley, Wang & Tron (2002)* [1,3]	Azevedo, Guthrie & Seibert (2004)[2]	Dabbagh & Denisar (2006) [1]
Azevedo, Cromley, Seibert, & Tron (2003)* [1,2]	Bauer & Koedinger (2006)[3]	MacGregor (1999)[1]
Azevedo, Cromley, Thomas, Seibert, & Tron, (2003)* [1,3]	Bell, (2007)[1]	
Azevedo et al. (2005)* [1]	Brush & Saye (2001)[3]	
Azevedo, Cromley & Seibert (2004)* [1]	Greene & Azevedo (2007)[1]	
Azevedo, Greene & Moos, (2007)* [2,3]	Greene et al. (2008)[1]	
Azevedo, Winters & Moos (2004)[1]	Greene & Azevedo (2009)[1]	
Azevedo et al. (2008)* [1]	Hadwin et al. (2007)[3]	
Cho (2004)* [1]	Jacobsen & Archididou (2000) [1]	
Eom & Reiser (2000)* [1,3]	Manlove, Lazonder & deJong (2009) [1]	
Gao (2003)[12]	McManus (2000)[3]	
Graessar et al. (2005) [1,2]	Moos &Azevedo (2008a)[1]	
Kauffman et al. (2004) [1]	Narciss, Proske & Koerndle (2007)[3]	
Kauffman et al. (2008)[1,2]	Nesbit et al. (2006)[3]	
Kramarski & Hirsch (2003)* [1]	Proske, Narciss & Koerndle (2007)[1,3]	
Kramarski & Gutman (2006)* [1,2]	Rui & Lui (2007)* [1]	
Kramarski, Mevarech & Liberman (2001) [1]	Wang & Lin (2007)[1,3]	
Kramarski & Mevarech (2003)[1]	Winters & Azevedo (2005)[1]	
Kramarski, Mevarech, Arami (2002)* [1,2]		
Kramarski & Ritkof (2002)* [1]		
Manlove, Lazonder & deJong (2006)* [1]		
Manlove, Lazonder & deJong (2007)* [1]		
Manlove, Lazonder & deJong (2008)* [2]		
McKendree (1990)[1,2]		
McNamara et al. (2006)[1,2]		
Moos &Azevedo (2006)[1]		
Moos & Azevedo (2008a)[1,2]		
Moos &Azevedo (2008b)* [1]		
Moos &Azevedo (2008c)* [1]		
van den Boom et al. (2004)* [1,2]		
White and Frederiksen (1998)[2]		
Yang (2006)[2]		

[1] – researched whether use of TELE affects knowledge gain
[2] – researched how TELEs can influence SRL behaviors
[3] – researched how pre-existing SRL tendencies influence TELE use
* – study makes comparisons between different types of TELEs

Noel, Code, Zhou, MacAllister, 2006). Tools to promote strategy use, monitoring and reflection are made available, but students must seek out these tools for their own use instead of being prompted or externally regulated by the environment. These designs lend themselves to exploratory methodologies which seek to identify naturally occurring trends in learners' SRL use. A third class of TELEs is simply a computerized representation of content that is not enhanced in any fashion. These are analogous to paper based learning environments which include identical content and no additional features. The main difference between these and paper-based learning environments is the presentation of content across multiple nodes. These lend themselves to comparisons of how readers navigate the text in both environments, and direct comparisons of the benefits of paper-based versus computer-based presentations.

As one might expect, the relationship between a learning environment and student's employment of SRL processes depends heavily on the nature of the learning environment itself. This will become evident in our study descriptions, as students behave differently in different types of TELEs. To illustrate the trends in utilization of different types of TELEs, we include Table 1 that groups studies by the structure of the learning environment. Each column contains studies that employ a didactic TELE (which uses external regulation and encourages SRL through tutoring, including human tutoring or by scaffolds and prompts), facilitative TELE (which supports self-regulation by providing tools but not prompting their use) and a computer-based learning environment (CBLE), which contains no enhancements. When multiple conditions are used, studies are grouped based on the condition that has been particularly foregrounded and an asterisk indicates comparison with other TELEs that serve as controls. Though argument can be made that a continuum exist in which TELEs are fully didactic through completely unenhanced, such a grouping is meant to enhance readers' understanding of the prevailing types of TELEs in use, and to draw connections between studies that employ similar procedures and arrive at similar or contrasting conclusions about how TELEs influence SRL. Further, to cue readers to the findings of each of these studies, superscripts have been applied that denote each study's content as it relates to our three research questions. These questions are answered in the next three sections of the chapter.

As can be seen from Table 1, considerable effort has been put forth to design and research TELEs which teach students to learn in self-regulated ways. Additional research has been conducted to determine how students will use TELEs and the tools which support SRL practices. The next three sections detail the findings of these studies. What we find is that different structures within a TELE, including the presence of tutors, prompts and tools, influence the learning process in unique ways.

QUESTION 3: WHEN INDIVIDUALS ENGAGE IN SRL BEHAVIORS IN A WELL-DESIGNED TELE, DO THEY SHOW GREATER LEARNING THAN THEIR PEERS WHO ENGAGE IN FEWER SRL BEHAVIORS?

Of the questions we posed in the beginning of this chapter, this question can be answered in the most straightforward manner. While learning with paper-based and identical but computer-based learning materials yield non-significant differences in achievement, TELEs do cause learners to acquire more knowledge than non-enhanced conditions. Training students to be more self-regulated causes them to be so, and to acquire more knowledge as a result (Azevedo & Cromley, 2004; Manlove, Lazonder & de Jong, 2006; 2007; 2008; 2009). Additionally, TELEs can successfully increase students' SRL behavior, though their impact on achievement is not definitive.

We answer this question in two parts. First, do TELEs seem to be superior to paper-based or computer-based learning tasks which are not technologically enhanced? The answer, as we detail below, seems to be "yes." Second, we address the question, "Do self-regulated learners actually learn more when using TELEs than less self-regulated peers?" Again, the answer is yes, and we can identify specific SRL processes that drive this phenomenon.

Knowledge Acquisition and TELE Type

Considering first the differences in learning outcome by TELE type, there is evidence that students' use of paper-based learning tasks and simple computer-based versions of the same task yield similar learning outcomes (Rui & Lui, 2007). Rui and Lui (2007) explored differences in a problem solving activity when students used different forms of data organization and found that students using the computerized database reported that it functioned as an organizational tool that decreased task difficulty. They also solved more science problems accurately than participants in both the paper database and no database conditions. Opening up our discussion to include learning environments that are technologically enhanced, evidence suggests that learners who use TELEs in lieu of or in addition to lecture and paper-based learning materials do experience increased knowledge acquisition (Dabbagh & Denisar, 2005; Manlove, Lazonder & de Jong 2009; Proske, Narciss & Koerndle, 2007; Rui & Lui, 2007).

Manlove, Lazonder, and de Jong (2009) found that when high school students interact with a TELE to facilitate problem-solving tasks, learners demonstrated increased accuracy while solving physics problems and constructed more accurate solution models than the control group. Dabbagh and Denisar (2006) provide a qualitative explanation of how TELEs can promote problem solving by comparing the students' solutions to problems whose scenarios were structured hierarchically or heterarchically. Content analysis indicated solutions derived from heterarchically structured problem were more cogent and comprehensive than the comparison structure. The TELE allowed students to approach the organization of the problem in different ways and influenced the process and quality of the students' solution.

Knowledge Acquisition and Engaging in SRL Actions in a TELE

Having documented the utility of TELEs for learners in general, we now turn our attention to how learners who self-regulate benefit from TELE use. Speaking generally about SRL and its impact on knowledge gain, it seems that, based upon the collected findings of individual studies across SRL classes, the tendency to self-regulate learning, is positively associated with knowledge acquisition in learners using TELEs (Azevedo & Cromley, 2004; Azevedo, Seibert, Guthrie, Cromley, Wang & Tron, 2002; Greene & Azevedo, 2007; Greene & Azevedo, 2009; Proske, Narciss, & Koerndle, 2007). These findings have been documented with a variety of student populations including high school students classified as low achieving (Azevedo, Winters & Moos, 2004), on grade level and gifted (Greene, Moos, Azevedo & Winters, 2008), as well as middle school (Greene & Azevedo, 2007; 2009) and college students (Azevedo & Cromley, 2004; Azevedo, Green, Moos, 2007; Kauffman, 2004; Kauffman, Gie, Xie & Chen, 2008; Manlove, Lazonder & de Jong, 2006; 2007; 2008; 2009).

Cited by many subsequent studies which attempt to answer this question, Azevedo and Cromley (2004) found that learners who receive 30 minutes of training before completing a science learning task in hypermedia experience greater declarative and conceptual knowledge gains than those who were not trained. Further, Wang and Lin (2007) found that self-efficacy moderates the benefits of TELE use. When using TELEs like Net-Ports, which support collaborative and individual learning, groups of students who felt efficacious about their TELE experience produced more high-quality ideas than less efficacious groups.

Individual macro-level SRL behaviors applied in TELEs such as monitoring (as compared to planning, strategy use, and others) have been found to predict knowledge gain (Greene & Azevedo, 2009). Dissecting these macro levels of SRL behavior into subcategories, specific behaviors within each category of SRL activity (planning, monitoring, strategy use and the handling of task difficulty and demands) have been identified as significant predictors of knowledge acquisition, as detailed in the studies below. As such, we can state that students who are more self-regulated tend to acquire more knowledge. This statement underscores the importance of identifying and building specific SRL behaviors into students' approach to a learning task.

SRL processes that predict larger gains in conceptual understanding include planning, monitoring and use of a higher proportion of effective strategies (Azevedo, Guthrie & Seibert, 2004). These include selecting new information sources, summarizing, re-reading, making inferences, hypothesizing, and elaborating. Greene and Azevedo (2007) analyzed trajectories of knowledge gains and patterns of SRL behaviors amongst adolescents and found that SRL microprocesses predict knowledge acquisitions in hypermedia learning tasks. SRL strategies like: coordinating information sources (text to diagram), making inferences, knowledge elaboration and monitoring activities like identifying the adequacy of information (assessing usefulness of content), and feeling of knowing (monitoring understanding) are associated with higher levels of knowledge gain. The tendency of students to focus on controlling conditions of the learning task (clicking to toggle settings features of a TELE such as zoom) was negatively associated with knowledge gain. We can say with some confidence that strategy use and monitoring are critical SRL behaviors for acquiring knowledge when using hypermedia.

While some TELEs provide students free reign to utilize or ignore features of a learning environment, others are more forceful and include regulative scaffolding to guide learning. Generally, there is less consistent evidence that TELEs that regulate (as compared to allowing student to self-regulate) learning improve knowledge acquisition. The forms of regulative support that have been studied include the effects of providing feedback (McKendree, 1990; Gao, 2003; Graessar, McNamara, & Van Lehn, 2005, prompting reflection (Kauffman, 2004; Kauffman, Gie, Xie & Chen, 2008; van den Boom, Paas, van Merrienboer, & van Gog, 2004), scaffolding metacognition (Aleven, McLaren, Roll & Koedginer, 2006; Kramarski, 2002; Kramarski & Gutman, 2006; Kramarski and Hirsch; 2003; Kramarski & Mizrachi, 2006; Kramarski & Ritkof, 2002), and scaffolding overall self-regulation (Jacobsen & Achididou, 2000; Manlove, Lazonder & de Jong, 2006; 2007; 2008; 2009; McNamara, O'Reilly, Best & Ozuru, 2006). Several studies have demonstrated that learning environments with immediate and elaborate feedback are superior to environments without these conditions in terms of knowledge acquisition (McKendree, 1990; Gao, 2003; Graesser, Lu, Jackson, Mitchell, Ventura, Olney, & Louwerse, 2004). However, in studies where feedback is conducted through e-messages and is less immediate, its benefits are less clear (van den Boom, Paas, van Merriënboer & van Gog, 2004). Prompting reflection has been shown to be effective at improving students' knowledge in mathematics, but also influences student behavior by encouraging students to allow themselves to become reliant on this external regulation.

Further demonstrating the efficacy of enhancing TELE with regulative scaffolds is the effect of metacognitive training on declarative and conceptual knowledge. Kramarski and Hirsch (2003) found that students who received math instruction with self-regulation skill lessons, when compared to their counterparts without self-regulation support, performed better on symbolic reasoning and patterns items, but not significantly better on manipulating algorithms and analysis change items. Reinforcing these conclusions, Kramarski and Mevarech (2003) found that middle school students who received metacognitive training in math outperformed their counterparts in overall reasoning including math explanations. These researchers also established that integrating metacognitive training into more than one domain yields higher learning gains than only including the support in one area (Kramarski, Mevarech & Lieberman, 2001). Additionally, their research indicated that peer emailing enhanced the effectiveness of these TELEs as evidenced by improved math explanations (Kramarski, 2002; Kramarski & Ritkof, 2002).

These findings are encouraging, yet they beg another question—are gains due primarily to this metacognitive training being embedded in the TELE or can a similar training provided by a human tutor produce similar results? Kramarski and

Mizrachi (2006) addressed this methodological concern, and found that middle school students who received metacognitive training online demonstrated heightened math literacy skills over and beyond that of their counterparts who received the same training from a human tutor. Azevedo and colleagues (Azevedo, Cromley & Seibert, 2004; Azevedo, Cromley, Seibert & Tron, 2003; Azevedo, Cromley, Thomas, Seibert & Tron, 2003; Azevedo, Cromley, Winters, Moos & Greene, 2005; Azevedo, Greene & Moos, 2007), however, find that when a human tutor trains learners to be self-regulated prior to the task and scaffolds their learning process, a change in behavior pattern accompanies increased learning. Learners who had a tutor more often engage in help seeking than other self-regulatory behaviors. Learners were utilizing resources and conducted co-regulation (with the tutor) instead of self-regulation of learning, which led to similar knowledge gains (Azevedo, Cromley & Seibert, 2004; Azevedo, Cromley, Winters, Moos & Greene, 2005). It seems then that if learners' metacognition is appropriately scaffolded by a TELE itself, or by a human tutor, learning gains will follow, though tutors who intend to also foster SRL should be wary of students' over-reliance on the tutor.

While adaptive scaffolding provided by a human tutor and computerized tutors have been shown to be beneficial for student learning, some TELEs provide automated scaffolding that is not adapted to student's needs. The benefit of these regulatory prompts is less clear. Specific TELE design features that prompt all students, regardless of SRL tendency, to conduct SRL behaviors have been shown to improve learning outcomes. Kauffman (2004) investigated whether web-based instructional prompts (to take notes, self monitor learning, and to consider one's self-efficacy) influenced SRL behavior and knowledge acquisition while completing a WebQuest. Findings suggest that those who took notes using the matrix tool provided and were prompted to reflect on learning gained more knowledge than those who were not prompted and those who took freeform notes. Significant main effects of note-taking condition and of monitoring were also found where each experimental group gained more knowledge than control groups, suggesting prompting does improve knowledge acquisition. These results indicate that TELEs that scaffold strategy use (note taking format) and metacognition (embedded prompts) primed learners to engage in SRL activities, which resulted in knowledge gain. However, additional study by Kauffman, Gie, Xie & Chen (2008) into how prompts impact problem solving and writing quality suggest that problem solving prompts, alone and in combination with self-reflection prompts, improved learning outcomes, while self-reflection prompts did not. We can conclude then that not all prompts are equally beneficial, and that additional research needs to be done to examine what effect different kinds of externally regulative objects employed in TELEs have on students, and how they might interact with SRL tendencies.

One final feature we have yet to review is the provision of tools that students may use to mark content in the TELE. These include highlighters, note taking tools, linking tools, and ways that students can build "information objects" onto pre-existing content from the learning task. These tools are available in TELEs such as gStudy (Nesbit, Winne, Jamieson-Noel, Code, Zhou, & MacAllister, 2006) and in Study 2000 (Proske, Narciss and Koerndle, 2007). Proske, Narciss and Koerndle (2007) found that use of marking and note taking tools is positively associated with improvements in knowledge gain.

Knowledge Acquisition and TELEs that Employ Multiple Features of Self-Regulation

While many basic TELEs include just one or two features aimed at regulating or encouraging self-regulation during learning, there have been successful interventions where multiple features of self-regulation were embedded in the instruction. Manlove, Lazonder, and de Jong (2006; 2007; 2008; 2009) conducted a series of intervention studies that consistently demonstrated science learning gains measured in scientific inquiry tasks (not mere declarative knowledge tasks) for students using multi-featured TELEs. The multiple features of SRL that were embedded in the treatment TELE were: goal lists, hints, prompts, cue and templates specifically designed for science inquiry. Similarly, Jacobson and Archididou (2000) successfully used scaffolds to assist students in transforming their naïve mental models of biology to more complete and advanced models. In light of the collected findings on different TELEs and their influence on student knowledge gain, we can conclude that students generally experience increased learning from TELEs.

Summary and Implications

To aggregate what we discussed about knowledge acquisition resulting from SRL strategy use, we can affirm that the following conditions are conducive to learning: (1) feedback must be immediate, elaborate, and in user friendly language (2) metacognitive training is effective with and without a human tutor; pre- task training amplifies this effect (3) tools that support annotation of TELE content enhance knowledge gain and (4) knowledge growth is also enhanced in TELEs where SRL training is intertwined with instruction.

In all of the studies, pretest scores ensured no differences in prior domain knowledge. These findings all apply to middle school to college students and math and science domains. Taken together, these results do emphasize the need for some type of regulative scaffolds, whether embedded in the task or provided as pre-task training, in order for students to take full advantage of the opportunity TELEs provide. Without regulative scaffolds, whether it is a human or software, most students do not have the propensity to flourish in a TELE. The regulative scaffolds are the keystone in structuring a TELE where the two necessary conditions of knowledge growth, opportunity and propensity, meet.

QUESTION 4: HOW HAVE TELES BEEN SHOWN TO PROMOTE SRLTENDENCIES IN LEARNERS?

Now that we have examined how didactic, facilitative, and unenhanced TELEs impact knowledge acquisition as well as how learners' characteristics mediate this interaction, we are compelled to determine how to use TELEs to improve learners' SRL tendencies themselves. In doing so, we begin a dialogue on how to build a student's propensity to maximize the full potential of the opportunities provided by a TELE.

Before discussing how TELEs have been shown to promote SRL tendencies in learners, it is helpful to first briefly review the tendencies in which a self-regulated learner engages. Following the summary of our current understandings, we discuss implications and areas that require further research related to the focal question of this section.

Recall from the descriptions of the theoretical models of Boekaerts and Zimmerman that appraisal and SRL interact and affect a learner's performance throughout the three recursive phases of SRL (Boekaerts & Niemivirta, 2000; Zimmerman & Tsikalas, 2005).

Researchers have considered how to improve the tendencies enlisted in the forethought phase and performance/ volition phases, but tendencies in the self-reflection phase, to the best of our knowledge, has received minimal attention. Moreover, promoting positive appraisals is an area of research that has received relatively scant attention. As will become evident in the following discussion, TELEs have been shown to promote SRL tendencies when their program features directly support a particular attribute. To date, researchers have implemented programs aimed at developing the following aspects of SRL: (a) the metacognitive skills of goal setting, planning, organizing, and monitoring learning, (b) strategy use, (c) adapting and (d) self-initiating.

Metacognitive Aspects of SRL

Goal Setting and Planning. Although several studies examine the role of goal structure in TELEs, none have been conducted to actually improve the metacognitive process of setting appropriate goals. To date, all of the TELEs have presented participants with some form of prescribed goals, subgoals and hints. As such, TELEs in the literature cannot be said to promote goal-setting. Perhaps TELEs in the future can be made to solicit goals rather than prescribe them. In regard to planning, however, one study (Manlove, Lazonder & de Jong, 2008) was designed to develop the process of planning through providing regulative support. Students who received cues, hints, goal lists, prompts and templates demonstrated more planning than those who did not receive the support.

Organizing. Researchers have examined the utility of computerized organizational tools and different forms of hypertext organization (Rui & Liu, 2007), but little or no research (to our knowledge) has been done to improve students' propensity to organize the material that is presented to them. All of the information provided in a TELE is pre-organized in some hierarchical or heterarchical form thereby precluding the opportunity for a student to self-initiate organizational tactics. Progress in this field of organizing information and resources is limited to the highly structured nature of TELEs.

Monitoring Learning. Two studies looked at improving metacognitive monitoring. White and Frederiksen (1998) used science inquiry simulations to enhance this attribute. Students demonstrated enhanced monitoring when they designed experimental plans and scenarios. Supplementing metacognition training with reflective assessment led to even higher learning gains. Kramarski and Gutman (2006) used metacognitive questions to increase monitoring of learning in the math domain. Math explanations seemed to lead to improved metacognition. Both of these programs proved efficacious in developing metacognition as measured by conceptual knowledge tasks.

Motivational Aspects of SRL

Empirical work has been conducted to ascertain the relationship between motivational beliefs such as self-efficacy and attitudes toward a particular domain. This relationship was analyzed in the previous section of this chapter. Without devaluing this progress, research has yet to address technology-based interventions that promote motivation. At present, we do know that learner motivation seems to increase over the course of a learning task in a TELE (Moos & Azevedo, 2008a), but additional investigation is warranted to determine why motivation increases, and if it increases similarly across all learners.

Strategic Aspects of SRL

Several studies have attempted to increase the quality and quantity of strategy use. The common denominator in the effective interventions is their specific nature; they focus on strategy use in a particular domain and/ or adapt the targeted instruction to the learner's needs. Aleven and his colleagues (2006) designed Cognitive Tutor Help Tutor to increase the frequency of appropriate uses of help-seeking in geometry. This program uses the individual student's problem solving actions to tailor feedback and guidance about the help tools (e.g. glossary) that the learner should use in order to understand the content. Students who utilized the Help Tutor decreased the frequency of inappropriate and inefficient use of help seeking strategies.

iStart (McNamara, O'Reilly, Best, & Ozuru, 2006), an interactive reading strategy trainer, aimed to improve comprehension of science text through developing the reading strategies of paraphrasing and making connections within text. Students who learned using iStart demonstrated improved reading strategy use and comprehension in their self-explanations of text including the science text comprehension questions.

Adaptive Aspects of SRL

The context in which TELEs have been shown to promote an adaptive tendency in learners is research on feedback. McKendree (1990) established that learners adapt from feedback that is both immediate and elaborate, that is, it provides an explanation for incorrect responses instantly. Gao (2003) extended this finding to conclude that a generative activity such as formulating an example or scenario will aid students in adapting to the demands of a TELE. Relatedly, Graessar and his colleagues' program, AutoTutor (2004), delivers immediate and elaborate feedback in the form of natural dialogue. The authors suggest that AutoTutor's conversational pedagogical agent is the feature that enables learners to adapt to the demands presented in their TELE. Kramarski, Zemira, Arami & Arami (2002) found that a learning environment enhanced with metacognitive training and email helped students adapt to the requirements of the math course.

Self-initiating Aspects of SRL

A multitude of studies have investigated different TELE conditions that encourage students to self-initiate strategy use. Azevedo and his colleagues (Azevedo, Cromley, Thomas, Siebert & Tron, 2003; Azevedo & Cromley; 2004, Azevedo, Greene & Moos, 2007) have contributed a great deal to our understanding of how to initiate SRL in students. One such condition involves incorporating SRL training before engaging in a TELE. Pre-task training has been shown to promote self-initiation of a variety of learning strategies such as activating prior knowledge and managing time (Azevedo & Cromley, 2004; Azevedo, Greene & Moos, 2007). Another condition, pre-task training along with co-regulation, has reliably shown students to activate prior knowledge, monitor their learning, self-question, gauge their progress toward goals, manage time and manage effort (Azevedo, Cromley, Thomas, Siebert & Tron, 2003). In addition, Kauffman and his colleagues (2008) identified a necessary condition for using prompts to initiate self-regulation—prompts must be embedded within the problem to be solved. Yang (2006) extended these efforts by presenting students with embedded prompts and found that this approach increased self-monitoring and self-instruction.

Summary and Implications

To synthesize what we know so far about how TELEs have been shown to promote SRL tendencies, two particular conditions seem to be necessary: (1) the objective of the TELE must be domain specific and strategy specific and (2) the instructional methods applied by the system must be in response to the learner's specific needs.

As with any empirical research, however, methodological shortcomings limit the implications that can be drawn from the findings. First, strategy-specific and adaptive TELEs have promoted strategy use in learners across age groups from middle school through college. It remains to be seen how these well-designed TELEs would influence elementary age students. Second, the interventions have been largely limited to just two content areas: math and science (with the exception of van den Boom and colleagues' (2004) study involving classroom scenarios). Third, many interventions seem to work best when participants are pre-trained in SRL, and to a greater extent, tutored through a task. This "other regulation" is somewhat at odds with the construct of student initiated self-regulation of learning that we would hope to see occur naturally in students. Nevertheless, the studies in this section have shown that TELEs can improve planning, monitoring of learning, help seeking, reading strategies, and adapting when they fit the criteria described above. That said, only a portion of the learning event has been modeled and examined in these studies. Several SRL attributes and processes that interact and affect learning have received minimal attention (to the best of our knowledge): positive appraisal, goal setting, organizing, interest, self-efficacy, motivation and self-reflective processes including causal attributions.

QUESTION 5: HOW DO PRE-EXISTING SRL TENDENCIES INFLUENCE THE WAYS IN WHICH LEARNERS INTERACT WITH TECHNOLOGY ENHANCED LEARNING ENVIRONMENTS?

In order to address this question, we first must define what we mean by pre-existing tendencies to self-regulate learning. Referring back to our definition of a self –regulated learner as being metacognitive, strategic, adaptive, engaged and self-initiating, a number of learner characteristics will influence learners' tendencies to self-regulate. Green and Azevedo (2009) conceptualize the SRL attributes we describe as including individuals' tendency to plan, monitor, use strategies, handle task demands and sustain interest. While they measured these tendencies during a task, such learner characteristics likely exist pre-task and across tasks, and could theoretically be measured a priori with the right tool. Additionally, other factors besides SRL attributes such as self-efficacy and prior knowledge can affect learners' interaction with a TELE and result in knowledge gain. In this section, we discuss how SRL attributes have been measured, how findings from studies using different research designs and measurement techniques might be interpreted, and then discuss how additional learner characteristics may affect (or interact with SRL to affect) TELE use.

Evolution of Measurement of SRL

Learners' tendencies to self-regulate have been measured in a variety of ways including content analyses of post-task interviews, pre-task assessment using self-report questionnaires and in-task assessment using think aloud and trace methodologies. These experimental design decisions have consequences in terms of the way we define learners as self-regulated, and color our ability to make causal or correlational attributions between SRL tendency and learner interaction with a TELE.

Qualitative Analysis. Early work exploring the relationship between TELE and SRL overlaps with research focusing on structures of hypertext and how learners use reading strategies to enhance their comprehension. One of the first studies to use SRL language with respect to TELEs was conducted by Balcytiene (1999), who identified three types of reading behaviors students employed when read a hypertext on architecture. From content analyses of videotaped interviews, Balcytiene determined that learners exhibit different reading strategies, and that readers could be categorized as cue-dependent or self-regulated. She concluded that TELEs are more beneficial to readers who exhibit SRL processes such as inference making, self-questioning and reflection.

While Balcytiene (1999) took a qualitative and post hoc approach to analyzing reading styles and connecting them to SRL behaviors, more recent studies of SRL tendency have evolved from quasi-experimental into experimental and exploratory designs. This trend has implications for the interpretation of findings based upon how we define a pre–existing SRL tendency.

Quasi experimental Designs. Scales to measure SRL tendency rely upon the self-report of the individual and include the Motivated Strategies for Learning Questionnaire, (MSLQ; Pintrich, Smith, Garcia & McKeachie, 1991), the Learning and Study Strategies Inventory (LASSI; Weinstein, Schulte & Palmer, 1987) and many derivatives and adaptations of these early SRL scales. Such a method of measurement follows from SRL theories which treat SRL as an aptitude which is consistent across contexts and is fixed within the individual (Boekaerts & Niemivirta, 2000). Some studies we reviewed to document the influence of learners' SRL tendency on their behaviors in TELEs employ these self-report measures and often categorize students as either high self-regulators or low self-regulators. Using self-report scales, early research (Eom & Reiser, 2000; McManus, 2000; Wang & Lin, 2007) suggested that highly self-regulated individuals did interact differently in TELEs than their lesser regulated peers.

Trace Methodologies in Exploratory and Experimental Designs. Studies by Eom & Reiser,(2000) and McManus (2000) which sought to determine a more quantitative causal impact of SRL tendency on TELE use represent a next step in methodological complexity, but their instruments suffered from a lack of validity, and their findings must be approached with some caution. SRL measurement through self-report methods has been shown to be poorly calibrated to true SRL tendency as evidenced by learner behaviors (Winne & Jamieson-Noel, 2002). Hadwin and colleagues (2007) used trace methods to further assess the relationship between self-reported SRL behaviors and representative behaviors. Even the most well informed students tend to accurately self-report their tendency to self-regulate only 40% of the time, while the mean is closer to a quarter of the time (Hadwin, Nesbit, Jamieson-Noel, Code, & Winne 2007). This discovery undermines what conclusions we can draw from studies that make use of self-report data. Instead, we choose to focus primarily on those studies that create experimental conditions by intentionally creating groups of students who are highly or minimally self-regulated in their learning behaviors through pre-task SRL training. Additional studies that employ self-report scales and traces of behavior (Nesbit, Winne, Jamieson-Noel, Code, Zhou, & MacAllister, 2006) use exploratory designs to examine patterns of SRL behaviors as they relate to SRL microprocesses (e.g. marking strategies like highlighting and note taking) might give insight into how learner characteristics influence TELE use.

Employing an experimental design and influencing SRL tendency by providing training in SRL to experimental groups (and withholding training from control groups) has demonstrated that highly self-regulated learners behave and achieve differently than minimally self-regulated learners (Azevedo, Seibert, Guthrie, Cromley, Wang & Tron, 2002; Azevedo, Cromley, Thomas, Seibert & Tron, 2003; Azevedo & Cromley, 2004). When learners are made to be more self-regulated by training, they tend to use a higher percentage of effective learning strategies than untrained peers (Azevedo, Seibert, Guthrie, Cromley, Wang & Tron, 2002; Azevedo, Cromley, Thomas, Seibert & Tron, 2003). Those who are trained in SRL prior to completing a task (high SRL) tend to spend more time coordinating sources of information, taking notes, drawing, and reading and reviewing notes than untrained (lower SRL) peers (Azevedo, Cromley, Thomas, Seibert & Tron, 2003). When not trained, learners (a) typically will vary greatly in their study tactics, (b) tend not to monitor their own learning, but do monitor the adequacy of information in TELE, (c) use strategies that include a mixture of effective and ineffective search tactics as well as repetition of goals in working memory, (d) typically do not plan and fail to integrate different sources within a TELE, (e) skip between instructional platforms (text, diagram, animation) and focus on a goal to memorize content by rereading passages and taking notes, but seldom reviewing notes and (f) will generally be performance-oriented, using only externally provided performance subgoals to guide action and will not plan their own learning. When trained however, learners behave in a more self-regulated fashion. They read, summarize, restate and activate prior knowledge. They monitor their understanding by determining if they know something, if they are learning and also conduct self-questioning. When a tutor is present, they engage the tutor for assistance, and when not, they are more planful, set specific goals and then engage in reading. They monitor by judging if learning is occurring, engage in self-question and reread to clarify misunderstandings. Training clearly improves performance in TELEs in terms of both SRL use and knowledge acquisition. Azevedo and Cromley (2004) found that learners who were made to be more self-regulated through training primarily summarized, made drawings and notes, read notes, elaborated on knowledge, coordinated sources and found location in the environment. Their untrained counterparts primarily conducted searches, both goal oriented and free, and selected new informational sources as means of learning. This set of behaviors is deemed less effective in producing knowledge gain, which makes such differences an important consideration for educators considering TELE use with students of varying SRL ability. In addition to the impact of pre-task SRL training, Azevedo, Greene and Moos (2007) found that tutored learners monitor learning more and use a greater percentage of effective SRL strategies compared to learners who are left to regulate their own learning without tutoring.

Exploratory Designs. A paradigm shift in measurement techniques resulted from Winne and Jamieson-Noel's (2002) findings regarding calibration of SRL self-reports, and researchers have, for the most part, begun to avoid labeling individuals as high or low SRL, as an enduring trait. Instead, advances in technology have enabled TELEs to trace individuals' behaviors in the environment using logs of their actions. A profile of learners' SRL behaviors has since replaced offline SRL scales as the preferred method of measuring student SRL tendency in context. This method lends itself more to the process model of SRL as described by Winne and Hadwin (1998) and Zimmerman (2000) but also undermines researchers' ability to make statements about an individual's degree of self-regulation across learning contexts unless SRL is intentionally enhanced pre-task. That is, if SRL can only be measured in context, it would require exposure to and measurement in multiple learning environments before an individual's behaviors indicate that he or she is highly or minimally self-regulated, in a trait-like sense. Thus, many later studies tend to take an associative approach and discuss learners' patterns of SRL behaviors when placed in a TELE. Examples of such research include studies by Bauer and Koedinger (2006) and Nesbit and colleagues (2006) who used software to trace students' tendency to select segments of text while studying as it relates to goal orientation and knowledge gain.

Additional Influence of Other Learner Characteristics

While our primary focus in this chapter is to discuss the interaction of SRL and TELEs, we must not ignore the role of other student and TELE characteristics that influence knowledge gain. These variables play a large mediating role in the way learner and TELE interact and what knowledge gain results. Mediating characteristics include learners' self-efficacy (Bell 2007), motivation (Moos & Azevedo, 2008c; Narciss, Proske & Koerndle, 2007; Nesbit, Winne, Jamieson-Noel, Code, Zhou, & MacAllister, 2006 2006), level of prior knowledge (Azevedo, Moos, Greene, Winters, & Cromley, 2008; Azevedo & Cromley, 2004; Balcytiene, 1999; Brusilovsky, 2004Cho, 2004; MacGregor, 1999; Moos & Azevedo, 2008a; Winters & Azevedo, 2005) and the TELE's features which allow for learner control for the task environment (Bauer & Koedinger, 2006; Eom and Reiser, 2000). Taken collectively, these findings, that highlight interactions between SRL tendency with other learner characteristics and with TELE characteristics, underscore the need to consider multiple characteristics of the learner as well as features of the TELE when considering the benefit of a TELE for specific users.

Having now answered our question about the influence of TELEs on student knowledge gain, we now turn our attention to the influence of TELEs on students' tendency to self-regulate their own learning.

Summary and Implications

These three very different research methodologies give us considerable evidence that does suggest that the propensity to learn in a self-regulated fashion is critical. Given the evolution of research designs to capture reliable data about learner's tendencies, this new wave of research using in-task measurement is presently in its infancy, which limits our understanding for the time being.

We can say with some certainty that learners' propensity can be intentionally increased through offline, pre-task training, and such increased propensity increases the degree to which learners take advantage of the opportunities provided by the technological enhancements of their learning environment. When we consider the individuals in typical learning situations, they look remarkably different from those who were trained. While learners are not significantly more mastery or performance oriented (Nesbit, Winne, Jamieson-Noel, Code, Zhou, & MacAllister, 2006), they are generally not very self-regulated, as evidenced by low utilization rates of TELE tools (Brush & Saye, 2001; Narciss, Proske & Koerndle, 2007; Nesbit, Winne, Jamieson-Noel, Code, Zhou, & MacAllister, 2006; Proske, Narciss, Koerndle, 2007). This low baseline level of self-regulation suggests that educators should aim to increase students' SRL propensity if they intend to instruct using TELEs. Without raising students' SRL acumen, the opportunities that TELEs provide are likely to be underutilized.

CONCLUSION

Having discussed a broad range of empirical findings as they relate to enhancing SRL and knowledge gain, we conclude by distilling these findings into a set of take home messages for educational practitioners and for those who intend to conduct further research into students' SRL and its interaction with TELE. We denote these messages as both opportunities to design TELEs that best meet the needs of learners, and as chances to increase the propensity of learners to benefit from using TELEs.

For Practitioners

Opportunity: The design of a TELE must be adapted to the learner. By and large, findings revealed that the students who increased their SRL tendencies and knowledge the most were the ones who had access to TELEs that gave them all they needed and nothing they did not need. If our goal is to provide students with the best learning opportunity possible, it is important that we design the TELE to provide them appropriate scaffolding with respect to tutors that answer their questions, but also allow students who do not need "other regulation" to opt out of it. This may require some pre-task screening for particular characteristics (e.g. prior knowledge) or constant monitoring of student use of TELEs by a teacher, but is likely to result in an improved educational experience. In terms of tool provision, we should strive to give students a full complement of tools they can use. Students who have the propensity to self-regulate learning know what they need, and, if given the option of using it, they will excel.

Opportunity: Teacher-led instruction and TELEs are not exclusionary, but complementary. Genuine opportunities to learn require well-designed teacher-led and technologically facilitated instruction. The promise for student achievement and self-regulation that our review has demonstrated can only be fulfilled if both necessary conditions of the learning environment are maintained: complementary use of teacher-led instruction and well-designed educational software. As the label (TELE) appropriately says, learning environments are only enhanced with computers and their software. Improvements in math, science and SRL strategies can only be replicated in a classroom if the practitioner is also providing instruction and monitoring the use of the TELE. To ensure this occurs, teachers must be trained fully in the technology before implementing the program. They must understand the theory, research and operations of the program in order to effectively integrate it into their learning environment. Otherwise, the software may be incorrectly used or not used at all. With proper training in use of TELEs, practitioners can diagnose and target the propensities that are needed for a task and ensure maximal fit between the technology and the student.

Propensity: Pre-task training is Key. Studies in which participants were left to their own devices to engage in SRL behaviors revealed that most participants did not engage in these behaviors very often. In contrast, studies in which participants were given pre-task training to increase their tendency to engage in SRL behaviors showed that training was effective and learning was enhanced. Thus, those who wish to create TELEs for learners would be wise to increase participants' propensity to learn via pre-task training. The need to focus on propensities prior to presenting opportunities to learn via TELEs is reminiscent of recent findings in the achievement literature which show that achievement gaps in particular grade levels (e.g., first grade in American schools, age 6) could be substantially reduced if interventions occurred prior to students' entry into that grade level (e.g., when students were 4 or 5 years old). When students enter a grade with the same proficiency, they show comparable levels of achievement at the end of that grade (e.g., Byrnes & Wasik, 2009).

Propensity: Practitioners must take advantage of the online monitoring that only TELEs can perform. To effectively diagnose, target and cultivate student propensities, practitioners need accurate data to make instructional decisions. Before TELEs, teachers could only observe individual students to collect data on their strategy use, reasoning and so on. Clearly, the amount of time this assessment consumed typically outweighed its perceived utility, therefore resulting in paper and pencil tests that are only intended to tap self-reported learning processes. However, TELEs have the capability to transform teachers' assessments of students' needs and abilities by improving the efficiency and accuracy of assessment. Trace methodologies track and log actual learning processes as they transpire, thereby allowing the teacher to review an organized, detailed file of a particular student's strategy use and decisions. There are many benefits to having access to how students regulate their learning. Chief among them is the opportunity to provide individualized instruction that can be designed to keep student learning progressing rather than stagnating. Properly trained teachers who are given the time and resources to evaluate trace data on student learning processes hold tremendous promise for improved instruction and student learning.

For Researchers

Opportunity: Student characteristics are numerous, interconnected and dynamic and should be measured as such. While this chapter focused on learners' interactions with TELEs, much attention was paid to discussing the contributing role of other student characteristics such as motivation, self-efficacy, prior knowledge and others. Because TELEs must be adapted to learners' needs, researchers who intend to accurately describe a learner-TELE interaction must also go to great lengths to describe the characteristics of the learner. Assessing these contributing factors known to influence learning is critical. The more we know about learners, the more accurately we can characterize their behaviors and adapt future TELEs to meet their needs.

Additionally, we see that learner characteristics can change throughout a learning task. Just as SRL is an iterative process, our measurement must be iterative, monitoring changes in the student over time that could affect the way the student interacts with the TELE. Such monitoring requires that researchers take a process approach to their research designs, and likely requires them to use trace or think aloud methodologies and reassess pre task characteristics (like motivation) in task periodically to monitor change.

Opportunity: The relationship between TELEs & SRL requires further study. In the process of mapping the relationship between TELEs, SRL and achievement, we identified gaps in our understanding that can be filled with intentionally designed future research. First, researchers have established a firm understanding of the mechanisms involved in enhancing particular self-regulatory attributes such as planning and help seeking. However, additional research is needed in the areas of: positive appraisal, motivation, goal setting, interest, self-efficacy, organizing and self-reflective processes. As these areas are explored and findings uncovered, a more comprehensive model of SRL with TELEs can be forged. Second, the generalizability of the findings we have reviewed is limited to the middle school, high school, undergraduate and graduate populations that have been studied. Researchers still need to chart these opportunity-propensity interactions in elementary school populations. Through understanding the nature and function of TELEs throughout all of the school years, designers can begin to develop programs that would scaffold content mastery as well as the necessary SRL strategies from elementary grades to high school and beyond. Third, our knowledge of the efficacy of TELEs is bound within the academic areas of science and math. TELEs that focus on improving writing, reading and social studies will broaden the scope of our inferences and implications. Certainly, it would be valuable to students to reap the benefits in conceptual knowledge and self-regulation in areas beyond math and science.

In sum, then, the literature on the relationship between SRL and TELEs has certainly grown over the years and much has been learned about this relationship. This increased insight is fortunate given the pervasiveness of TELEs in contemporary society. However, the full power of TELEs can only be harnessed if researchers and practitioners work together and share insights.

ACKNOWLEDGMENT

All authors contributed equally to this chapter.

REFERENCES

Aleven, V., McLaren, B., Roll, I., & Koedinger, K. (2006). Toward meta-cognitive tutoring: A model of help seeking with a Cognitive Tutor. *International Journal of Artificial Intelligence in Education, 16*(2), 101–128.

Azevedo, R. Cromley, J.G., Thomas, L., Seibert, D., & Tron, M. (2003, April). *Online process scaffolding and students' self-regulated learning with hypermedia.* Paper presented at the annual meeting of the American Educational Research Association, Chicago, IL.

Azevedo, R., & Cromley, J. G. (2004). Does training on self-regulated learning facilitate students' learning with hypermedia? *Journal of Educational Psychology, 96*(3), 523–535. doi:10.1037/0022-0663.96.3.523

Azevedo, R., Cromley, J. G., & Seibert, D. (2004). Does adaptive scaffolding facilitate students' ability to regulate their learning with hypermedia? *Contemporary Educational Psychology, 29*(3), 344–370. doi:10.1016/j.cedpsych.2003.09.002

Azevedo, R., Cromley, J. G., Seibert, D., & Tron, M. (2003). *The role of co-regulated learning during students' understanding of complex systems with hypermedia.* Paper presented at the Annual meeting of the American Educational Research Association.

Azevedo, R., Cromley, J. G., Winters, F. I., Moos, D. C., & Greene, J. A. (2005). Adaptive human scaffolding facilitates adolescents' self-regulated learning with hypermedia. *Instructional Science*, *33*(5-6), 381–412. doi:10.1007/s11251-005-1273-8

Azevedo, R., Greene, J. A., & Moos, D. C. (2007). The effect of a human agent's external regulation upon college students' hypermedia learning. *Metacognition and Learning*, *2*(2), 67–87. doi:10.1007/s11409-007-9014-9

Azevedo, R., Guthrie, J. T., & Seibert, D. (2004). The role of self-regulated learning in fostering students' conceptual understanding of complex systems with hypermedia. *Journal of Educational Computing Research*, *30*(1), 87–111. doi:10.2190/DVWX-GM1T-6THQ-5WC7

Azevedo, R., Moos, D. C., Greene, J. A., Winters, F. I., & Cromley, J. G. (2008). Why is externally-facilitated regulated learning more effective than self-regulated learning with hypermedia? *Educational Technology Research and Development*, *56*(1), 45–72. doi:10.1007/s11423-007-9067-0

Azevedo, R., Seibert, D., Guthrie, J. T., Cromley, J. G., Wang, H., & Tron, M. (2002, April). *How do students regulate their learning of complex systems with hypermedia?* Paper presented at the annual meeting of the American Educational Research Association, New Orleans, LA.

Azevedo, R., Winters, F. I., & Moos, D. C. (2004). Can students collaboratively use hypermedia to learn science? the dynamics of self-and other-regulatory processes in an ecology classroom. *Journal of Educational Computing Research*, *31*(3), 215–245. doi:10.2190/HFT6-8EB1-TN99-MJVQ

Balcytiene, A. (1999). Exploring individual processes of knowledge construction with hypertext. *Instructional Science*, *27*(3), 303–328. doi:10.1007/BF00897324

Bauer, A., & Koedinger, K. (2006). Pasting and encoding: note taking in online courses. *Proceedings of the 6th IEEE Conference on Advanced Learning Technologies*, Kerkrade, The Netherlands, July 2006, pp 789-793.

Bell, P. D. (2007). Predictors of college student achievement in undergraduate asynchronous web-based courses. *Education*, *127*(4), 523–533.

Boekaerts, M., & Niemivirta, M. (2000). Self-regulated learning: Finding a balance between learning goals and ego-protective goals. In Boekaerts, M., Pintrich, P. R., & Zeidner, M. (Eds.), *Handbook of self-regulation* (pp. 417–450). San Diego, CA: Academic Press. doi:10.1016/B978-012109890-2/50042-1

Brush, T., & Saye, J. (2001). The use of embedded scaffolds with hypermedia-supported student-centered learning. *Journal of Educational Multimedia and Hypermedia*, *10*(4), 333–356.

Brusilovsky, P. (2004). Adaptive navigation support in educational hypermedia: The role of student knowledge level and the case for meta-adaptation. *British Journal of Educational Technology*, *34*(4), 487–497. doi:10.1111/1467-8535.00345

Byrnes, J. P. (2001). *Minds, Brains and Learning: Understanding the Psychological and Educational Relevance of Neuroscientific Research*. New York: The Guilford Press.

Byrnes, J. P. (2003). Factors predictive of mathematics achievement in White, Black, and Hispanic 12th graders. *Journal of Educational Psychology*, *95*(2), 316–326. doi:10.1037/0022-0663.95.2.316

Byrnes, J. P., & Miller, D. C. (2007). The relative importance of predictors of math and science achievement: An opportunity-propensity analysis. *Contemporary Educational Psychology*, *32*(4), 599–629. doi:10.1016/j.cedpsych.2006.09.002

Byrnes, J. P., & Wasik, B. A. (2009). Factors predictive of mathematics achievement in kindergarten, first, and third grades: An opportunity-propensity analysis. *Contemporary Educational Psychology*, *34*(2), 167–183. doi:10.1016/j.cedpsych.2009.01.002

Carroll, J. B. (1989). Intellectual abilities. In Lesgold, A., & Glaser, R. (Eds.), *Foundations for a psychology of education* (pp. 137–197). Hillsdale, NJ: Lawrence Erlbaum Associates.

Cho, M. (2004, Oct). *The effects of design strategies for promoting students' self-regulated learning skills on students' self-regulation and achievements in online learning environments.* Association for Educational Communications and Technology, 27th, Chicago, IL, October 19-23, 2004.

Corno, L., Cronbach, L. J., Kupermintz, H., Lohman, D. F., Mandinach, E. B., Porteus, A. W., & Talbert, J. E. (2002). *Remaking the concept of aptitude: Extending the legacy of Richard E. Snow*. Mahwah, NJ: Erlbaum.

Dabbagh, N., & Denisar, K. (2005). Assessing team-based instructional design problem solutions of hierarchical versus heterarchical web-based hypermedia cases. *Educational Technology Research and Development*, *42*(2), 71–82.

Eom, W., & Reiser, R. (2000). The effects of self-regulation and instructional control on performance and motivation in computer-based instruction. *International Journal of Instructional Media*, *27*(3), 247–260.

Gao, T. (2003). The effects of different levels of interaction on the achievement and motivational perceptions of college students in a web-based learning environment. *Journal of Interactive Learning*, *14*(4), 367–386.

Graesser, A., McNamara, D., & Van Lehn, K. (2005). Scaffolding deep comprehension strategies through Point&Query, AutoTutor, and iStart. *Educational Psychologist*, *40*(4), 225–234. doi:10.1207/s15326985ep4004_4

Graesser, A.C., Lu, S., Jackson, G. T., & Mitchell, H, H., Ventura, M., Olney, A. & Louwerse, M. M. (2004). AutoTutor: A tutor with dialogue in natural language. *Behavior Research Methods, Instruments, & Computers*, *36*(2), 180–192.

Greene, J. A., & Azevedo, R. (2007). Adolescents' use of self-regulatory processes and their relation to qualitative mental model shifts while using hypermedia. *Journal of Educational Computing Research*, *36*(2), 125–148. doi:10.2190/G7M1-2734-3JRR-8033

Greene, J. A., & Azevedo, R. (2009). A macro-level analysis of SRL processes and their relations to the acquisition of a sophisticated mental model of a complex system. *Contemporary Educational Psychology*, *34*(1), 18–29. doi:10.1016/j.cedpsych.2008.05.006

Greene, J. A., Moos, D. C., Azevedo, R., & Winters, F. I. (2008). Exploring differences between gifted and grade-level students' use of self-regulatory learning processes with hypermedia. *Computers & Education*, *50*(3), 1069–1083. doi:10.1016/j.compedu.2006.10.004

Hadwin, A. F., Nesbit, J. C., Jamieson-Noel, D., Code, J., & Winne, P. H. (2007). Examining trace data to explore self-regulated learning. *Metacognition and Learning*, *2*(2), 107–124. doi:10.1007/s11409-007-9016-7

Jacobson, M., & Archididou, A. (2000). The design of hypermedia tools for learning: Fostering conceptual change and transfer of complex scientific knowledge. *Journal of the Learning Sciences*, *9*(2), 149–199. doi:10.1207/s15327809jls0902_2

Jones, K. K., & Byrnes, J. P. (2006). Characteristics of students who benefit from high-quality mathematics instruction. *Contemporary Educational Psychology, 31*(3), 328–343. doi:10.1016/j.cedpsych.2005.10.002

Kauffman, D. F. (2004). Self-regulated learning in web-based environments: Instructional tools designed to facilitate cognitive strategy use, metacognitive processing, and motivational beliefs. *Journal of Educational Computing Research, 30*(1), 139–161. doi:10.2190/AX2D-Y9VM-V7PX-0TAD

Kauffman, D. F., Ge, X., Xie, K., & Chen, C. (2008). Prompting in web-based environments: Supporting self-monitoring and problem solving skills in college students. *Journal of Educational Computing Research, 38*(2), 115–137. doi:10.2190/EC.38.2.a

Kramarski, B. (2002). Enhancing math discourse: The effects of email conversation on learning graphing. *Educational Media International, 39*(1), 100–107. doi:10.1080/09523980210131169

Kramarski, B., & Gutman, M. (2006). How can self-regulated learning be supported in mathematical E-learning environments? *Journal of Computer Assisted Learning, 22*(1), 24–33. doi:10.1111/j.1365-2729.2006.00157.x

Kramarski, B., & Hirsch, C. (2003). Using computer algebra systems in mathematical classrooms. *Journal of Computer Assisted Learning, 19*(1), 35–45. doi:10.1046/j.0266-4909.2003.00004.x

Kramarski, B., & Mevarech, Z. (2003). Enhancing mathematical reasoning in the classroom: The effects of cooperative learning & metacognitive training. *American Educational Research Journal, 40*(1), 281–310. doi:10.3102/00028312040001281

Kramarski, B., Mevarech, Z., & Lieberman, A. (2001). Effects of multilevels versus unilevel metacognitive training on mathematical reasoning. *The Journal of Educational Research, 94*(5), 292–300. doi:10.1080/00220670109598765

Kramarski, B., Mevarech, Z. R., & Arami, M. (2002). The Effects of Metacognitive Instruction on Solving Mathematical Authentic Tasks. *Educational Studies in Mathematics, 49*(2), 225–250. doi:10.1023/A:1016282811724

Kramarski, B., & Mizrachi, N. (2006). Online discussion and self-regulated learning: Effects of instructional methods on mathematical literacy. *The Journal of Educational Research, 99*(4), 218–230. doi:10.3200/JOER.99.4.218-231

Kramarski, B., & Ritkof, R. (2002). The effects of metacognition and email interactions on learning graphing. *Journal of Computer Assisted Learning, 18*(1), 33–43. doi:10.1046/j.0266-4909.2001.00205.x

Kramarski, B., Zemira, R., Arami, M., & Arami, M. (2002). The effects of metacognitive instruction on solving mathematical authentic tasks. *Educational Studies in Mathematics, 49*(2), 225–250. doi:10.1023/A:1016282811724

MacGregor, S. K. (1999). Hypermedia navigation profiles: Cognitive characteristics and information processing strategies. *Journal of Educational Computing Research, 20*(2), 189–206. doi:10.2190/1MEC-C0W6-111H-YQ6A

Manlove, S., Lazonder, A., & de Jong, T. (2006). Regulative support for collaborative scientific inquiry learning. *Journal of Computer Assisted Learning, 22*(2), 87–98. doi:10.1111/j.1365-2729.2006.00162.x

Manlove, S., Lazonder, A., & de Jong, T. (2007). Software scaffolds to promote regulation during scientific inquiry learning. *Metacognition and Learning, 2*, 144–155. doi:10.1007/s11409-007-9012-y

Manlove, S., Lazonder, A., & de Jong, T. (2008). Trends and issues of regulative support use during inquiry learning: Patterns from three studies. *Computers in Human Behavior, 25*(4), 795–803. doi:10.1016/j.chb.2008.07.010

Manlove, S., Lazonder, A., & de Jong, T. (2009). Collaborative versus individual use of regulative software scaffolds during scientific inquiry learning. *Interactive Learning Environments, 17*(2), 105–117. doi:10.1080/10494820701706437

Mayer, R. E. (2005). Introduction to multimedia learing. In Mayer, R. E. (Ed.), *The Cambridge Handbook of Multimedia Learning.* New York: Cambridge University Press.

McKendree, J. (1990). Effective feedback content for tutoring complex skills. *Human-Computer Interaction, 5*(4), 381–413. doi:10.1207/s15327051hci0504_2

McManus, T. F. (2000). Individualizing instruction in a web-based hypermedia learning environment: Nonlinearity, advance organizers, and self-regulated learners. *Journal of Interactive Learning Research, 11*(2), 219–251.

McNamara, D. S., O'Reilly, T. P., Best, R. M., & Ozuru, Y. (2006). Improving adolescent students' reading comprehension with iSTART. *Journal of Educational Computing Research, 34*(2), 147–171. doi:10.2190/1RU5-HDTJ-A5C8-JVWE

Moos, D. C., & Azevedo, R. (2006). The role of goal structure in undergraduates' use of self-regulatory processes in two hypermedia learning task. *Journal of Educational Multimedia and Hypermedia, 15*(1), 49–86.

Moos, D. C., & Azevedo, R. (2008a). Exploring the fluctuation of motivation and use of self-regulatory processes during learning with hypermedia. *Instructional Science: An International Journal of the Learning Sciences, 36*(3), 203–231.

Moos, D. C., & Azevedo, R. (2008b). Monitoring, planning, and self-efficacy during learning with hypermedia: The impact of conceptual scaffolds. *Computers in Human Behavior, 24*(4), 1686–1706. doi:10.1016/j.chb.2007.07.001

Moos, D. C., & Azevedo, R. (2008c). Self-regulated learning with hypermedia: The role of prior domain knowledge. *Contemporary Educational Psychology, 33*(2), 270–298. doi:10.1016/j.cedpsych.2007.03.001

Narciss, S., Proske, A., & Koerndle, H. (2007). Promoting self-regulated learning in web-based learning environments. *Computers in Human Behavior, 23*(3), 1126–1144. doi:10.1016/j.chb.2006.10.006

Nesbit, J. C., Winne, P. H., Jamieson-Noel, D., Code, J., Zhou, M., & MacAllister, K. (2006). Using cognitive tools in gStudy to investigate how study activities covary with achievement goals. *Journal of Educational Computing Research, 35*(4), 339–358. doi:10.2190/H3W1-8321-1260-1443

One Laptop Per Child. (2009). *One Laptop Per Child.* Retrieved April 24, 2009, from http://laptop.media.mit.edu/

Opdenakker, M., Van Damme, J., De Fraine, B., Van Landeghem, G., & Onghena, P. (2002). The effects of schools and classes on mathematics achievement. *School Effectiveness and School Improvement, 13*(4), 399–427. doi:10.1076/sesi.13.4.399.10283

Pea, R. D., Wulf, W. A., Elliot, S. W., & Darling, M. A. (2003). (Eds.) *Planning for two transformations in education and learning technology: Reports from a workshop.* Washington, DC: National Academies Press.

Pintrich, P. R. (2000). The role of goal orientation in self-regulated learning. In Boekaerts, M., Pintrich, P. R., & Zeidner, M. (Eds.), *Handbook of self-regulation* (pp. 451–502). San Diego, CA: Academic Press. doi:10.1016/B978-012109890-2/50043-3

Pintrich, P. R., Smith, D. A. F., Garcia, T., & McKeachie, W. J. (1991). *A Manual for the use of the Motivated Strategies for Learning Questionnaire (MSLQ)*. Ann Arbor, MI: National Center for Research to Improve Postsecondary Teaching and Learning, University of Michigan.

Pressley, M., Wharton-McDonald, R., & Rafael, L. M. (2002). Exemplary first-grade teaching. In Taylor, B., & Pearson, P. D. (Eds.), *Teaching reading: Effective schools, accomplished teachers* (pp. 73–88). Mahwah, NJ: Lawrence Earlbaum.

Proske, A., Narciss, S., & Koerndle, H. (2007). Interactivity and learners' achievement in web-based learning. *Journal of Interactive Learning Research, 18*(4), 511–531.

Reynolds, A. J., & Walberg, H. J. (1991). A structural model of science achievement. *Journal of Educational Psychology, 83*(1), 97–107. doi:10.1037/0022-0663.83.1.97

Rui, L., & Liu, M. (2007). Understanding the effects of databases as cognitive tools in a problem based multimedia learning environment. *Journal of Interactive Learning, 18*(3), 345–363.

Sloan Consortium. (2008). *Staying the Course: Online Education in the United States.* Retrieved March 27, 2009, from http://www.sloanc.org/publications/survey/pdf/staying_the_course.pdf

Tate, W. F. (1995). Returning to the root: A culturally relevant approach to mathematics pedagogy. *Theory into Practice, 34*(3), 166–173. doi:10.1080/00405849509543676

van den Boom, G., Paas, F., van Merriënboer, J. J. G., & van Gog, T. (2004). Reflection prompts and tutor feedback in a web based learning environment: Effects on students' self-regulated learning competence. *Computers in Human Behavior, 20*(4), 551–567. doi:10.1016/j.chb.2003.10.001

Waits, T., & Lewis, L. (2003). *Distance Education at Degree-Granting Postsecondary Institutions: 2000–2001 (NCES 2003-017)*. Washington, DC: U.S. Department of Education, National Center for Education Statistics.

Wang, S., & Lin, S. S. J. (2007). The application of social cognitive theory to web-based learning through NetPorts. *British Journal of Educational Technology, 38*(4), 600–612. doi:10.1111/j.1467-8535.2006.00645.x

Weinstein, C. E. Schulte, A. C. & Palmer, D. R., (1987). *LASSI user's manual*. Clearwater, FL: H & H Publishing.

White, B. Y., & Frederiksen, J. R. (1998). Inquiry, modeling, and metacognition: Making science accessible to all students. *Cognition and Instruction, 16*(1), 3–118. doi:10.1207/s1532690xci1601_2

Wigfield, A., Byrnes, J. P., & Eccles, J. S. (2005). Adolescent development. In Alexander, P. A., & Winne, P. (Eds.), *Handbook of Educational Psychology* (2nd ed.). Mahwah, NJ: Erlbaum.

Winne, P. H. (2005). A perspective on state-of-the-art research on self-regulated learning. *Instructional Science, 33*(5-6), 559–565. doi:10.1007/s11251-005-1280-9

Winne, P. H., & Hadwin, A. F. (1998). Studying as self-regulated learning. In Hacker, D. J., Dunlosky, J., & Graesser, A. C. (Eds.), *Metacognition in educational theory and practice* (pp. 277–304). Mahwah, NJ: Lawrence Erlbaum Associates.

Winne, P. H., & Jamieson-Noel, D. (2002). Exploring students' calibration of self reports about study tactics and achievement. *Contemporary Educational Psychology, 27*(4), 551–572. doi:10.1016/S0361-476X(02)00006-1

Winne, P. H., & Perry, N. E. (2000). Measuring self-regulated learning. In Boekaerts, M., Pintrich, P., & Zeidner, M. (Eds.), *Handbook of self-regulation* (pp. 531–566). Orlando, FL: Academic Press. doi:10.1016/B978-012109890-2/50045-7

Winters, F. I., & Azevedo, R. (2005). High-school students' regulation of learning during computer-based science inquiry. *Journal of Educational Computing Research*, *33*(2), 189–217. doi:10.2190/F7HM-9JN5-JUX8-4BM9

Yang, Y. (2006). Effects of embedded strategies on promoting the use of self-regulated learning strategies in an online learning environment. *Journal of Educational Technology Systems*, *34*(3), 257–269. doi:10.2190/9472-TU0X-1M7J-3Y8Q

Zimmerman, B. (2000). Attaining self-regulation: A social cognitive perspective. In Boekaerts, M., Pintrich, P., & Zeidner, M. (Eds.), *Handbook of self-regulation* (pp. 13–39). San Diego, CA: Academic Press. doi:10.1016/B978-012109890-2/50031-7

Zimmerman, B., & Tsikalas, K. (2005). Can computer-based learning environments (CBLEs) be used as self-regulatory tools to enhance learning? *Educational Psychologist*, *40*(4), 267–271. doi:10.1207/s15326985ep4004_8

KEY TERMS AND DEFINITIONS

Adaptive Scaffold: In task, supportive feature of a TELE that adjusts its responses according to the learner's performance.

Automated Scaffold: In task, supportive feature of a TELE that provides pre-determined, non-adaptive responses to the learner's performance.

Causal Attribution: Process by which a learner links a performance outcome with a controllable cause or uncontrollable cause.

Computer-Based Learning Environment: Learning environment designed only to present content in a computerized form; there are no enhancements or features that promote SRL.

Co-Regulation: In task form of SRL scaffolding where a human tutor assists in prompting self-regulatory behaviors.

Didactic TELE: TELE designed to teach students how to self-regulate.

External regulation: In task form of SRL scaffolding where features of the TELE assist in prompting self-regulatory behaviors.

Facilitative TELE: TELE designed to allow learners to self-initiate self-regulatory behaviors; provides tools that support self-regulatory behaviors.

Forethought Phase: Phase of SRL where the learner engages in task analysis, goal setting, and strategic planning.

Immediate & Elaborate Feedback: TELE's response to a learner's action with minimal time delay & helpful information about the learner's actions.

Metacognitive Processes: Evaluation and use of one's cognitive processes and resources.

Negative Appraisal: Evaluation of a context as being unfavorable to one's well-being.

Opportunity: Culturally defined context in which an individual is presented with content to learn or given favorable setting(s) in which to practice skills.

Opportunity-Propensity Framework: Framework that asserts that learners will attain high levels of achievement with in a particular domain if: (1) they are given authentic contexts to learn and practice skills (2) they are willing and able to take advantage of these contexts.

Parsimony: Problem of multiple, partially overlapping theories in the self-regulation literature which generally explain the same phenomenon with slightly different terminology.

Performance Phase: Phase of SRL where the learner engages in metacognitive and strategic actions.

Positive Appraisal: Evaluation of a context as being favorable to one's well-being.

Pre-Existing SRL Tendencies: Metacognitive and strategic tendencies that a learner was trained in pre-task or that a learner spontaneously demonstrated pre-task.

Propensity: Any factors that relate to the ability or willingness to learn content once it has been exposed or presented in particular contexts.

Relevance: Presumption that SRL may be germane to the learner attaining optimal benefit from TELEs.

Self-Efficacy: Perceptions of one's capabilities to attain a designated outcome.

Self-Reflection Phase: Phase of SRL where the learner evaluates and reacts to their performance outcome.

Self-Regulatory Behaviors: Metacognitive and strategic actions that learners engage in while performing a task.

Self-Regulated Learning (SRL): Adaptive process by which a learner self-initiates metacognitive and strategic actions to perform a task.

Self-Regulatory Tendencies: Metacognitive and strategic actions that learners engage in while performing a task; used interchangeably with the term *self-regulatory behaviors*.

Trace Methodology: Method of measuring SR behaviors in task using a TELE's ability to track the learner's metacognitive and strategic actions.

Utility: This terms refers to the benefit of an Opportunity-Propensity Framework for understanding the relations between SRL and TELEs.

Chapter 2
Measuring and Profiling Self-Regulated Learning in the Online Environment

Lucy Barnard-Brak
Baylor University, USA

William Y. Lan
Texas Tech University, USA

Valerie Osland Paton
Texas Tech University, USA

ABSTRACT

While the presence of technology-enhanced learning environments (TELEs) will only increase in higher education, this book chapter examines current literature concerning the measurement of online SRL behaviors and the application of this online SRL measurement with regard to profiling SRL behaviors in TELEs. The methodologies and issues associated with the measurement of SRL behaviors in TELEs is discussed in view of extant research. The organization of SRL behaviors into five, distinct profiles is then discussed in view of a social cognitive perspective concerning the development of SRL (e.g. Zimmerman & Schunk, 2001). The book chapter concludes with recommendations for future research concerning the presence of SRL profiles and their relationship to other metacognitive factors and academic achievement.

INTRODUCTION

The latest U.S. report on development of distance education (Parsad & Lewis, 2008) describes the rapid growth of distance education at the post-secondary level. Data from the Postsecondary Education Quick Information System (PEQIS) survey indicate that:

...during the 2006-07 academic year, 66% of 2-year and 4-year Title IV degree-granting post-secondary institutions reported offering online, hybrid/blended online, or other distance education courses for any level or audience. Sixty-five

DOI: 10.4018/978-1-61692-901-5.ch002

percent of the institutions reported college-level credit-granting distance education courses, and 23% of the institutions reported noncredit distance education courses. (Parsad & Lewis, 2008, p. 2)

The report estimates that during the academic year of 2006-2007, 12.2 million students enrolled in college-level credit granting distance education courses, and 77% of these students were enrolled in online courses. Compared to an estimate of 2.8 million students who were enrolled in distance education in the PEQIS report six years ago in 2000-2001 (Waits & Lewis, 2003), it is apparent that the growth of distance education, especially online education, is exponential. For the purposes of the current book chapter, online learning refers to the delivery of typical course curriculum via the medium of the Internet.

Among the modalities used for distance education, online instruction was reported as the most popular mode of course delivery: 61% of the institutions that offered distance education courses reported offering online courses, 35% reported offering hybrid/blended courses that at least had an online component. Sixty-two percent of the institutions that offered online courses required the online courses be 100% delivery online (Parsad & Lewis, 2008). Asynchronous internet-based technology was cited as the most widely used technology for delivery of distance education courses: 75% of institutions reported that they used the technology to a large extent and 17% used it to a moderate extent. Compared with only 43% of the institutions that used online delivery four years ago (Waits & Lewis, 2003), online instruction and learning is playing an increasingly significant role in distance education.

One of the characteristics of online learning, or TELEs in general, is the autonomy students experience in the learning environment. Online instruction eliminates the limitation of place, time, and physical materials and to a great degree gives students the control over when, what, and how to study (Cunningham & Billingsley, 2003). Some researchers believed that the online environment allowed instructors to present information in a nonlinear fashion that gave students the freedom to unrestrictedly move from one topic to another "without concern for predetermined order or sequence" (McManus, 2000, p. 221). These researchers (McManus, 2000) believed information obtained from online instruction was more personally relevant than what they learned from traditional classroom. Other researchers (Bowen, 1996) found the autonomous online environment was the most beneficial for the students with an internal locus of control who believed they had control over things that happened to them: students with internal accountability beliefs performed better than students with an external locus of control in online courses. When having options, students chose online courses over traditional face-to-face courses mostly because they valued the autonomy to determine the pace and timing of learning (Roblyer, 1999). A considerable body of literature also supported the importance of SRL behaviors in TELEs (e.g. Ally, 2004; Barnard, Lan, Crooks, & Paton, 2008; Barnard, Lan, To, Paton, & Lai, 2009; Barnard, Paton, & Lan, 2008; Fisher & Baird, 2005).

Because the online environment is characterized with autonomy, self-regulation becomes a critical success factor for online learning. This book chapter asserts that it is important not only to measure self-regulated learning in the online learning environment but to profile learners according to the self-regulated skills and strategies that they endorse. While the reliable and valid measurement of self-regulated learning skills and strategies is indeed important because it depicts the current developmental level of self-regulation of students, it is just one half of the challenge of helping learners to succeed in the online learning environment. The other half of this challenge is to understand the trajectory of the development of online self-regulation and profile these learners in order to help them become self-regulated in the online learning environment. All learners are not the same or similar in the development of these self-regulated learning skills and strategies (Barnard-Brak, Lan, & Paton, 2009). By profiling learners according to the self-regulated learning skills and strategies that they endorse, we can more effectively focus interventions on where learners are in development of these self-regulated learning skills and strategies.

BACKGROUND

Self-regulated learning (SRL) refers to students' self-generated thoughts, feeling, and actions, which are systematically oriented toward attainment of their goals (Zimmerman & Schunk, 2001). Self-regulated learning is carried out with skills and strategies that can include but are not limited to: goal setting, environment structuring, self-monitoring, help seeking, and task strategies (Zimmerman & Schunk, 2001). In view of social cognitive theory, these SRL behaviors develop as a function of the reciprocal interaction of personal, environmental, and behavioral factors (Bandura, 1986; 1997; Schunk, 2001; Zimmerman, 1994). In this sense, the development of SRL strategies may be considered cyclical, suggesting an iterative process. Thus, each of these personal, environmental, and behavioral factors interact, adjusting and modifying the developmental cycle of SRL strategies, including affective, cognitive and adaptations. In view of this social cognitive perspective, Zimmerman (1998) suggested a three-phase model through which SRL strategies develop. These three phases consist of a *forethought* phase, *performance/volitional control* phase, and *self-reflection* phase. In the *forethought* phase, those SRL strategies that precede learner performance, such as goal-setting, occur. In the *performance/volitional control* phase, those SRL strategies, such as environment structuring, self-monitoring, help seeking, and task strategies, that occur during learning are emphasized. In the *self-reflection* phase, those SRL strategies, such as self-evaluation and attribution, that influence learners to monitor and react to the outcomes associated with their performance are utilized and alter the elements of the forethought phase for a next learning task. It should be noted that motivational and affective factors can play an important role in the development of SRL skills and strategies, especially in view of social cognitive theory (Bandura, 1986; 1997). Social cognitive theory provided the framework in which Zimmerman and Schunk (2001) developed their SRL model, which includes the triadic reciprocal causation that occurs between the person, the behavior, and the environment.

SELF-REGULATED LEARNING

Measuring Self-Regulated Learning

Self report measures of SRL have tended to dominate research examining this important meta-cognitive construct. The most notable of these measures has been the Motivated Strategies for Learning Questionnaire (MSLQ) (Pintrich, Smith, Garcia & McKeachie, 1993). The MSLQ is a domain-general, 81-item questionnaire with a 7-point Likert-type response format consisting of values ranging from "not at all true of me" to "very true of me." From these eighty-one items, the MSLQ measures the two primary constructs of learning strategies and motivation. Both of these primary constructs are further composed of subscales to measure lower-order constructs. The primary construct of learning strategies has two of these subscales: cognitive-meta-cognitive and resource management. While the primary construct of motivation has three of these subscales: valuing, expectancy, and affect.

Two more notable domain-general measures of SRL have been the Learning and Study Strategies Inventory (LASSI) (Weinstein, Schulte & Palmer, 1987) and the Self-Regulated Learning Interview Scale (SRLIS) (Zimmerman & Martinez-Pons,

1986; 1988). The LASSI is an 80-item questionnaire with a 5-point Likert-type response format consisting of values ranging from "not at all typical of me" to "very much typical of me." The LASSI is composed of ten scales: concentration, selecting main ideas, information processing, motivation, attitude, anxiety, time management, study aids, self-testing, and test strategies. These latter four scales of the LASSI appear to be directed to measuring SRL strategies. The SRLIS is a structured interview instrument where individuals respond to open-ended questions, which are subsequently coded according to fourteen SRL categories. The SRLIS may be considered a multi-domain-specific instrument to measure SRL behaviors across multiple domain-specific learning contexts in order to measure and capture a domain-general view of an individual's SRL behaviors.

While domain-general, self-report measures of SRL have been the most notable given their applicability to a wide breath of learning contexts, many self-report measures have been developed that are entirely context-specific to the domain of interest. These self-report SRL measures that have been contextualized to a particular domain of interest may be considered as preferred given that the development of SRL skills and strategies have been indicated as being context-specific (Schunk, 2001). Schunk (2001) has indicated that the development and execution of self-regulated learning behaviors is "highly context dependent" (p. 125). Given this domain specificity of self-regulated learning skills and strategies, the measurement of SRL behaviors would appear to be best measured in a manner that is specific to the learning context. The three aforementioned SRL instruments were designed to measure SRL strategies in regular, face-to-face classrooms and may not be appropriate to be used to measure SRL strategies in TELEs. When attempting to explain a significant and negative relationship between SRL and performance in an online learning task reported in McManus' study (2000), researchers suspected the mismatch between the measurement of SRL and the learning context might be the reason (Lan, Bremer, Stevens, & Mullen, 2004).

To measure SRL in both the online and blended learning environments the Online Self-regulated Learning Questionnaire (OSLQ) has been developed (Barnard, Lan, To, Paton, & Lai, 2009). The OSLQ is a 24-item questionnaire with a five-point Likert-type style response format. Items included in the instrument were generated from a structured interview where students who had online learning experiences were asked to share their strategies to regulate their own learning activities to accomplish learning goals. A copy of the items in this instrument may be obtained from the appendix of Barnard, Lan, To, Paton, and Lai (2009). Developed in view of Schunk and Zimmerman's (2001) model, the OSLQ assesses six sub-processes of SRL: goal setting, environment structuring, help seeking, task strategies, time management, and self-evaluation. In utilizing the OSLQ, research has indicated sufficient psychometric properties (Barnard, Lan, Crooks & Paton, 2008; Barnard, Paton & Lan, 2008) along with its validation across two study samples of online and blended learners based upon reliability and confirmatory factor analyses (Barnard, Lan, To, Paton, & Lai, 2009). This research using the OSLQ has also suggested evidence towards the ecological validity of the scale. Barnard, Lan, Crooks, and Paton (2008) found that SRL as measured by the OSLQ was positively associated with academic achievement. Barnard, Paton, and Lan (2008) echoed this finding also indicating a positive association between SRL as measured by the OSLQ and academic achievement. The work of Barnard, Lan, To, Paton, and Lai (2009) provides a measurement analysis of the OSLQ, which revealed its sound psychometric properties across two samples. Below are two examples of items from the OSLQ from the goal setting and help seeking scales respectively:

- I set standards for my assignments in online courses.

- I am persistent in getting help from the instructor through e-mail.

While self report measures can provide an important means of measuring the construct, other means of measuring SRL have also been employed. For instance, the examination of messages exchanged between learners and instructors may be considered another self-report form of measuring SRL activities that may be considered more interactional over typical self-report measures (Dettori, Giannetti, & Persico, 2006; Dettori & Persico, 2008). Thus, other methods, such as trace methods and think-aloud protocols, to measure observable self-regulated learning behaviors have been suggested. Traces refer to those observable and measurable behaviors on the part of the individual that may be considered as indicative of the cognition of the individual (Winne & Perry, 2000). Methods for measuring these trace behaviors or traces include recording the frequency and/or pattern of observable and measurable SRL behaviors through ICT. The ICT utilized in recording traces can arrange from the low to high tech. For instance, mouse-tracking software can be employed to record the number of hyperlinks that a learner accesses in studying materials presented online and where the mouse cursor was placed with respect to the content displayed on the computer monitor. Eye-tracking equipment software may be considered another ICT used to record traces of SRL behaviors. Eye-tracking equipment and software can follow the gaze of a learner in studying materials presented on a computer monitor. Think-aloud protocols measure observable and unobservable SRL behaviors *in temporis* (in real or present time) as learners report their thoughts about the materials they are studying and how they are studying these materials. These think-aloud protocols are directed by learners thus permitting open-ended responses that can later be coded. Promising research has been conducted utilizing think-aloud protocols to measure SRL behaviors (e.g. Azevedo & Cromley, 2004; Greene & Azevedo, 2007).

Issues Associated with Measuring Self-Regulated Learning

Much of the research concerning SRL behaviors has utilized self-report measures. These self-report measures of SRL, however, have been criticized as being biased and inaccurate. Winne and Jamieson-Noel (2002) found that learners tended to over-estimate their SRL activities in self-report measures as compared to those SRL activities measured by trace methods. This positive bias in the self-report of SRL behaviors, however, may be viewed as a form of measurement error that accompanies the measurement of many *self*-related constructs such as self-estimates of intelligence. The *above average* effect, also known as the Lake Wobegon Effect (Kruger, 1999), is a positive bias where individuals tend to estimate their own intelligence and ability as *above average* as compared to the group at large. Interestingly, the prevalence of the *above average* effect has been previously noted as possibly, "...pos[ing] issues of self-regulation for students who overestimate their abilities" (Barnard & Olivarez, 2007). Despite self-reported measures tending to be a biased over-estimate of SRL activities, we adopted a self-report measure as this bias over-estimate would appear to translate to all self-report measures, thus may be considered an issue of measurement error.

Trace methods, where ICT record observable SRL activities such as through a computer software program, have been suggested as a better means of measuring SRL activities. Zimmerman (2008) discusses the use of trace methods in the measurement of SRL and describes their potential as "extraordinary" (p. 170). Trace methods are, indeed, precise and eliminate the aspect of human error in the measurement of SRL activities. The disadvantage of trace methods is that all observable behaviors by the learner, involuntary and voluntary, are recorded. For instance, if a student lingers on a particular set of materials and stares off into space or day-dreams, then that time may be recorded as studying. In considering trace methods in measuring SRL activities, there is no differentiation between those behaviors that are a function of voluntary self-regulation and those that are not. Martin (2004) posited that "agency" or "...the capability of individual human beings to make choices and to act on these choices in ways that make a difference in their lives" is an important assumption of SRL (p. 135). Involuntary behavior, however, are not a function of human agency; therefore, trace methods can be viewed as problematic given that these methods record all observable behaviors, both voluntary and involuntary.

Another issue surrounding trace methods for the measurement of SRL is that these methods measure only activities and do not include measurement of cognitive and affective domains. This limitation of trace methods may be one of the reasons why students are appearing to overestimate their self-regulated learning abilities (Winne & Jamieson-Noel, 2002). Trace methods do not record a learner rehearsing material in their head, self-quizzing, affective or attitudinal SRL strategies. While trace methods are helpful, these methods must be used in conjunction with a self-report measure to capture those unobservable and voluntary SRL strategies. Think-aloud protocols have been criticized for similar reasons in terms of unobservable SRL strategies. Zimmerman (2008) noted that think-aloud protocols have yet been developed to measure planning, goal setting, and other *forethought* phase SRL strategies that precede the process of studying on the part of the learner.

Profiling Self-Regulated Learning

This overview of issues related to measuring SRL activities sets the stage for our discussion of profiling learners' online SRL strategies. Based on reliable and valid measurement of online SRL strategies, we can address even more important tasks of identifying the trajectory of development of online SRL strategies and profiling characteristics of online SRL learners. Online SRL behaviors are multi-faceted and complex as a function of both *skill* and *will* on the part of learners (Woolfolk, Winne & Perry, 2000). In accurately characterizing the online SRL strategies of learners, we can develop profiles of different kinds or types of self-regulated learning behaviors exhibited by learners who are in different phases of the development of online SRL behaviors. In a study conducted by Barnard-Brak, Lan and Paton (2010) using standardized scores derived from the OSLQ, the researchers indicated the presence of five, distinct profiles of online self-regulated learning using mixture modeling techniques (Nagin, 2005). Mixture modeling techniques permit unobserved heterogeneity to be examined to reveal distinct latent groups or classes among a data set (see Nagin (2005) for more information). These five, distinct profiles of online SRL were replicated across two different study samples of online learners, which may be considered as evidence towards the cross-validation of mixture model results. The first profile of the five, revealed across the two samples, consisted of learners who appeared to be the least self-regulated in their learning. These *non-self-regulators* or minimal self-regulators endorsed the least online SRL skills and strategies as measured across all subscales of the OSLQ (Barnard-Brak, Lan, & Paton, 2010).

The second and third profiles consisted of learners who appeared to be disorganized in their self-regulation. Learners belonging to the second profile appeared to more highly endorse goal setting and environment structuring skills, strategies used in the forethought phase in Zimmerman's (2001) model, while endorsing other online SRL skills and strategies to a lesser extent. Thus, learners belonging to the second profile were characterized as *forethought-endorsing self-regulators* given their emphasis on strategies associated with the *forethought* phase of online SRL development

(Zimmerman 1998; Zimmerman & Schunk, 2001). Conversely, learners belonging to the third profile appeared to more highly endorse task strategies, time management, and self-evaluation, strategies utilized in the performance/volitional control and self-reflection phases in Zimmerman's (2001) model, while endorsing other online SRL skills and strategies to a lesser extent. Thus, learners belonging to this third profile were characterized as *performance/reflection-endorsing self-regulators* in keeping with the *performance control* or *self-reflection* phases of the development of online SRL (Zimmerman, 1998; Zimmerman & Schunk, 2001).

The remaining two online SRL profiles were more distinguishable than the second and third profiles of disorganized self-regulators. The fourth profile consisted of learners who appeared to be highly self-regulated, which were referred to as *super self-regulators* (Barnard-Brak et al., 2009). These super self-regulators endorsed highly the skills and strategies associated with SRL across all subscales of the OSLQ. The fifth, and final, profile consisted of learners who appeared to moderately to highly endorse SRL skills and strategies but not to the same extent of those learners belonging to the *super self-regulators* profile class. These learners were referred to as *competent self-regulators* in that these learners may be considered as doing what it takes in order to achieve in their learning environment (Barnard-Brak, Lan, & Paton, 2010).

After discerning the presence of five, distinct profiles of online SRL across two, separate study samples, Barnard-Brak, Lan, and Paton (2010) then examined the association of academic achievement with profile membership. As "self-regulated learning is seen as a mechanism to help explain achievement differences among students and as a means to improve achievement..." (Schunk, 2005, p. 85), the association between online SRL and academic achievement becomes all the more relevant. In conducting their analyses, Barnard-Brak, Lan, and Paton (2010) indicated that *super* and *competent self-regulators* did not differ significantly in their academic achievement as measured by grade point average (GPA). This result suggests that learners do not necessarily have to be exceptionally self-regulated in their learning yet must be self-regulated enough to achieve in their learning environment. Competent self-regulators may represent a savvy set of learners who are able to navigate their learning environment and figure out how much and when to be self-regulated in their learning in order to achieve. Non-self-regulators (Profile 1) did not differ significantly in their academic achievement from disorganized self-regulators (e.g. Profile 2 -*forethought* and Profile 3 - *performance/reflection endorsing self-regulators*). This result would indicate that a learner being disorganized in their online SRL strategies is just as disadvantaged as being a non-self-regulator (Barnard-Brak, Lan, & Paton, 2010). Super and competent self-regulators did have significantly higher academic achievement as measured by GPAs than individuals belonging to the other three profiles with a Cohen's f value of.65. This value of Cohen's f indicates a large effect size with values of.10,.25, and.40 or larger indicating small, medium, and large effect sizes respectively (Cohen, 1988).

These SRL profiles appear to be powerful in highlighting how learners self-regulate in view of theory regarding their development across time (e.g. Zimmerman & Schunk, 2001). Additionally, these SRL profiles appear to significantly differ in their association with academic achievement outcomes as measured by GPA. Thus, these SRL profiles become even more efficacious as relating to academic achievement outcomes. While the results of Barnard-Brak et al. (2009) are promising, we suggest that the presence of these SRL profiles be examined across other study samples to provide further evidence as to the generalizability of these SRL profiles beyond the online learning environment and other TELEs. Additionally, the presence of these SRL profiles should be examined across other study samples employing a different means of measurement besides the OSLQ. A key limitation of these profiles as developed was the absence of data on characteristics relating to intelligence or motivation. Thus, future research should consider controlling these variables by the prior achievement or some motivation scale.

FUTURE RESEARCH DIRECTIONS

Future research should explore how these online SRL profiles may be utilized to provide early assessment and intervention for learners in TELEs. Previous research has indicated that learners do not automatically develop these self-regulated learning skills and strategies (Barnard-Brak, Paton, & Lan, *in press*). Thus, developing specific interventions according to the learner's online SRL profile may improve academic outcomes. We hypothesize that learners belonging to the disorganized SRL profiles (e.g. Profile 2 - *forethought* and Profile 3 - *performance/reflection endorsing self-regulators*) may be considered the most promising for intervention, in that these learners do appear to endorse at least some of the online SRL skills and strategies. Schunk (2001) has noted that self-regulated learning behaviors are "highly context dependent" (p. 125). Therefore, future research should replicate the results of Barnard-Brak, Lan, and Paton (2010) across several domains in order to cross-validate findings of indicating the presence of these five, distinct online SRL profiles. In addition, future research should also consider examining the presence of these five profiles across time. The need for research to examine the longitudinal stability of self-regulated learning over time has been noted (Meece, 1994). Vermetten, Vermunt, and Lodewijks (1999) have additionally noted the scarcity of longitudinal research concerning the online SRL of students in higher education. Thus, latent transition analyses should be employed to examine online SRL profile membership across time. For instance, we hypothesize that a learner may belong to a disorganized profile (e.g. Profile 1 - *forethought* or Profile 3 - *performance/reflection endorsing self-regulators*) at the beginning of his/her college matriculation but develop into a *competent* or *super self-regulator* (Profiles 4 and 5) by the end of their studies in higher education. Alternatively, some learners could stagnate in the development of the online SRL skills and strategies and remain disorganized.

Additionally, future research should consider examining the relationship between other metacognitive factors such as epistemological beliefs and SRL (e.g. Barnard, 2007; Barnard, Lan, Crooks, & Paton, 2008; Pintrich & Zusho, 2002). A learner's SRL online profile membership may be associated with other metacognitive factors such as epistemological beliefs. We hypothesize that individuals with more sophisticated, availing, or constructivist-oriented epistemological beliefs would appear to be more likely characterized as *super* or *competent self-regulators* in view of the five, distinct profiles discerned by Barnard-Brak, Lan, and Paton (2010) using the OSLQ. This hypothesis appears to have some initial evidence to support its claim in view of extant literature. In another study of online learners, Barnard, Lan, Crooks, and Paton (2008) found a statistically significant and highly positive relationship between more sophisticated epistemological beliefs and online SRL skills and strategies. The association between these epistemological beliefs and SRL skills and strategies, however, has yet to be examined in view of the five, distinct profiles discerned by Barnard-Brak, Lan, and Paton (2010).

CONCLUSION

This chapter has examined current literature concerning the measurement and profiling of the presence of SRL strategies in TELEs, specifically the online learning environment. While there is much research to be conducted concerning the measurement and profiling of online SRL strategies, we suggest that current literature provides much direction regarding future research with specific attention to online SRL profiles that have emerged (Barnard-Brak, Lan, & Paton, 2010). We suggest that methodological advancements such as log-analysis (or tracing) and mixture modeling techniques in software packages such as M*Plus* (*v.* 5.20; Muthén & Muthén, 2008) make this research more than possible to conduct. We also hope researchers will design and conduct longitudinal studies to enrich our understanding of the trajectory of development of online SRL. We anticipate that, based on findings with regard to trajectory and profiles of online SRL development, individualized interventions may be developed to help learners become skillful in online self-regulation and successful in online learning tasks. Thus, the practical implication of these profiles is that we may begin to help individuals based upon their specific profile be better able to allocate time and resources towards interventions.

REFERENCES

Ally, M. (2004). Foundations of educational theory for online learning. In Anderson, T. (Ed.), *The Theory and Practice of Online Learning* (pp. 15–44). Edmonton, AB: Athabasca University Press.

Azevedo, R., & Cromley, J. G. (2004). Does training on self-regulated learning facilitate students' learning with hypermedia? *Journal of Educational Psychology*, *96*(3), 523–535. doi:10.1037/0022-0663.96.3.523

Bandura, A. (1986). *Social Foundations of Thought and Action: A Social Cognitive Theory*. Englewood Cliffs, NJ: Prentice-Hall.

Bandura, A. (1997). *Self-efficacy: The exercise of control*. New York, NY: Freeman.

Barnard, L. (2007). The expert ceiling in epistemological beliefs. *Essays in Education*, *19*(1), 85–94.

Barnard, L., Lan, W. Y., Crooks, S. M., & Paton, V. O. (2008). The relationship of epistemological beliefs with self-regulatory skills in the online course environment. *Journal of Online and Learning Teaching*, *4*(3), 261–266.

Barnard, L., Lan, W. Y., To, Y. M., Paton, V. O., & Lai, S. L. (2009). Measuring self-regulation in online and blended learning environments. *The Internet and Higher Education*, *12*(2), 1–6. doi:10.1016/j.iheduc.2008.10.005

Barnard, L., & Olivarez, A. (2007). Self-estimates of multiple, *g* factor, and school-valued intelligences. *North American Journal of Psychology*, *9*(3), 501–510.

Barnard, L., Paton, V. O., & Lan, W. Y. (2008). Online self-regulatory learning behaviors as a mediator in the relationship between online course perceptions with achievement. *International Review of Research in Open and Distance Learning*, *9*(2), 1–11.

Barnard-Brak, L., Lan, W. Y., & Paton, V. O. (2010). Profiles in self-regulated learning in the online learning environment. *International Review of Research in Open and Distance Learning*, *11*(1), 61–80.

Barnard-Brak, L., Paton, V. O., & Lan, W. Y. (2010). Self-regulation across time of first-generation online learners. *Journal of Association of Learning and Technology*, *18*(1), 61–70.

Bowen, V. S. (1996). The relationship of locus of control and cognitive style to self-instructional strategies, sequencing, and outcomes in a learner-controlled multimedia environment. *Dissertation Abstracts International Section A: Humanities & Social Sciences*, *56*(10-A), 3922.

Cohen, J. (1988). *Statistical power analysis for the behavioral sciences* (2nd ed.). Hillsdale, NJ: Lawrence Erlbaum & Associates.

Cunningham, C. A., & Billingsley, M. (2003). *Curriculum Webs: A practical guide to weaving the Web into teaching and learning*. Boston: Allyn and Bacon.

Dettori, G., Giannetti, T., & Persico, D. (2006). SRL in online cooperative learning: implications for pre-service teacher training. *European Journal of Education, 41*(3/4), 397–414. doi:10.1111/j.1465-3435.2006.00273.x

Dettori, G., & Persico, D. (2008). Detecting self-regulated learning in online communities by means of interaction analysis. *IEEE Transactions on Learning Technologies, 1*(1), 11–19. doi:10.1109/TLT.2008.7

Fisher, M., & Baird, D. E. (2005). Online learning design that fosters student support, self-regulation, and retention. *Campus-Wide Information Systems, 22*(5), 88–107. doi:10.1108/10650740510587100

Greene, J. A., & Azevedo, R. (2007). Adolescents' use of self-regulatory processes and their relation to qualitative mental model shifts while using hypermedia. *Journal of Educational Computing Research, 36*(2), 125–148. doi:10.2190/G7M1-2734-3JRR-8033

Kruger, J. (1999). Lake Wobegon be gone! The 'below-average effect' and the egocentric nature of comparative ability judgment. *Journal of Personality and Social Psychology, 77*(2), 221–232. doi:10.1037/0022-3514.77.2.221

Lan, W., Bremer, R., Stevens, T., & Mullen, G. (2004, April). Self-regulated learning in the online environment. Paper presented at AERA 2004 annual meeting, San Diego, CA.

Martin, J. (2004). Self-regulated learning, social cognitive theory, and agency. *Educational Psychologist, 39*(2), 135–145. doi:10.1207/s15326985ep3902_4

McManus, T. F. (2000). Individualizing instruction in a Web-based hypermedia learning environment: Nonlinearity, advance organizers, and self-regulated learners. *Journal of Interactive Learning Research, 11*(3), 219–251.

Meece, J. L. (1994). The Role of Motivation in Self-regulated Learning. In Schunk, D. H., & Zimmerman, B. J. (Eds.), *Self-regulation of learning and performance: Issues and educational applications* (pp. 25–44). Mahwah, NJ: Lawrence Erlbaum.

Muthén, L. K., & Muthén, B. O. (2008). *MPlus User's Guide*. Los Angeles, CA: Muthén & Muthén.

Nagin, D. S. (2005). *Group-Based Modeling of Development*. Cambridge, MA: Harvard University Press.

Parsad, B., & Lewis, L. (2008). Distance education at degree-granting postsecondary institutions: 2006-07. Retrieved August 16, 2009, from http://nces.ed.gov/pubs2009/2009044.pdf.

Pintrich, P. R., Smith, D. A. F., Garcia, T., & McKeachie, W. J. (1993). Reliability and predictive validity of the motivated strategies for learning questionnaire (MLSQ). *Educational and Psychological Measurement, 53*(2), 801–813. doi:10.1177/0013164493053003024

Pintrich, P. R., & Zusho, A. (2002). The development of academic self-regulation: The role of cognitive and motivational factors. In Wigfield, A., & Eccles, J. S. (Eds.), *Development of achievement motivation* (pp. 249–284). San Diego, CA: Academic Press. doi:10.1016/B978-012750053-9/50012-7

Roblyer, M. D. (1999). Is choice important in distance learning? A study of student motives for taking Internet-based courses at the high school and community college levels. *Journal of Research on Computing in Education, 32*(1), 157.

Schunk, D. H. (2001). Social cognitive theory and self-regulated learning. In Zimmerman, B. J., & Schunk, D. H. (Eds.), *Self-regulated learning and academic achievement* (2nd ed.). Mahwah, NJ: Lawrence Erlbaum.

Schunk, D. H. (2005). Self-regulated learning: The educational legacy of Paul R. Pintrich. *Educational Psychologist, 40*(2), 85–94. doi:10.1207/s15326985ep4002_3

Vermetten, Y. J., Vermunt, J. D., & Lodewijks, H. G. (1999). A longitudinal perspective on learning strategies in higher education-different viewpoints towards development. *The British Journal of Educational Psychology, 69*(2), 221–242. doi:10.1348/000709999157699

Waits, T., & Lewis, L. (2003). *Distance education at degree granting postsecondary institutions: 2000-2001.* Retrieved November 19, 2008, from http://nces.ed.gov/surveys/peqis/publications/2003017/

Weinstein, C. E., Schulte, A. C., & Palmer, D. R. (1987). *LASSI: Learning and Study Strategies Inventory.* Clearwater, FL: H. & H.

Winne, P. H., & Jamieson-Noel, D. (2002). Exploring students' calibration of self reports about study tactics and achievement. *Contemporary Educational Psychology, 27*(1), 551–572. doi:10.1016/S0361-476X(02)00006-1

Winne, P. H., & Perry, N. E. (2000). Measuring self-regulated learning. In Boekaerts, M., Pintrich, P., & Zeidner, M. (Eds.), *Handbook of self-regulation* (pp. 532–566). Orlando, FL: Academic Press. doi:10.1016/B978-012109890-2/50045-7

Woolfolk, A. E., Winne, P. H., & Perry, N. E. (2000). *Educational psychology*. Scaborough, ON: Allyn and Bacon.

Zimmerman, B. J. (1994). Dimensions of academic self-regulation: A conceptual framework for education. In Schunk, D. H., & Zimmerman, B. J. (Eds.), *Self-regulation of learning and performance* (pp. 3–21). Hillsdale, NJ: Lawrence Erlbaum.

Zimmerman, B. J. (1998). Academic studying and the development of personal skill: A self-regulatory perspective. *Educational Psychologist, 33*(2), 73–86. doi:10.1207/s15326985ep3302&3_3

Zimmerman, B. J. (2008). Investigating self-regulation and motivation: Historical background, methodological developments, and future prospects. *American Educational Research Journal, 45*(1), 166–183. doi:10.3102/0002831207312909

Zimmerman, B. J., & Martinez-Pons, M. (1986). Development of a structured interview for assessing students' use of self-regulated learning strategies. *American Educational Research Journal, 23*(1), 614–628.

Zimmerman, B. J., & Martinez-Pons, M. (1988). Construct validation of a strategy model of student self-regulated learning. *Journal of Educational Psychology, 80*(3), 284–290. doi:10.1037/0022-0663.80.3.284

Zimmerman, B. J., & Schunk, D. H. (2001). *Self-regulated learning and academic achievement: Theoretical perspectives* (2nd ed.). Mahwah, NJ: Lawrence Erlbaum.

ADDITIONAL READING

Allen, I. E., & Seaman, J. (2006). *Making the grade: Online education in the United States, 2006.* Needham, MA: Sloan-C.

Arbaugh, J. B. (2004). Learning to learn online: A study of perceptual changes between multiple online course experiences. *The Internet and Higher Education*, *7*(3), 169–182. doi:10.1016/j.iheduc.2004.06.001

Barnard, L., Paton, V. O., & Rose, K. (2007). Perceptions of online course communications and collaboration. *Online Journal of Distance Learning Administration, 10*(4). Available online from: http://www.westga.edu/~distance/ojdla/winter104/barnard104.html

Boekaerts, M., & Cascallar, E. (2006). How far have we moved toward the integration of theory and practice in self-regulation? *Educational Psychology Review*, *18*(1), 199–210. doi:10.1007/s10648-006-9013-4

Kramarae, C. (2001). *The third shift: Women learning online*. Washington, DC: American Association of University Women Educational Foundation Press.

Lynch, R., & Dembo, M. (2004). The relationship between self-regulation and online learning in a blended learning context. *International Review of Research in Open and Distance Learning*, *5*(2), 1–16.

Muthén, B. O. (2002). Beyond SEM: General latent variable modeling. *Behaviormetrika*, *29*(1), 81–117. doi:10.2333/bhmk.29.81

Nota, L., Soresi, S., & Zimmerman, B. J. (2004). Self-regulation and academic achievement and resilience: A longitudinal study. *International Journal of Educational Research*, *41*(3), 198–215. doi:10.1016/j.ijer.2005.07.001

Pintrich, P. R. (2000). The role of goal orientation in self-regulated learning. In Boekaerts, M., Pintrich, P. R., & Zeidner, M. (Eds.), *Handbook of self-regulation* (pp. 451–502). San Diego, CA: Academic Press. doi:10.1016/B978-012109890-2/50043-3

Severiens, S., Ten Dam, G., & Wolters, B. V. H. (2001). Stability of processing and regulation strategies: Two longitudinal studies on student learning. *Higher Education*, *42*(4), 437–453. doi:10.1023/A:1012227619770

KEY TERMS AND DEFINITIONS

Information and Communication Technology (ICT): Technology that has applications pertaining to the dissemination of information and the communication of individuals and organizations across time and/or space.

Online Self-Regulated Learning Questionnaire (OSLQ): A self-report scale that measures the self-regulated learning skills and strategies that learners endorse in the online and blended learning environments (for more information and the complete scale including its psychometric properties, see Barnard, Lan, To, Paton & Lai, 2009).

Self-Regulated Learning (SRL): The skills and strategies invoked by individuals in order to achieve in their learning environment.

Technology-Enhanced Learning Environment (TELE): Learning contexts which are enhanced or enriched by the use of one or more technology.

Chapter 3
Design of the SEAI Self-Regulation Assessment for Young Children and Ethical Considerations of Psychological Testing

Jesús de la Fuente
University of Almería, Spain

Antonia Lozano
University of Almería, Spain

ABSTRACT

As knowledge in the area of self-regulated learning has progressively expanded, there is a perceived need for new methods and assessment instruments that are in line with the construct and with the subject. Computer-assisted assessment has been proposed as an excellent means for responding to these demands for new types of measurement. Nonetheless, new instruments and assessment processes must be submitted to the same ethical standards required elsewhere, whether in aspects relating to design or to usage. Development of the SEAI program was guided by a psychological model as well as a model for designing computer-aided assessment. This chapter presents the SEAI program design, and explains how both its design and use seek to meet ethical standards related to computer-aided assessment.

INTRODUCTION

In the last two decades, the process of self-regulated learning has been the object of numerous studies in Educational Psychology research (Elliot & Dweck, 2007; De la Fuente & Mourad, 2010; Post, Boyer & Brett, 2006; Schunk & Zimmerman, 2008). This fact has contributed to the development of increasingly complex and complete models which attempt to incorporate the diverse cognitive, metacognitive, affective, motivational and contextual elements, as well as the interaction

DOI: 10.4018/978-1-61692-901-5.ch003

and synergy produced between them (Lozano, 2009). Without a powerful framework it is very difficult to give meaning to the empirical evidence that has been supplied, referring to reliability and validity aspects of measuring instruments (Pintrich, 2000; 2004). However, more than a few difficulties remain for the advancement of knowledge in this field: development of theory requires empirical support for its validation, and this means having the necessary assessment instruments. From this perspective, assessment encompasses every systematic method that can be used for collecting information and evidence about the student's progress, process or outcomes (Brinke, Van Bruggen, Hermans, Giesbergs, Koper & Latour, 2007).

This chapter has two basic objectives. The first is to present the design of an assessment instrument which conforms to a psychological model of self-regulated learning and also meets the guidelines of a model for designing computer-aided assessment. The second objective is to establish how this software, as a computerized platform for assessment, meets ethical assessment standards.

The chapter is organized as follows: we begin with an examination of theoretical models which form the basis for the software design, and proceed with an explanation of how ethical standards for assessment instruments have been taken into account.

ISSUES ASSOCIATED WITH ASSESSING SELF-REGULATED LEARNING

In the area of self-regulated learning assessment, several authors point to a series of issues that should be addressed (De la Fuente & Lozano, 2010; Butler, 2002; Torrano & González, 2004; Winne & Perry, 2000):

- There is a need for further data triangulation through different protocols and measurement comparison rules. More research is needed to coordinate different measurements, making it possible to characterize the complete spectrum of self-regulated learning.
- There has been little research on self-regulated learning strategies in small children (6 years and younger), so that practically no measurement protocols have been developed which are appropriate to this developmental stage.
- There is a need to devise new methods and designs for coordinated assessment of self-regulation in its two expressions, as an aptitude and as an event. So far this construction has almost always been measured through self-report measures; additional methods and measuring instruments should be created and validated, complementary to the use of self-reports, so as to allow self-regulated learning to be assessed as a dynamic, continuous process that unfolds over time and in a specific context.

Self-regulated learning is by definition a response behavior (Hadwin, Winne, Stockley, Nesbit & Woszcyna, 2001). Numerous researchers have reached the conclusion that, in order to achieve an adequate, comprehensive model of this construct, the self-regulation process should be investigated while it is being produced (Boekaerst & Cascallar, 2006; Boekaerst & Corno, 2005). According to Pintrich (2004), self-report assessment only shows us the pupil's predisposition to use self-regulated learning strategies. Other procedures, such as think-aloud measures, only show us processes which are in working memory, but not automatic processes (Prins, Veenman & Elshout, 2006). These difficulties are further aggravated when seeking to assess self-regulated learning in small children.

Pintrich (2004) affirms that use of self-regulation strategies should be "captured" when they are being used in an activity; online recording is one of the most important measuring processes for doing so. The record of how the subject interacted with a virtual activity gives us much more precise information about his or her learning strategies than do self-reports of any type (Hadwin, Winne & Nesbit, 2005). These authors summarize the advantages of using the computer for research in the field of psychology:

- It can collect information which is otherwise impossible to gather.
- Software can be programmed to interact with the learner's behavior and to do so in a reliable fashion.
- It can collect information at the minute level, reliably, without overlooking anything and without biases (beyond those that may form part of the program itself).

As for Educational Psychology, in particular, Hadwin, Winne and Nesbit (2005) consider that the following aspects should be taken into account when using computer programs:

- Measuring constructs: subjects should have the possibility of modifying their answers.
- Allowing creation of new conditions for learning and interaction.
- Allowing new methods to operationalize the variables that are to be observed.
- Allowing new ways to communicate and interact at different levels.
- Making it possible to investigate the processes while they are under way, to record data, make calculations, combine variables, etc., over long periods. This will facilitate a reformulation of learning and motivation theories as constructs in transition rather than as a state or trait.

USEFUL THEORETICAL MODELS FOR ASSESSING SELF-REGULATED LEARNING

The Psychological Model

We have designed an online assessment program assuming the Pintrich (2000) model as theoretical model from Educational Psychology. His global model of self-regulated learning proposes four phases or processes (planning, monitoring, control and reflection); at each phase the model considers regulation activities within four separate areas (cognitive, affective-motivational, behavioral and contextual). The model is presented as a guide for thinking, since not all academic learning necessarily involves explicit self-regulation. Each of the areas includes aspects such as the following:

- Regulating *cognition*: this area encompasses different cognitive strategies that subjects can use in the academic context, as well as metacognitive strategies for controlling and regulating their own cognition. The planning phase includes processes such as establishing goals, activating prior knowledge and metacognitive knowledge. The monitoring phase involves awareness and follow-up of cognitive and metacognitive aspects. The control and regulation phase includes actions the subject uses to adapt and change his or her cognition. The reaction and reflection phase includes self-assessment and attributions about what was done.
- Regulation of *motivation and affect*. The planning phase involves processes such as self-efficacy judgments, and judgments about the difficulty, value and interest of the activity. The observation phase means that the subject is aware of his or her affective states. The control and regulation phase involves using strategies such as self-directed dialog, promising yourself

awards, etc. The final phase takes in emotional reactions such as justification of what was carried out.

- Regulation of *behavior:* this area includes the subject's attempts to control his or her behavior over the length of the activity. The first phase refers to planning and administering one's time and effort. The monitoring phase involves making adjustments in the latter in order to accomplish the objective. During the control and regulation phase, one's time and effort are adjusted according to the difficulty of the task.
- Regulation of the *context:* this area does not form part of the individual, but is external to him or her. In this case it is not the area itself that leads us to use the term "self-regulated", rather, it is whether the subject becomes involved in trying to observe, control and regulate the learning context through the use of strategies. The planning phase includes activating the subject's perception of the classroom context and atmosphere. In the monitoring phase, the subject must become aware of and come under the restrictions and opportunities of the classroom social system. The control phase includes all the subject's strategies that seek to regulate the activity and the context: negotiating assessment criteria, establishing norms for discussion and reasoning, etc. In the reflection phase, the subject is likely to make an overall evaluation of the activity and of the class environment.

The *Pro&Regula* Program is an intervention program based on this model, for improving self-regulation strategies (De la Fuente & Martínez, 2004); it is currently available in print, with an online version under development.

The Model for Design of Computer-Assisted Assessment

Mislevy, Steinberg and Almond (1999; 2001) draw up a broad conceptualization of the assessment process, and they establish two general areas of educational assessment: a substantive area and an area of evidence-based reasoning. The substantive area refers to the concrete field of knowledge, to how the students learn it and how they use what they know. The area of evidence-based reasoning refers to what and how much we can learn about students' knowledge, based on what they say and do. These authors propose integrating the two areas: although the most visible element in educational assessment is the task, its purpose is to provide evidence about aspects that cannot be directly observed at all, such as what the subject thinks or what he or she can do. This conceptual framework is made up of different levels which they call models:

- The *student* model incorporates knowledge, strategies and other attributes that we wish to assess, since it refers directly to the assessment objectives themselves.
- The *evidence* model describes how to extract the key items of evidence from what the pupil does or says in the context of the task, and how such evidences relate to the student model. It therefore involves behaviors or actions that should reveal the constructs that are to be evaluated.
- The *task* model provides the framework for constructing and describing the situation within which the subject acts. The task should incorporate the particular circumstances that give the subject a chance to act in a way that reveals what he or she knows or is able to do.
- The *assembly* model defines the combination of actions that will constitute the students' assessment.

- The *simulation* model includes the description and requirements of the environment in which the assessment task will run.
- The *environmental* model describes the global environment for whatever is needed to carry out the assessment.

DESIGN OF THE SEAI SELF-REGULATION ASSESSMENT FOR YOUNG CHILDREN

The *SEAI* (acronym from its Spanish name) is a software program for assessing self-regulation strategies in five-year-olds (Lozano & De la Fuente, 2009). It is made up of three activities which are independent of each other and contain curriculum content that the schoolchildren have already worked on. This program addresses one of the big gaps in research on self-regulated learning, that of strategy assessment in young children.

The *SEAI* (op. cit) design was accomplished by incorporating the Pintrich (2000) psychological model into a computer program that follows the assessment model guidelines proposed by Mislevy, Steinberg, Almond, Haertel and Penuel (1999; 2001). Assessment is done through inferring the appropriateness of strategies observed in the subject's performance of three assigned tasks. A summary of the assessment process follows, structured according to the different levels proposed in the assessment model.

The Student Model

The student model includes all variables related to self-regulated learning that we wish to assess in the subjects. Variables were selected by considering:

- Whether they could be captured through the subject-computer interaction.
- Whether they could be easily recorded and interpreted.
- Whether the subject needed to make use of them as he or she progressed into the activity.
- That they would be the same for all three tasks.

In line with the guiding psychological model, variables selected for the student model were self-regulation strategies such as comprehension, planning, self-efficacy, help seeking, persistence, self-assessment and attribution. These variables are explained in more detail below:

- Comprehension. This variable refers to awareness of one's own understanding of the task statement and awareness of what has to be done. Poor comprehension of the task and the subject's failure to recognize this lack of comprehension may be at the root of many academic problems (Hadwin & Winne, 2001).
- Planning. This strategy refers to the subject's choice, appropriate or inappropriate, of objects he believes he can do the activity with. Planning is understood as deliberate organization of the action to be followed in accomplishing a goal (Prevost, Bronson & Casey, 1995)
- Self-efficacy. Involves the subject's perception about his or her own ability to correctly solve the task. Beliefs about one's own efficacy are positively related to self-regulated learning (Pintrich, 1999; Schunk & Ertmer, 1999). In order to be effective, the perception of self-efficacy should be derived from the subject's estimation of the difficulty of the task, something which five-year-olds are already capable of (Winne, 1997).
- Help seeking. This strategy refers to the subject asking for help when he or she feels unable to carry out the task alone. According to Pintrich (2000), help seeking can be an adaptive or a maladaptive strate-

gy. It is considered to be adaptive when the subject is fundamentally seeking to understand and to learn, but needs help to overcome an especially difficult aspect of the activity. It is a maladaptive strategy when the subject asks for help in order to get the correct answer without making much effort, or to finish the task quickly without much understanding or learning.

- Persistence. Defined as constancy in sticking with the task until reaching the end, regardless of any assessment of performance. One failure in regulation is not having enough motivation or not knowing how to control it in order to persist when difficulties appear (De la Fuente, 2008; De la Fuente & Cardelle-Elawar, 2009; Wolters, 2003).
- Self-assessment. Refers to the subject's awareness of how well he or she has performed and of the results obtained (Cleary & Zimmerman, 2004; Schunk & Ertmer, 1999).
- Attribution. This type of strategy involves the justification and motives to which the subject attributes his or her performance on each task (Cleary & Zimmerman, 2004).

Evidence Model

This level consists of two components, an assessment component and a statistical component. The assessment component records and evaluates all that the subject does in his or her interaction with the activity. In the SEAI, the subject's actions are recorded numerically and include the following types of actions: answer to a question, choice of an option, decision making (set of diverse actions that the program groups together as indicators of certain variables). This component also records performance achieved on each of the three activities. The measurement or statistical component, expresses how the recorded variables relate in probabilistic fashion with the variables from the student model. The psychometric model followed in the SEAI is based on factorial analysis.

The Task Model

In the SEAI, the subject is presented with three activities that we can analyze from different perspectives. From a psychological perspective, the tasks were selected based on the intent to assess certain cognitive and affective actions that would be required in performing the activities. It must be emphasized that the task design will determine the degree of self-regulation that the subject is capable of demonstrating (Parsons, in press). From an instructional perspective, task selection took into account both the curricular content of the tasks and their previous use in similar research. From the technological perspective, the program acts as a virtual tutor that presents the activity, asks questions, offers help and scores the activity. Finally, from the perspective of the subject-SEAI interaction, the design was guided by the principle of "maximum interaction with the software with the minimum use of hardware"; in addition, small animations were included and the language employed was simple and age-appropriate for small children.

The Assembly Model

The section is highly important since it manages the interaction between the pupil, the task and the evidence model. Each activity in the SEAI was developed based on a previous pilot study with five-year-olds and with manual versions of the same tasks. Several refinements and adjustments were subsequently carried out, concluding with the beta version of the program.

The Simulation Model

The SEAI program attempts to simulate the conditions in which the usual academic activity of a five-year-old takes place, including the activity itself as well as the presence of a virtual tutor that offers similar assistance to that of a real teacher, such as explaining and helping. The help which the child requires is adjusted to the demands of the situation as detected by the program, based on what the subject does at each point in the activity.

The Environmental Model

Since we wish to see the child's performance on an academic task and within its context, the SEAI assessment process must take place in an environment as similar as possible to the one of an ordinary classroom.

ETHICAL ASPECTS OF COMPUTER-ASSISTED ASSESSMENT AND THE SEAI PROGRAM

In a recent review of ethical standards for psychological assessment, Schulenberg and Yutrzenka (2004) offer a synthesis of ethical criteria most relevant for computer-assisted assessment. In order to make this synthesis, they base their work on the AERA *Standards for Educational and Psychological Testing* (1999), the APA *Ethical Principles of Psychologists and Code of Conduct* (2002), and the *Code of Conduct* (2001) of the ASPPB. Their review of ethical aspects related to this type of assessment encompass elements regarding competency, interpretation and use of computer-produced reports, characteristics of the subject to be evaluated, equivalency procedures between paper-and-pencil versions and computer versions, and finally, confidentiality.

Each of these standards is described below, as well as how each one affects the design and use of the SEAI Program.

Competency

In the sphere of professional practice it is an accepted fact that the professional should act within the limits of his or her area(s) of competency. Consequently, the American Psychological Association (APA) states that those practitioners who perform traditional assessments and computer-assisted assessments must have the necessary knowledge about assessment techniques, as well as about the instruments and the measurements that they make use of. The following competencies are involved: (1) knowledge of the research which endorses the measurement instruments and procedures, including their strengths and weaknesses; (2) appropriate interpretation and presentation of assessment results; (3) implementation of the assessment process which is consistent with the test being used; (4) performing the assessment only after receiving the required training, and protecting the confidentiality of the subjects being assessed; and finally, (5) a certain amount of computer experience is essential. When professionals make use of computer-assisted assessments, this implies that they themselves and their assistants who administer the test are responsible for their use.

The SEAI is an online assessment instrument whose design and validation were the object of research for a doctoral thesis (De la Fuente & Lozano, 2009; Lozano, 2009; Lozano & De la Fuente, 2009). Its benefits include those which are shared by all computerized psychological assessment instruments, and the benefits of computer-assisted assessment of self-regulated learning. According to Carlson and Smith Harvey (2004) and O'Neill (2002), general benefits include the following:

- ease and increased security in the collection and treatment of information;
- reduced time involved in data collection, storage and treatment;
- orderly collection of information and flexibility in data treatment;
- the possibility to quickly produce a subject's profile;
- reduced error in data processing and protection.

It is important to note that many of the drawbacks of human assessors are eliminated, such as oversights, moods, fatigue, hunger, boredom, etc. (Epstein & Klinkerberg, 2001). Other noteworthy advantages with a computerized assessment include reduced storage space for materials, assurance of proper use of instructions and standards, and improved test security.

The SEAI, as a computer-assisted instrument for assessing self-regulated learning, enjoys benefits mentioned by different researchers. According to Pintrich (2004), on-line records make it possible to capture self-regulated learning strategies in an activity as they are put into practice:

1. Ease of recording and interpreting actions indicative of self-regulated learning (Chung & Baker, 2003; Hadwin & Winne, 2001; Vrugt & Oort, in press).
2. Information obtained about learning strategies is much more precise than through self-reports of any type (Hadwin, Winne & Nesbit, 2005).
3. It is possible to observe the pupil's cognitive activity without having to interrupt him or her (Chung & Baker, 2003; Van Biljon, Tolmie & Du Plessis, 1999; Winne & Stockley, 1998).
4. It is possible to collect reliable information about what subjects actually do when learning, as compared to what they say they do (Winne & Jamieson-Noel, 2003).
5. Possible language-related deficiencies are avoided (O'Neill, 1999; Yeh & Lo, 2005).

For the specific case of strategy assessment in small children, the SEAI Program gets around several age-related difficulties: it does not interrupt or overload the child's cognitive activity with questions, it does not depend on the child's awareness or memory of strategy use, it does not depend on the child's answers, which are usually quite vague and irrelevant at this age, it does not require explanations from the children about their own cognitive processes.

As for disadvantages, most have to do with instrument use: some subjective elements are unavoidable, since the instrument was programmed by human beings (Pardeck, 1997); it may be used by subjects who lack proper training (Carlson & Smith Harvey, 2004); the computer-generated report may be accepted without questioning (Epstein & Rotunda, 2000); the test may be applied to a sample different from what it was intended for; measurement errors may arise due to interaction between test content, test design and the user interface (Bennet, 1999); test results may be influenced by aspects such as computer experience, familiarity and attitudes toward the computer (McDonald, 2002). Some assessment problems specific to the SEAI are as follows: because of its content it can only assess self-regulated learning strategies in children from 4 to 6 years of age; during the assessment process the child must be constantly accompanied by an adult; it does not take any measurement of the time that the child spends on any complete activity or on its different parts.

Interpretation and Use of Computer-Generated Reports

A frequent error of professionals untrained in computer-assisted assessment is uncritical confidence in the computerized interpretation when making their diagnosis. This is in addition to the established weakly validity of many of these assessment programs, as mentioned above. The assessor must balance these possible risks and understand the reliability and validity of the test that is being applied; one must also take into account any discrepancies between the computer-generated assessment report and the characteristics of the subject being assessed. It is recommended that the user be familiar with: the instrument's psychometric criteria, studies and research about the instrument or using the instrument, the user guide, possible effects that should be taken into account in interpreting results, information that he or she may contribute to interpretation of the report, and any other existing information (e.g. statistical information, the profile's consistency with prototypical patterns).

The isolated use of computer-assisted assessment is only justified if it has been demonstrated to be more effective than traditional assessment. Psychometric characteristics of the SEAI are discussed in previous studies (De la Fuente & Lozano, 2009; Lozano, 2009; Lozano y De la Fuente, 2009). Moreover, we believe that this type of virtual instrument is most appropriate in the assessment of self-regulated learning strategies in young children, since it is the only type of instrument that overcomes the deficiencies associated with assessment at this age, as discussed earlier.

Characteristics of the Subject Being Assessed

During the assessment process it is necessary to consider characteristics of the subject which may interact with the use of the computer and which may affect the reliability and validity of the results. On one hand, we refer to attitudes toward and familiarity with the computer, on the other hand, anxiety and/or aversion to the computer. These aspects were reviewed in some depth in the previous section. It is thus recommended that the subject be informed beforehand that the assessment will be done with a computer.

In the SEAI validation process (Lozano, 2009), one of the variables considered was the child's previous experience with the computer. Research has shown that this is not an important variable in the regulation process or in the results of a given subject (Lozano, 2009). The program design takes into account the psychomotor particularities of small children and their difficulties in using computer peripherals. The task is executed using the "click" function of the mouse, it is not necessary to drag or to use the right button.

Other aspects considered when designing the interface were the use of small animations; limiting the number of words and explanations, both oral and written; and the size of the icons. Following the criteria of authors such as Markopoulos and Bekker (2003), animations are used to draw attention to different parts of the interface and to parts of the task. The animation is always accompanied by brief verbal instructions, this way the activity interface is more effective in guiding the pupil, and the amount of verbal instructions is reduced. With regard to icon size, we followed research contributions such as those from Hourcade, Bederson, Druin and Guimbretière (2004), using a minimum icon size of 32 pixels in the activity interface, considered appropriate in order for the child to perform at the same level as an adult (in an activity involving straight paths from one icon to another).

Equivalence Procedures between Pencil-and-Paper Versions and Computer Versions

Despite numerous advantages in computer-assisted assessment, there are still aspects that must be carefully considered by the assessor. One of these is the need to understand the empirical basis of validation; when using the alternative form of a certain assessment instrument, the most important aspect is to determine whether both versions are measuring exactly the same construct. When using the computerized version of an instrument, it is advisable to be familiar with studies that have been done to establish equivalence between the two versions, as well as the degree of equivalence established.

The SEAI is not a computerized version of a previous instrument; rather, it was developed directly in an online format in an attempt to overcome the difficulties previously discussed. The three activities proposed in the program, however, have been used previously; tasks 1 and 3 were used in non-computerized versions for similar purposes. Task 1, a model which the child must copy exactly, was verified as valid for assessing self-regulated learning strategies in small children in two previous studies (Amate 2003; 2004). Task 2 is an adaptation of activities proposed in Building Blocks (Sarama & Clements, 2004) in order to reflect the child's cognitive activity through manipulation and actions with objects that involve mathematical activity. Task 3 is an adaptation of an activity designed by Muñoz (2003) to assess self-regulation strategies in five-year-olds; it consists of putting together a simple puzzle.

Confidentiality

Security of information collected through computer-assisted assessment is the responsibility of the assessor; confidentiality breaches may occur if access to assessment results is not properly protected, for example, if the information is stored on the hard drive or on a network. It is advisable to use some kind of codified system to access the information, using numbers instead of the subjects' names, as well as assigning system administration to a trusted individual.

Addressing the school psychologist in particular, Carlson and Harvey (2004) give practical recommendations for using assessment software:

- Before using any assessment software, first consult several sources (such as specialized journals and software directories) that may offer useful information about the resource.
- Be familiar with critical reviews, from subject experts, about specific software programs.
- In the absence of publications or critiques about the assessment instrument, the practitioners themselves ought to carry out an evaluation of this type.
- It is important to understand one's own competence in handling assessment software and to attempt to minimize one's limitations before using this type of instrument.
- The program should never totally replace the assessment task of the professional, the assessor's judgment is not only irreplaceable, but it is also key in making decisions later on.
- Training of future professionals in educational psychology should include curriculum content regarding computer-assisted assessment.

CONCLUSION

Assessment is a central element in research processes of any kind. As new research needs to arise, instruments are developed that seek to meet the demands. These new tools and instruments must be submitted to the same demands and requirements, in design and in usage, that previously existing tools were subject to. The parallel world of virtual applications involves certain particularities that deserve special mention. Despite existing literature on this topic (Epstein & Klinkerberg, 2001; Garb, 2000; Korukonda, 2005), the ethical aspects of computer-assisted assessment require further study and differentiation from assessment in its traditional format: Are the results of measuring the same construct substantially different? Which results should be given greater importance when both are measuring similar aspects? How much weight should be given to computer-generated reports when we are making decisions? In the specific field of computer-assisted assessment of self-regulated learning, since research has scarcely begun to take shape in this area, the questions are even more numerous: Do subjects self-regulate in quantitatively similar fashion, or are there differences as a function of the activity format or the assessment instrument? Is exactly the same construct being measured? What amount of systematic error is introduced in the computerized version? Authors such as Clauser, Kane and Swanson (2002) assert that this format may produce an increase in systematic error, but also a decrease in random error, which may be a good exchange for many contexts.

Having noted the increasing interest in recording the subject's cognitive processing through his or her browsing history (Shih, Feng & Tsai, 2007; Vrugt & Oort, in press), and the timely increase in instruments and research techniques in this field, it is worth inquiring into the inclusion of this area in study plans for future psychologists and education professionals. We defend the need for computer-assisted assessment to receive explicit treatment, not only anecdotal mention, in the training of future practitioners, and not to leave this initial preparation up to one's personal choice. We consider that the fundamental progress of research in this field of educational psychology is at issue.

ACKNOWLEDGMENT

This work was carried out under the auspices of:

1. RD&I Project ref. BSO2003-06493. *Improving self-regulated learning in university students through online regulatory teaching strategies*. Ministry of Science and Technology and European Social Fund (2003-2006), Spain.
2. RD&I Project ref. SEJ2007-66843/educ. *Assessing improvement in the teaching-learning process and assessment of competencies in the European Higher Education Area: Model and Protocols.* Ministry of Education and Science, European Social Fund (2007-2010). Spain.

REFERENCES

Amate, J. (2003). Procedure for Evaluating Self-regulation Strategies during Learning in Early Childhood Education. *Electronic Journal of Research in Educational Psychology, 1*(1), 19-42. Available online at http://www.investigacion-psicopedagogica.org/revista/articulos/1/english/Art_1_3.pdf

Amate, J. (2004). *Evaluación de la autorregulación y relaciones con la ejecución en Educación Infantil.* [Self-regulation assessment and relationship to performance in Early Childhood Education.] A study carried out for obtaining the Spanish qualification as a researcher and diploma of advanced studies. Pending publication.

American Educational Research Association. (1999). *Standards for educational and psychological testing*. Washington, DC.

American Psychological Association. (2002). *Ethical principles of psychologists and code of conduct.* Washington, DC.

Association of State and Provincial Psychology Boards. (2001). *ASPB code of conduct*. Montgomery, AL: American Psychological Association.

Bennet, R. E. (1999). Using new technology to improve assessment. *Educational Measurement: Issues and Practice*, (Fall): 5–12.

Boekaerst, M., & Cascallar, E. (2006). How far have we moved toward the integration of theory into practice. *Educational Psychology Review, 18*(3), 199–210. doi:10.1007/s10648-006-9013-4

Boekaerst, M., & Corno, L. (2005). Self-regulation in the classroom: a perspective on assessment and intervention. *Applied Psychology: an International Review, 54*(2), 199–231. doi:10.1111/j.1464-0597.2005.00205.x

Brinke, D., Van Bruggen, J., Hermans, H., Burgers, J., Giesbers, B., Koper, R., & Latour, I. (2007). Modeling assessment for re-use of traditional and new types of assessment. *Computers in Human Behavior, 23*(6), 2721–2741. doi:10.1016/j.chb.2006.08.009

Butler, D. L. (2002). Qualitative approaches to investigating self-regulated learning: Contributions and challenges. *Educational Psychologist, 37*(1), 59–63.

Carlson, J. F., & Smith Harvey, V. S. (2004). Using computer-related technology for assessment activities: ethical and professional practice issues for school psychologists. *Computers in Human Behavior, 20*(5), 645–659. doi:10.1016/j.chb.2003.10.010

Chung, G. K. W. K., & Baker, E. L. (2003). An exploratory study to examine the feasibility of measuring problem-solving processes using a click-through interface. *Journal of Technology, Learning and Assessment, 2*(2). Available online at http://www.jtla.org

Clauser, B. E., Kane, M. T., & Swanson, D. B. (2002). Validity issues for performance-based tests scored with computer-automated scoring systems. *Applied Measurement in Education, 15*(4), 413–432. doi:10.1207/S15324818AME1504_05

Cleary, T. M., & Zimmerman, B. J. (2004). Self-regulation empowerment program: A school-based program to enhance self-regulated and self-motivated cycles of student learning. *Psychology in the Schools, 41*(5), 537–550. doi:10.1002/pits.10177

De la Fuente, J. (2008). Action-emotion style as a characteristic of achievement motivation in university students. In Valle, A., & Núñez, J. C. (Eds.), *Handbook of Instructional Resources and their applications in the classroom* (pp. 279–310). New York: NOVA Science Publishers.

De la Fuente, J., & Cardelle-Elawar, M. C. (2009). Research on action-emotion style and study habits: Effects of individual differences on learning and academic performance of undergraduate students. *Learning and Individual Differences, 19*(4), 567–576. doi:10.1016/j.lindif.2009.07.009

De la Fuente, J., & Lozano, A. (2009). Validación inicial de Software para Evaluar la Autorregulación Infantil [Initial validation of software to evaluate children self-regulation]. *Revista Mexicana de Psicología, 26*(2), 175–183.

De la Fuente, J., & Lozano, A. (2010). Assessing Self-Regulated Learning in Early Childhood Education: Difficulties, Needs, and Prospects for the Future. *Psicothema, 22*, 277–283.

De la Fuente, J., & Martínez, J. (2007). *Scales for Interactive Assessment of the Teaching-Learning Process (IATLP)*. Almería, ES: Education & Psychology I+D+i, e-Publishing Series. Available at www.education-psychology.com/e-publishing

De la Fuente, J., & Martínez, J. M. (2004). *Programa Pro&Regula* [*The Pro&Regula Program*]. Archidona, Málaga (Spain): Algibe

De la Fuente, J., & Mourad, A. (Eds.). (2010). *International Handbook on Applying Self-Regulated Learning in Different Settings*. Almería, ES: Education & Psychology I+D+i, e-Publishing Series. Available at www.education-psychology.com/e-publishing

Elliot, A. J., & Dweck, C. S. (2007). *Handbook of Competence and Motivation*. New York, NY: Guilford Press.

Epstein, J., & Klinkenberg, W. D. (2001). From Eliza to Internet: a brief history of computerized assessment. *Computers in Human Behavior, 17*(3), 259–314. doi:10.1016/S0747-5632(01)00004-8

Epstein, J., & Rotunda, R. J. (2000). The utility of computers versus clinician-authored assessment in aiding the prediction of patient symptomatology. *Computers in Human Behavior, 16*(5), 519–536. doi:10.1016/S0747-5632(00)00023-6

Garb, H. N. (2000). Computers will become increasingly important for psychological assessment: not that there's anything wrong with that! *Psychological Assessment, 12*(1), 31–39. doi:10.1037/1040-3590.12.1.31

Hadwin, A. F., & Winne, P. H. (2001). CoNoteS2: a software tool for promoting self-regulation. *Educational Research and Evaluation, 7*(2-3), 313–334. doi:10.1076/edre.7.2.313.3868

Hadwin, A. F., Winne, P. H., & Nesbit, J. C. (2005). Roles for software technology in advancing research and theory in educational psychology. *The British Journal of Educational Psychology, 75*(1), 1–24. doi:10.1348/000709904x19263

Hadwin, A. F., Winne, P. H., Stockley, D., Nesbit, J., & Woszcyna, C. (2001). Context moderates students' self-reports about how they study. *Journal of Educational Psychology, 93*(3), 477–487. doi:10.1037/0022-0663.93.3.477

Hourcade, J. P., Bederson, B., Druin, A., & Guimbretiére, F. (2004). Differences in pointing task performance between preschool children and adults using mice. *ACM Transactions on Computer-Human Interaction, 11*(4), 357–386. doi:10.1145/1035575.1035577

Korukonda, A. R. (2005). Personality, individual characteristics and predisposition to technophobia: some answers, questions and points to ponder about. *Information Sciences, 170*(2-4), 309–328. doi:10.1016/j.ins.2004.03.007

Lozano, A. (2009). *Diseño y Validación de Software para evaluar las estrategias de aprendizaje autorregulado en Educación Infantil.* [Design and validation of a software program to assess self-regulated learning strategies in Early Childhood Education.] Doctoral Thesis. Almería, ES: University Publications Service.

Lozano, A., & De la Fuente, A. (2010). Assessment of self-regulated learning in early childhood education: the SEAI application. In J. de la Fuente & A. Mourad (Eds.). *International Handbook on Applying Self-Regulated Learning in Different Settings* (pp._101-116). Almería, ES: Education & Psychology I+D+I, e-Publishing Series.

Lozano, A., & De la Fuente, J. (2009). *Software para Evaluar la Autorregulación Infantil, SEAI.* [SEAI: Software program for Early Childhood Self-regulation Assessment]. Almería, ES: Education & Psychology I+D+i, e-Publishing Series.

Markopoulos, P., & Bekker, M. (2003). Interaction design and children. *Interacting with Computers, 15*(2), 141–149. doi:10.1016/S0953-5438(03)00004-3

Martínez, J. M., & De la Fuente, J. (2007). Self-regulation of learning through the Pro&Regula Program. *Electronic Journal of Research in Educational Psychology, 2*(1), 145-156. Available online at http://www.investigacion-psicopedagogica.org/revista/articulos/3/english/Art_3_34.pdf

McDonald, A. (2002). The impact of individual differences on the equivalence of computer-based and paper-and-pencil educational assessments. *Computers & Education, 39*(3), 299–312. doi:10.1016/S0360-1315(02)00032-5

Mislevy, R. J., Steinberg, L., & Almond, R. (1999). *On the roles of task model variables in assessment design.* (CSE Technical Report 500) [online article]. Los Angeles, CA: University of California, National Center for Research on Evaluation, Standards and Student Testing (CRESST). Available at http://www.cse.ucla.edu/products/reports_set.htm

Mislevy, R. J., Steinberg, L., Almond, R., Haertel, G. D., & Penuel, W. (2001). *Leverage points for improving educational assessment.* (CSE Technical Report 534) [online article]. Los Angeles, CA: University of California, National Center for Research on Evaluation, Standards and Student Testing (CRESST). Available at http://www.cse.ucla.edu/products/reports_set.htm

Muñoz, J. (2003). *Enseñanza-aprendizaje de estrategias metacognitivas en niños de Educación Infantil.* [The teaching-learning of metacognitive strategies in Early Childhood Education.] Unpublished thesis. University of Burgos.

O'Neil, H. F. Jr. (1999). Perspectives on computer-based performance assessment of problem solving. *Computers in Human Behavior, 15*(3-4), 255–268. doi:10.1016/S0747-5632(99)00022-9

O'Neil, H. F. Jr. (2002). Perspectives on computer-based assessment of problem solving. *Computers in Human Behavior, 18*(6), 605–607. doi:10.1016/S0747-5632(02)00044-4

Pardeck, J. T. (1997). Computer technology in clinical practice: a critical analysis. *Social Work & Social Sciences Review, 7*, 101–111.

Parsons, S. (2010). Using the ACCES framework to design tasks that promote students' self-regulated learning. In J. De la Fuente & A. Mourad (Eds.), *International Handbook on Applying Self-Regulated Learning in Different Settings* (pp. 167-182). Almería, ES: Education & Psychology I+D+I, e-Publishing Series.

Pintrich, P. (1999). The role of motivation in promoting and sustaining self-regulated learning. *International Journal of Educational Research, 31*(6), 459–470. doi:10.1016/S0883-0355(99)00015-4

Pintrich, P. R. (2000). The role of goal orientation in self-regulated learning. In Boekaerst, M., Pintrich, P. R., & Zeidner, M. (Eds.), *Handbook of self-regulation* (pp. 452–502). San Diego, CA: Academic Press. doi:10.1016/B978-012109890-2/50043-3

Pintrich, P. R. (2004). A conceptual framework for assessing motivational and self-regulated learning in college students. *Educational Psychology Review, 16*(4), 385–407.

Post, Y., Boyer, W., & Brett, L. (2006). A historical examination of self-regulation: helping children now and in the future. *Early Childhood Education Journal, 34*(1), 5–14. doi:10.1007/s10643-006-0107-x

Prevost, R., Bronson, M., & Casey, M. (1995). Planning processes in preschool children. *Journal of Applied Developmental Psychology, 16*(4), 505–527. doi:10.1016/0193-3973(95)90002-0

Prins, F. J., Veenman, M. V. J., & Elshout, J. J. (2006). The impact of intellectual ability and metacognition on learning: New support for the threshold of problematicity theory. *Learning and Instruction, 16*(4), 374–387. doi:10.1016/j.learninstruc.2006.07.008

Sarama, J., & Clements, D. H. (2004). Building blocks for early childhood mathematics. *Early Childhood Research Quarterly, 19*(1), 181–189. doi:10.1016/j.ecresq.2004.01.014

Schulenberg, S. E., & Yutrzenka, B. (2004). Ethical issues in the use of computerized assessment. *Computers in Human Behavior, 20*(4), 477–490. doi:10.1016/j.chb.2003.10.006

Schunk, D. H., & Ertmer, P. A. (1999). Self-Regulatory processes during computer skill acquisition: Goal and self-evaluative influences. *Journal of Educational Psychology, 91*(2), 251–260. doi:10.1037/0022-0663.91.2.251

Schunk, D. H., & Zimmerman, B. J. (2008). *Motivation and Self-Regulated Learning*. New York: Routledge.

Shih, M., Feng, J., & Tsai, C. (2008). Research and trends in the field of e-learning from 2001 to 2005. A content analysis of cognitive studies in selected journals. *Computers &. Education, 51*(2), 955–967.

Torrano, F., & González, M. C. (2004). Self-Regulated Learning: current and future directions. *Electronic Journal of Research in Educational Psychology, 2*(1), 1-34. available online at http://www.investigacion-psicopedagogica.org/revista/articulos/3/english/Art_3_27.pdf

Van Biljon, J. A., Tolmie, C. J., & Du Plessis, J. P. (1999). Magix: an ICAE system for problem-based learning. *Computers & Education, 32*(1), 65–81. doi:10.1016/S0360-1315(98)00051-7

Vrugt, A., & Oort, F. (in press). Effective self-regulated learning of university students. In J. De la Fuente & A. Mourad (Eds.), *International Handbook on Applying Self-Regulated Learning in Different Settings*. Almería, ES: Education & Psychology I+D+I, e-Publishing Series.

Winne, P. H. (1997). Experimenting to bootstrap self-regulated learning. *Journal of Educational Psychology*, *89*(3), 397–410. doi:10.1037/0022-0663.89.3.397

Winne, P. H., & Jamieson-Noel, D. (2003). Self-regulating studying by objectives for learning: Students' reports compared to a model. *Contemporary Educational Psychology*, *28*(3), 259–276. doi:10.1016/S0361-476X(02)00041-3

Winne, P. H., & Perry, N. E. (2000). Measuring self-regulated learning. In Boekaerst, M., Pintrich, P. R., & Zeidner, M. (Eds.), *Handbook of self-regulation* (pp. 531–566). San Diego, CA: Academic Press. doi:10.1016/B978-012109890-2/50045-7

Winne, P. H., & Stockley, D. (1998). Computing technologies as sites for developing self-regulated learning. In Schunk, D. H., & Zimmerman, B. J. (Eds.), *Self-regulated learning. From teaching to self-reflective practice* (pp. 106–136). New York: Guilford Press.

Wolters, C. A. (2003). Regulation of motivation: Evaluating an underemphasized aspect of self-regulated learning. *Educational Psychologist*, *38*(4), 189–205. doi:10.1207/S15326985EP3804_1

Yeh, S., & Lo, J. (2005). Assessing metacognitive knowledge in web-based CALL: a neural network approach. *Computers & Education*, *44*(2), 97–113. doi:10.1016/j.compedu.2003.12.019

KEY TERMS AND DEFINITIONS

Computer-Assisted Assessment: Models where software utilities are implemented in the assessment of self-regulated learning.

Ethical Aspects: Ethical and deontological safeguards that should be incorporated in the assessment of self-regulated learning as a human psychological process.

Psychological Testing: The assessment of psychological processes, where methodology requirements and techniques for reliability and validity must be fulfilled.

Self-Regulation Assessment: The object of this research, having to do with issues in evaluating the self-regulated behavior of subjects, especially in learning situations.

Young Children: Children in a stage of psychological development typical of preschool or early childhood education.

Chapter 4
Self-Regulated Strategies and Cognitive Styles in Multimedia Learning

Barbara Colombo
Catholic University of the Sacred Heart, Italy

Alessandro Antonietti
Catholic University of the Sacred Heart, Italy

ABSTRACT

An experiment was carried out to investigate how participants self-regulate their access to explanatory pictures that were designed to facilitate learning. Participants learned from two multimedia presentations, one in audio, and the other in video format. Participants were given the opportunity to ask for an explanatory picture when they felt they needed more information to better understand the text. Recording the requests for pictures assessed self-regulation of strategies that promote picture use. Before completing comprehension questions, participants explained why they asked for pictures and were asked to express their level of awareness of the cognitive processes involved in learning from pictures. Two questionnaires were administered to measure the right/left thinking styles and the spontaneous tendency to use mental images. Results showed that participants, without full awareness, self-regulated their cognitive strategies according to presentation complexity. Judgments of picture utility were internally coherent. Finally, cognitive styles played a minor role in self-regulating learning, but tended to influence the metacognitive awareness of the strategies applied.

DOI: 10.4018/978-1-61692-901-5.ch004

INTRODUCTION

Multimedia learning materials can provide learners with rich educational environments where concepts can be learned in multiple formats: written texts, oral narratives, static pictures, animated videos, etc. Numerous theoretical and empirical questions come to mind when considering multimedia-learning materials. While considerable research has attempted to understand when and why presenting the same content both visually and verbally can foster learning, this chapter, on the other hand, is concerned with whether participants are able to identify optimal strategies for using multimedia learning materials when learning a new concept.

Paivio's (1986) dual-coding theory has provided a rich foundation for research on multimedia learning. Dual-coding theory contends that verbal and non-verbal information are processed in parallel and therefore normally do not compete for resources. Verbal representations are composed of words for objects, events and ideas while non-verbal representations are embedded in non-verbal representations with some resemblance to the perceptions that give rise to them.

Mayer's (2001; 2005) Cognitive Theory of Multimedia Learning is an applied model that has tested Paivio's dual-coding theory with multimedia learning materials. Mayer's theory is an empirically supported model inspired by a learner-centered approach. According to the theory, people learn better if information is learned via both systems, that is, when learning occurs with verbal and non-verbal information, than with verbal information alone. Such a general principle was specified by a set of sub-principles experimentally tested by Mayer and his colleagues (Mayer, Moreno, Boire & Vagge, 1999; Mayer, Dow & Mayer, 2003; Mayer & Moreno, 2003). These principles include:

- Spatial contiguity: Students learn better when corresponding words and pictures are presented near rather that far from each other on the page or on the screen.
- Temporal contiguity: Students learn better when corresponding words and pictures are presented simultaneously rather than successively.
- Coherence: Students learn better when extraneous material is excluded rather than included in the presentation.
- Modality: Students learn better from animation and narration than from animation and on-screen text.
- Redundancy: Students learn better from animation and narration than from animation, narration, and text.

A body of research suggests that individual differences in cognitive style might significantly influence how one would learn from a multimedia-learning environment. In this chapter we tested how participants, tested for individual differences in cognitive style, would differentially self-regulate their access to relevant pictures (that were designed to facilitate learning) in a multimedia learning environment.

We will first review cognitive styles, then we will discuss relevant questionnaires for measuring individual differences in cognitive style, and before we discuss the study, we will briefly turn our attention to self-regulation, which is crucial for any learning environment.

Cognitive Styles

The efficacy of the principles drawn from Cognitive Theory of Multimedia Learning varies depending on individual differences. Antonietti (2003) argues that cognitive styles (e.g., visualizers, verbalizers, etc.), which can be defined as the way an individual perceives, remembers and re-elaborates information, modulate people's way of dealing with information, effectiveness of cognitive processing, and learning in general. For instance, those who prefer to process information in a visual format (visualizers) should benefit from the concurrent presentation of texts both verbally and non-verbally (excluding situations of working memory overload) more than people who prefer to process information in a verbal format (verbalizers). Furthermore, the integration of texts and pictures, which is needed when learning from a multimedia presentation, should occur easily for people who tend to process information from different stimuli (holistic-intuitive style) but not for people who tend to process information separately (analytical-systematic style).

While the relationship between cognitive styles and learning from hypertexts has been investigated in the past (e.g., Calcaterra, Antonietti & Underwood, 2005; Fiorina, Antonietti, Colombo & Bartolomeo, 2007), multimedia learning itself has been rarely studied from a cognitive styles standpoint. Chen and Macredie (2004) attempted to determine the relationship between learners' cognitive styles and their perceptions and attitudes toward the features of a Web-based multimedia instructional program. Results indicated that cognitive styles influenced participants' reactions to the hypertext nature of the program and to the possibility to self-regulate their behavior while using the program. In another study, Chen, Magoulas, and Dimakopoulos (2005) adopted an individual difference approach to explore user's attitudes towards various interface features provided by existing Web directories. For their research, cognitive style was considered to influence the effectiveness of information seeking. For example, cognitive style influenced participants' reactions to the organization of subject categories, presentation of results, and screen layout. In related research, Chen, Ghinea and Macredie (2006) tested 132 undergraduates, balanced as for gender and somewhat for style (46 verbalizers, 54 visualizers and 32 bimodals, i.e., individuals failing to demonstrate a preference for either) attending Brunel University's Department of Information Systems and Computing. Results demonstrated that participants didn't show a preference toward either an analytic or an intuitive way of processing information preferred to draw on visual sources for informational purposes and that the presence of text in multimedia clips had a detrimental effect on the knowledge acquisition irrespective of a participant's cognitive style.

Special care should be used in choosing the instrument to measure participants' cognitive styles. In fact, while evaluating the effects of cognitive learning style on learning probability and statistics, Miller (2005) compared two instruments: The Gregorc Style Delineator and the Kolb Learning Style Inventory.

The Gregorc Style Delineator was "designed to aid an individual to recognize and identify the channels through which he/she receives and expresses information efficiently, economically, and effectively" (Gregorc, 1982a, p. 1). There are two dimensions within the model: Perception and Ordering (Gregorc, 1982b). The Perception dimension is a learner's preference for grasping information either abstractly or concretely while the Ordering dimension is a learner's preference to arrange and refer to information either sequentially or randomly. In Gregorc's (1982a, 1982b) model, four combinations are described to define a learner's preferences: Concrete Sequential, Abstract Sequential, Abstract Random, and Concrete Random.

The Kolb Learning Style Inventory is based on Kolb's Experiential Learning Model (Kolb, 1984). In this model, knowledge is created from grasping and transforming one's experience. There are two modes of grasping experience: Concrete Experience and Abstract Conceptualization and two modes of transforming experience: Reflective Observation and Active Experimentation. This results in four learning styles: assimilators favor Abstract Conceptualization and Reflective Observation, convergers favor Abstract Conceptualization and Active Experimentation, divergers favor Concrete Experience and Reflective Observation and accommodators favor Concrete Experience and Active Experimentation. Miller (2005) found that participants identified as Concrete Sequential, learned significantly less than participants identified as Concrete Random with the Gregorc Style Delineator, while no differences were found with the Kolb Learning Style Inventory.

According to the cognitive style perspective, learners should adapt their behavior, based on their cognitive style, while studying a multimedia presentation. This perspective has two implications. First, learners are consciously aware of distinctive features of the multimedia presentation as well as their cognitive style preferences. Second, learners can exert control over what they are doing, thus choosing the best strategy. In other words, self-regulation is needed to learn efficiently from a multimedia presentation. Returning to the question of this chapter, we test whether learners are able to identify and apply the best strategies when they are trying to learn new topics through multimedia presentations.

Self-Regulation

Self-regulation, in general, is needed when one is faced with a complex and challenging task, and different modes of dealing with it are available (Carver & Scheier, 1998). According to Boekaerts, Pintrich & Zeidner (2000) learners have to identify optimal learning strategies in order to succeed (where optimality means learners have to take into consideration their goals, the conditions under which they are learning, and their own personal learning style). Winne and Hadwin (1998) maintain that self-regulated learning involves four components: 1) definition of the task; 2) setting the goals and planning the strategies to reach them; 3) application of the strategies; and 4) adaptation to future similar instances. Similarly, Zimmerman (1989; 1998) argued that self-regulated learning implies: 1) self-evaluation and monitoring; 2) goal setting; 3) strategic planning; and 4) implementation of the strategies. For both Zimmerman (1989; 1998) and Winne and Hadwin (1998), strategies are linked to how goals and the learning situation are conceived as well as to learners' cognitive preferences/resources and to the feedback they receive.

Task evaluation, goal setting, planning, monitoring, retrospective judgment and self-perception are concepts that stress the close relationship between self-regulation and metacognition. A literature review on metacognition and multimedia (Antonietti & Colombo, in press) highlighted that research has converged in acknowledging the important role of metacognition in learning. This review highlighted that much research has shown how metacognition helps learners to be aware of personal cognitive styles and the interaction between these styles and multimedia learning via self-regulated learning. That is, the more the learner is proficient in monitoring and controlling his/her cognitive strategies while using a multimedia presentation, the more the multimedia environment will meet users' needs and allow them to learn proficiently. It is worth mentioning that Moreno (2005) recently proposed an extended version of Mayer's model in which motivation and metacognition were included to stress the importance of self-regulation.

Present Study

The experiment reported in this chapter investigated the role of self-regulation in multimedia learning. Multimedia environments are rich, complex, and challenging environments, which prompt self-regulation because the learner has to manage materials in various formats (text, pictures, etc.,) and decide which material should be attended to, and check whether task demands and the personal cognitive styles were met. In previous research by Mayer (2005), participants were presented with "fixed" multimedia materials (in that participants used presentations that were completely pre-determined in their format), built to be coherent or incoherent with the multimedia principles. Lack of flexibility of the learning materials meant that Mayer's was more interested in the learner's behavior and learning outcomes and less interested in learner self-regulation. In contrast, in this chapter we explored how learners, presented with "flexible" multimedia materials, self-regulate their approach to those materials. We assessed to what extent the learners were aware of what they were doing with the materials (i.e., metacognitive awareness) and tested if cognitive styles interacted with self-regulation. In detail, we investigated:

- Level of self-regulation for learning strategies applied to multimedia materials in audio and video format;
- Level of metacognitive awareness underlying such strategies;
- Influence of self-regulation and metacognitive awareness on learning outcomes;
- Influence of individual cognitive styles on self-regulation, metacognitive awareness, and learning outcomes.

METHOD

Participants and Design

Twenty-four university students from a large Italian university (20 women, 4 men), aged between 20 and 34 years (mean = 23) participated for partial course credit. The experiment was a within-subjects design with 4 experimental conditions (2 topics x 2 formats). Each participant viewed two presentations, one audio, and the other video. The order of presentation by topic (invisibility cloak/Renaissance lute) and format (audio/video) was counterbalanced. The order of the questionnaires (i.e., SOLAT and USOIM77, to be described below) was also counterbalanced.

Materials

In a control study, the multimedia materials, that were based in part on those used by Mayer (2001), were tested in order to ensure that both were equally "intelligible" and were equally "difficult." Sixteen participants browsed two multimedia presentations (8 for each presentation). Participants rated them and expressed their opinions regarding topic difficulty, clearness, and adequacy of pictures. Our materials, while on different topics than Mayer's (to avoid a potential gender bias) were similar in terms of text length, text structure, text complexity, as well as the explanatory/procedural structure. The topic selected for our multimedia materials were on how an invisibility cloak works and on how to change frets on a Renaissance lute.

Each multimedia presentation was divided into 16 sections. A picture was associated with each section. The audio format consisted of text presented orally while the video format consisted of text shown on-screen. Both formats were controlled for length. The audio format was read at the same rate as a typical undergraduate would read text. The rate was determined in a pilot test, where we obtained the mean reading times of five undergraduate students.

Picture Requesting

Participants were free to ask for the corresponding picture while listening/watching the presentation. The software, compiled in Visual Basic, used for the multimedia presentation recorded participants' requests for pictures automatically. Participants' requests for the pictures were proposed as evidence of self-regulation promotion. Picture requesting was therefore considered to temporally indicate when the learner felt that the picture would assist comprehension. Picture viewing durations provided insight regarding the metacognitive awareness and control participants exert on the learning process. Thus, we propose that good self-regulated learners will vary the time they spend in looking at pictures depending on the perceived degree of difficulty of the passage.

Participant Interview

Participants were interviewed following the presentation. Participants were asked to estimate the complexity of the materials, the quality and utility of the pictures retrieved, and their evaluation of their behavior while browsing the multimedia presentations. To understand participants' metacognitive judgments, they were asked to justify their answers.

Questionnaire of Cognitive Style

Following the interview, the SOLAT (Style Of Learning And Thinking) questionnaire (Torrance, 1988) was employed to assess the analytic-systematic vs. holistic-intuitive thinking style. Torrance's (1988) theory is inspired by previous work on brain hemispheric dominance that contends that right-thinkers (i.e., analytic-systematic) tend to use verbal-abstract code linked to analytic and sequential procedures, while left-thinkers (i.e., holistic-intuitive) tend to use visual-motor code linked to intuition and innovative procedures (see Torrance et al., 1978). Torrance's (1988) theory also includes personality traits, right-thinkers are supposed to be imaginative, inventive, enterprising, change seeking, and non conformists; while on the contrary, left-thinkers are supposed to be realistic, repetitive, settled, planners, and conformist. Bimodal thinkers tend to be balanced along the personality trait dimensions and therefore tend to utilize them when contextually appropriate. The distinction between right and left thinking styles, even though supported by neurobiological evidence (Martindale, 1999), is controversial (see Antonietti, Fabio, Boari & Bonanomi, 2005). Nevertheless, SOLAT has satisfactory psychometric properties, and succeeds in distinguishing analytic-systematic strategies from holistic-intuitive strategies.

The USOIM77 (USe Of IMagery, 77 items) questionnaire (Antonietti & Colombo, 1996-1997) was used to assess the tendency for spontaneous mental imagery. Each item concerned a situation in which people may mentally visualize an image. The items concern different mental functions (e.g., memorizing, problem-solving, daydreaming, etc.), different kinds of mental images (static and dynamic, single and interactive, etc.), different situations (e.g., study activities, leisure time, etc.) and have different contents (e.g., objects, persons, places, etc.).

Learning Outcomes

Comprehension of the multimedia presentation was tested following Mayer's (2001) methodology. The methodology included three questions, one was a memory question (i.e., retention) and the other two were problem-solving questions, (i.e., trouble-shooting and redesign). Participants were first asked a retention question, in that they were asked to explain either how to change frets on a Renaissance lute or how an invisibility cloak works (according to presentation order) to someone who does not know anything about either. Next, participants were asked a transfer (or trouble-shooting) question, in that they were told that either the lute or cloak were not working (again, according to presentation order) and participants were asked to discover the reason why it was not working. For the cloak trouble-shooting question, participants were told that, after wearing the invisibility cloak a person did not disappear and participants were asked to provide possible explanations. For the fret trouble-shooting question, participants were told that after changing the frets, the instrument did not sound good and participants were asked to provide possible explanations. Finally, responders were asked to redesign the original system according to a new specific request. For the cloak redesign question, participants were asked to describe a new procedure for using the invisibility cloak to be seen as wearing something different instead of being invisible. For the lute redesign question, participants were asked to describe a new procedure for changing a fret as they were

Table 1. Mean number (SD in parentheses) of pictures spontaneously requested during the presentation for the audio and video conditions

Presentation	Condition		
	Audio	Video	Total
Lute	13.54 (3.67)	14.23 (3.03)	13.88
Cloak	13.73 (4.80)	11.55 (4.72)	12.64
Total	13.63	13.00	

Table 2. Mean time (in sec; SD in parentheses) of picture requesting during the presentation for the audio and video conditions

Presentation	Condition		
	Audio	Video	Total
Lute	92.973 (78.401)	13.898 (10.46)	11.597
Cloak	84.161 (74.949)	115.923 (84.063)	100.042
Total	88.934	128.410	

told to imagine that just before a concert a fret broke and they did not have the time to following the 16-step procedure they had learned.

RESULTS AND DISCUSSION

Self-Regulation During the Multimedia Presentation

We conducted a 2 (format) x 2 (topic) ANOVA for each of the dependent variables to test whether the audio vs. video format and the two different topics (cloak and lute) affected image requesting. Table 1 reports descriptive statistics for the number of pictures requested.

The main effect for format $F(1,22) = 1.13$, $p = .30$) and topic $F(1,22) = 0.76$, $p = .39$) was not significant. The interaction between format and topic was however significant $F(1,22) = 4.22$, $p < .05$, $\eta^2 = .23$). Pictures were requested less frequently for the invisibility cloak presentation. Difficulty ratings, described below, are offered as a possible mediator. Table 2 reports descriptive statistics for picture requesting (in seconds) following the text presentation.

An ANOVA was performed to explore the possible effect of image request timing in the audio and video condition with respects to the two different topics. The main effect for format approached significance $F(1,22) = 4.00$, $p = .06$) (in the video condition participants were somehow slower in requesting the images) while the main effect for topic was not significant $F(1,22) = 0.29$, $p = .60$).

Metacognitive Judgments

Participants' evaluations regarding level of difficulty of presentation content and their evaluations regarding the level of usefulness for the pictures were coded on a 4-points scale (not at all = 0, a little = 1, fairly = 2, a lot = 3). No main effect for difficulty $F(1,22) = 0.81$, $p = .38$), topic $F(1,22) = 2.96$, $p = .10$) was found, and the interaction between format and topic $F(1,22) = 2.36$, $p = .14$) was also not significant. Participants rated the cloak presentation in the video format as the easiest to understand. Perhaps the difference in difficulty ratings was due to differences in familiarity for the topics, though this was not tested in our study. Therefore, for the invisibility cloak the elements of the topic, e.g., a cloak, computer, video projector, etc., could have been more familiar then the elements from the Renaissance lute topic, e.g., a lute, frets, etc. The video format probably reinforced participant' perception of ease with the materials since they could read and re-read the text of each passage according to his/her own inner timing. The analysis of usefulness for the pictures revealed no main effect for format $F(1,22) = 0.01$, $p = 0.98$, topic $F(1,22) = 2.26$, $p = .15$, and the interaction between format and topic $F(1,22) = 0.12$, $p = .74$ was also not significant. Images were considered most useful for the cloak presentation in the video condition, which was also rated as the easiest to understand. We infer that images were perceived to be the most useful when the text was perceived as easy to comprehend, maybe because participants had more cognitive resources available to process the pictures.

Justifications for the Metacognitive Judgments

During the post-experiment open-format interview, participants explained their rated level of complexity and image usefulness. Two independent judges categorized responses. Categories were identified by asking two independent judges to read all responses and propose their own classification system. The two judges then compared the resulting categories and cases of disagreement were resolved. Once a set of shared categories was devised, responses were then re-coded by two other independent judges.

Categories derived from the explanations of text complexity were: "The descriptions are detailed," "Images are useful," "The procedure is clear," "The topic is easy," "I was allowed to read again" – when the text was rated "easy/clear"; "The topic is new"; "The topic is too complex"; "The topic is not easy to visualize" – when the text failed to be rated as clear/easy. Categories derived from the explanations of image usefulness were: "Images promote mental visualization"; "Image accelerate/integrate comprehension" – for the positive role of images; "Images are not intelligible"; "Images are put in a bad position"; "The text is already clear even without images" – for the negative role of images. Cross tabulations aimed at testing possible association between explanations given by participants and experimental conditions were computed. No connections emerged. However, we observed that in the audio condition, the multimedia presentations were rated as less familiar than in the video condition (the cloak story was rated as the least familiar. Images in the audio condition for the invisibility cloak story were most likely to be regarded as an aid to comprehension. In the video condition, the pictures appeared to promote mental imagery in the presentation about the cloak, whereas in the presentation about the lute they were perceived as an aid to comprehension.

When participants were asked what they would have changed and why, they answered that they would have asked for more images in the presentation where they had previously rated them to be more useful (cloak presentation). Where pictures had been perceived as less useful (lute presentation), participants would have changed their position.

Learning Outcomes

Table 3 reports the descriptive statistics for the retention question.

There was no main effect for format $F(1,22) = 3.75$, $p = .07$ nor of the topic $F(1,22) = 1.09$, $p = .31$, and the interaction between topic $F(1,22) = 2.21$, $p = .14$, was also not significant. Overall, the audio condition tended to promote the best retention and this tendency was strongest for the Renaissance lute presentation. These trends, though not statistically significant, confirm Mayer's (2001) theory in that since the audio condition was assessed, on the average, as being more demanding, promoted a higher level of attention resulting in better retention.

Answers to the trouble-shooting question were divided into 5 categories: 1) no answer, 2) irrelevant hypothesis (namely, concerning aspects which are not the focus of the problem), 3) correct identification of the problem (that is, the core issue was caught), 4) correct identification of the problem and one or more relevant (even though not exact) hypotheses, and 5) correct answer. Cross tabulations were computed to assess possible relationships between participants' answers and the experimental conditions. No significant associations emerged.

Table 3. Mean retention scores (SD in parentheses) for each presentation and condition

Presentation	Condition		
	Audio	Text	Total
Lute	8.83 (3.41)	6.58 (3.50)	7.70
Cloak	6,18 (5.02)	5.91 (4.49)	6.04
Total	7.57	6.26	

Answers given for the redesign question were classified according to 5 categories: 1) no answer, 2) ineffective solution, 3) ineffective but creative solution (i.e., solution which cannot be actually implemented but which is an original attempt to face the problem), 4) effective solution, 5) effective and creative (i.e., unique) solution. Cross tabulations between answers and experimental conditions failed to demonstrate any significant connections between each pair of variables.

In order to verify whether learning outcomes depended on behavior exhibited during the multimedia presentation, the number of images requested and the timing of requesting them were related to the answers given in the learning test. On the basis of the number of images requested and of the duration of the exposure to pictures (timing), participants were split into two sub-samples: high and low in image requesting and high and low in timing request. To explore the possible effect of the retention questions in the audio vs. video conditions based on the sub-samples (low vs. high in image requesting), a 2 x 2 ANOVA was computed by collapsing performance in the two presentations. No main effect for format $F(1,22) = 1.71$, $p = .20$) nor of the low vs. high level of image requesting $F(1,22) = 1.25$, $p = .31$ was found and the interaction was not significant $F(1,22) = 1.71$, $p = .28$. Examining the interaction between image requests and retention, we noticed that participants who were high in image requesting scored better in the retention test and this was especially true for the audio condition. A further 2 x 2 ANOVA

Table 4. Cross tabulation (percentages in parentheses) between participants' answers for the re-design question for the two sub-samples of high and low in timing for picture requesting

Re-design	Request timing		
	Low	High	Total
Don't answer/can't answer	4 (36.4)	7 (58.3)	11 (47.8)
Ineffective solution	5 (45.5)	0 (0.0)	5 (21.7)
Ineffective but creative solution	0 (0.0)	1 (8.3)	1 (4.3)
Effective solution	1 (9.1)	4 (33.3)	5 (21.7)
Effective and creative solution	1 (9.1)	0 (0.0)	1 (4.3)
Total	11 (100)	12 (100)	23 (100)

$\chi^2 = 9.40$; df = 4; $p < .05$

was computed assuming low vs. high timing in image requesting as the independent variable. The format $F(1,22) = 0.58$, $p = .72$, and timing of image requesting $F(1,22) = 0.11$, $p = .75$, as well as the interaction $F(1,22) = 0.31$, $p = .58$, failed to yield significant results. Participants who were classified as high in requesting time scored higher in the retention question, and this was especially true for the audio condition. We interpret this as meaning that participants who take more time with the pictures are reflecting more on the topic and therefore learning better.

To assess possible connections between answers given to the trouble-shooting and re-design questions and the different sub-samples of participants, cross tabulations were computed. Significant associations failed to emerge. An exception was the cross tabulation between participants' answers to the re-design question for the two sub-samples of high and low in request timing (see Table 4).

Participants classified as high in time requesting did not tend to answer more often than low participants, whereas those who replied more gave more answers that were incorrect. We interpret this as meaning that taking more time to understand the message meant being more aware of what was ineffective (and hence participants did not answer if they thought their solution to be ineffective).

Right vs. Left Thinking Style

To analyze the relationship between right vs. left cognitive style on self-regulation, metacognitive dimension, and learning outcomes, we classified the participants into three categories: right thinkers, left thinkers, and bimodal thinkers. To do so we considered the relative weight of each of them. If the score for a single modality was highest, the categorization was immediate; where we had similar scores, we proceed to compare them. If the right and the left scores were similar, the participant was categorized as Bimodal; if the right or the left score was similar to the bimodal score, the student was classified as Bimodal.

ANOVAs were computed on the total number of images requested. No significant effect for the total number of images requested $F(2,22) = 1.36$, $p = .28$ and the image timing $F(2,22) = 0.27$, $p = .77$ was found. No effect of the right vs. left style was detected on the perception of the complexity of the multimedia presentations, respectively $F(2,21) = 0.21$, $p = .81$, $F(2,22) = 0.30$, $p = .74$. The same was true for the perception of image usefulness, respectively $F(2,22) = .73$, $p = .49$, $F(2,22) = 0.51$, $p = .61$. Inspection of the responses showed a tendency for the right thinkers to rate the pictures as being more useful and to facilitate comprehension. Moreover, according to right thinkers, pictures were useful because they accelerated or integrated comprehension, while according to the left thinkers they were helpful because they promoted mental visualization.

To investigate the effect of right vs. left style on learning, an ANOVA was computed on the number of recalled passages (retention question). No effect of cognitive style was detected on the number of passages reported by participants in the two presentations, respectively $F(2, 22) = 0.23$, $p = .98$, $F(2,22) = 0.35$, $p = .71$. Table 5 and Table 6 report the cross tabulations between participants' categorized answers to the trouble-shooting question and participants' cognitive style.

We noticed some tendencies. In the first presentation right and bimodal thinkers tended to identify correctly the problem, while left thinkers

Table 5. Cross tabulation (percentages in parentheses) between participants' categorized answers for the trouble-shooting question, first presentation and right vs. left thinking style

Trouble-shooting	Style of thinking			
	Right	Left	Bimodal	Total
No answer	2 (18.2)	0 (0.0)	1 (11.1)	3 (12.5)
Irrelevant hypothesis	1 (9.1)	0 (0.0)	3 (33.3)	4 (16.7)
Identify correctly the problem	4 (36.4)	0 (0.0)	3 (33.3)	7 (29.2)
Identify correctly the problem – more than one correct hp	2 (18.2)	4 (100.0)	1 (11.1)	7 (29.2)
A right and a wrong answer	2 (18.2)	0 (0.0)	1 (11.1)	3 (12.5)
Total	11 (100)	4 (100)	9 (100)	24 (100)

$\chi^2 = 13.90$; $df = 8$; $p = .08$

Table 6. Cross tabulation (percentages in parentheses) between participants' categorized answers for the trouble-shooting question, second presentation and right vs. left thinking style

Trouble-shooting	**Style of thinking**			
	Right	Left	Bimodal	Total
No answer	2 (20.0)	0 (0.0)	0 (0.0)	2 (8.7)
Irrelevant hypothesis	1 (10.0)	1 (25.0)	2 (22.2)	4 (17.4)
Identify correctly the problem	4 (40.0)	2 (50.0)	2 (22.2)	8 (34.8)
Identify correctly the problem – more than one correct hp	2 (20.0)	0 (0.0)	4 (44.4)	6 (26.1)
A right and a wrong answer	1 (10.0)	1 (25.0)	1 (11.1)	3 (13.0)
Total	10 (100)	4 (100)	9 (100)	23 (100)

$\chi^2 = 13.90$; df = 8; p = .08

gave more than one correct solution. In the second presentation, the bimodal thinkers gave more than one right answer. Table 7 and Table 8 show the cross tabulations between participants' categorized answers to the re-design question and participants' style.

Associations were not significant, but an interesting difference could be noticed in the cloak presentation. For this presentation, the right thinkers tended to give more solutions that were ineffective. In the lute presentation, bimodal participants tended not to answer while right thinkers were the ones who gave more answers that were correct.

By analyzing self-regulation during the two presentations by considering together the number of images requested and the timing of the request and the responses given to the metacognitive questions, it was possible to divide the sample into two sub-samples: coherent and incoherent with

Table 7. Cross tabulation (percentages in parentheses) between participants' categorized answers for the re-design question in the cloak presentation and right vs. left thinking style

Re-design	**Style of thinking**			
	Right	Left	Bimodal	Total
Don't answer/can't answer	0 (0.0)	0 (0.0)	2 (22.2)	2 (8.3)
Ineffective solution	6 (54.5)	1 (25.0)	0 (0.0)	7 (29.2)
Ineffective but creative solution	0 (0.0)	1 (25.0)	3 (33.3)	4 (16.7)
Effective solution	5 (45.5)	2 (50.0)	4 (44.4)	11 (45.8)
Total	11 (100)	4 (100)	9 (100)	24 (100)

$\chi^2 = 11.93$; df = 6; p = .06

Table 8. Cross tabulation (percentages in parentheses) between participants' categorized answers for the re-design question, lute presentation and right vs. left thinking style

Re-design	Style of thinking			
	Right	Left	Bimodal	Total
Don't answer/can't answer	3 (30.0)	2 (50.0)	6 (66.7)	11 (47.8)
Ineffective solution	3 (30.0)	1 (25.0)	1 (11.1)	5 (21.7)
Ineffective but creative solution	1 (10.0)	0 (0.0)	0 (0.0)	1 (4.3)
Effective solution	3 (30.0)	1 (25.0)	1 (11.1)	5 (21.7)
Total	10 (100)	4 (100)	9 (100)	23 (100)

$\chi^2 = 5.79$; df = 8; p = .67

respects to the cognitive styles as emerged from SOLAT questionnaire. Coherent participants were those who showed correspondence among number of images requested and timing of requests with respect to their cognitive style (e.g. right thinkers would have asked for the images quickly while left thinkers should have been slower). Such classification was used as a new independent variable for t-tests (Table 9).

Participants who were coherent with their style (being right or left) asked on the whole for more pictures $t = 2.58$, $p < .05$, $\eta 2 = .31$. Being "cognitively coherent" allows one to be more adequate self-regulation while using multimedia tools.

Visualization Style

To explore the effect of the spontaneous tendency to use mental visualization on self-regulation,

Table 9. Mean number of pictures requested and SDs according to cognitive style coherence

Cognitive style coherence	Total number of pictures requested		
	M		SD
Coherent	31.43		0.97
In-coherent	25.25		0.69

metacognition and learning outcomes, participants were classified, according to the score obtained in the USOIM77. The two categories were low and high visualizers. An ANOVA demonstrated that there was no effect of individual differences in visualizing on the total number of images requested ($F(1,22) = 0.02$, $p = .90$) or on image request timing ($F(1,22) = 0.12$, $p = .73$). The preference for spontaneous mental visualization failed to influence the evaluation of the level of complexity of the presentations ($F(1,22) = 1.22$, $p = .28$) and the perception of the image efficacy ($F(1,22) = 1.14$, $p = .30$). As far as retrospective metacognitive evaluations were concerned, low visualizers admitted they would have changed pictures or coherently with their cognitive style, they would have asked for less pictures, while high visualizers would have changed picture position on the computer screen.

To investigate the effects of the spontaneous use of visualization on learning, an ANOVA was computed on the number of recalled passages. No significant effect was found, $F(2,22) = 1.31$, $p = .26$. To investigate the effects of the tendency toward visualizing on trouble-shooting and re-design questions cross tabulations were computed. No significant connections emerged, even though high visualizers tended to give just a correct answer while low visualizers were more likely to identify correctly the problem and to give more solutions that were correct. Cross tabulations on the answers to the re-design questions, and participants' tendency to use mental visualization failed to show any significant connection.

GENERAL DISCUSSION AND CONCLUSION

We observed that fewer images were requested in the video condition, especially for the invisibility cloak presentation. Participants were slowest in requesting pictures and in exploring the text presentation. Metacognitive responses confirmed these results as participants regarded the invisibility cloak presentation to be easier in general than the Renaissance lute condition, and the invisibility cloak presentation easier in the video condition. The general conclusion that the cloak presentation was easier than the lute is not surprising because it consisted of elements, (e.g., a computer, a video projector, etc.) which were more familiar to participants than the elements of the Renaissance lute presentation (e.g., lute, frets, etc.). While this difference was not significant we believe that this difference is potentially important and could reach significance with a larger sample size. These data are not consistent with Mayer's (2005) modality principle. We wonder whether written texts provided the participants with a strong sense of "control" that could provide the right environment for the learner to be more confident in self regulating their cognitive processes.

An interesting result was that the images were considered to be more useful in the video condition, which was also considered to be easier. We expected that images usefulness would increase as presentation content complexity increased. We also expected that images would have been rated as most useful in the audio condition, since that was the condition where images were requested more often. This counterintuitive result can be explained if one considers the images can be used vicariously, that is, instead of reading carefully the passage, when the passage is perceived to be easy and familiar.

Across presentation topic, we found that the audio condition was rated to be more unfamiliar than the video condition. Rather surprisingly, the presentation concerning the invisibility cloak was rated as more unfamiliar than the lute presentation and this result highlights the lack of metacognitive awareness in the participants, since they previously rated this condition as very easy and their interactions with the learning materials were consistent with this judgment. The images were regarded as more of an aid to comprehension in the audio condition (consistent with Mayer's (2005) multimedia theory). Participants seemed to be coherent when asked what they would have changed about the images. They said that they would have asked for more images, when they were interpreted as being most useful (i.e., the invisibility cloak presentation) and when participants had been perceived the images as less useful (i.e., the lute presentation), they would change their disposition towards the images, in that participants said that they understood a posteriori the actual importance of images. This increase in metacognitive awareness linked to this specific presentation leads us to wonder whether an increase in complexity and unfamiliarity for materials is associated with more accurate a posteriori metacognitive understanding of behaviors.

Concerning learning outcomes, results were not 100% consistent with Mayer's (2005) theory. Consistently with Mayer's theory, the audio condition promoted better results for the retention test than the video condition, but we wonder whether the more demanding condition promoted a higher utilization of attentional resources which resulted in better retention performance. Improved performance for the trouble-shooting questions in the audio condition stressed this point. Consistently with previous findings, an increase in image requesting for the audio condition underscored how requesting more images promoted better learning outcomes (retention test). Concerning the trouble-shooting questions, images seemed to promote a better understanding since participants who were high in image requesting tended to give more correct answers. Answers to the re-design problem, instead, stressed that images can result to be both an aid or a distracter, maybe because a strong visual memory of the images proposed can prevent, or slow down, the effective restructuring of the problem or can generate self-imposed ties.

The role of timing in image requesting stresses the importance of individual difference in a trouble-shooting situation, since taking less or more time to reflect turned out to be a gain or a loss. Instead, when facing a re-design problem, using more time appeared to bring people to be more aware of what solutions are ineffective, so that the number of ineffective solution decreases. It is interesting to highlight that different problem structures (trouble-shooting and re-design) evoked different cognitive strategies even if starting from the same multimedia presentation, so stressing the relevance of individual self-regulation strategies.

Cognitive style appeared to have indirect effects. Right thinkers generally rated the images to be more useful and to accelerate comprehension, coherently with their style. Bimodal and left thinkers, conversely, maintained that images promoted mental visualization. For the trouble-shooting questions right and bimodal thinkers were the more effective ones – and this is not surprising since they should be the ones able to explore more strategic options. Left thinkers, though, were able to find more than one correct answer. Answers to the re-design questions returned once more a coherent picture: right thinkers were the ones to give answers that are more effective.

In addition, the tendency toward the spontaneous use of mental visualization was taken into account. We found participants' explanations of images functions to be coherent with such a style. High visualizers said that they would had changed only images' disposition, while low visualizers would had asked less images, focusing only on the most useful images. Trouble-shooting and re-design questions stressed once again that images can result to be both an aid or a distracter, since often low visualizers tended to be more creative and to give more corrects answers: this can be explained hypothesizing that a too strong visual memory can prevent from an effective restructuring of the problem field.

In conclusion, participants appear to be able to self-regulate their learning while they are trying to learn from a multimedia presentation: they change strategies according to the features of the multimedia presentation they are facing (for instance, in the audio condition participants asked for more pictures). Yet, people's metacognitive awareness fails to reflect the different feature of multimedia presentations. Nevertheless, individuals' judgments about such features tend to be internally coherent. Hence, people appear not to be aware of the actual potentiality of multimedia, even if they are able to use them effectively. Cognitive styles play only a minor role in the spontaneous fruition of multimedia tools, but they influence metacognitive judgments. Hence, we can interpret cognitive style as a promising path to promote a metacognitive awareness and, consequently, self-regulation. It is possible that teaching people to be aware of the cognitive peculiarities of their thinking style and training them to use such characteristics could help them to become more aware of the matching characteristics of a multimedia presentation and, hence, to self-regulate better the use of multimedia tools.

This study aimed to explore the role and the interaction of many variables in multimedia learning. The focus was on learners' self-regulation and their metacognitive awareness of their actions in learning environments where they were free to devise their own strategies to browse the presentation. The learning environment designed as such challenged participants to find for themselves the best way to interact in that environment. Our learning environment therefore differed significantly from that used by Mayer (2001) since Mayer's environment was static and only allowed learners to follow a pre-determined approach. In this multimedia environment, participants, therefore, had to self-regulate the selection of the pictures which were designed to improve their understanding of the presentation topic and be integrated with the text. The strategies applied by the learners, even if they fail to mirror the cognitive principles identified by previous researchers, appear to be functionally relevant and internally coherent. Learners, though not completely aware of what they are doing and why, behave in a way which is for the most part consistent with their perception of the task and partially consistent with their own cognitive style. Metacognitive awareness and cognitive style appear to be important variables for future research on self-regulated learning.

REFERENCES

Antonietti, A. (2003). Cognitive styles assessment. In *Encyclopaedia of psychological assessment* (*Vol. I*, pp. 248–253). London: Sage.

Antonietti, A., & Colombo, B. (1996-1997). The spontaneous occurrence of mental visualization in thinking. *Imagination, Cognition and Personality*, *16*(4), 415–428.

Antonietti, A., & Colombo, B. (in press). Metacognitive beliefs about learning from multimedia artifacts. In Antonietti, A., Confalonieri, E., & Marchetti, A. (Eds.), *Cognitive and social development in educational settings: Recent issues in theory, research, and application*. New York: Cambridge University Press.

Antonietti, A., Fabio, R. A., Boari, G., & Bonanomi, A. (2005). Il questionario "Style of Learning and Thinking" (SOLAT): dati psicometrici per una validazione e standardizzazione della versione italiana. *TPM. Testing Psicometria Metodologia*, *12*(4), 299–316.

Boekaerts, M., Pintrich, P. R., & Zeidner, M. (Eds.). (2000). *Handbook of self-regulation*. San Diego, CA: Academic Press.

Calcaterra, A., Antonietti, A., & Underwood, J. (2005). Cognitive style, hypermedia navigation and learning. *Computers & Education*, *44*(4), 441–457. doi:10.1016/j.compedu.2004.04.007

Carver, C. S., & Scheier, M. F. (1998). *On the self-regulation of behavior*. New York: Cambridge University Press.

Chen, S. Y., Ghinea, G., & Macredie, R. D. (2006). A cognitive approach to user perception of multimedia quality: An empirical investigation. *International Journal of Human-Computer Studies*, *64*(12), 1200–1213. doi:10.1016/j.ijhcs.2006.08.010

Chen, S. Y., & Macredie, R. D. (2004). Cognitive modelling of student learning in web-based instructional programmes. *International Journal of Human-Computer Interaction*, *17*(3), 375–402. doi:10.1207/s15327590ijhc1703_5

Chen, S. Y., Magoulas, G. D., & Dimakopoulos, D. (2005). A flexible interface design for web directories to accommodate different cognitive styles. *Journal of the American Society for Information Science and Technology*, *56*(1), 70–83. doi:10.1002/asi.20103

Fiorina, L., Antonietti, A., Colombo, B., & Bartolomeo, A. (2007). Thinking style, browsing primes and hypermedia navigation. *Computers & Education*, *49*(3), 916–941. doi:10.1016/j.compedu.2005.12.005

Gregorc, A. F. (1982a). *Gregorc Style Delineator: Development, technical and administration manual*. Columbia, CT: Gregorc Associates.

Gregorc, A. F. (1982b). *An adult's guide to style*. Columbia, CT: Gregorc Associates.

Kolb, D. A. (1984). *Experiential learning*. Englewood Cliffs, NJ: Prentice-Hall.

Martindale, C. (1999). Biological bases of creativity. In Sternberg, R. J. (Ed.), *Handbook of creativity*. Cambridge, MA: Cambridge University Press.

Mayer, R. E. (2001). *Multimedia learning*. Cambridge, UK: Cambridge University Pres.

Mayer, R. E. (2005). *The Cambridge handbook of multimedia learning*. New York, NY: Cambridge University Press.

Mayer, R. E., Dow, G. T., & Mayer, S. (2003). Multimedia learning in an interactive self-explaining environment: What works in the design of agent-based microworlds? *Journal of Educational Psychology*, *95*(4), 806–812. doi:10.1037/0022-0663.95.4.806

Mayer, R. E., & Moreno, R. (2003). Nine ways to reduce cognitive load in multimedia learning. *Educational Psychologist*, *38*(1), 43–52. doi:10.1207/S15326985EP3801_6

Mayer, R. E., Moreno, R., Boire, M., & Vagge, S. (1999). Maximizing constructivist learning from multimedia communications by minimizing cognitive load. *Journal of Educational Psychology*, *91*(4), 638–643. doi:10.1037/0022-0663.91.4.638

Miller, L. M. (2005). Using learning styles to evaluate computer-based instruction. *Computers in Human Behavior*, *21*(2), 287–306. doi:10.1016/j.chb.2004.02.011

Moreno, R. (2005). Instructional technology: Promise and pitfalls. In Pytlik Zillig, L., Bodvarsson, M., & Bruning, R. (Eds.), *Technology-based education: Bringing researchers and practitioners together* (pp. 1–19). Greenwich, CT: Information Age Publishing.

Paivio, A. (1986). *Mental representations: A dual coding approach*. Oxford, UK: Oxford University Press.

Torrance, E. P. (1988). *Style of Learning and Thinking: Administrator's manual*. Bensenville, IL: Scholastic Testing Service.

Torrance, E. P., Reynolds, C. R., Ball, O. E., & Riegel, T. (1978). *Revised norms technical manual for your Style of Learning and Thinking Forms A and B*. Athens: Georgia Studies of Creative Behaviour.

Winne, P. H., & Hadwin, A. F. (1998). Studying as self-regulated learning. In Hacker, D. J., Dunlosky, J., & Graesser, A. C. (Eds.), *Metacognition in educational theory and practice* (pp. 277–304). Mahwah, NJ: Lawrence Erlbaum Associates.

Zimmerman, B. J. (1989). Models of self-regulated learning and academic achievement. In Zimmerman, B. J., & Schunk, D. H. (Eds.), *Self-regulated learning and academic achievement. Theory, research and practice* (pp. 1–25). New York: Springer.

Zimmerman, B. J. (1998). Academic studying and the development of personal skills: A self-regulatory perspective. *Educational Psychologist*, *33*(2-3), 73–86. doi:10.1207/s15326985ep3302&3_3

KEY TERMS AND DEFINITIONS

Cognitive Style: Cognitive styles refer to a person's habitual, prevalent, or preferred mode of perceiving, memorizing, learning, judging, making decisions, and solving problems. Individual differences about how people carry out tasks involving these functions may constitute a style if they appear to be pervasive (that is, they emerge consistently in different contexts, independently of the particular features of situation) and stable (namely, they are always the same at different times).

Metacognition: Metacognition is defined as the knowledge and control of cognitive objects and cognitive processes. Such a wide definition can be broadened to anything psychological (e.g. emotions), rather than anything merely cognitive. Similarly any kind of monitoring might be seen as part of metacognition processes.

Multimedia Learning: Multimedia learning is an active, student-centered approach in which learners can select relevant words and images, organizing them into coherent verbal and visual models, and integrating them into whole conceptual structures.

Chapter 5
Re-Conceptualizing Calibration Using Trace Methodology

Rylan G. Egan
Simon Fraser University, Canada

Mingming Zhou
Simon Fraser University, Canada

ABSTRACT

In this chapter, the authors challenge the traditional differentiation between metacognitive monitoring and control in text-based self-regulated learning (SRL). Building on Pieshl (2009), the authors presented a case for conceptualizing and measuring calibration as the interaction between metacognitive monitoring and control under the assumption that learners adjust metacognitive judgments as they monitor and control their learning both within and between trials. To this end they describe three separate but related measures of calibration – assessment, internal, and strategic calibration – to address such questions as what kind of test will be given; how will I perform on such a test; and what can I do to improve my performance, respectively. Each type of calibration is mutually exclusive; however, overall calibration accuracy relies on the hierarchical interplay among all three types. Finally, they provide examples of how trace data for each type of calibration may be collected in a multimedia-learning environment.

INTRODUCTION

Relationships between the accuracy of metacomprehension judgments and effective self-regulated learning (SRL) are a pivotal feature of the SRL literature (e.g., Glenberg, Sanocki, Epstein & Morris, 1987; Maki & Serra, 1992; Winne, 1997). Metacomprehension judgments are learners' relative (socially or norm-referenced) and absolute (criterion-referenced) judgments of text comprehension. All SRL models are fundamentally concerned with metacognitive control as this is the mechanism by which actions, based on predictions about metacognitive performance, forge the self-regulation process (e.g., Boekaerts, Pintrich & Zeidner, 2000; Pintrich, 2000; Winne & Perry, 2000; Zimmerman, 2000). Thiede, Anderson and Therriault (2003) illustrate the importance of metacomprehension to SRL by providing evidence of a statistically detectable positive correlation

DOI: 10.4018/978-1-61692-901-5.ch005

between metacomprehension judgment accuracy and SRL performance. Unfortunately, it is equally clear that learners, in general, are notoriously poor judges of the accuracy and extent of their learning (for a review, see Thiede, Griffin, Wiley & Redford, 2009). For example, a recent study by Thiede, Griffin, Wiley and Redford (2009) found that average judgment accuracy measured by gamma coefficients in more than 34 studies and 57 experiments over the last 28 years was .27.

This paper provides a unique perspective on the formulation and measurement of metacomprehension judgments. We posit such fine grained measurement of multiple intervals in the judgment process allows educators to pinpoint and correct heuristic biases resulting in judgment inaccuracy and thus provide effective support and guidance to improve students' self-evaluation and learning process.

Multiple contextual and cognitive factors associated with metacognitive judgment have been proposed in the metacomprehension literature including: test-judgment grain size alignment (Dunlosky, Rawson & Middleton, 2005), immediate vs. delayed recall (Thiede & Anderson, 2003), prior assessment experiences (Moore, Lin-Agler & Zabrucky, 2005), deceptively simple text (Lin, Zabrucky & Moore, 2002; Weaver & Bryant, 1995) and ineffective study strategies (Griffin, Wiley & Thiede, 2008; Thiede, Griffin, Wiley & Redford, 2003). As well, multiple factors outside of the learners' direct control have been indentified, such as self-esteem and locus of control (Garner & Alexander, 1989), individual interest (Lin & Zabrucky, 1998), test anxiety (Miesner & Maki, 2007), and working memory span (Griffin, Wiley & Thiede, 2008). In line with Winne and Perry (2000) we label the former factors "state" and the latter "trait". State factors are differentiated from trait by the possibility for relatively simple and short term manipulation. In contrast, individual traits are expected to be stable within context and resistant to researcher manipulation. In the following chapter we concentrate primarily on state factors, as this data is most relevant to the learning technology we introduce and discuss in this chapter. However, we acknowledge the interdependence of learners' traits and states and intend to focus future research on measuring both factors. The scope and intent of the current chapter does not allow us to comprehensively review all state correlates of metacomprehension inaccuracy; instead, we concentrate on a selected review of interventions in the cognitive and educational psychology literatures theorized to improve calibration within our proposed theoretical model and traditional measures of calibration. Specifically, we investigate the links between heuristic cues and metacomprehension calibration (e.g., Linderholm, Zhao, Therriault, & Cordell-McNulty, 2008; Rawson & Dunlosky, 2002; Rawson, Dunlosky, & Thiede, 2000).

We propose the trichotomous calibration model (TCM) wherein intermediate judgments made throughout the learning process are considered as requisite for metacomprehension calibration. Specifically, in line with Pieschl (2009) we argue that traditional calibration can be further deconstructed into three parts: (1) assessment calibration; (2) strategic calibration; and (3) internal calibration. Assessment calibration refers to the match between learners' ability to interpret the learning context to estimate the depth, complexity, and coverage of an upcoming assessment and actual qualities of a task. Strategic calibration concerns the match between learning strategies a learner surveys and then chooses, relative to needs for developing knowledge and skills perceived to be required by the task. Strategies are defined here as the aggregate of two or more actions (tactics) enacted to achieve a predefined end. Lastly, internal calibration refers to the accuracy of metacomprehension judgments of future performance *based on* learners' assessment prediction. Contingent on assessment prediction accuracy, learners accurately calibrated in all three forms will be better positioned to maximize metacomprehension judgment accuracy. Over time, feedback and instruction on each of these forms will allow instructors and researchers to detect the causes of judgment inaccuracy and thus, through facilitation, learners can be able to more accurately monitor and predict their learning.

In a recent review Thiede, Griffin, Wiley and Redford (2009) concluded that metacomprehension accuracy is constrained by factors "that arise from the inherent complexity of monitoring learning from text" (p. 101). We acquiesce and expand by proposing that learners are able to manage monitoring complexity through self-regulation of metacomprehension processes. In line with general notions of SRL, we believe metacomprehension judgments are regulated by a) setting goals or standards, b) using appropriate strategies to achieve mastery, and c) evaluating products with standards developed from (a) (see Winne, 1997). Through self-regulation, learners can learn to manipulate the level(s) and processes of text comprehension in accordance with goals and strategies aligned with future assessment. Although ground breaking work in the field of metacomprehension calibration cannot be overstated, by relying on experimental manipulation there is a danger that the centrality of the learner will be overshadowed so that the autonomous *self* in SRL could be lost. Granted, learners may be able to manipulate their environment to align with experimental findings. However, without specific feedback and metacognitive monitoring on aspects of processing that support accurate judgments (e.g., level of text processing, strategies used to improve judgments etc.), learners may be blind to the nature of their judgments and, as a result, have little basis for improving. In this chapter we aim to provide a method whereby metacomprehension research can move beyond manipulations of external factors that improve judgment accuracy and turn toward helping learners develop more accurate judgment processes.

In the next section we briefly review the concept of calibration. For interested readers more comprehensive reviews are available by Schraw (2009) and Maki, Shields, Wheeler, and Zacchilli (2005). Then we briefly examine how heuristic cues implicated in metacomprehension judgment function within the TCM. Moreover, we discuss how training for improved TCM calibration may render more accurate heuristic judgments. Next, using scenarios and descriptions we explain how innovative computing software, called nStudy (Winne & Hadwin, 2009), could be used to non-invasively explore metacomprehension calibration for each proposed form. Finally, we conclude our discussion and offer recommendations for future research.

METACOMPREHENSION CALIBRATION

Calibration, as we use it, refers to absolute and relative measures of accuracy; as well as bias. To calculate absolute accuracy predicted or post-dicted scores for each unit of information assessed are subtracted from actual item scores on those assessments and then squared ($\frac{1}{N}\sum_{i=1}^{N}\left(C_i - P_i\right)^2$ where *Ci* represents a confidence rating and *Pi* represents a performance score, see Schraw, 2009, p. 36). Mean absolute accuracy is the sum of squared differences between the learner's prediction of the score for each item and actual item score divided by the total number of items. Similarly, directional inaccuracy or bias is obtained by computing the value of each difference between estimated and actual outcome, then summing those results ($\frac{1}{N}\sum_{i=1}^{N}\left(C_i - P_i\right)$ where C_i represents a confidence rating and P_i represents a performance score, see Schraw, 2009, p. 37). Positive or negative bias are represented by the sign and magnitude of the deviation from zero for differences between the learners' predicted score and an item's actual score (Maki, Shields, Wheeler, & Zacchilli, 2005; for a review see Schraw, 2009).

A second method for determining metacomprehension judgment accuracy is to calculate relative accuracy or the correlation between judgments and performance (e.g., Schraw, 2009; Maki, 1998; Maki & Serra, 1992). Relative accuracy is usually gauged by the gamma coefficient as suggested by Nelson (1984) but is also possible with Pearsons correlation (see Maki, 2005). We do not distinguish between relative and absolute accuracy or bias. Instead, we follow Pieschl (2009) and apply the term calibration to "all situations in which congruence, alignment or match between students' metacognitive judgments and their performance on a criterion task are diagnosed" (p. 4). In other words, although the reliability of absolute and relative calibration has been questioned (see Maki, 2005), we assume that both measures gauge learners' metacomprehension judgment accuracy. Moreover, we expect the TCM will be relevant to all metacomprehension accuracy measures.

Heuristics Decisions and the TCM

Manipulations of learners' states through contextual or cognitive (strategic) manipulations have a long history. Examples include

- manipulations of text coherence (Begg, Duft, Lalonde, Melnick & Sanvito, 1989; Maki, Foley, Kajer, Thompson & Willert, 1990; Thomas & McDaniel, 2007),
- reading strategies (Dunlosky & Rawson, 2005; Griffin, Wiley & Thiede, 2008; Nietfeld & Schraw, 2002; Rawson, Dunlosky, & Thiede, 2000),
- delayed summaries and keyword generation (Thiede, Anderson, & Therriault, 2003; Thiede, Dunlosky, Griffin, & Wiley, 2005),
- text difficulty (Maki, Shields, Wheeler, & Zacchilli, 2005),
- researcher feedback (Dunlosky, Rawson & Middleton, 2005; Glenberg, Sanocki, Epstein & Morris, 1987; Hacker, Bol, Horgan & Rakow, 2000; Magliano, Little, & Graesser, 1993; Rawson & Dunlosky, 2007; Moore, Lin-Agler & Zabrucky, 2005),
- adjunct questioning (Maki, Foley, Kajer, Thompson, & Willert, 1990; Pressley, Snyder, Levin, Murray & Ghatala, 1987; Walczyk & Hall, 1989; Weaver & Bryant, 1995).

In this section we discuss how state factors such as those listed above interact with ease of cognitive processing and anchoring heuristics (Linderholm, Zhao, Therriault & Cordell-McNulty, 2008). We then explore the effect of heuristics on calibration within the TCM.

Ease of Cognitive Processing

We define ease of cognitive processing as the relationship between the speed and salience of text based information contemplated during initial reading and/or upon retrieval from long or short term memory and judgments about the nature of that information (e.g., Rawson & Dunlosky, 2002; Maki, 1998; Thiede, Griffin, Wiley & Redford, 2009). In the cognitive psychology literature, cognitive processing difficulty has been induced by limiting information availability, and has consistently resulted in biased judgments (e.g., Lichtenstein, Slovic, Fischhoff, Layman, & Combs, 1978; Reyes, Thompson & Bower, 1980; Kahneman, Slovic & Tversky, 1982). For example, Schwarz, Bless, Strack, Klumpp, Rittenaurer-Schatka and Simons (1991) found that participants who retrieved 6 instances of self-assertive behavior rated assertiveness statistically detectably higher than those who effortfully retrieved 12 instances. Similarly, using associative word pair judgments Koriat and Bjork (2005) found that judgments of learning were statistically detectable overconfident when word pairs were presented with high semantic (e.g., kitten-cat), but low forward associative (e.g., cat-kitten) relationships. Koriat and Bjork concluded that "when the to-be-learned materials trigger associations [processing ease] during study that are weak or absent during subsequent test, participants are prone to illusions of competence when predicting their own future recall" (p. 193).

Congruent to related fields, findings in the metacomprehension literature also provide support that judgments are biased by ease of cognitive processing (e.g., Rawson & Dunlosky, 2002; Maki, Shields, Wheeler & Zacchilli, 2005; Dunlosky, Rawson, & Middleton, 2005). Ease of processing in this field has been manipulated at the text-based and situation model levels. The text-based level refers to "coherence relations among the propositions in a text and their organization" (Kintsch, 1986, p. 89). In contrast, the situation model level refers to the connections made between text level propositions that create "mental representation of the situation found in the text" (Kintsch, 1986, p. 89). Text-based factors such as letter deletion, sentence cohesiveness, and text difficulty have been shown to increase the difficulty of initial text processing and act as cues for metacomprehension judgments (e.g., Griffin, Wiley & Thiede, 2008; Maki, Shields, Wheeler, & Zacchilli, 2005; Rawson & Dunlosky, 2002). For example, by deleting letters and rearranging sentences Rawson and Dunlosky (2002) found that metacomprehension ratings were lower and slightly more accurate than those who read altered text (see also Maki, Foley, Kajer, Thompson, & Willert, 1990). Similarly, a number of researchers have found that judgments are more accurate for texts that are challenging, but not to the extent that they consume the entirety of learners' cognitive resources (Dunlosky & Rawson, 2005; Griffin, Wiley & Thiede, 2008; Maki, Shields, Wheeler, & Zacchilli, 2005; Weaver & Bryant, 1995). Griffin and colleagues (2008) conjectured that cognitive resources must be available to metacogntively monitor comprehension alongside text level decoding. They supported their hypothesis that participants with poor reading comprehension could equal the calibration of more effective readers when rereading alleviated processing constraints. Put together, research indicates that when resources required for cue monitoring are available, learners use difficult cognitive processing of text as a cue to lower metacomprehension judgments.

It is clear that learners use text based heuristics to make metacomprehension judgments but the accuracy of these judgments has been inconsistent (e.g., Rawson & Dunlosky, 2002; Maki, Foley, Kajer, Thompson. & Willert, 1990; Thiede, Griffin, Wiley, & Redford (2009). For example, Thiede, Griffin, Wiley, & Redford (2009) argue that although processing difficulty at the text level is "one possible cue that a reader might use, it might not actually predict comprehension all that well" (p. 102). They go on to explain that "accurate metacomprehension depends on cues produced by accessing and utilizing one's situation model" (p. 102). A program of research over the last six years conducted by Thiede and colleagues (Thiede & Anderson, 2003; Thiede, Anderson & Therriault, 2003; Thiede, Dunlosky, Griffin, & Wiley 2005) has provided support for these contentions. Experiments follow a common procedure whereby participants read text, and either recall information (summaries or key words) immediately or after a delay. Consistently, those retrieving information after a delay have enjoyed metacomprehension calibration between .5 and .7. In contrast, those recalling information immediately score below .3. Thiede, Griffin, Wiley, and Redford (2009) reconcile large increases in metacomprehension accuracy with delayed recall by contending that text level information is no longer available to learners after a delay, therefore forcing learners to evaluate metacomprehension cues at the situation model level. Given that assessments are also provided after a delay, and often require inference level understanding, these cues tend to be more accurate. The benefit of contextual congruence between delayed tests and judgments is also predicted by test-appropriate monitoring hypothesis that assumes "the accuracy of people's judgments of memory is a direct function of the match between the properties of the judgment context and properties of the test context" (Dunlosky, Rawson, & Middleton, 2005, p. 552).

It is plausible that impressive increases in judgment accuracy are a function of judgment cues at the level of the situation model. However, thus far, findings are restricted to situations where learners recall text from memory that has been either read or reread. It is less clear how ease of cognitive processing cues influence learners who use complex study strategies to process text information (in vivo or from memory). Consider a situation where learners use elaborative study strategies such as concept mapping or argumentation. Situation model level strategies such as these may allow learners to monitor the ease or difficulty of constructing interconnections between concepts. As a result, assessments focused on testing interconnections should be relatively more accurate. In contrast, performance on assessments requiring rote memorization or recognition recall may be less accurately predicted. In this case, processing difficulty of text studied using mnemonics, rehearsal, or method of loci may be more informative. Although empirical research is required to verify our contentions, following from recent research we propose that congruence between predicted assessment objectives and study strategies may increase the predictive validity of cognitive processing cues. Importantly, if our contentions are verified, learners who are taught the importance of attending to relevant contextual assessment cues, and aligning these cues to more appropriate strategies, will be able to take ownership over factors influencing metacomprehension calibration, and hence be better able to discriminate between texts that are mastered and those that need to be revisited. Thus, by measuring and facilitating the three forms of calibration introduced in the TCM, ease of cognitive processing cues may be more predictive of assessment performance. Next we provide evidence that calibration in these forms can moderate the anchoring heuristics and provide more valid "starting points" for the judgment process.

Anchoring

Anchoring is a cognitive heuristic proposed by Tversky and Kahneman (1974) that refers to "the assimilation of numeric judgment to a previously considered standard" (for a review see Mussweiler, Englich & Strack, 2004). In the cognitive psychology literature Tversky and Kahneman (1974) provided a salient example of the anchoring heuristic by asking participants to estimate the percentage of African nations represented in the United Nations. On average, participants' estimates differed by 20% when given a high (65%) or a low (25%) initial anchor. Building upon the same thought, Linderholm, Zhao, Therriault, and Cordell-McNulty (2008) argue that "readers typically anchor their predictions based on perceived ability and/or their initial exposure to the task, but insufficiently adjust away from that anchor as a function of text difficulty, topic knowledge and/or interest in the topic" (p. 184). Multiple findings in the literature support the notion that learners anchor judgments to these factors prior to engaging with text. For example, Glenberg, Sanocki, Epstein and Morris (1987) found that students with experience in a domain tended to overestimate comprehension within that domain regardless of text difficulty. Moore, Lin-Agler and Zabrucky (2005) found in a path analysis that, after 12 learning trials with text of varying complexity, participants tended to use past assessment experience instead of study experiences to form metacomprehension judgments. Moreover, in Hacker, Bol, Horgan, and Rakow's (2000) study of undergraduate educational psychology students during a 15-week course, previous performance predictions accounted for more calibration variance than reported study time. Interestingly, feedback from actual performance did not statistically detectably account for any variance in judgment accuracy in their study.

We support Linderholm, Zhao, Therriault, and Cordell-McNulty's (2008) view on the anchoring and adjustment heuristic. Both evidence in the literature and intuitive logic attest to the notion that judgments must start somewhere. However, we disagree that accessibility of information and the ease of processing that results necessarily "represents [the] bottom-up processes that readers use to make estimates (Linderholm, Zhao, Therriault, & Cordell-McNulty, 2008, p. 185). Thiede and colleagues have illustrated that after a delay recall of information from memory may provide readers a means to infer understanding from a more global (situation model) perspective. Moreover, we argue that through facilitation test predictions may serve as the basis for judgment anchors. As Hacker, Bol, Horgan, & Rakow (2000) claimed, metacomprehension judgments are based on predictions of "how well one will perform *given the nature of the test*, the kinds of *items on the test*, and the *difficulty of the items*" (p.161; emphases ours, see also Maki, 1998). If these predictions are accurate, anchors based on specific learning goals congruent to the nature of a future assessment may mitigate unrealistic a priori anchors. Thus, by accurately estimating the nature of an upcoming assessment, learners may be able to appropriately anchor their judgments to future assessment. Most importantly, anchors based on cues associated with actual future assessment can replace (or at least temper) *potentially* irrelevant anchoring cues such as perceived ability, prior test experiences, and previous metacomprehension judgments. In this way measuring and facilitating assessment judgments in the TCM provides specific data relevant to judgment inaccuracy and may provide a more accurate starting point for the judgment process.

Trichotomous Calibration Model

In this section we further explain the three distinctive forms of metacomprehension calibration. As noted in Figure 1, the three forms of calibration are dependent on learners' perceptions of future assessment, which necessarily aligns with the task's context; including assessment instructions, learning objectives, and instructional techniques. However, basing metacomprehension judgments on warranted assessment predictions may not be well developed in many learners (Linderholm, Zhao, Therriault, & Cordell-McNulty, 2008) and may require facilitation. Based on assessment perceptions learners can make future performance judgments and decide upon appropriate study strategies. The use of metacognitive monitoring is required when learners make these judgments (as shown in grey boxes). The discrepancy between perceptions and the standards (as shown in white boxes) constitutes different forms of calibrations (as shown in black boxes). The validity of these standards is a function of: 1) actual assessments corresponding to specifically stated learning objectives and instructional methods, and 2) theoretically appropriate strategies being warranted as the "best strategies" for stated learning objectives and instructional techniques. It is only when these two conditions used to make metacomprehension judgments (in the TCM) are met, that internal calibration can align with traditionally measured calibration. The solid black arrows in Figure 1 indicate judgments based on specific assessment items (i.e., item specific). The dotted lines refer to recursive relationships between feedback and metacomprehension judgments. The nature of feedback may range from intensive intervention facilitating assessment prediction, adopting strategies, and evaluating indicators of comprehension to simply providing actual assessments, strategies, and outcomes. Lastly, the unidirectional dotted lines emanating from *perceived assessment* to *perceived performance on predicted items* and *adopted strategies* denote adjustments required when assessment perceptions are altered by feedback.

Based on the TCM model we argue that metacomprehension is an amalgamation of judgments about the: (a) complexity level of information processing required by future assessment(s) (as-

Figure 1. The Trichotomous Calibration Model

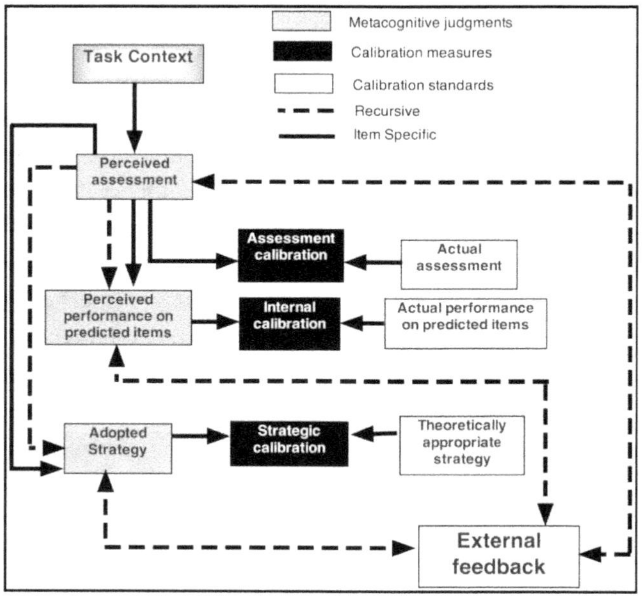

sessment calibration, see the revised Bloom's taxonomy in Anderson & Krathwohl, 2001; Kintsch, 1986); (b) matches between learning strategies and assessment requirements (strategic calibration); and (c) predictions about kinds of or levels of outcomes that can be achieved in relation to (a) (internal calibration). Moreover, we hypothesize that learners who are inaccurate at any form of calibration are challenged to self-regulate study activities effectively (Thiede, Anderson, & Therriault, 2003; Winne, 1997). Therefore, a mechanism for tracking calibration for a given assessment (*item specific*) or over multiple assessments (*recursive*) represents an important supplement to designing research on metacomprehension. In the scenarios we present later, we suggest methods for gathering data on these three forms of judgments to improving metacomprehension calibration as measured by the TCM and traditional measures.

Assessment Calibration

In line with Pieshl (2009), assessment calibration requires the ability to anticipate defining features of an upcoming assessment (item specific) and self-regulate predictions based on feedback (recursive). To accurately predict future assessment learners must focus on relevant indicators of assessment format and complexity. We define assessment complexity through Anderson and Krathwohl's (2001) revised Bloom's taxonomy. Specifically, learners may be required to process materials in terms of retrieving, understanding, applying, analyzing, evaluating, and/or creating. We posit that when instructors or textbook authors align vocabulary and instructional materials with assessment, learners can develop the capacity to determine levels of assessment. For example, a first year anatomy student who is provided with lists of vocabulary attached to parts of the body should expect an assessment at the retrieving level. Similarly, when verbs like "define, recite, or retrieve" are used in the vocabularies of learning outcomes, assessment instructions, or rubrics use, learners may expect retrieve level assessment. In contrast, terms such as "generate, plan, and produce" indicate the creation of knowledge (Anderson & Krathwohl, 2001; Bloom, 1956). Accurate self-regulation requires that learners minimize discrepancies between assessment judgments and attributes of the topic of those judgments. If learners have not accurately predicted the scope, depth (Anderson & Krathwohl, 2001) and fluency (i.e., automaticity) of information processing required to prepare for and work on an assessment task, or have not predicted the nature of the assessment task at all, they will not be able to set appropriate learning goals and learning strategies. To dissect assessment inaccuracies researchers must evaluate discrepancies between learners' assessment perception and reality, and how discrepancies are reflected in their subsequent metacomprehension judgments.

Strategic Calibration

Strategic calibration requires congruence between features of knowledge and skills predicted to be needed for future assessment, and tactics and strategies used to acquire required knowledge and skills. Investigations of strategic calibration may concern strategic compatibility with particular assessment(s) (item specific), or the development of more appropriate strategies through continuous feedback from multiple study-assessment intervals (recursive). Unfortunately, poor strategic calibration may obscure the benefits of accurate assessment predictions as learners lack strategies for success.

The importance of aligning strategies with the nature of future assessment has been well supported in the literature. For example, self-questioning, rereading, and delayed summaries have evidenced large statistically detectable improvements in metacomprehension judgment accuracy (e.g., Griffin, Wiley, & Thiede, 2008; Rawson, Dunlosky, & Thiede, 2000; Thiede & Anderson, 2003; Thomas & McDaniel, 2007). However, traditional metacomprehension measurement techniques do not allow researchers to detect and further explain poor strategic calibration. For example, learners may base metacomprehension judgments on heuristic indicators derived from poorly chosen strategies. Traditional calibration measures are not able to distinguish this influence from poor assessment or internal calibration. To understand why learners are unable to judge the extent of their learning, researchers must collect data about strategic learning decisions and calculate their congruence with assessment requirements.

Internal Calibration

Internal calibration refers to the match between learners' judged success on future assessment tasks and actual achievement (Pieshl, 2009). Internal calibration can also be measured at item specific and recursive levels. Calibration measurements based on single text interaction are termed item specific, and those based on accuracy across assessments are termed recursive. As previously discussed, learners are poor judges of comprehension even when accurate assessment exemplars are provided (e.g., Dunlosky, Rawson, & Middleton, 2005; Rawson & Dunlosky, 2002). Unfortunately, using traditional metacomprehension measures the origin of poor calibration is unclear. By evaluating metacomprehension judgments based on learners' own predictions, and measuring assessment and strategic calibration, researchers can evaluate if learners are simply poor at evaluating their own comprehension, unable to align strategies to assessment properties, or unable to predict upcoming assessment correctly.

Measuring Trichotomous Calibration

In this section we outline a methodology for collecting data on the three forms of calibration specified by the TCM using modern technology. Moreover, we present new software called nStudy used to collect on-the-fly behavioral artifacts (traces) without inserting potentially disruptive manipulations into the learner's environment (Winne, 1997). Finally, we provide scenarios to operationalize measurement of each type of calibration in a naturalistic setting.

Traditionally metacomprehension studies have investigated calibration variance as a function of experimental manipulation(s) or learner trait(s) (Maki, 1998; Thiede, Griffin, Wiley, & Redford, 2009). Recently Linderholm and colleagues (Linderholm, Zhao, Therriault & Cordell-McNulty, 2008; Zhao & Linderholm, 2008) collected learners' descriptions of factors considered in the judgment process. We believe such efforts to investigate cognitive interactions with affective states and environmental manipulation is essential to furthering research in the field. However, ongoing concerns about self-reported data (see Winne, Jamieson-Noel, & Muis, 2002 for a review) must be considered. Descriptive analysis captures the process(es) that generate judgments but suffers from subjectivity. Learners' descriptions of their judgment-making experiences can be restricted by their vocabulary and knowledge. Also, the dynamic nature of calibration *in vivo* requires gathering data about learning "on-the-fly". Trace methodology can ameliorate those concerns. The discrete nature of this method enables researchers to track some learning activities without disrupting navigation and adding to the learners' cognitive load. We introduce nStudy, a software system capable of recording fine-grained activities that reflect learners' assessment expectations, internal judgments, and strategic choices. Using nStudy, metacomprehension research can be done in natural environments and the three forms of metacomprehension judgments can be obtained without intruding on learner's activities.

nStudy (see Figure 2) is a software system for collecting data on many aspects of self-regulated learning (Winne & Hadwin, 2009). In nStudy, learners express agency as they use different tools available in nStudy while processing information. nStudy's tools include: making notes based on a choice of schemas (e.g., question and answer, summary, etc.), tagging selected content to classify its properties (e.g., important, review this, don't understand, etc.), hyperlinks that expose new information, constructing new glossary terms, drawing and manipulating concept maps to assemble information within and across information objects (e.g., selections in a text; among notes, terms, etc.), a powerful multi-faceted search tool, and a chatting tool. These tools afford multiple and varied options for learners to exercise and express agency as they construct knowledge. As learners select and use tools in nStudy, the system records fine-grained traces about these choices, such as changing focus from one web page to a note's window, clicking a button to add a note to a concept map, reviewing a glossary term or tagging selected text. The actions traced during the student's studying session are logged to the millisecond and can be examined to make inferences about student activity. These records provide empirical grounds for interpretations about learners' decision-making during knowledge construction. By separating decisions based on calibration types, teachers are able to more precisely target interventions to improve metacomprehension judgment accuracy.

In the following section each type of calibration is described in detail, and a hypothetical scenario is provided as an example:

Measuring Assessment Calibration with Trace Data

Assessment calibration is operationalized as the relationship between predicted and actual assessment; the stronger the relationship, the better the calibration. It can be decomposed into two functional levels. At the item specific level, assessment calibration can be measured by asking learners to predict defining features of future assessments, such as levels of assessment complexity (Anderson & Krathwohl, 2001), or writing assessment items (researchers code responses). By recording predicted assessment complexity along an ordinal scale, the complexity of expected assessment (couched in a specific context) can be compared to actual assessment complexity and assessment calibration can be calculated. At the recursive

Figure 2. nStudy interface

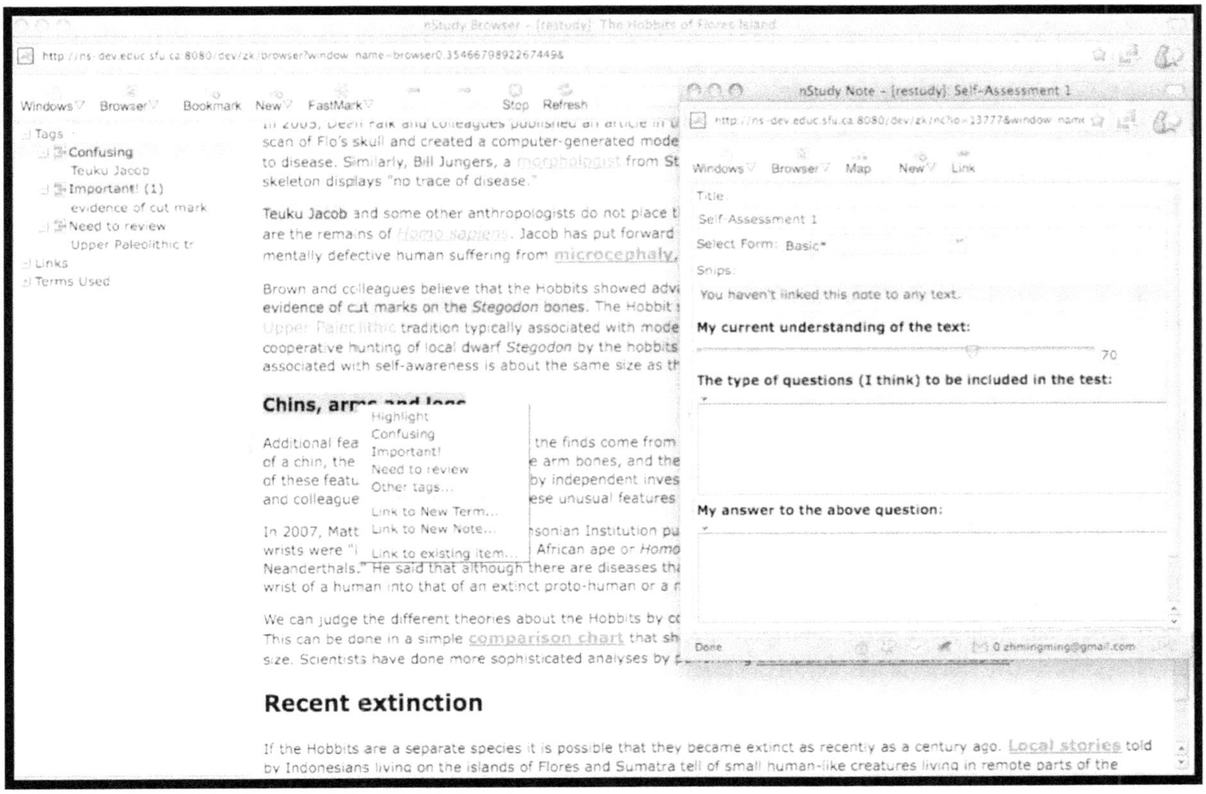

level assessment calibration will be calculated based on the difference between predicted and actual assessments across trials.

Scenario I

Cathy, a second-year undergraduate student is asked to study a chapter on cognitive development using nStudy in her *Educational Psychology* course. The instructor provides a learning objective: *Understand and apply the theoretical stages of cognitive development*. Cathy studies and marks up the chapter using nStudy's tagging tool. She is then asked to use the note tool to draft a short-answer question corresponding to the learning objective: she writes "what are the four stages of cognitive development according to Piaget"? Next, she is provided with three short-answer questions created by the instructor (at different processing depths) and is told to select questions most likely to occur on the quiz. Cathy picks the one similar to the question she just created. The written, selected, and actual test questions are coded into numeric values (e.g., according to the six levels of the revised Bloom taxonomy in Anderson and Krathwohl, 2001). Finally, assessment calibration is calculated by correlating the complexity of her questions and the question she has selected with actual assessment complexity.

Later Cathy discovers that the actual question reads: According to Piaget, what typical cognitive development characteristics would a 2-year-old boy display (*apply*) and why (*understand*)? Thus, Cathy's assessment calibration (at the item specific level) is low. Cathy notes discrepancies between the depth of actual and expected assessment, and her teacher provides additional instructions by asking her to elaborate her reasoning and emphasizing the importance of *understand* and *apply* in the assessment instructions. As a result, Cathy is more aware of the type of assessment to expect given an "understand and apply" learning objective and attempts to create questions at this level of complexity in the future. When studying for the rest of the chapter and receiving similar instructions, Cathy is able to make more accurate predictions of the future assessment. By comparing differences between actual and predicted assessment items over two or more trials, researchers assess Cathy's recursive assessment calibration.

Measuring Strategic Calibration with Trace Data

Strategic calibration is operationalized as relationship between strategies used to prepare for a perceived assessment and strategies theoretically appropriate to the complexity of assessment item(s) and learning goals. Although study strategies are considered control (not monitoring) processes, metacognitive monitoring is required to predict strategy outcomes based on products of study tactics employed in the past. For example, learners may monitor discrepancies between their current level of understanding, the level predicted for a future assessment, and strategies expected to minimize the difference. At the item specific level, strategic calibration deals with strategies chosen to regulate learning at perceived levels of an assessment. Over multiple trials, strategy changes associated with shifting assessment predictions can provide recursive strategic calibration data.

Unlike conventional strategy data collected via probes (self-reports, interviews, concurrent think-aloud protocols) (Azevedo, 2009; Winne, Jamieson-Noel & Muis, 2002), nStudy captures fine grained tactic traces without alerting the learner. By analyzing strategies in this way a number of confounding variables such as reliance on potentially inaccurate recall, reports conforming to expectations, and increased cognitive load are avoided. To define strategies based on traces of enacted tactics, pattern-based analysis can be conducted. Pattern-based analysis (for a review see Winne, Jamieson-Noel, & Muis, 2002) refers to a process where the nature and order of study tactics are recorded, tallied, and calculated to reveal tactic patterns. Reoccurring patterns within a similar context can then be defined as strategies. Enacted strategies can then be compared with learners' perceptions of future assessment to calculate strategic calibration.

Scenario II

As in scenario I, when studying a new chapter Cathy creates and selects questions most likely to be tested. She then predicts her performance for each item (as described in scenario III below). Finally, she completes the instructors' quiz with questions provided at different levels of complexity. After receiving quiz scores, she repeats the same procedure for a second chapter. From quiz feedback and additional instructions, she finds that she does a fairly good job on questions with predicted complexity, but not on questions at other complexity levels. Moreover, she begins to notice that understanding materials is difficult when only simple highlighting tactics are used. To compensate she modifies her strategy from simply highlighting to using an assortment of nStudy tools: she creates a concept map to uncover relations between major concepts in the chapter, writes mini-summaries on notes she has created, and links notes to the concept map for future reference.

On-the-fly data gathered by nStudy during study sessions records Cathy's study tactics, the text she chooses, questions she predicts, actual quiz items, metacognitive judgments of future quiz performance, and performance indicators such as quiz marks. Pattern-based analysis of study tactic data shows that Cathy uses simple highlighting strategies while reading the first chapter, yet in the second chapter, she chooses to add definitions to her glossary for future recall. Also, based on her observations in scenario one that the assessment will require application, she begins to link glossary items to specific scenarios presented in the text. At the item specific level, Cathy's strategic calibration is low during the first study session, but improves greatly in the second session. For example, one of her patterns shifts from a simple "select text →add label" tactic pattern referred to as a "tagging" strategy to "select text→add to glossary→ link to large text (qualitatively coded as a scenario)" tactic pattern referred to as a "term application" strategy, which evidences improved strategic calibration.

Measuring Internal Calibration with Trace Data

Internal calibration is operationalized by the correlation between learners' metacomprehension judgment and actual performance on assessment questions as predicted by learners. Learners who misidentify the complexity of future assessment may be internally calibrated but perform poorly on traditional calibration measures (Pieshl, 2009). By separating internal calibration from assessment and strategic calibration researchers can determine which of these three processes are responsible for notoriously poor metacomprehension calibration found in the literature (see Maki, 1998). Internal calibration can be measured at item specific and recursive levels. Calibration measurements based on single text interaction are termed item specific, and those based on accuracy across assessments are termed recursive. To trace internal calibration researchers must evaluate performance on items in an actual assessment mapped to items predicted by learners (see Scenario I).

Scenario III

When Cathy studies each chapter, she also uses a judgment slider to make predictions about her performance on each question made or selected. nStudy records Cathy's predicted assessment items, the actual assessment items, her future assessment performance judgments, as well as her actual performance. Researchers can calculate item specific internal calibration, by correlating Cathy's judgments with her performance on items at predicted levels of complexity, assuming Cathy's assessment calibration is not zero. By repeating this procedure multiple times, Cathy receives feedback after each assessment (e.g., her actual score on each item) and can adjust her assessment expectations appropriately. Tracking these changes over time researchers can calculate her recursive internal calibration.

CONCLUSION

Within the traditional paradigm, metacomprehension calibration has been defined as the relationship between judgments of comprehension and outcomes of assessments developed by researchers. A wealth of insightful research has identified a multitude of factors influencing calibration such as text complexity, working memory span, and delayed retrieval. In this chapter, we offer the TCM as a means for elucidating the metacomprehension process. We posit that evaluating calibration in the assessment, strategic, and internal form provides a means for a) creating interventions targeted at specific factors associated with poor calibration, b) providing specific feedback to enable learners to self-regulate judgment accuracy, and c) identifying instances where curricular and instructional assessment incongruence acts as a barrier to calibration. In contrast to Linderholm et al., who view the role of ambiguous assessments as an "anchoring trigger", we posit that assessment prediction (with experience) may serve as the anchor itself. If this is the case, assessment calibration within the TCM may provide a valid starting point for adjustment based on comprehension judgments. We also propose that, when strategies are appropriate for an upcoming assessment, processing ease provides a better indication of comprehension. If these assumptions can be verified by empirical research, benefits to metacomprehension due to emphasizing strategic calibration within the TCM become clear. Moreover, our exploration of modern trace technology couched in the TCM framework leverages current metacomprehension research to provide precise, accurate, and on-the-fly feedback. By tracing learners studying independently in a naturalistic environment nStudy can provide activity descriptions with far greater accuracy and detail than think aloud protocols or self-reports. Thus, feedback and resulting interventions may provide learners with process data needed to self-regulate their own comprehension judgments. In this way, we believe that future research can provide learners with tangible means for self-regulating metacomprehension calibration.

RECOMMENDATIONS FOR FUTURE RESEARCH

The TCM framework expands our understanding of the (meta)cognitive processes involved in metacomprehension and presents tracing technology as a means to enhance traditional methods. However, it also reveals several metacomprehension constituents requiring to be addressed in future research.

First, adopting the TCM requires researchers to ensure learners are provided with a) sufficient assessment instruction, b) text/instruction that mirror assessment and c) learning objectives clearly representative of future assessment.

Second, to date few attempts have been made to collect data about learners' predictions (or lack thereof) concerning the nature of future assessments (Zhao & Linderholm, 2008). Recently, Zhao and Linderholm (2008) presented evidence indicating that learners generally do not set goals/standards based on future assessment expectations but rather anchor judgments on prior experiences. This persists even when researchers have provided practice tests that are identical (e.g., Maki & Serra, 1990) or similar (e.g., Glenberg, Sanocki, Epstein, & Morris, 1987) to future assessments. Future research should investigate why learners may not attend to salient contextual assessment cues, and design interventions to encourage full use of provided information (cues) to make effective assessment and comprehension judgments.

Third, learning is an ongoing process that can be seen as a cycle of continual review and revision. As discussed previously, each form of calibration in TCM can be item specific or recursive. Given the large number of computer-lab based experiments involving a single study session, longitudinal research is needed to measure recursive calibration (Pieshl, 2009). By tracking metacomprehension judgments over time, researchers are able to address such issues as (a) the form of calibration most influenced by feedback and/or interventions; and (b) the relationship between internal calibration measured by the TCM and traditional calibration measures. Fourth, nStudy's fine grained activity traces allow researchers to explore cognitive triggers or consequences of changes like those described above. For example, does a certain learning tactic (e.g., making notes) lead to higher metacomprehension ratings or better calibration within a particular context? Does the use of a particular study tactic change as a function of (in) accurate internal calibration?

Finally, metacomprehension research has traditionally focused on in-laboratory learning settings. It is significant to extend current research by studying how contextual variables in authentic learning situations influence metacomprehension processes and accuracy as measured by the TCM. More specifically, future research should be conducted to investigate how fine-grained feedback in all three forms of calibration provided by trace analysis in natural settings will improve learners' ability to self-regulate their judgment accuracy.

REFERENCES

Anderson, L. W., & Krathwohl, D. R. (Eds.). (2001). *A taxonomy for learning, teaching and assessing: A revision of Bloom's Taxonomy of Educational Objectives*. New York: Longman.

Azevedo, R. (2009). Theoretical, methodological, and analytical challenges in the research on metacognition and self-regulation: A commentary. *Metacognition and Learning*, *4*(1), 87–95. doi:10.1007/s11409-009-9035-7

Begg, I., Duft, S., Lalonde, P., Melnick, R., & Sanvito, J. (1989). Memory predictions are based on ease of processing. *Journal of Memory and Language*, *28*(5), 610–632. doi:10.1016/0749-596X(89)90016-8

Boekaerts, M., Pintrich, P. R., & Zeidner, M. (Eds.). (2000). *Handbook of self-regulation*. San Diego, CA: Academic Press.

Dunlosky, J., & Rawson, K. A. (2005). Why does rereading improve metacomprehension accuracy? Evaluating the levels-of-disruption hypothesis for the rereading effect. *Discourse Processes*, *40*(1), 37–55. doi:10.1207/s15326950dp4001_2

Dunlosky, J., Rawson, K. A., & Middleton, E. L. (2005). What constrains the accuracy of metacomprehension judgments? Testing the transfer-appropriate-monitoring and accessibility hypothesis. *Journal of Memory and Language*, *52*(4), 551–565. doi:10.1016/j.jml.2005.01.011

Garner, R., & Alexander, P. A. (1989). Metacognition: Answered and unanswered questions. *Educational Psychologist*, *24*(2), 143–158. doi:10.1207/s15326985ep2402_2

Glenberg, A. M., Sanocki, T., Epstein, W., & Morris, C. (1987). Enhancing calibration of comprehension. *Journal of Experimental Psychology. General*, *116*(2), 119–136. doi:10.1037/0096-3445.116.2.119

Griffin, T. D., Wiley, J., & Thiede, K. W. (2008). Individual differences, rereading, and self-accuracy. *Memory & Cognition*, *36*(1), 93–103. doi:10.3758/MC.36.1.93

Hacker, D. J., Bol, L., Horgan, D. D., & Rakow, E. A. (2000). Test prediction and performance in a classroom context. *Journal of Educational Psychology*, *92*(1), 160–170. doi:10.1037/0022-0663.92.1.160

Kahneman, D., Slovic, P., & Tversky, A. (Eds.). (1982). *Judgment under uncertainty: Heuristics and biases*. Cambridge, UK: Cambridge University Press.

Kintsch, W. (1986). Learning from text. *Cognition and Instruction*, *3*(2), 87–108. doi:10.1207/s1532690xci0302_1

Koriat, A. (2008). Easy comes, easy goes? The link between learning and remembering and its exploitation in metacognition. *Memory & Cognition*, *2*(2), 416–428. doi:10.3758/MC.36.2.416

Koriat, A., & Bjork, R. A. (2005). Illusions of competence in monitoring one's knowledge during study. *Journal of Experimental Psychology. Learning, Memory, and Cognition*, *31*(2), 187–194. doi:10.1037/0278-7393.31.2.187

Lichtenstein, S., Slovic, P., Fischhoff, B., Layman, M., & Combs, B. (1978). Judged frequency of lethal events. *Journal of Experimental Psychology. Human Learning and Memory*, *4*(6), 551–578. doi:10.1037/0278-7393.4.6.551

Lin, L., & Zabrucky, K. M. (1998). Calibration of Comprehension: Research and Implications for Education and Instruction. *Contemporary Educational Psychology*, *23*, 345–391. doi:10.1006/ceps.1998.0972

Lin, L., Zabrucky, K. M., & Moore, D. (2002). Effects of text difficulty and adults' age on relative calibration of comprehension. *The American Journal of Psychology*, *115*(2), 187–198. doi:10.2307/1423434

Linderholm, T., Zhao, Q., Therriault, D. J., & Cordell-McNulty, K. (2008). Metacomprehension effects situated within an anchoring and adjustment framework. *Metacognition and Learning, 3*(3), 175–188. doi:10.1007/s11409-008-9025-1

Magliano, J. P., Little, L. D., & Graesser, A. C. (1993). The impact of comprehension instruction on the calibration of comprehension. *Reading Research and Instruction, 32*(3), 49–63.

Maki, R. H. (1998). Test predictions over text material. In Hacker, D. J., Dunlosky, J., & Graesser, A. C. (Eds.), *Metacognition in educational theory and practice* (pp. 117–144). Mahwah, NJ: Erlbaum.

Maki, R. H., Foley, J. M., Kajer, W. K., Thompson, R. C., & Willert, M. G. (1990). Increased processing enhances calibration of comprehension. *Journal of Experimental Psychology. Learning, Memory, and Cognition, 16*(4), 609–616. doi:10.1037/0278-7393.16.4.609

Maki, R. H., & Serra, M. (1992). The basis of test predictions for text material. *Journal of Experimental Psychology. Learning, Memory, and Cognition, 18*(1), 116–126. doi:10.1037/0278-7393.18.1.116

Maki, R. H., Shields, M., Wheeler, A. E., & Zacchilli, T. L. (2005). Individual differences in absolute relative metacomprehension accuracy. *Journal of Experimental Psychology, 97*(4), 723–731.

Miesner, M. T., & Maki, R. H. (1997). The role of test anxiety in absolute and relative metacomprehension accuracy. *The European Journal of Cognitive Psychology, 19*(4), 650–670. doi:10.1080/09541440701326196

Moore, D., Lin-Agler, L. M., & Zabrucky, K. M. (2005). A source of metacomprehension inaccuracy. *Reading Psychology, 26*(3), 251–265. doi:10.1080/02702710590962578

Mussweiler, T., Englich, B., & Strack, F. (2004). Anchoring effect. In Pohl, R. F. (Ed.), *Cognitive illusions* (pp. 183–200). New York: Psychology Press.

Nelson, T. O. (1984). A comparison of current measures of the accuracy of feeling-of-knowing predictions. *Psychological Bulletin, 95*(1), 109–133. doi:10.1037/0033-2909.95.1.109

Nietfeld, J. L., & Schraw, G. (2002). The effect of knowledge and strategy training on monitoring accuracy. *The Journal of Educational Research, 95*(3), 131–142. doi:10.1080/00220670209596583

Pieschl, S. (2009). Metacognitive calibration-an extended conceptualization and potential applications. *Metacognition and Learning, 4*(1), 3–31. doi:10.1007/s11409-008-9030-4

Pintrich, P. R. (2000). The role of goal orientation in self-regulated learning. In Boekaerts, M., Pintrich, P. R., & Zeidner, M. (Eds.), *Metacognition in educational theory and practice* (pp. 452–494). San Diego, CA: Academic Press.

Pressley, M., Snyder, B. L., Levin, J. R., Murray, H. G., & Ghatala, E. S. (1987). Perceived readiness for examination performance (PREP) produced by initial reading of text and text containing adjunct questions. *Reading Research Quarterly, 22*(2), 219–236. doi:10.2307/747666

Rawson, K., & Dunlosky, J. (2007). Improving students' self-evaluation of learning for key concepts in textbook materials. *The European Journal of Cognitive Psychology, 19*(4), 559–579. doi:10.1080/09541440701326022

Rawson, K. A., & Dunlosky, J. (2002). Are performance predictions for text based on ease of processing? *Journal of Experimental Psychology. Learning, Memory, and Cognition, 28*(1), 69–80. doi:10.1037/0278-7393.28.1.69

Rawson, K. A., Dunlosky, J., & McDonald, S. L. (2002). Influences of metamemory on performance predictions for text. *The Quarterly Journal of Experimental Psychology*, *55*(2), 505–524.

Rawson, K. A., Dunlosky, J., & Theide, K. W. (2000). The rereading effect: Metacomprehension accuracy improves across reading trials. *Memory & Cognition*, *28*(6), 1004–1010.

Reyes, R. M., Thompson, W. C., & Brower, G. H. (1980). Judgmental biases resulting from differing availabilities of arguments. *Journal of Personality and Social Psychology*, *39*(1), 2–12. doi:10.1037/0022-3514.39.1.2

Schraw, G. (2009). A conceptual analysis of five measures of metacognitive monitoring. *Metacognition and Learning*, *4*(1), 33–45. doi:10.1007/s11409-008-9031-3

Schwarz, N., Bless, H., Strack, F., Klumpp, G., Rittenauer-Schatka, H., & Simons, A. (1991). Ease of retrieval as information: Another look at the availability heuristic. *Journal of Personality and Social Psychology*, *61*(2), 195–202. doi:10.1037/0022-3514.61.2.195

Thiede, K. W., & Anderson, M. C. M. (2003). Summarizing can improve metacomprehension accuracy. *Contemporary Educational Psychology*, *28*(2), 129–160. doi:10.1016/S0361-476X(02)00011-5

Thiede, K. W., Anderson, M. C. M., & Therriault, D. (2003). Accuracy of metacognitive monitoring affects learning of texts. *Journal of Educational Psychology*, *95*(1), 66–73. doi:10.1037/0022-0663.95.1.66

Thiede, K. W., Dunlosky, J., Griffin, T. D., & Wiley, J. (2005). Understanding the delayed keyword effect on metacomprehension accuracy. *Journal of Experimental Psychology. Learning, Memory, and Cognition*, *31*(6), 1267–1280. doi:10.1037/0278-7393.31.6.1267

Thiede, K. W., Griffin, T. D., Wiley, J., & Redford, J. (2009). Metacognitive monitoring during and after reading. In Hacker, D., Dunlosky, J., & Graesser, A. (Eds.), *Handbook of metacognition in education* (pp. 85–106). New York: Taylor & Francis.

Thomas, A. K., & McDaniel, M. A. (2007). The negative cascade of incongruent generative study-test processing in memory and metacomprehension. *Memory & Cognition*, *35*(4), 668–678.

Tversky, A., & Kahneman, D. (1974). Judgment under uncertainty: Heuristics and biases. *Science*, *185*(4157), 1124–1130. doi:10.1126/science.185.4157.1124

Walczyk, J. J., & Hall, V. C. (1989). Effects of examples and embedded questions on the accuracy of comprehension self-assessments. *Journal of Educational Psychology*, *81*(3), 435–437. doi:10.1037/0022-0663.81.3.435

Weaver, C. A., & Bryant, D. S. (1995). Monitoring of compehension: The role of text difficulty in metamemory for narrative and expository text. *Memory & Cognition*, *23*(1), 12–22.

Winne, P. H. (1997). Experimenting to bootstrap self-regulated learning. *Journal of Educational Psychology*, *89*(3), 397–410. doi:10.1037/0022-0663.89.3.397

Winne, P. H., & Hadwin, A. F. (2009). *nStudy: A web application for researching and promoting self-regulated learning* (version 1.01) [computer program]. Simon Fraser University, Burnaby, BC, Canada.

Winne, P. H., Jamieson-Noel, D. L., & Muis, K. (2002). Methodological issues and advances in researching tactics, strategies, and self-regulated learning. In Pintrich, P. R., & Maehr, M. L. (Eds.), *Advances in motivation and achievement: New directions in measures and methods* (*Vol. 12*, pp. 121–155). Greenwich, CT: JAI Press.

Winne, P. H., & Perry, N. E. (2000). Measuring self-regulated learning. In Boekaerts, M., Pintrich, P., & Zeidner, M. (Eds.), *Handbook of self-regulation* (pp. 531–566). Orlando, FL: Academic Press. doi:10.1016/B978-012109890-2/50045-7

Zhao, Q., & Linderholm, T. (2008). Adult metacomprehension: Judgment processes and accuracy constraints. *Educational Psychology Review*, *20*(2), 191–206. doi:10.1007/s10648-008-9073-8

Zimmerman, B. J. (2000). Attaining self-regulation: A social cognitive perspective. In Boekaerts, M., Pintrich, P. R., & Zeidner, M. (Eds.), *Handbook of self-regulation* (pp. 13–39). San Diego: Academic Press. doi:10.1016/B978-012109890-2/50031-7

KEY TERMS AND DEFINITIONS

Absolute Judgment Accuracy: Actual performance viewed as a function of predicted performance.

Anchoring: The assimilation of judgments to a previously considered standard.

Assessment Calibration: Learners' ability to interpret the learning context to estimate the depth, complexity, and coverage of an upcoming assessment.

Criterion Referenced: (measures): Performance that is compared to a pre-specified standard.

Ease of Cognitive Processing: The relationship between the speed and salience of text based information contemplated during initial reading and/or upon retrieval from long or short term memory and judgments about the nature of that information.

Internal Calibration: The accuracy of metacomprehension judgments of future performance based on learners' assessment prediction.

Locus of Control: Outcomes perceived to be controllable by an individual's own characteristics or actions (internal control) or by external forces such as luck, fate or others (external control). (see Rotter, 1966)

Metacognition: A general term used to describe awareness of our own learning, memory, and thought processes. (see Flavell, 1979)

Metacomprehension Calibration: The accuracy of metacomprehension judgments.

Metacognitive Control: Regulation of cognitive processing to achieve a desired outcome.

Metacomprehension Judgment: Judgment of comprehension or comprehension performance over text materials.

Metacognitive Monitoring: Evaluation of the progress and outcome of cognitive processing.

Norm Referenced: (measures): An individual's performance is compared to the (usually) normally distributed performance of other students.

Relative Calibration Accuracy: The correlation between judgments and performance.

Self-Esteem: A judgment consisting of feelings of worth and acceptance (see Panteleev, 1991).

Self-Regulated Learning: An active, constructive process whereby students set goals for their learning and then attempt to monitor, regulate, and control their cognition, motivation, and behavior guided and constrained by their goals and the contextual features in the environment. (see Pintrich, 2000)

Strategic Calibration: The match between learning strategies a learner surveys and then chooses relative to needs for developing knowledge and skills perceived to be required by the task.

Chapter 6
Using Student Assessment Choice and eAssessment to Achieve Self-Regulated Learning

Cath Ellis
University of Huddersfield, UK

Sue Folley
University of Huddersfield, UK

ABSTRACT

This chapter explores how we can harness technology to foster self-regulated learning in assessment practices. Traditionally innovation in assessment lags behind that in other areas of teaching and learning, however, it is important to make sure that assessment methods and practices are aligned with learning objectives. For assessment to be a beneficial learning experience for students it is important that they are afforded more autonomy and agency over what, when and how they are assessed. This chapter reflects on the 'problem' that assessment and feedback presents and on what the research is showing academics need to concentrate. Secondly, it considers how eAssessment tools can provide a way forward to achieving these objectives and helping students to develop more self-regulated learning strategies. Finally the authors will explore how the use of these tools can allow students greater autonomy over the whole assessment process, and the essential role that technology may play in achieving this.

INTRODUCTION

In a fast-paced world it is becoming increasingly likely that a significant proportion of what students learn as undergraduate students will be out of date or obsolete by the time they finish their professional working careers. In such a world, which Barnett (2008) describes as both 'uncertain' and 'super-complex,' it is no longer enough for students to learn 'facts' and 'information'. As

DOI: 10.4018/978-1-61692-901-5.ch006

Dochy et al. (2008) put it: 'today's knowledge community expects graduates not only to have a specific knowledge base but to be able to apply this knowledge to solve complex problems in an efficient way' (p. 87). Being able to find, manage and evaluate information and process it into knowledge is becoming an increasingly important graduate attribute across all disciplines. As many scholars have pointed out, empowering students to become self-regulated learners is vital for their achieving this (M Boekaerts, 1999; M. Boekaerts & Simons, 1995; Kurtz & Weinert, 1989; Weinert, Schrader, & Helmke, 1989).

Self-Regulated Learning (SRL) has become an increasingly important concept in education studies in recent years. Zimmerman & Schunk (1989) define SRL as self-generated thoughts, feelings, and actions, that students systematically orient toward attaining goals they have developed or value. The growing interest in SRL has occurred partly in response to the changing nature of knowledge, information and professional work, and partly because of increasingly successful attempts by governments to widen participation in post-compulsory education. Self-regulated learning, therefore, marks a significant shift away from the traditional *modus operandi* of Higher Education which has been dominated by instructional pedagogical paradigms (characterised by 'chalk-and-talk' and end-loaded, summative assessment). There is considerable literature exploring and outlining the means by which these changes may come about but achieving them will be futile if they are not matched with changes in assessment and feedback strategy that also both encourage and reward SRL.

As many leading scholars have pointed out, assessment is the single most important factor influencing student behaviour and attitude (Bloxham & Boyd, 2007; see for instance Boud & Falchikov, 2007; Dochy et al., 2008; Scouller, 1998; Snyder, 1971). The influence is so strong that the negative 'backwash effect' of assessment that Elton (cited in Biggs & Tang, 2007, p. 169) identified and Biggs has developed further, whereby assessment sends undesirable messages to students, means that even the most brilliant teaching and learning can be negated by weak assessment design. It is this principle which is at the heart of Biggs' very influential theory of 'constructive alignment' whereby teaching and learning activities and assessment tasks are systematically aligned with the intended learning outcomes according to the learning activities required in the outcomes (Biggs & Tang, 2007, p. 7). What this means for self-regulated learning is that it is not enough to design learning environments, activities and outcomes that encourage and empower it without appropriately designed assessment tasks that also encourage SRL.

This chapter is interested in exploring the key aspects of SRL in terms of assessment and feedback. As Boekaerts has pointed out, choice is one of the hallmarks of SRL, alongside accessibility and adaptability (M Boekaerts, 1999, p. 451). This chapter will focus, therefore, on the role that student assessment choice can play in terms of empowering students to become self-regulated learners and how eAssessment tools can be used to achieve this. For the purposes of this paper, we use the term eAssessment to mean electronic and/or online tools which can be used for student assessment. This is not limited to Computer Marked Assessment (such as automatically assessed multiple choice questions) and includes Tutor Marked Assessment which makes use of communication and information technology in some form. The context of this discussion is limited to the Higher Education sector, so focussing on adult learners.

BACKGROUND

As any student will tell you, assessment and feedback is important and has a huge influence on their perceptions and behaviour. Orsmond et al (2002) state that 'Assessment tends to shape every part of the student learning experience' (p.24).

Keppell et al (2006) claim that students define the curriculum or module according to the assessment, and that it sends both explicit and implicit messages to them about what is considered important. This is, of course, a double edged sword. While the 'backwash effect' can be negative, as Gibbs recently remarked, assessment is teachers' main lever 'to change the way students study and get them to put effort into the right things' (Gibbs, reported by Attwood, 2009). Taras (2001) suggests that in a fees-based system:

students as paying customers have invested in higher education and their returns are seen to materialise in the form of assessment grades. Without pandering to these negative perceptions, we do need to assure them that assessment is fair and above reproach. (p. 606)

Getting assessment and feedback right is not just important to students but to individual institutions and, indeed, the industry as a whole. But, as Boud (1995) has pointed out, in an open-access system with increasing student numbers and diversity this is, if anything, getting harder to achieve.

The 'Problem' of Assessment and Feedback in Higher Education

Across the Higher Education sector, the practice of assessment and feedback is attracting increasing attention to the extent that it is now frequently discussed in terms of being a 'problem'. Graduate experience and satisfaction surveys show that assessment and feedback frequently score low results in terms of student satisfaction (see for instance the NSS in the United Kingdom where Assessment and Feedback has consistently registered the lowest level of student satisfaction). It is also the area in which Higher Educational Institutions (HEIs) most frequently struggle in terms of Quality Assurance Audits. Student Unions have also voiced their concerns about assessment and feedback. In the UK the National Union of Students (NUS) has initiated a targeted campaign, led by the Great Feedback Amnesty briefing paper (NUS, 2008) which fed into a briefing statement listing ten principles for good practice on assessment and feedback. The implication was clear: the dissatisfaction reported by students indicates that across the sector these principles are simply not being adhered to. In combination these factors have meant that HEIs are newly motivated to tackle the issue of assessment and feedback.

The scholarship on assessment and feedback agrees that assessment in Higher Education is indeed a 'problem'. Boud and Falchikov (2007) open their book Rethinking Assessment in Higher Education with the simple statement: 'Assessment affects people's lives'(p.3). They argue that the 'high-risk' nature of assessment has meant that it has been subject to less innovation and change than other aspects of University teaching and learning. They argue:

We face a system of assessment that has been subject to slow incremental change, to compromise and inertia. We are afraid to change the system because of the risks, but we also avoid looking at it because doing so might entail major effort. (p.3)

Bloxham and Boyd (Bloxham & Boyd, 2007, p. 3), in the introduction to their book Developing Effective Assessment in Higher Education: A Practical Guide, agree. They bemoan the fact that 'University assessment lags well behind its equivalent in the school sector [...] relying largely on a limited range of tried (but not always tested) methods' (Bloxham & Boyd, 2007, p. 3). Assessment is, they argue, 'dealt with in an ad hoc way' by most academics who have learned the craft 'informally through being assessed [them] selves and through being part of a community of practice'(p.3). They assert that 'the contemporary environment of higher education means that assessment cannot carry on unaltered' (p. 4). Phil Race, (2005) in a chapter section titled 'What's wrong with assessment?' asserts that in terms of teaching, learning and assessment 'assessment is the weakest link'. He goes on:

it's easier (and safer) to fiddle around with the quality of teaching or learning than to tackle the big one: assessment. [...T]here are significant shortfalls in the extent to which many of the most common assessment practices measure up to bringing the qualities [of validity, reliability, transparency and authenticity] to bear on assessment. (p. 74-5)

Clearly there is a problem with how HEIs manage assessment and feedback across the sector.

While assessment and feedback can reasonably be considered a 'problem', there has been a considerable amount of excellent research conducted on how to do assessment and feedback well. Most notable has been the REAP report led by David Nicol at the University of Strathclyde (Nicol, 2007). There are also a group of highly respected scholars working in the field, including Graham Gibbs, Sally Brown, Phil Race, Chris Rust and David Boud to name a few (see for example Boud, 1995; Gibbs, 1992; Rust, 2002). Their research and the literature in general tends to agree on the fundamental principles of good assessment and feedback design.

Our survey of the research on assessment and feedback shows that in the current academic climate the following objectives are priorities for HEIs:

- Achieving higher levels of student participation, empowerment and motivation by improving the diversity, flexibility and authenticity of assessment tasks;
- Increasing student satisfaction by offering greater involvement and control over assessment and by better aligning assessment with learning objectives and student-centred pedagogies;
- Supporting the widening participation agenda with assessment that accommodates a larger range of learning needs and learning styles;
- Developing stronger life-long learning skills by harnessing the power of self- and peer-evaluation in assessment;
- Encouraging good scholarly conduct by designing out plagiarism;
- Maximising both quality improvements *and* staff workload efficiencies.

In terms of designing assessment tasks the research highlights the following. Assessment must:

- be *for* learning not *of* learning (i.e. assessment that benefits student learning rather than simply measuring learning achievement);
- measure understanding not just memory, which is especially important in an information-rich world where being able to remember things is becoming increasingly unimportant but being able to understand and interpret things increasingly more important;
- be fair, reliable, consistent, innovative, inspiring, motivating, regular, reflective and manageable;
- be authentic (i.e. asking students to do something that they can understand that someone in the real world would realistically be asked to do or want to do);
- be transparent, so that it is clear to students what is expected of them, how their work will be evaluated and how their grade was arrived at;
- be empowering, so that the students feel motivated and involved.

The hallmarks of SRL, to do with choice and autonomy, are clearly evident here.

It stands to reason that if we want students to regulate their own learning that this should include regulating their own assessment. On this point Taras (2001) makes an important contribution. She asserts: 'if we want students to take responsibility we must allow them to do so' (p.612 emphasis ours). This observation is very telling. It follows that one of the key aspects of empowering students to become self-regulated learners necessarily involves academic staff relinquishing some of the regulation they have traditionally controlled. There is little doubt that assessment (with its strict submission deadlines and formats accompanied by harsh penalties for non-compliance) is one of the most heavily regulated aspects of the student learning experience, no doubt further evidence of both its importance and 'high-risk' status. After all, there is undoubtedly a clear contradiction between academic staff wanting to maintain control over assessment regulation while at the same time bemoaning students' lack of self-regulation. Taras (2001) argues convincingly for increased power-sharing in assessment practices and Hafner & Hafner (2003) call for a wholesale move away from what they call the 'semi-secret' devices of traditional university assessment. It is clear that for SRL to be achieved, academic teaching staff must give up at least some of their control over assessment choices.

Allowing students to make more decisions regarding their assessment brings other important benefits. Taras (2001) suggests that finding ways students can be involved in the process of their assessment can also reduce the likelihood of their complaining about it. If they are more involved in the process, she argues, they are better able to judge for themselves if assessment of their work is justified, and this, therefore, should result in increased levels of student satisfaction. Because decision making necessarily involves choice, emerging from this is the growing consensus that increasing levels of student assessment choice should have a long-term beneficial impact on student learning and achievement and will be crucial to their development as self-regulated learners.

STUDENT ASSESSMENT CHOICE

While increasing levels of student assessment choice may be ideal in theory, getting it to work and be beneficial for student learning in practice is another thing altogether. For one thing, there are many different aspects of assessment over which students could be given more choice and control, so distinguishing between them is helpful. The research undertaken for the REAP report (Nicol, 2007) makes it clear that assessment choice covers five main areas:

- **Format**: the format in which they present their learning achievements and/or research findings;
- **Subject**: the topics, questions and/or problems students address and/or engage with, in their work;
- **Criteria**: against which their achievement will be measured;
- **Timing**: when they submit their work and the penalty they will incur if this deadline is not met;
- **Result**: the grade they receive for their work.

Importantly, these different aspects are not equal in terms of the ease with which they may be implemented and the anxieties they may generate in students.

The implications of such anxiety generation on students' learning needs careful consideration. While the relationship between assessment and emotion is both complex and, as Boud & Falchikov (2007) point out, 'underinvestigated and undertheorised' (2007, p. 147), it is, they argue, nevertheless important, not least because 'assessment experiences can be long lasting and influential on personal and academic development' (Boud & Falchikov, 2007, p. 152). The relationship between student autonomy and the emotional experience of being assessed is, they argue, a central concern. They make the important point that while it is clear that autonomy is important for learners, as Krapp's research shows, 'not all learners are prepared for full autonomy', (Krapp, 2005, as cited in Boud & Falchikov, 2007, p. 152) and that 'learners desire and enjoy freedom of action only when they believe themselves to be capable of learning and carrying out the tasks involved' (p. 152). So while autonomy, control and choice are important for SRL, students must be offered appropriate and timely support in their decision making and it is unlikely that giving absolute choice and power to students is going to be helpful or effective.

It is useful therefore to consider in more detail these different aspects of assessment choice in terms of student anxiety. While all of these aspects of student assessment have traditionally been regulated by academic staff, there is already a certain amount of student involvement and choice in some. For instance, during their course of study most students will experience some degree of choice in the subject on which they are to be assessed (which could include a choice of essay topic or exam question, and such things as open design briefs or performance choices). This is therefore something with which students and staff are both already likely to be accustomed, which consequently is likely to cause little anxiety and is therefore considered relatively low risk. In contrast, most academics would agree there are other aspects of student assessment choice over which it is almost unthinkable to allow student choice. For instance, few students will have any say at all on the result they receive for their work within the normal assessment procedures (i.e. outside a formal 'complaints' or 'appeals' processes). Most academic staff would argue, quite stridently, that deciding and regulating student results sits clearly within the bounds of their professional judgement and practice and University regulations would tend to agree. This aspect of student assessment choice, being unfamiliar and outside normal regulated practise, therefore has the very real potential to generate significant anxiety for students, staff and the institution itself and is consequently seen as 'high-risk'. It stands to reason then, that for staff relinquishing and students taking up control over some of these aspects of assessment choice is, in both a conceptual and a practical sense, less risky and therefore easier to achieve than others.

We have represented this diagrammatically in Figure 1 below.

The relationship between anxiety and learning is an important but complex one and it is worth dwelling on this here in some more detail. Pekrun, Goetz, Titz & Perry's (2002) research shows that anxiety is the emotion reported most often by students in relation to academic settings. But, as they point out, anxiety 'is not necessarily the most detrimental negative academic emotion' (p.100) and anxiety doesn't always equate to poor learning outcomes. To put it simply: 'Some students may [...] profit motivationally from their anxieties, whereas others are handicapped' (p.100). Similarly, the relationship between anxiety and self-regulated learning is not straightforward. Their findings suggest that

> *positive emotions foster students' self-regulation, whereas negative emotions lead to reliance on external guidance. However, the reverse direction of causation may play a role here as well: Self-regulating one's own learning may induce positive feelings, whereas external control may induce anger, anxiety or boredom. (p. 99)*

On the whole, however, they found that emotions are closely tied to students' 'self-appraisals of competence and control in the academic domain, [...] and to classroom instruction and social environments affecting control, values, and goals' (p.103). It is clear therefore, that affording students more control and regulation over their assessment can only be helpful to their learning both in the short and long term but only if they are confident in their competence to make such decisions. It follows that it is unhelpful and perhaps even dangerous to simply give control, choice and

Figure 1. Anxieties induced by assessment choices

	Choice	Anxieties		Current Frequency	Institutional barriers
		Staff	Students		
Format	How they present their learning for assessment	Moderate	Moderate	Occasional	Moderate
Subject	Topic, question and/or problem students address and/or engage with, in their work	Low	Low	Common	low
Criteria	Measures against which their work is judged	High	High	Rare	High
Timing	When their work is to be submitted	High	Moderate	Rare	High
Result	Grade received for work	Very High	High	Almost never	Very high

regulation to students without guidance, encouragement and support that is appropriate for their developing confidence in their competencies.

Concomitant to this is the fact that each aspect of student assessment choice brings with it a different range of barriers in terms of administrative practicalities. Offering students a choice of four or five different essay topics is one thing; allowing all students to submit whatever assessment they want, whenever they want is another thing altogether and something that institutional regulations are unlikely to condone. While handing some aspects of assessment regulation to students is important, and perhaps even long overdue, it is clear that introducing wholesale and widespread student assessment choice across all of these areas all at once is neither feasible nor wise. As such, by making the distinction between different aspects of student assessment choice in this way, alongside the anxieties they are likely to generate and the practical difficulties they bring is helpful in that it allows a more informed, strategic approach to a successful and sustainable introduction of increased student assessment choice. This thereby allows for better development and support of SRL in such a way that risks are minimised for academic staff, students and institutions alike.

The Role of Technology e-Assessment

Teaching technologies and e-Assessment tools can help manage the pragmatic aspects of increasing student assessment choice. Technology has made increasing student assessment choice viable in two key ways.

First, online environments, particularly the vast array of web 2.0 resources and the accessibility and affordability of multi-media production tools, have given students an unprecedented choice in how to collaborate, demonstrate their learning achievement and share their assessment outputs. Students can now realistically produce films, collaboratively authored reports, high quality publications (such as posters and brochures) and websites from their PCs which can be viewed and commented on by their peers. One of the key benefits of using these kinds of resources is the shareable nature of the assessment produced. Many of these tools also allow students to comment on each other's work from within the resource (such as the comments tool embedded in YouTube and Flickr) and/or to collaborate (for instance Voicethread allows multiple people to comment on the same image or film). As we shall argue, the shareable nature of web2.0 resources brings a new dimension to student assessment which has considerable potential for the development of SRL.

Secondly, computer software, and particularly Virtual Learning Environments (VLEs), have given academic staff powerful tools with which to manage complicated assessment and feedback strategies efficiently and effectively. Mobile computer hardware and increasingly powerful and integrated software has given academic staff more flexibility and freedom in terms of when, where and how they process and store student assessment and offer students feedback/forward. One of the key benefits that these tools bring is the capacity for academic staff to automate many of the repetitive processes involved with assessment administration allowing for increases in both efficiency and quality.

Together, the range of eAssessment tools now available to staff, students and the institution have much to contribute to the realisation of greater degrees of student assessment choice and thereby SRL.

Student Assessment Choice

We now turn to consider each aspect of student assessment choice in more detail. The purposes of this consideration is to offer an overview of each aspect while touching on some of the potential benefits and risks, and also considering, the impact that new technologies can have on their feasibility.

Format

Until fairly recently, students were restricted in terms of how they could present their learning achievements for assessment. For the most part paper-based assessment such as exams and coursework essays formed the bulk of the student assessment 'diet' occasionally supplemented with portfolios of practical work, presentations, etc. Technology has now changed all of that. Students can now realistically be expected to build a web site, publish a brochure or make a film as part of their assessment. Web2.0 resources are a key part of this with sites like Voicethread, UStreamTV, YouTube, flickr, Jing, Qik and countless others making it possible for students to publish their assessment as sophisticated, shareable audio visual objects.

Choice of format brings some important benefits to students in terms of their being able to regulate their own learning. By choosing their method, students can find a medium that best suits their skills and best presents their learning/findings. Text-based tools such as blogging, micro-blogging and webpage authoring might suit students with strong writing skills whereas screencasting, podcasting and film making resources allows students with strong oral communication skills to play to their strengths. It is highly likely that some if not all of these media could offer significant benefits to the large proportion of students in HE living with cognitive conditions such as dyslexia for whom written communication presents a barrier.

This wide range of formats also allows students to choose the media which best accommodates the material or argument they want to put. For instance, YouTube is well suited to students presenting learning as a film while VoiceThread is much better for commenting on a film. Giving choice to students, thereby, adds another important skill: choosing the medium that best communicates the message they want to send.

The 'shareability' of web2.0 objects brings a valuable new dimension to student assessment work. At a time when sharing information and publication is easier and cheaper than ever, students sharing or 'publishing' their assessment outputs remains relatively rare. Assessment is still considered a 'private affair' with the vast majority of student work read only by the student and their marker(s). Looked at purely objectively, this constitutes a monumental waste of effort. Knowing that their assessment work may be of value to other students in a peer-learning context and/or that it contributes to a body of student learning materials to be reused by future student cohorts is a significant motivating factor for students – something that is evidenced in the significant body of work on peer-learning and the Contributing Student Approach (Collis & Moonen, 2001).

Shared assessment improves the opportunity for students to compare their work with others. This may be beneficial to students across the spectrum of achievement. Low and middle achieving students can see where they need to be aspiring and high-achieving students can get a better sense of why their work is valued at the level it is. Importantly, this therefore better supports self-assessment and self-evaluation – both important aspects of self-regulated learning. It also reduces student collusion and plagiarism because any evidence of similarity between students' work is clearly displayed. It can also encourage collaboration as students can more easily see who shares their interests and/or can match themselves up with peers with complementary skills sets to theirs. Anecdotally, students report taking more care over their work if they know their peers are going to see and use it. Clearly in an SRL environment, where students are actively encouraged and supported in choosing their own paths of inquiry, the shareable nature of using web2.0 resources to present assessment outputs brings significant benefits.

While 'publishing' assessment work clearly brings benefits, students tend to perceive it as being more 'risky' than simply submitting work to a tutor for marking. The anecdotal evidence which suggests that students are more likely to take more care over work which they know their peers will see indicates that sharing their work is also more anxiety inducing. Supporting students through this process is vital and again emerging and established eAssessment tools are vital in this process. Using incremental and iterative developmental steps whereby students move gradually from low-risk, short, reflective publications (making regular blog or discussion-board entries on a VLE for instance) towards a larger-scale, public output that is submitted for summative assessment (such as a film or web-published essay) can help students develop confidence over a period of time. Encouraging students to seek and provide formative feedback on each other's developmental work can also help them develop a shared sense of trust and thereby increase their confidence.

The use of technology is vital in terms of helping students and teachers manage increased choice in assessment method. Anything that can be uploaded or linked to the VLE can very easily be managed through the VLE operating as a kind of portal. Something as simple as a blog, discussion board or wiki can be used as a point for students to use to submit their work and view each others' work. This can also enable the students to share their work without it being made available publicly outside of the student cohort.

Importantly, the increased choice and inherent shareability that web2.0 resources offer, brings with it a single, important assessment design issue. If assessment work is to be shared and if this sharing is going to be useful to students in terms of peer-learning, any assessment task for which there is only one or a limited number of 'right answers' simply will not work. This brings with it the second in our list of student assessment choices: subject.

Subject

Offering students a choice in assessment topics is certainly nothing new, but the vast majority of assessment tasks offer only a limited choice. Opening up the choice more widely is seen as more risky but can bring important benefits. Key amongst these is the possibility of making assessment much more personal and thereby more responsive to SRL. Asking students to choose the subject of their assessment based on their own personal experiences, environments and interests is immediately more motivating and has the capacity to make their learning experience more authentic by anchoring it to their lives.

Offering increased choice in subject is also more likely to be recognised by students as closer to real-life activities i.e. work that someone in the real world would reasonably want or be expected to do (Rust, Price, & O'Donovan, 2003). This kind of task has been shown to motivate students to undertake deep-learning and make them less likely to cheat. Offering choice on the subject of assessment is more in line with constructivist pedagogies which allow students to build on their prior knowledge and to discover and/or pursue their own interests and passions. The benefits of increased assessment choice in terms of subject are not just for the students. Having a wider range of topics addressed in student assessment can also make marking much less repetitive – something academic staff everywhere are likely to welcome!

Opening up the student choice of subject in assessment brings with it risks for students and academic staff alike. Students can feel overwhelmed when faced with a 'blank slate' thus making the experience more anxiety inducing. As any scholarly researcher will know, coming up with a topic can sometimes seem like half the work. Providing students with appropriate support is vital to reduce their anxieties. Given that much of this process is very generic to all research in any given discipline area, regardless of the specific topic, it is possible then to design and develop a suite of learning activities for students to work through which allow them to develop relevant skills. Tasks focussed on, for instance, writing a thesis-statement, building an annotated bibliography, evaluating resources and writing an abstract, can support students in the generation of their own essay topic. Similar tools for corresponding steps in response to an open design brief or applied research project could also be developed. For such learning activities to be beneficial to students in an SRL context, they need to be both on-demand and self-paced, thus allowing students to work through them when and where they are ready to move onto the next stage of the process. Again, the flexibility that asynchronous learning tools, embedded in a VLE for instance, offer are vital to support students through this learning process. This strategy gives students a more open scope on the product they generate from their work but has the added benefit of emphasising the generic, shared elements of the process of their scholarly endeavour. So just as the learning they achieve has considerable elements of self-regulation, it also enables students to develop valuable lifelong learning skills which they can reapply in different contexts.

Allowing a more open scope in the topic of student assessment can, however, bring significant risks for those managing and marking the work. The requirement for those grading the assessment to have enough knowledge and expertise to 'cover' all of the subjects addressed by students in their work can be unfeasible, especially in large classes. In addition, having completely unlimited scope from which students can choose the topic of their assessment makes a mockery of them working from a shared syllabus. It is logical, therefore, to have some agreed and proscribed limits on this choice and/or to make it a requirement that students submit proposals for their choice of topic for approval at an early stage in the process. Again, tools embedded in VLEs such as blogs, discussion boards and wikis can be useful in this process.

Encountering increased choice in method and subject will be, for most students for some time to come, an unfamiliar experience. If students are able to choose both what they are going to engage with and how they are going to present their findings, having a clear set of criteria which is fair and equitable to everyone is vital. Therefore it is important that students understand and have some sense of ownership of the criteria against which their learning achievement will be measured. It is here that increasing student choice of criteria comes into play.

Criteria

A vast body of research on assessment and feedback design demonstrates that the use of criteria-referenced marking systems is widely beneficial to students. This research agrees, however, that providing students with the criteria isn't enough to trigger beneficial learning. Students must also be supported in their engagement with it. The tacit knowledge of marking is an important part of this.

Tacit knowledge has been defined as: 'we can know more than we can tell' (Polanyi cited in Rust et al., 2003, pp. 151-152). As Rust, Price and O'Donovan explain, tacit knowledge is

deeply rooted in action and often in an individual's commitment to a profession, [it] consists partly of technical skills based on professional experience, and in a more cognitive dimension, in our ingrained mental models, beliefs and perspectives. (p.152)

As many scholars have pointed out, marking is a kind of tacit knowledge. As Saunders and Davis put it: 'over time discussion and shared experiences of marking and moderation among staff enable the sharing of tacit knowledge, resulting in more standardised marking' (Saunders & Davis cited in Rust et al., 2003, p. 152). This tells us that the 'professional judgement' academics use in their marking and share with each other is something that comes with practice. It stands to reason that for criteria-based marking to be useful to students as they build their assessment they too need to share in this tacit knowledge. Taras (2001) describes this as students becoming their own double markers. She argues that all the benefits of double marking for teaching accrue to students if they too are engaged with it. This includes making the process more fair and helping coordinate the understanding of criteria and standards. She draws on the finding of Boud (1995) that 'all assessment is questioned to some degree' (p.608) and speculates that: 'perhaps self-assessment should develop the confidence and independence of students sufficiently for them to be able to judge for themselves if other people's assessment of their work is justified' (p. 608). Obviously, just as for academic staff, this isn't something in which students can be expected to develop expertise quickly. So including them, over time, in the assessment culture and allowing and encouraging them to take responsibility for being one of their own markers is strategically useful and important.

Having said this, if the learning criteria against which student work is to be measured has been devised solely by academic teaching staff, it is entirely possible that students who achieve well against these measures have done little more than, as Race (2005) puts it, succeed in getting 'their minds into our assessment culture' (p. 78). Allowing students to have *more* to say about the criteria against which their work will be measured can mitigate against this by immediately making the assessment culture theirs. This is not to say that students should be allowed to be assessed completely on what they want. Students who have *some* involvement in the construction of the assessment criteria against which their work will be measured, however, are more like to have a greater sense of ownership over it and thereby a deeper engagement with it.

Obviously allowing students completely free reign on the construction of learning criteria is undesirable and would not satisfy even the most lenient quality assurance requirements for validation. However, offering guidance and a framework which students collaboratively 'fill-in', discussing and negotiating changes and weightings, can be both an empowering and motivating experience for students. While academic staff may feel uncertain about giving up this element of control, those who have attempted it indicate that these anxieties are unfounded. In a feature article on assessment and feedback in the Times Higher Education, Race reported on his experience of having students formulate assessment criteria themselves:

The first two or three times I did this I had in my briefcase what I thought was a good set of criteria we could use if the students didn't come up with a suitable one. But every time, I walked out humbled: they'd come up with something better. (Race, reported in Attwood, 2009)

Race is adamant that having students compose their own assessment criteria assists in the value and quality of both self- and peer-evaluation, both of which can bring measurable benefits to student learning and achievement.

Again, technology can offer valuable assistance in this endeavour. Collaborative authoring tools, such as wikis and document sharing, allow multiple people to contribute to and edit documents simultaneously. Many such tools also keep historical records of alterations so that alterations can be tracked (in terms of both who made them and when they were made) and easily reinstated. There is also a growing body of e-Assessment tools, such as Turnitin's suite of assessment management software, (http://turnitin.com) which include rubric calculators, banks of common comments and powerful automated peer-assessment tools, which allow students and staff to evaluate assessment against the same criteria. These tools can bring other administrative benefits, such as plagiarism checking, the backing up of student work and electronic submission and return, which make the management of student assessment both more secure and efficient.

Timing

Across the sector, institutions place great significance in the timing of assessment, ensuring that due dates are clearly published in advance and that students suffer significant consequences if their work is late. These rules are so well accepted that few think to challenge or even question them. But, when considered objectively, the idea of set deadlines and such harsh penalties can be difficult to justify and out of step with constructivist pedagogical principles. Barr and Tagg (1995) in their influential paper "From Learning to Teaching – a new paradigm for undergraduate education" outline the characteristics of their paradigmatic shift from an instruction paradigm to a learning paradigm. Under the heading of teaching/learning structures they include:

Time held constant, learning varies → Learning held constant, time varies

End of course assessment → Pre, during and post assessments

In this kind of new paradigm, it makes sense that students should be able to present their learning achievement when they are ready to, that is, when they consider that they have achieved their learning and not by when their tutor thinks they should have achieved it. One of the most common arguments put for the maintenance and defence of due dates and harsh penalties is that it helps students develop time management skills. When looked at objectively, however, enforcing due dates does nothing of the sort. Any academic will tell you, and many students will admit, that a great many students routinely leave their work to the last minute regardless of when the deadlines are set. Time management, in terms of such things as effective strategies and the use of electronic tools, is not taught through set deadlines, rather students are more often than not left to figure out time management skills and strategies for themselves. Allowing students to set their own deadlines, would allow students greater agency in managing their work and at the very least allow them to submit their work when it suits them, rather than, which is currently the case, having work due when it suits the tutor and/or institution best.

Similarly, when looked at objectively, the relatively harsh penalties for late submission of student work seems hard to justify. A penalty for lateness which automatically caps a result at, say, the pass grade is fundamentally inequitable. Furthermore, if a grade is to give an indication of student learning achievement, the fact that a student receives a lower grade simply for submitting work late without an extension does not necessarily mean that their learning achievement is also reduced. Arguably such penalties can be justified if, and only if, submitting work on time is an important and heavily weighted assessment criteria for every piece of assessment work. Whichever way you look at it, making penalties so harsh is difficult to justify. Most academics would argue that some kind of incentive (whether it be a 'stick' or a 'carrot') is essential to encourage students to manage their time and submit their work otherwise everything would be left to the last minute. While this may be true for some students, surely self-regulation is part and parcel of this: if students are to be expected to manage the timing of their assessment deadlines and penalties they must be trusted and supported and guided through this process. Supporting students in the development of their time management skills would appear to be a vital learning experience. It would seem logical, then, that students have at least some say in the timing of their assessment deadlines and the penalties they will incur for not meeting them.

There is a range of options available for increasing student choice of assessment timing, from removing all deadlines and allowing individual students to submit when they are ready to, to allowing individual students to negotiate their timing, to allowing a student cohort to democratically decide amongst themselves the timing of their assessment and the penalties incurred for not meeting the deadlines. Either way, electronic tools are useful to both students and staff in the management of this process. Again, collaboration tools, such as wikis, and online voting systems can assist students in the management of democratic processes for deciding assessment deadlines. Similarly, electronic assessment management tools, such as those embedded within proprietary VLEs, assist academic staff in the management and tracking of student assessment which is submitted over staggered or multiple deadlines as they provide a safe, secure and backed-up repository where student course work can be submitted, marked, double marked/moderated, returned and archived. Similarly, tools such as RSS feeds and the early warning systems embedded into VLEs can be set to alert academic staff to the submission of coursework.

Result

Of the five aspects of student assessment choice, giving students some control over the result they receive for their own work is likely to be the most controversial. The accepted wisdom is that students cannot be trusted with this kind of judgement and offering this kind of choice makes a mockery of the principles of academic standards. There is, however, considerable potential for self-evaluation to be beneficial to students for both formative and summative reasons. For formative assessment, as Taras (2001) points out, it seems fair and logical to have students redo their work after they have undertaken self-evaluation. This is, sadly, often prohibited by workload pressures and university assessment regulations. However it is possible to involve self-evaluation on formative tasks (Taras, 2001, insists it should always be formative) or to design summative tasks so that they work cumulatively and iteratively together. Offering students the option of negotiating their result is a valuable addition to the exercise. Again, using the tools available within a VLE is helpful in supporting students through this process.

The novelty of such an option is, in the first instance, likely to be a strong motivation to undertake self-evaluation, something that as Andrade & Valcheva (2009), quoting Goodrich, have shown students can be reluctant to undertake without support and direct instructions, even if they are aware of the benefits of it to their learning. Again, shareability is key here. Being able to revisit their own work in direct comparison to the work of all others in their cohort and even from previous years, allows students to get a clearer sense of how their achievement rates against others, and therefore are better able to place themselves and their work more accurately in the marks range. Offering students a private and secure forum in which to evaluate their own work, compare their evaluation against a tutor's evaluation of their work and then to present a claim for a change in grade (referenced against their original work and the assessment criteria) offers a valuable learning experience for students. For this kind of activity, a private blog, with an RSS feed, is ideally suited to the purpose. Motivating and supporting students through this process, whereby they undertake self-evaluation then compare their findings with the tutor's evaluation, makes it much more likely that students actively engage with their feedback and feedforward in a way which many tutors fear is both important and rare.

Further, having the option to negotiate their result involves students much more fully in the assessment process and improves the levels of power-sharing and transparency. As Taras (2001) points out, the sense of involvement given to students through the option of negotiating their result, makes them more likely to accept their result as justified and therefore less likely to complain, even if they do not take the offer up. In our experience of using result negotiation, supported by self-evaluation using a scored rubric, only a very small percentage of students took up the offer, but many more reported finding the option to do so a good source of motivation to self-evaluate. Again, electronic assessment management tools and private means of communication (both standard within proprietary VLEs) are vital to the effective and efficient management of such a process. In practice, collecting self-evaluation data systematically and comparing it to the tutor-evaluation data can also provide valuable diagnostic information about the cohort as a whole and individual students, particularly in terms of their understanding of the assessment criteria.

CONCLUSION

To foster self-regulated learning in students we have to allow and empower them with more choice over the way in which the study and what they learn. But this will be rendered effectively meaningless unless we also allow them to have more control and choice over how they are assessed. With recent developments in technology, and the mainstreaming of Technologically Enhanced Learning Environments, we now have the tools at our disposal to both allow and manage increased student assessment choice.

REFERENCES

Andrade, H., & Valtcheva, A. (2009). Promoting Learning and Achievement Through Self-Assessment. *Theory into Practice*, *48*(1), 12–19. doi:10.1080/00405840802577544

Attwood, R. (2009). Well, what do you know? [Electronic Version]. *Times Higher Education*. Retrieved 25/08/09 from http://www.timeshighereducation.co.uk/story.asp?sectioncode=26&storycode=405152.

Barnett, R. (2008). Assessment in higher education: An impossible mission? In D. J. Boud & N. Falchikov (Eds.), *Rethinking Assessment in Higher Education: Learning for the longer term* (pp. 29-40). Abingdon: Routledge.

Barr, R. B., & Tagg, J. (1995). From Teaching to Learning. In D. DeZure (Ed.), *Learning from Change*. London: Kogan Page.

Biggs, J., & Tang, C. (2007). *Teaching for Quality Learning at University* (3rd ed.). Maidenhead: Open University Press.

Bloxham, S., & Boyd, P. (2007). *Developing Effective Assessment in Higher Education: A Practical Guide*. Maidenhead: Open University Press.

Boekaerts, M. (1999). Self-regulated learning: where we are today - Theory, research, and practice. *International Journal of Educational Research*, *31*(6), 445–457. doi:10.1016/S0883-0355(99)00014-2

Boekaerts, M., & Simons, P. R. J. (1995). *Leren en instructie: Psychologie van de leerling en het leerproces [Learning and instruction: The psychology of the student and the learning process]. (2nd, revised version ed.)*. Assen: Van Gorcum.

Boud, D. J. (1995). *Enhancing Learning through Self-assessment*. London: Kogan Page.

Boud, D. J., & Falchikov, N. (2007). Introduction: Assessment for the longer term. In Boud, D. J., & Falchikov, N. (Eds.), *Rethinking Assessment in Higher Education: Learning for the longer term. Abingdon: Routledge*.

Collis, B., & Moonen, J. (2001). *Flexible Learning in a digital world*. London: Kogan Page Ltd.

Dochy, F., Segers, M., Gijbels, D., & Struyven, K. (2008). Assessment engineering: breaking down barriers between teaching and learning, and assessment In D. J. Boud & N. Falchikov (Eds.), *Rethinking Assessment in Higher Education: Learning for the longer term* (pp. 87-100). Abingdon: Routledge.

Gibbs, G. (1992). *Improving the Quality of Student Learning*. Bristol: Technical and Educational Services Ltd.

Hafner, J., & Hafner, P. (2003). Quantitative analysis of the rubric as an assessment tool: an empirical study of student peer-group rating. *International Journal of Science Education*, *25*(12), 1509–1528. doi:10.1080/0950069022000038268

Keppell, M., Au, E., & Chan, C. (2006). Peer learning and learning-oriented assessment in technology-enhanced environments. *Assessment & Evaluation in Higher Education*, *31*(4), 453–464. doi:10.1080/02602930600679159

Kurtz, B. E., & Weinert, F. E. (1989). Metamemory, memory performance, and causal attributions in gifted and average children. *Journal of Experimental Child Psychology*, *48*(1), 45–61. doi:10.1016/0022-0965(89)90040-4

Nicol, D. (2007). *Re-engineering Assessment Practices in Scottish Higher Education*. JISC.

NUS. (2008). *The great nus feedback amnesty - Briefing Paper*.

Orsmond, P., Merry, S., & Reiling, K. (2002). The Use of Exemplars and Formative Feedback when Using Student Derived Marking Criteria in Peer and Self-assessment. *Assessment & Evaluation in Higher Education*, *27*(4), 309–323. doi:10.1080/0260293022000001337

Pekrun, R., Goetz, T., Titz, W., & Perry, R. (2002). Academic Emotions in Students' Self-Regulated Learning and Achievement: A Program of Qualitative and Quantitative Research. *Educational Psychologist*, *37*(2), 91–105. doi:10.1207/S15326985EP3702_4

Race, P. (2005). *Making Learning Happen: A Guide for Post-Compulsory Education*. London: Sage Publications.

Rust, C. (2002). The Impact of Assessment on Student Learning: How Can the Research Literature Practically Help to Inform the Development of Departmental Assessment Strategies and Learner-Centred Assessment Practices? *Active Learning in Higher Education*, *3*(2), 145–158. doi:10.1177/1469787402003002004

Rust, C., Price, M., & O'Donovan, B. (2003). Improving Students' Learning by Developing their Understanding of Assessment Criteria and Processes. *Assessment & Evaluation in Higher Education*, *28*(2), 147–164. doi:10.1080/02602930301671

Scouller, K. (1998). The influence of assessment method on students' learning approaches: Multiple choice question examination versus assignment essay. *Higher Education*, *35*(4), 453–472. doi:10.1023/A:1003196224280

Snyder, B. (1971). *The Hidden Curriculum*. Cambridge, MA: MIT.

Taras, M. (2001). The Use of Tutor Feedback and Student Self-assessment in Summative Assessment Tasks: towards transparency for students and for tutors. *Assessment & Evaluation in Higher Education*, *26*(6), 605–614. doi:10.1080/02602930120093922

Weinert, F. E., Schrader, F. W., & Helmke, A. (1989). Quality of instruction and achievement outcomes. *International Journal of Educational Psychology*, *13*(8), 895–912.

Zimmerman, B. J., & Schunk, D. H. (1989). *Self-regulated learning and academic achievement: Theory, research, and practice*. New York: Springer.

KEY TERMS AND DEFINITIONS

Backwash Effect: A situation in which assessment design sends undesirable messages to students.

Computer Marked Assessment: assessment which is marked automatically by a computer

Constructive Alignment: A situation whereby teaching and learning activities and assessment tasks are systematically aligned with the intended learning outcomes according to the learning activities required in the outcomes.

Criteria Referenced Marking: measures student achievement against predetermined criteria rather than against the marks of a student population as a whole.

eAssessment: electronic and/or online tools which can be used for student assessment. This is not limited to Computer Marked Assessment (such as automatically assessed multiple choice questions) and includes Tutor Marked Assessment which makes use of communication and information technology in some form

Tacit Knowledge: knowing more by experience than can be easily explained to others.

Tutor Marked Assessment: assessment which is marked by a tutor

Chapter 7
The Role of SRL and TELEs in Distance Education:
Narrowing the Gap

Maureen Snow Andrade
Utah Valley University, USA

Ellen L. Bunker
Brigham Young University Hawaii, USA

ABSTRACT

Self-regulated learning (SRL), defined as learners taking responsibility for their own learning (Dembo & Eaton, 2000), is a critical component for success in distance education. Distance education contexts, typically TELEs (Technology Enhanced Learning Environments), also have the potential to foster SRL. This chapter focuses on the importance of SRL in distance education, specifically in higher education and lifelong learning contexts, and how SRL can mediate the gap between the learner and instructor and decrease the distance that may be created by Information and Communication Technology (ICT). The chapter reviews the use of ICT in distance education, explicates key terms related to SRL, presents a model for course design, and illustrates how behaviors of key stakeholders can support development of SRL.

INTRODUCTION

Study is necessary for all students to find knowledge but to study successful is difficult. - English language learner in Cambodia

DOI: 10.4018/978-1-61692-901-5.ch007

The value of higher education to individuals and society is well-established. Benefits not only include increased earning power over a life time but less easily measured societal benefits such as political stability, decreased poverty, lower crime rates, more social capital, an increase of new ideas, and a better quality environment

(McMahon, 2009). Given such advantages, rather than limiting opportunities for higher education to an elite group as has been the case traditionally, movements referred to as widened participation, equity, and the democratization of higher education are occurring in countries such as England, Australia, and Japan.

European nations are cooperating on educational initiatives such as the Bologna Process and the Lisbon Agenda to educate their citizenry, strengthen economies, and increase global competitiveness. The Bologna Process has established a system of higher education that supports student mobility across nations in terms of transferability of credits, transparency of academic records, and efficient degree structures (NAFSA, 2007). The Lisbon Agenda aims to create "a learning society" (Caneiro & Steffens, 2006, p. 374) comprised of learning centers and diverse methods of instruction and delivery to support lifelong learning. Even those who have completed formal degrees and training may need to retool multiple times within their lives. The U.S. Department of Labor (2008) reports that Americans born between 1957 and 1964 have held an average of 10.8 jobs. The popular YouTube video "Did You Know 3.0" (http://www.youtube.com/watch?v=jpEnFwiqdx8) claims that "We are preparing students for jobs that don't yet exist."

Diverse learners and the on-going demand for re-tooling throughout life require innovative educational approaches. Brick and mortar institutions cannot meet the increasing demand for higher education (Gourley, 2009) nor are they generally flexible enough to accommodate learners of varied age groups, learning goals, and personal life situations. This need for new approaches, particularly using media, is reflected in educational research and theory. For example, Jonassen served as an early leader in looking at the learning environment and the effect of technology on this environment (Jonassen, 1993; 1999; 2000; Jonassen, Campbell & Davidson, 1994; Jonassen & Land, 2000). Distance education is increasingly a common means of providing educational opportunities for diverse learners. But what does it take for learners to be successful in distance education contexts and navigate TELEs?

Those who have the capacity to learn independently and are internally motivated are more likely to succeed than those without these characteristics. Learners must be self-regulatedto monitor their learning processes and achieve their goals. But learner characteristics are only part of the equation. To expect that learners who enroll in higher education or lifelong learning opportunities will already possess SRL skills or develop them on their own is naïve. Providers of educational opportunities must not only create programs for learners with various academic backgrounds, levels of technological expertise, and purposes for learning, but also courses that support the growth of SRL and strengthen the likelihood of learner success.

ICT is enabling institutions to reach more students and approach learning in new ways, but it also requires more expertise on the part of learners who must manage the learning process in a technology-based environment. Because of this, TELEs create a challenge for institutions, course designers, instructors, and learners; however, they can also foster the development of self-regulation, paramount to success in this new learning paradigm. As the introductory quote from our Cambodian student acknowledges, the need to study is readily apparent; the key is learning to study successfully.

The objective of this chapter is to identify how distance education contexts can support the development of SRL through course design, instructor feedback, and institutional support, and by so doing, mediate the gaps created by the distance and the TELE. We present an overview of ICT in distance education, explore the concept of SRL, and introduce a model for developing SRL in distance learners. Then we share applications of the model for key stakeholders and identify future research directions.

BACKGROUND

To understand the role of TELEs in distance education and how it requires and potentially fosters SRL, we first provide an introduction to the use of ICT in distance education. We then explore the concept of SRL and related terminology.

Distance Education and TELEs

Throughout its long history, distance education courses have, by their nature, been required to incorporate various means of mediated instruction, beginning with the penny post and continuing through the use of each type of ICT as it was developed (Bunker, 1998; 2003; Mason, 1999; Watkins & Wright, 1991). The field of distance education also has an established record of recognizing and addressing the need for learner independence.

The purpose of mediated instruction in both correspondence and distance education was and is to close the gap between the instructor and learner. Early theories, such as Holmberg's (1983; 2007) theory of guided didactic conversations and Moore's (1972; 2007) theory of transactional distance incorporated elements of communication and interaction to address this challenge. Moore's theory includes three key variables: dialogue (the interaction between the learner and teacher), structure (the degree to which the course accommodates learners' preferences and needs), and learner autonomy (the learner's ability to create learning plans, find resources to support study, and self-evaluate).

Moore (1989) later addressed more specifically the interaction occurring in distance education, outlining three types of interaction: learner-content, learner-instructor, and learner-learner. Yet, ICT sometimes create an additional gap due to learners' unfamiliarity with the TELE. Hillman, Willis, and Gunawardena (1994) added a fourth component, learner-interface interaction, specifically addressing the role of mediated communication from the learner's perspective. They write that the "inability to achieve learner-interface interaction successfully" becomes a major problem for learners unfamiliar with ICT protocols (p. 33). This situation creates a need for both student support in the use of ICT (Peters, 2003) and for stronger SRL behaviors on the part of learners.

As noted, Moore's theory of transactional distance includes the dimension of (learner) autonomy. Moore (1972), conducting his early research in the area of independent study, noted that correspondence educators failed to recognize the ability of students to manage responsibility for their learning and that successful learners used some level of control to achieve their goals. Within Moore's theory, learner autonomy ranged from full autonomy—determining study goals, planning how to accomplish goals, and determining how much to learn, to limited or no autonomy—lacking decision-making power related to the course. Moore (2007) has recently used the term self-management as a synonym for autonomy and other researchers refer to autonomy with a variety of terms, as discussed below.

AUTONOMY AND SRL

Autonomy encompasses learner choice and involvement (Moore, 1972). It also refers to learner capacity in the sense of taking responsibility, active learning, self-awareness, self-direction, and self-reflection (Holec, 1981; Hurd, 1998; Garrison, 2003). Other characteristics associated with the term are metacognition, motivation, strategic competence, behavior, time management, self-direction, and goal setting (Hurd, 1998; Hurd, Beaven & Ortega, 2001; Peters, 1998). Although autonomy generally consists of two dimensions—pedagogical (taking control of learning) and psychological (metacognition) (Peters, 1998), definitions vary and reflect a broad range of characteristics. Choice and involvement, key aspects of autonomy, are not sufficient to improve learning. Although the term also reflects capacity for learning, we posit that the concept of SRL is more conducive to facilitating learner success in TELEs even though it has not typically been applied to distance education research.

The term SRL, predominantly used in the field of educational psychology, also reflects a variety of related concepts, specifically motivation, affect, cognition, metacognition, social context (Butler, 2002), active learning (Zimmerman, 1994), and strategic action (Perry, 2002; Winne, Jamieson-Noel & Muis, 2002). However, the term consistently refers to "the ability of learners to control the factors or conditions affecting their learning" (Dembo, Junge & Lynch, 2006, p. 188), and to four major components: cognitive (learning strategies to understand and remember information), metacognitive (planning, setting goals, monitoring, evaluating), motivation (ability to self-motivate, taking responsibility for success and failures; developing self-efficacy), and behavior (seeking help, creating a positive learning environment) (Dembo, Junge & Lynch, 2006; Zimmerman & Kitsantis, 1997).

One of the advantages of conceptualizing distance learning in terms of SRL rather than autonomy, is that the former consists of descriptive components—cognition, metacognition, motivation, behavior—as well as processes such as how to approach learning, the use of strategies, managing performance, and evaluating. This makes SRL a more usable concept for purposes of distance course design and instructor and institutional support than autonomy. The four primary components have been further broken down into six dimensions, specifically motive (*why*), methods (*how*), time (*when*), physical environment (*where*), social environment (*with whom*), and performance (*what*) (Zimmerman, 1994). These six dimensions of SRL provide a framework and process that guides the application of SRL to distance learning contexts and assists in identifying behaviors that support learner self-regulation.

DEVELOPING SRL IN DISTANCE LEARNERS

We next explore the challenge of developing SRL in distance learners and present a model to guide key stakeholders in this process. Before the advent of ICT, learner support and feedback in distance education was time consuming and slow, making it necessary to put most support into the learning materials. However, the increasing availability of ICT allows stakeholders to seriously (and feasibly) address the issue of increasing SRL in distance learners. But aside from the challenges related to terminology mentioned, how to develop SRL in learners can be problematic. Can characteristics such as motivation or locus of control be changed? Can ICT have a positive effect on the development of SRL? Although in theory ICT can enhance education and the development of SRL (an important component to lifelong learning) supporting evidence for this is limited (Banyard, Underwood & Twiner, 2006; Steffens, 2006). ICT can enable different kinds of teaching and learning, but cannot ensure that learning goals are achieved (Kirkwood & Price, 2005) or that SRL is increased. The latter takes deliberate, focused efforts on the part of course designers, instructors, learners, and institutions.

Model of Self-Regulated Distance Learning

To realize the goal of fostering SRL in distance education TELEs, we introduce the model of self-regulated distance learning (*Figure 1*), which guides course designers and instructors and has implications for learners and institutions. The model is based on Moore's (1972, 2007) theory of transactional distance and Zimmerman's (1994) six dimensions of SRL. The model demonstrates how interaction with structure and dialogue contributes to self-regulation (Andrade & Bunker, 2009). Learners approach the learning context with initial levels of self-regulation, commitment,

Figure 1. Model of self-regulated distance learning (Adapted from [Andrade & Bunker, 2009])

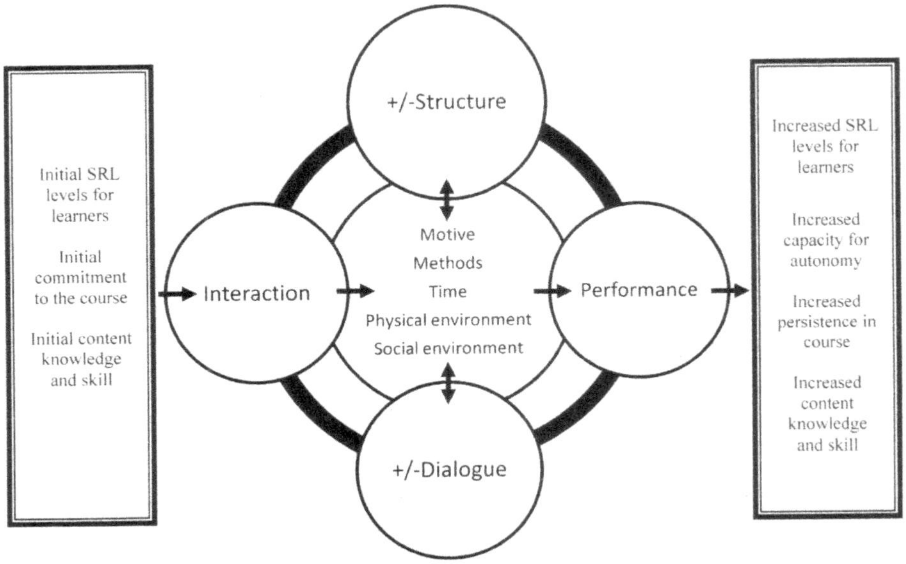

and knowledge. Structure is provided through interaction with course components designed to build SRL. More structure means less individual choice (autonomy), but builds self-regulation and the capacity for autonomy, and decreases distance. A high level of dialogue (e.g., tutorials, conferences, feedback) similarly provides support for SRL although it decreases autonomy in the sense of choice. It also decreases transactional distance and helps increase self-regulation, capacity for autonomy, course persistence, and knowledge. In other words, teachers use dialogue to direct students (i.e., to tell them what to do). This potentially limits students' freedom to choose (autonomy), but ultimately the direction guides them in becoming more independent.

To illustrate further, we use an example from courses we have developed in distance language learning. At the beginning of an online intermediate level English language course, learners take a self-assessment survey designed to give them an initial understanding of their strengths and weaknesses in the components of SRL. Using their personal results from the survey, learners then choose from a set of activities related to the dimensions of SRL and submit a weekly written response (as a learner journal) to the teacher. Teachers can then interact with each student, giving support and feedback as needed based on the learner's understanding of these dimensions. In addition to this teacher/student interaction, use of the model has application to other key stakeholders, as discussed next.

The Six Dimensions: The Behavior of Stakeholders

Examined carefully, each of the six dimensions of SRL has implications for the behavior of key stakeholders in the learning process. Stakeholders include course designers, instructors, learners, and educational institutions. Distance education in TELEs requires behavior changes for these primary actors as there is no face to face contact in a true distance education mode. This has implications for the way learners learn, teachers teach, and institutions interact with learners (Harlow, 2007).

In this section, we suggest ways that the principal actors in the learning process can help develop and support the various dimensions of

SRL as it relates to TELEs in distance education courses and within the context of our model. Due to space limitations, we are not able to discuss all six dimensions for each stakeholder, but for each section we first provide illustrative examples from the literature and then from our own English language courses.

Course Design

Support systems for distance language courses can "encourage language learners to develop strategies that work for them personally, and . . . lead to more effective learning methods and enhanced learning outcomes" (Hurd, 2000, p. 37). These support systems should be built into the course design. The key elements of our model—dialogue, structure, the dimensions of SRL, and autonomy—have been applied to course design in TELEs with positive results. We next present relevant research from the literature and then share our own course design work.

Dialogue and structure, the basis of Moore's theory (1972, 2007), are evident in distance course design and can be linked to the development of SRL. A number of studies in the literature support this relationship. For example, a structural means of helping students enrolled in foreign language classes become more self-regulated is student marked assignments in which students assess their knowledge of grammar and semantics (Hurd, 2004). Dialogue is provided through detailed tutor feedback to help learners correct mistakes and analyze errors so that they develop self-correction and self-monitoring abilities. Dimensions of SRL evident in this design are *performance* and *methods*; learners reflect on their progress and make necessary adjustments in strategies to attain their goals.

In online Biblical language courses (Harlow, 2007), dialogue is created through web lectures with accompanying notes; phone calls and e-mail exchanges between students and teachers; the availability of online answers to exercises and exams for purposes of immediate feedback; additional instructor-generated e-mail feedback that explains what was missed on exams and why; regular instructor-generated e-mail to encourage, provide progress updates, and offer help; and personal tutoring by phone or live online conferencing. Structural components include a course introduction, study guide, weekly schedule, answer keys, and a CD for aural practice. Redundancy in materials—texts, lectures, notes, CDs, charts, drills, reviews, and practice exams—occurs so that students see concepts multiple times before testing. These elements support SRL dimensions such as *methods* for learning (e.g., text, lectures, notes, study guide), *social environment* (e.g., e-mail, live tutoring, online conferencing), and *performance* (e.g., feedback, online answers, progress updates, practice exams). Based on these research findings, these SRL components lead to learner satisfaction and satisfactory performance in higher level Biblical language courses

Other courses have reported gains in SRL as measured by standardized instruments and interviews. Students enrolled in a web-based course monitored their progress by completing a study time chart, writing reflecting summaries of assigned readings, and keeping a learner journal (Chang, 2005). They examined how much time they spent on each reading, considered the score they received on their reflective summaries, and reviewed their journals to identify the strategies used. In this way, several SRL dimensions were supported and learners actively monitored their learning process (i.e., the *performance* dimension of SRL) as they self-observed, self-evaluated, self-monitored, and adjusted their strategies. The impact of ICT and SRL components led to students finding value and use in course materials, recognizing that learning outcomes depended on their effort, viewing learning as their responsibility, and increasing *motivation* (another SRL dimension).

Research has also focused on how different types of technology support different aspects of SRL. For example, tools for collaboration and communication such as online discussion groups aided help seeking, self-monitoring, and reading comprehension (Dabbagh & Kitsantas, 2005). In the same study, posting writing drafts and the use of rubrics supported self-evaluation and goal setting. Questionnaires examining how TELEs affected SRL for teacher trainees enrolled in a blended course determined that social aspects such as seeking help, communication, and collaboration were more positively affected by ICT than cognitive, motivational, and emotional aspects (Dettori, Giannetti, & Persico, 2006). These studies demonstrate positive effects on dimensions such as *methods*, *social environment*, and *performance* through the structure and dialogue in the courses.

Media is a key means of closing the gap in distance education "for it provides the vehicle for the exchange of information between the distance teacher and the distance learner" (Harlow, 2007, p. 15). However, course design must consider student abilities and access to ICT. Anticipating variations in hardware and software, the amount of time students can work online, the possibility of shared computers and facilities, downloading time, fluency with ICT, and competency across ICT applications are important issues (Kirkwood & Price, 2005).

We next turn to examples from our own courses that illustrate the effectiveness of the model. In our course design, we considered the learner—international students, some in remote areas, who spoke English as a second language and had varying levels of FITness (fluency in technology; i.e., see Caneiro & Steffens, 2006). Accordingly, we provided a CD-Rom, printed study guide, and printed textbook in addition to our online course materials. This enabled learners to work on the course when they did not have access to ICT. We also provided "technology how-to" tutorials to introduce students to the online course management system, and learn how to upload assignments, create their own Web pages, use live interactive tutoring, and participate in discussion groups.

As a key aspect or "backbone" of the course, a Manage Your Learning (MYL) component was formed to function in the place of a language learner journal. MYL activities allowed learners to focus on one dimension of SRL each week, choosing from a pool of activities that included all dimensions (see Table 1). For example, sample activities include: *motivation*—goal setting, developing positive self-talk, and analyzing strengths and weaknesses as a language learner; *social environment*—making the most of tutoring sessions and teacher conferences, getting help when needed, and interacting with classmates at a distance; *time management* —keeping a daily 24-hour schedule and prioritizing daily activities.

For the most part, activities could be accomplished away from or on the computer, allowing flexibility. However, submitting the learner journal and receiving feedback from the instructor requires the use of ICT. Initial evaluation of student learner journals produced following completion of the MYL activities focused on SRL dimensions shows increased language production compared to learner journals from similar assignments in traditional face-to-face courses. In addition, students completing these activities showed a heightened understanding of their own SRL behaviors and an ability to use this understanding in completing assignments (Bunker & Takashima, 2009). For example, in a one-page journal response, a student wrote the following about how an activity designed to improve use of study materials helped her learn about the structure in her textbook. "I found a lot of new things in the book which I did not learn when I was in ... school years ago ... My textbook is my best friend at the moment" (English language learner in Indonesia).

Instructor

Face-to-face teaching is spontaneous, emotionally motivating, and involves communication through the human voice and body language (Moore, 1972) whereas the separation of learner and teacher in

Table 1. Example manage your learning activities for a high intermediate level English language writing course

Motivation	Methods of learning	Time management	Social environment	Physical environment
Week One: Beginning survey				
Strengths/weakness as language learner	Use your study guide/ course materials	Record activities 24 hours	Seek help	Evaluate study environment
Goal setting I Goal setting II	Use textbook	Prioritize activities	Make best use of tutoring session	Classify distractions
Midterm: Revisit beginning survey: performance evaluation, self-observation, and evaluation				
Develop positive self-talk, Part I, II	Language learning strategies, Part I	Evaluate time use	Make best use of teacher conference	Restructure physical environment
Evaluate progress on goals	Language learning strategies, Part II	Organize information/ time	Interact w/ classmates at distance	Create learning environment
Final Week: Re-do survey. Write final performance evaluation and response				

distance education creates a communication gap (Moore & Kearsley, 2005). Distance education instructors have been referred to as "contingent tutors" (Banyard, Underwood & Twiner, 2006, p. 485), as interpreters, guides, advisers, and supporters (Hurd, 2000), and as coaches (Harlow, 2007). "Encouraging learners to develop the metacognitive strategies that enable them to become autonomous in their learning should be the prime aim for all those involved in course delivery" (Hurd, 2000). For this to occur, learners must experiment with a variety of strategies and discover what works for them; teachers can help learners develop awareness of strategies that fit particular tasks (Hurd, 2000).

The role of the distance education instructor is to help learners develop SRL and increase success. When designers follow the model of self-regulated distance learning, the structure for SRL exists in the course, leaving the instructor to guide, encourage, and provide feedback—elements of dialogue that support the idea of teacher as facilitator. The model offers instructors a systematic guide for promoting the dimensions of SRL. A key role of the instructor is to provide greater or lesser amounts of dialogue and structure as needed to encourage students. Students cannot raise their hands to ask a question, so dialogue must be maximized and clear, and questions answered promptly (Harlow, 2007). Teachers should use different types of media to communicate, check on progress, encourage, and give feedback such as e-mail, chat rooms, and web pages, and structure must be provided through clear expectations, objectives, assignments, and redundancy (Harlow, 2007).

In our experience, teachers working with the model of self-regulated distance learning have clear direction in terms of not only helping students gain content knowledge, in our case language proficiency, but helping them increase their levels of SRL, capacity for autonomy, and persistence in the course. Teachers know that course goals are more likely to be accomplished if they can guide students to develop SRL skills. Our student population is typically accustomed to passive learning practices, and many of them work full time, provide financial support for extended family, and have limited experience with and access to technology.

Teachers have found that the elements of the model help them direct students in becoming self-regulated as does the TELE of the course. Teachers create dialogue by posting announcements on the course management system to remind learners of assignment due dates, and by sending e-mail reminders about late or missing assignments. They post models of student work, provide feedback through rubrics, and make direct comments on student work, which is uploaded to the management system. The management system allows teachers to make comments when they allocate or report grades on assignments and tests. Learners are given the opportunity to respond to learner journal questions and teachers' responses to journals encourage and prompt them to reflect more deeply about their learning approaches. Finally, new ICT allow teachers to speak with learners in a conference using Internet voice connections.

In other words, the model and the ICT used in the course provide a framework for developing the dimensions of SRL such as *methods* of learning, reflection on *performance*, *motive*/goal setting, and so forth. Teachers need a clear understanding of the components and how the course elements and their own behaviors fit within them. This insures that they are consistently approaching all of the dimensions as needed by individual learners.

The TELE of our courses also supports teacher voice (Anderson, 2007; Hurd, Beaven & Ortega, 2001; Moore & Kearsley, 2005; White, 2005), which is the means of establishing dialogue with distance learners. Although most distance courses are pre-packaged for purposes of efficiency and consistency, teachers can make a course their own through the voice they use to communicate with students about their learning. This voice is also evident in course materials such as the study guide, instructions for assignments, and multimedia presentations. To provide learners with a connection to their teachers, we post a brief introductory video of the teacher on the course management system. The teacher's address is a mixture of teaching philosophy, personal trivia, and motivation-building talk. This makes the teacher more real as the student commences what is likely a new experience in distance learning.

In a face-to-face class, teachers can easily give feedback to an entire class. Teachers new to distance education miss this direct approach. However, after some experience in online courses, our teachers report that ICT allow them to give immediate, concrete feedback to individual learners beyond what can be offered without this support. While the constant demand of in-coming messages can be an adjustment, teachers used e-mail and comment features to address student language learning needs. One teacher created a picture roster to view each student as she worked.

The SRL activities in our courses become critical in helping teachers know their students; as students share experiences in their journals about their development of SRL, though far from the teacher geographically, they become close in social presence terms. For example, a student from a remote area in China (who at times herded goats as she studied) wrote with some detail how positive and negative self-talk affected her life as a student. The teacher gave an extensive response showing understanding and supporting the student's efforts to improve her SRL behavior.

Learner

The behaviors of the course designer and the instructor are important, but the focus of their efforts is on helping learners develop SRL. The learner ultimately plays the key role in this development process. Successful learners are able to construct knowledge, reflect critically, be actively involved, and make choices for effective learning, but some question if these skills can be enhanced (Murphy, 2005). Studies have shown that less successful distance learners cannot tolerate ambiguity and believe the teacher should direct learning (Bown, 2006; White, 1999). This suggests the need for those involved in distance education to find ways to change student beliefs and approaches to learning. This has occurred successfully through the use of ICT and an emphasis on SRL.

When students in a TELE utilized course components that helped them identify needed skills for tasks, form an action plan, reflect on their work, discuss performance with a tutor, summarize feedback, and locate sources of help, they were able to prioritize, change their study approaches, question their assumptions, and develop metacognitive strategies (Murphy, 2005). Students who reflected on their learning processes and received feedback from a tutor experienced positive gains in SRL (van den Boom, Paas & van Merrienboer, 2007). Weblogs helped students develop metacognitive skills as they assisted each other, shared knowledge, solved problems, raised questions, reflected on issues, and linked to related topics and content (Baggetun & Wasson, 2006). Other social networking tools such as instant messaging, e-mail, and online forums helped learners self-assess and boosted motivation (Carneiro & Steffens, 2006).

Although these findings demonstrate that specific types of ICT have been successfully applied to developing various dimensions of SRL in distance learners, in some cases, students may demonstrate the potential to apply SRL skills but not use them, particularly if they are not embedded into the TELE (Porras-Hernandez, 2000). Also, competing demands for students' time may interfere with the use of effective study strategies or students may be only extrinsically motivated (Thang, 2005). Training is needed to help students become aware of useful approaches and strategies, and change their attitudes about learning (Thang, 2005). This could be accomplished by revising study guides to focus on these areas and allow more flexibility and choices to support a variety of learning styles (Thang, 2005).

Using our model, we adopted a direct approach to training students in SRL. Not only do our course study guides contain learning tips, pose reflective questions, and offer opportunities for interaction with teachers, tutors, and peers through live interactive tutoring, web pages, and online discussion groups, but the courses are designed to increase students' English language proficiency and SRL skills simultaneously. Hence, using the model to guide both the design/development phase and the delivery/teaching phase, the SRL dimensions are integral to the course and the language and SRL activities are folded together. One student made thoughtful and fairly lengthy responses for a language learner to the MYL activities each week.

I've learned so much from each MYL activity. It always surprises me, because it gives me many of new methods to become a better learner. The beginning survey has played a key role in this activity. According to the survey, I've got a chance to know myself better. I also found out which part that I could work on and I can become a better learner. This activity has been helping me a lot during this online course. I've been trying to apply all of the principles that I've learned from this MYL into my live too. I plan for my next day before hand, figure out what I should do to make the next better, make sure that everything that I'm doing or I'll do will help me to accomplish my goals, etc. I've also shown many principles from this activity to my friends. I love MYL, and I'll keep applying it into every phrase of my learning journey (English language learner in Taiwan).

Institution

Many nations are widening access to higher education and expanding training opportunities to support lifelong learning. Accordingly, collaborative multi-national initiatives such as the Bologna Process, Lisbon Agenda, and the New Lifelong Learning Programme, which integrates lifelong learning programs in Europe under one structure (Caneiro & Steffens, 2006), are establishing innovative educational approaches. These include the use of ICT. TELEPEERS, a European effort to measure how TELEs support SRL (Caneiro & Steffens, 2006), is a particular example of these efforts.

The vision and breadth exemplified by these movements is commendable. SRL in TELEs can be encouraged through large, multi-national projects and fostered by individual institutions. The need for institutions to support SRL is clear. "Self-directed study, which is prevalent at the graduate level, is being pushed down to the undergraduate level because of online learning" (Institute for Higher Education Policy, 2000, p. 17). The extent to which ICT is incorporated into learning contexts and the degree to which the use of ICT is combined with traditional approaches affects the level of self-regulation learners need. The primary role for institutions is to provide the necessary structures and support for this to occur successfully.

In distance learning, institutions must ensure convenient access to faculty and staff and ensure that learners will receive prompt answers to questions about course materials and assignments (Harlow, 2007). This interaction can occur through a virtual student service office, which offers personal service in the form of giving feedback, answering questions, encouraging, providing suggestions for effective study and time management, and regularly calling and e-mailing students (Harlow, 2007). Institutions must recognize the lifestyles of adult learners and their multiple roles, and work to meet students' needs by being available, providing ready access to support, removing barriers, and minimizing isolation (Harlow, 2007; Moore & Kearsley, 2005). Access to institutional support can be provided through various kinds of ICT, even those as simple as a 24-hour response time turnaround for e-mail, discussion boards for students, technology support call centers, and interactive live chat. Although these are straightforward solutions, they enable appropriate self-regulated behavior on the part of learners.

In the model of self-regulated distance learning, the kinds of support described relate predominantly to the element of dialogue, and encourage development of the *social environment* dimension of SRL in that learners are enabled to seek help from others and exercise responsibility for their learning. Such support also potentially assists with *motive* in that it encourages learners, with *methods* and *time management* depending on the types of suggestions provided, and with *performance* as learners reflect on their behaviors and adjust their goals. These kinds of support signify a learner-centered institution whose goal is to produce learning and create an environment that helps students create knowledge (Harlow, 2007).

In our experience applying the model to distance English language courses, we provide weekly online live sessions with staff from our campus Language Center for the purpose of speaking and listening practice. Learners utilize this opportunity to ask logistical and practical questions about assignments, due dates, the course management system, and study guide, and course content. Another support service we offer to learners is access to staff from the campus Writing Center who use e-mail and software tracking tools to give writing feedback. Course evaluation surveys and follow-up interviews with distance students now on campus show overwhelmingly that these tutoring support sessions are a favored part of the course. Such sessions would be impossibly expensive without ICT and students would miss opportunities to both interact and develop their language abilities and get logistical needs met without the technologies now available.

An element of self-regulation is knowing where to get help and demonstrating positive help-seeking behaviors; this is much more easily accomplished in face-to-face learning than in a distance context. However, when institutions provide transparent access to existing campus support systems through ICT, they are enabling SRL.

Solutions and Recommendations

Great strides are being made in improving accessibility to higher education and lifelong learning. For these efforts to be successful, educators, institutions, and nations must work collectively to share best practices related to ICT. The diversity of learners wanting to be successful in obtaining education and training requires that learning styles, strategies, and lifestyles receive prominent attention in the delivery and support of educational opportunities. Increasingly, those involved in the use of ICT in education are showing commitment to embedding SRL into course work and training with positive results.

In the fields of distance education and language learning as well as education itself, overlapping definitions and a proliferation of terms related to the concepts of autonomy and SRL do little to direct course designers, instructors, and researchers to focus on how to embed, support, and measure SRL in TELEs through the use of ICT. Our approach has been to develop a model that guides key stakeholders in the development of SRL in distance learning contexts. We contend that the model leads designers to consider how various forms of ICT can be utilized in helping students make gains in the six dimensions of SRL. It also directs teachers in how they interact with students. The model helps both designers and instructors determine how varying levels and types of structure and dialogue can enhance learners' SRL, the capacity for autonomy, persistence in the course, and content knowledge. Institutions play a key role in establishing appropriate support structures, providing resources, and demonstrating leadership in order for such initiatives to succeed.

FUTURE RESEARCH DIRECTIONS

Future research needs to provide empirical data related to various aspects of the model of self-regulated distance learning for different subject areas, learner populations, contexts, and TELEs. Similar to the TELEPEERS project in Europe (Caneiro & Steffens, 2006), further investigation must occur related to the success of various types of ICT incorporated into course design to advance the six dimensions of SRL. The model serves as a framework for building SRL components into a TELE course and for the measurement of SRL.

Additionally, increases in SRL need to be measured. Measuring SRL is somewhat problematic, however, and recent work in this area must be given careful consideration. Although self-report and standardized measures have predominated in SRL research, the field is increasingly using alternate means of assessment. These include qualitative measures such as observations, interviews and focus groups, think aloud protocols, learner journals, and teacher judgments (De Groot, 2002; Patrick & Middleton, 2002; Winne & Perry, 2000). Multiple data collection approaches allow for triangulation and involve the consideration of social settings, observations, and interviews, to capture the attitudes, beliefs, and perceptions of learners about their experiences (Patrick & Middleton, 2002). Trace data, examining behaviors such as note taking, underlining, and highlighting (Winne & Perry, 2000) further expand our knowledge of SRL. Another area of interest is how SRL evolves as the result of feedback (Winne, Jamieson-Noel & Muis, 2002). Many of these measures have the potential to capture changes in SRL behavior over time.

In sum, SRL measures should examine what learners actually do when they are engaged in an academic task and how the use of SRL varies depending on the context (Perry, 2002) and type of ICT. Our work in distance language learning TELEs demonstrates how courses can be designed to develop SRL, decrease transactional distance, increase the capacity for autonomy, improve persistence, and increase content knowledge/skills.

CONCLUSION

Global calls for education, specifically lifelong learning and higher education opportunities to improve the quality of life and the economic vitality of nations, require new approaches. This chapter has outlined how the field of distance education has pioneered alternate delivery methods and the use of ICT. It has also established the importance of autonomy in learner success in distance contexts. Our contribution has focused on synthesizing research in distance education and SRL to produce a model by which courses can be developed and learners supported through interaction with instructors and sponsoring institutions. The model serves as a basis for future research and measurement of SRL. We have also demonstrated how course designers, instructors, learners, and institutions can work within the constructs of the model to support SRL. We close with a student voice describing the benefit of the MYL activities.

At first, I don't why I need to do the 'manage your learning activities'. It is because I think that I just need to finish the writing process assignment. . . . I find out that it helps me become a better learner. I learnt many useful learning skills. The most effective activity is 'strengths and weakness as an English language learner'. This following activity helped me to think about my strengths and weaknesses. Then, I adopted a strategy of balancing my strengths and weaknesses. Before this activity, I would like to shirk my weaknesses. But now I welcome to facing them. Also, I learnt many things from other 'manage your learning activities', it helped me become a better student. The most important thing is I can use these learning skills into my new school life. I can build a better study environment for myself. From these activities, I also find out which is my suitable study style. Having a suitable study style, it affects our study progress immediately (English language learner in Hong Kong).

REFERENCES

Anderson, T. (2007). Modes of interaction. In Moore, M. G. (Ed.), *Handbook of distance education* (2nd ed., pp. 298–309). Mahwah, NJ: Lawrence Erlbaum.

Andrade, M. S., & Bunker, E. L. (2009). Language learning from a distance: A new model for success. *Distance Education, 30*(1), 47–61. doi:10.1080/01587910902845956

Baggetun, R., & Wasson, B. (2006). Self-regulated learning and open writing. *European Journal of Education, 41*(3/4), 453–472. doi:10.1111/j.1465-3435.2006.00276.x

Banyard, P., Underwood, J., & Twiner, A. (2006). Do enhanced communication technologies inhibit or facilitate self-regulated learning? *European Journal of Education, 41*(3/4), 473–489. doi:10.1111/j.1465-3435.2006.00277.x

Bown, J. (2006). Locus of learning and affective strategy use: Two factors affecting success in self-instructed language learning. *Foreign Language Annals, 39*(4), 640–659. doi:10.1111/j.1944-9720.2006.tb02281.x

Bunker, E. L. (1998). *An historical analysis of a distance education forum: The International Council for Distance Education world conference proceedings, 1938 to 1995.* Unpublished doctoral dissertation, The Pennsylvania State University, State College.

Bunker, E. L., & Takashima, K. (2009). Formative evaluation and interview data. Unpublished raw data.

Butler, D. L. (2002). Qualitative approaches to investigating self-regulated learning: Contributions and challenges. *Educational Psychologist, 37*(1), 59–63.

Caneiro, R., & Steffens, K. (2006). Editorial. *European Journal of Education, 41*(3/4), 345–352.

Chang, M. M. (2005). Applying self-regulated learning strategies in a web-based instruction—An investigation of motivation perception. *Computer Assisted Language Learning, 18*(3), 217–230. doi:10.1080/09588220500178939

Dabbagh, N., & Kitsantas, A. (2005). Using web-based pedagogical tools as scaffolds for self-regulated learning. *Instructional Science, 33*(5), 513–540. doi:10.1007/s11251-005-1278-3 de Groot, E. V. (2002). Learning through interviewing: Students and teachers talk about learning and schooling. *Educational Psychologist, 37*(1), 41–52.

Dembo, M. H., & Eaton, M. J. (2000). Self-regulation of academic learning in middle-level schools. *The Elementary School Journal, 100*(5), 473–490. doi:10.1086/499651

Dembo, M. H., Junge, L. G., & Lynch, R. (2006). Becoming a self-regulated learner: Implications for web-based education. In O'Neil, H. F., & Perez, R. S. (Eds.), *Web-based learning: Theory, research, and practice* (pp. 185–202). Mahwah, N. J: Lawrence Erlbaum Associates.

Dettori, G., Giannetti, T., & Persico, D. (2006). SRL in online cooperative learning: Implications for pre-service teacher training. *European Journal of Education, 41*(3/4), 397–414. doi:10.1111/j.1465-3435.2006.00273.x

Garrison, R. D. (2003). Self-directed learning and distance education. In Moore, M. G., & Anderson, W. G. (Eds.), *Handbook of distance education* (pp. 161–168). Mahwah, NJ: Lawrence Erlbaum.

Gourley, B. (2009, June). *Higher education for a digital age*. Paper presented at the meeting of International Council for Open and Distance Learning, Maastricht, The Netherlands.

Harlow, J. (2007). Successfully teaching Biblical language online at the seminary level: Guiding principles of course design and delivery. *Teaching Theology and Religion, 10*(1), 13–24. doi:10.1111/j.1467-9647.2007.00302.x

Hillman, C. A., Willis, D. J., & Gunawardena, C. N. (1994). Learner-interface interaction in distance education: An extension of contemporary models and strategies for practitioners. *American Journal of Distance Education, 8*(2), 30–42. doi:10.1080/08923649409526853

Holec, H. (1981). *Autonomy and foreign language learning: Council of Europe*. Oxford, UK: Pergamon Press.

Holmberg, B. (1983). Guided didactic conversation in distance education. In Sewart, D., Keegan, D., & Holmberg, B. (Eds.), *Distance education: International perspectives* (pp. 114–122). London, UK: Croom Helm.

Holmberg, B. (2007). A theory of teaching-learning conversations. In Moore, M. G. (Ed.), *Handbook of distance education* (2nd ed., pp. 69–76). Mahwah, NJ: Lawrence Erlbaum.

Hurd, S. (1998). Too carefully led or too carelessly left alone? *Language Learning Journal, 17*(1), 70–74. doi:10.1080/09571739885200121

Hurd, S. (2000). Helping learners to help themselves: The role of metacognitive skills and strategies in independent language learning. In M. Fay & D. Ferney (Eds.), *Current trends in modern language provision for non-specialist linguists* (pp. 36-52). London, UK: The Centre for Information on Language Teaching and Research (CILT) in association with Anglia Polytechnic University (APU).

Hurd, S. (2004). Autonomy and the distance language learner. In Holmberg, B., Shelly, M., & White, C. (Eds.), *Distance education and languages: Evolution and change* (pp. 1–19). Clevedon, UK: Multilingual Matters.

Hurd, S., Beaven, T., & Ortega, A. (2001). Developing autonomy in a distance language learning context: Issues and dilemmas for course writers. *System, 29*(3), 341–355. doi:10.1016/S0346-251X(01)00024-0

Institute for Higher Education Policy. (2000, April). *Quality on the line: Benchmarks for success in internet-based distance education*. Washington, DC. Retrieved May 28, 2009, from http://www.ihep.org/assets/files/publications/m-r/QualityOnTheLine.pdf

Jonassen, D. (1993). The trouble with learning environments. *Educational Technology*, *33*(1), 35–37.

Jonassen, D. (1999). Designing constructivist learning environments. In Reigeluth, C. (Ed.), *Instructional-design theories and models* (2nd ed., pp. 215–240). Mahwah, NJ: Lawrence Erlbaum Associates.

Jonassen, D. (2000). *Computers as mindtools for schools: Engaging critical thinking*. Columbus, OH: Prentice-Hall.

Jonassen, D., Campbell, J., & Davidson, M. (1994). Learning with media: Restructuring the debate. *Educational Technology Research and Development*, *42*(2), 31–19. doi:10.1007/BF02299089

Jonassen, D., & Land, S. (2000). *Theoretical foundations of learning environments*. Mahwah, NJ: Lawrence Erlbaum Associates.

Kirkwood, A., & Price, L. (2005). Learners and learning in the twenty-first century: What do we know about students' attitudes towards and experiences of information and communication technologies that will help us design courses? *Studies in Higher Education*, *30*(3), 257–274. doi:10.1080/03075070500095689

Mason, R. (1999). The impact of telecommunications. In Harry, K. (Ed.), *Higher education through open and distance learning* (pp. 32–47). London, UK and New York: Routledge.

McMahon, W. W. (2009). *Higher learning, greater good: The private and social benefits of higher education*. Baltimore, MD: Johns Hopkins Press.

Moore, M. G. (1972). Learner autonomy: The second dimension of independent learning. [from http://www.ajde.com/Documents/theory.pdf]. *Convergence*, *5*(2), 76–88. Retrieved August 4, 2008.

Moore, M. G. (1989). Three types of interaction. *American Journal of Distance Education*, *3*(2), 1–6. doi:10.1080/08923648909526659

Moore, M. G. (2007). The theory of transactional distance. In Moore, M. G. (Ed.), *Handbook of distance education* (2nd ed., pp. 89–105). Mahwah, NJ: Lawrence Erlbaum Associates.

Moore, M. S., & Kearsley, G. (2005). *Distance education: A systems view* (2nd ed.). Belmont, CA: Wadsworth Publishing Company.

Murphy, L. (2005). Critical reflection and autonomy: A study of distance learners of French, German and Spanish. In Holmberg, B., Shelley, M., & White, C. (Eds.), *Distance education and languages: Evolution and change* (pp. 20–39). Clevedon, U.K: Multilingual Matters.

NAFSA. Association of International Educators (2007). *The Bologna process. International Educator supplement*. Washington, DC: NAFSA.

Patrick, H., & Middleton, M. J. (2002). Turning the kaleidoscope: What we see when self-regulated learning is viewed with a qualitative lens. *Educational Psychologist*, *37*(1), 27–39.

Perry, N. E. (2002). Introduction: Using qualitative methods to enrich understandings of self-regulated learning. *Educational Psychologist*, *37*(1), 1–3.

Peters, O. (1998). *Learning and teaching in distance education. Analysis and interpretation from an international perspective*. London, UK: Kogan Page.

Peters, O. (2003). Learning with new media. In Moore, M. G., & Anderson, W. G. (Eds.), *Handbook of distance education* (pp. 87–112). Mahwah, NJ: Lawrence Erlbaum Associates.

Porras-Hernandez, L. H. (2000). Student variables in the evaluation of mediated learning environments. *Distance Education*, *21*(2), 385–403. doi:10.1080/0158791000210211

Steffens, K. (2006). Self-regulated learning in technology-enhanced learning environments: Lessons of a European peer review. *European Journal of Education*, *41*(3/4), 353–379. doi:10.1111/j.1465-3435.2006.00271.x

Thang, S. M. (2005). Investigating Malaysian distance learners' perceptions of their English proficiency courses. *Open Learning*, *20*(3), 243–256. doi:10.1080/02680510500298683

U.S. Department of Labor. (2008, June). *Number of jobs held, labor market activity, and earnings growth among the youngest baby boomers: Results from a longitudinal survey summary.* Retrieved from http://www.bls.gov/news.release/nlsoy.nr0.htm van den Boom, G., Paas, F., & van Merrienboer, J. J. G. (2007). Effects of elicited reflections combined with tutor or peer feedback on self-regulated learning and learning outcomes. *Learning and Instruction, 17*(6), 532-548.

Watkins, B. L., & Wright, S. J. (1991). *The foundations of American distance education: A century of collegiate correspondence study.* Dubuque, IA: Kendall/Hunt.

White, C. (1999). Expectations and emergent beliefs of self-instructed language learners. *System*, *27*(4), 433–457. doi:10.1016/S0346-251X(99)00044-5

White, C. (2005). Towards a learner-based theory of distance language learning: The concept of the learner-context interface. In Holmberg, B., Shelley, M., & White, C. (Eds.), *Distance education and languages: Evolution and change* (pp. 55–71). Clevedon, UK: Multilingual Matters Ltd.

Winne, P. H., Jamieson-Noel, D., & Muis, K. R. (2002). Methodological issues and advances in researching tactics, strategies, and self-regulated learning. In Pintrich, P. R., & Maehr, M. L. (Eds.), *New directions in measures and methods* (pp. 121–155). Ann Arbor, MI: University of Michigan.

Winne, P. H., & Perry, N. E. (2000). Measuring self-regulated learning. In Boekaerts, M., Pintrich, P. R., & Zeidner, M. (Eds.), *Handbook of self-regulation* (pp. 531–566). San Diego, CA: Academic Press. doi:10.1016/B978-012109890-2/50045-7

Zimmerman, B. J. (1994). Dimensions of academic self-regulation: A conceptual framework for education. In Schunk, D. H., & Zimmerman, B. J. (Eds.), *Self-regulation of learning and performance* (pp. 3–21). Hillsdale, NJ: Lawrence Erlbaum Associates.

Zimmerman, B. J., & Kitsantis, A. (1997). Developmental phases in self-regulation: Shifting from process to outcome goals. *Journal of Educational Psychology*, *89*(1), 29–36. doi:10.1037/0022-0663.89.1.29

ADDITIONAL READING

Carneiro, R., & Steffens, K. (Eds.). (2006). [Special issue]. *European Journal of Education*, *41*(3/4).

Dembo, M. H., & Seli, H. (2008). *Motivation and learning strategies for college success: A self-management approach* (3rd ed.). Mahwah, NJ: Lawrence Erlbaum.

Garrison, R. D. (2003). Self-directed learning and distance education. In Moore, M. G., & Anderson, W. G. (Eds.), *Handbook of distance education* (pp. 161–168). Mahwah, NJ: Lawrence Erlbaum.

Gibson, C. C. (1998). The distance learner in context. In Gibson, C. C. (Ed.), *Distance learners in higher education: Institutional responses for quality outcomes* (pp. 113–126). Madison, WI: Atwood Publishing.

Holmberg, B., Shelley, M., & White, C. (Eds.), *Distance education and languages: Evolution and change* (pp. 55–71). Clevedon, UK: Multilingual Matters Ltd.

Hurd, S. (2001). Managing and supporting language learners in open and distance learning environments. In Mozzon-McPherson, M., & Vismans, R. (Eds.), *Beyond language teaching towards language advising* (pp. 135–149). London, UK: CILT.

Hurd, S. (2006). Towards a better understanding of the dynamic role of the distance language learning: Learner perceptions of personality, motivation, roles, and approaches. *Distance Education, 27*(3), 303–329. doi:10.1080/01587910600940406

Little, D. (1991). *Learner autonomy 1: Definitions, issues, and problems*. Dublin, EI: Authentik.

O'Neil, H. F., & Perez, R. S. R. S. (Eds.). (2006). *Web-based learning: Theory, research, and practice.* Mahwah, NJ: Lawrence Erlbaum Associates.

Oxford, R. (1994). *Language learning strategies: What every teacher should know*. New YorkNewbury House.

Perry, N. E. (Ed.). (2002). [Special issue]. *Educational Psychologist, 37*(1).

Wedemeyer, C. A. (1981). *Learning at the back door: Reflections on nontraditional learning in the lifespan*. Madison, WI: University of Wisconsin Press.

White, C. (2003). *Language learning in distance education*. Cambridge, UK: Cambridge University Press. doi:10.1017/CBO9780511667312

Zimmerman, B. J. (1986). Development of self-regulated learning: Which are the key subprocesses? *Contemporary Educational Psychology, 11*(4), 307–313. doi:10.1016/0361-476X(86)90027-5

Zimmerman, B. J. (1990). Self-regulated learning and academic achievement: An overview. *Educational Psychologist, 25*(1), 3–17. doi:10.1207/s15326985ep2501_2

Zimmerman, B. J. (1998). Academic studying and the development of personal skill: A self-regulatory perspective. *Educational Psychologist, 33*(2/3), 73–86. doi:10.1207/s15326985ep3302&3_3

KEY TERMS AND DEFINITIONS

Autonomy: The ability to learn independently; involves choice in terms of how, what, when, and where to study as well as capacity related to the degree to which a learner can successfully monitor and regulate the process of learning.

Behavior: Related to the ability of the learner to seek help, establish effective study conditions, manage time, and control other factors related to learning.

Cognition: The use of various learning strategies and methods employed to learn and retain information.

Dialogue: Within distance education courses, this refers to interactions among teachers and learners and occurs through e-mail, live online chat, assessment feedback, discussion boards, and forms of communication.

Metacognition: The ability to reflect on the learning process; involves preparing, setting goals, planning, monitoring, reflecting, and evaluating performance.

Motivation: Includes reasons for learning and the ability to motivate oneself; also involves taking responsibility for positive and negative learning outcomes.

Structure: Provided within distance education courses through course components such as assignments, activities, study guides, calendars, deadlines, learning objectives, and texts.

Theory of Transactional Distance: Developed by Michael G. Moore (1972); refers to the learning gap between the learner and the instructor in distance education contexts. The gap increases or decreases with varying levels of structure and dialogue. Learner autonomy increases when structure and dialogue are low and decreases when they are high.

Chapter 8
Strategies to Promote Self–Regulated Learning in Online Environments

Bruce R. Harris
Western Illinois University, USA

Reinhard W. Lindner
Western Illinois University, USA

Anthony A. Piña
Sullivan University System, USA

ABSTRACT

The primary purpose of this chapter is to present techniques and strategies that can be incorporated in online courses to promote students' use of self-regulated learning strategies. In addition, the authors discuss why self-regulated learning skills are particularly critical in online learning environments, present a model of self-regulated learning, discuss issues related to measuring self-regulated learning, address the issue of whether or not self-regulated learning can be taught, and discuss why online learning environments are ideal environments to scaffold self-regulation. The authors present several strategies and techniques they have found successful for promoting self-regulated learning that can be readily incorporated and implemented in online courses. These strategies are organized by the three main components of the Self-Regulated Learning Model: Executive Processing, Cognitive Processing, and Motivation. The chapter concludes with a scenario that represents an idealized model of how to promote self-regulated learning in an online learning environment by employing an intelligent tutoring component as a tool to support students' use and development of self-regulated learning tactics and strategies.

DOI: 10.4018/978-1-61692-901-5.ch008

INTRODUCTION

As a result of recent advances in web-based technologies, online learning has become a major form of learning and teaching around the world. More and more instructors are converting their traditional face-to-face classes to an online course environment (Allen & Seaman, 2007). From our experiences in converting face-to-face classes to online courses over the last 15 years, we have generally found that online learning environments require learners to take greater responsibility for their learning than face-to-face courses. Without some of the characteristics and abilities that exist in a face-to-face class (e.g., regular class meeting times and locations, ability of the instructor to respond to non-verbal cues, ability of students to initiate and maintain verbal social interaction with other students, etc.), learners in an online course environment typically find that they must more effectively learn how to monitor their own learning processes to accomplish their learning goals.

Others have also come to this conclusion. For example, Schunk and Zimmerman (1998) state "Self-regulation seems critical due to the high degree of student independence [referring to distance learning] deriving from the instructor's physical absence" (p. 230). Kauffman (2004) explains "The nature of many Web-based instructional tasks, however, involves independent learning that requires students to be highly self-regulated" (p. 140). Successful online learners must generally be more self-regulated than in traditional face-to-face courses because the nature of online courses involves more independent learning. As Dabbagh and Bannan-Ritland (2005) have observed: "Helping students become self-directed is critical to their success in online learning environments" (p. 224).

In short, students who are academically successful in online courses tend to be self-regulated. Although there is some variance among theories and models of self-regulated learning, it is generally held that the construct of self-regulated learning consists of three key components: metacognitive, cognitive/behavioral, and motivational processes and strategies (Zimmerman & Martinez-Pons, 1986). Metacognitive processes and strategies include setting goals and planning, monitoring actions, evaluating progress, etc. Cognitive/behavioral processes and strategies include managing the learning environment, using rehearsal, organizational, and elaboration learning strategies, etc. Motivational processes and strategies include high self-efficacy, self-attributions, self-motivation, volition, etc.

Our experiences in teaching online courses have also shown that students often lack the necessary self-regulated learning skills to be successful in reaching their goals in an online learning environment. As Graesser, McNamara and VanLehn (2005) observe: "It is rare to find a student who spontaneously and skillfully enacts self-regulated learning" (p. 225). In addition, Kauffman (2004) states "Unfortunately, not all students are self-regulated.... This may be particularly relevant in Web-based environments where students are often asked to complete complex academic tasks with little or no support from classmates or teachers" (p. 140). This raises the very real probability that a significant number of students are not as successful as they could be in online learning environments because they do not have adequate proficiencies in using self-regulated learning strategies.

Consequently, the primary purpose of this chapter is to discuss several techniques and strategies instructors can incorporate in their online courses to promote students' use of self-regulated learning strategies. We will first discuss why self-regulated learning skills are critical in online learning environments. We next present a model of self-regulated learning developed by the authors, discuss issues related to measuring self-regulated learning, and address the issue of whether or not, and to what degree, self-regulated learning can be taught. Lastly, we discuss why online learning environments are ideal settings to promote self-regulation, and then discuss several strategies for promoting self-regulated learning strategies in online courses, including a scenario that represents an idealized model of how to promote self-regulated learning in an online learning environment by employing an intelligent tutoring component.

BACKGROUND

Self-Regulated Learning Skills are Critical in Online Learning Environments

In a previous paper, we identified several characteristics of the online learning environment that makes possession of self-regulated learning skills necessary for success (Harris, Piña & Lindner, 2002). These characteristics include a) the inability of the instructor to receive and process non-verbal cues indicating that the learner may not be understanding or may be having problems, b) the necessity for learners to inform their instructors when they are experiencing difficulties, c) the difficulty of initiating and maintaining social interaction between learners, and d) the managing of busy schedules to include sufficient time for course activities and assignments.

Loomis (2000) and other researchers have observed that traditional learning environments often do not adequately prepare students to develop self-regulated learning skills, nor do they provide sufficient opportunities to apply these skills (Eastmond, 1996). In a typical face-to-face classroom setting, the primary role for many learners has been to receive information, absorb and memorize what is deemed to be of most importance, and then recall the information on a subsequent examination. Although effective learner-content, learner-instructor and learner-learner interaction can occur in a face-to-face class, the regulation of learning often tends to be controlled by the instructor. A high degree of learner control in most traditional classrooms is minimal (Chang, 2005).

Chang (2005) notes that, in contrast, online learning environments place demands upon learners that exceed those encountered in traditional classrooms. "For students, web-based learning is a suitable environment for them to take charge of their own learning since they can control their own learning process. However, providing students with opportunities to integrate their knowledge through web-based instruction may not be effective if they lack the skills needed to regulate their learning. Thus, strategies that prepare students for the rigors of learning at a distance and increase the probability of retention and success must be put into practice" (p. 217). These challenges can be exacerbated for first-generation college students or those new to online learning (Williams & Hellman, 2004) or when the online course has not been well-designed (Harris, Piña & Lindner, 2002).

Online attrition is another area that may be influenced by the learner's level of self-regulation. Although retention and attrition rates for online courses vary widely between and within institutions, the literature is consistent in the observation that online courses tend to have higher attrition rates than face-to-face courses (e.g. Moody, 2004; Patterson & McFadden, 2009; Tyler-Smith, 2006). First-generation college students and those new to online learning appear to have a particularly difficult time (Williams & Hellman, 2004).

The nature of online learning environments--particularly those that are asynchronous--is such that variables previously under control of the instructor must now be controlled by the learner (Dettori, Gianetti, & Persico, 2006; Williams & Hellman, 2004). These include the hour of the instructional delivery, the length of time that it takes to deliver instruction, the amount of practice and review time, and the location where the learner receives the instruction (Puzziferro, 2008). Jonassen, Davidson, Collins, Campbell, and Haag (1995) observe that this changing of roles and tasks between online instructors and learners makes self-regulated learning skills more important in online learning environments than they are in the typical traditional classroom environment. Williams and Hellman (2004) state that those who are highly self-regulated tend to set proximal goals, which supports self-regulated learning for online environments, given that learners tend to have more freedom of choice with online instruction than they do with face-to-face instruction. Puzziferro (2008) studying the affect of self-regulation/self-actualization behaviors in 815 community college students, found that students engaging in self-regulation displayed increased academic performance and higher satisfaction in online courses. Other recent studies (e.g. Chang, 2005; Whipp & Chiarelli, 2004) have confirmed the necessity of self-regulated learning skills for students taking online courses. Given the significance of self-regulated learning, particularly in online learning environments, the critical question seems to be: what exactly does it mean for a learner to be self-regulated?

A Model of Self-Regulated Learning

Models of self-regulated learning come in a variety of forms, rooted in different theoretical orientations (Puustinen & Pulkkinen, 2001). Common to the various approaches researchers have put forward, as Paris and Paris (2001) note, are "autonomy and control by the individual who monitors, directs, and regulates actions toward goals of information acquisition, expanding expertise, and self-improvement" (p. 89). Our own approach to self-regulated learning views it as a type of complex skill carried out by the general cognitive (information processing) system. Figure 1 provides a basic representation of our current working model of self-regulated learning. Our argument is that the executive or metacognitive level in the information processing system is the central player in the self-regulation of academic performance. The primary functions of the executive in our model is to: a) focus attention on the critical components and conditions of a learning task; b) suppress, or inhibit, automatic cognitive processing; c) determine a conditional plan for accomplishing whatever goals the system selects relative to the task; and d) monitor and evaluate progress toward such goals, adapting the plan and accompanying strategies to specific task demands as they arise in the course of learning or problem solving.

Whether or not the individual sets a goal to accomplish some task, rather than to procrastinate or avoid it altogether, is, of course, also dependent on the motivational and affective dynamics the task evokes. However, we do not, as in some approaches to self-regulated learning (see, for example, Pintrich, 2004) assume that motivation and related affective factors are separate components of self-regulation. Since self-regulation involves decision making, we agree with Kunda (1990) that motivation, particularly in this case, is thoroughly "cognitively mediated' (p. 480).

We contend that once engaged in the planning phase where key decisions are made, an individual is more likely to set a goal to accomplish a task if, relative to the demands of the task, one has high self-efficacy, makes constructive attributions, and has a learning or mastery goal orientation. Once the intention to learn or accomplish a learning task is activated, the cognitive system is brought into play. That is, any relevant knowledge, declarative or procedural, domain specific or general, the system possesses is then activated in the service of accomplishing the task at hand. More specifically, tactics and strategies specific to the nature of the learning task or challenge are retrieved, assembled, and activated. The executive component of the system subsequently integrates, monitors and evaluates the efficacy of the strategy and tactics employed and makes, via corrective feedback, any necessary adjustments (cognitive and/or affective) as the need arises.

Our own research has repeatedly demonstrated a significant relationship between self-regulated learning and academic performance (Lindner & Harris, 1992; 1998; 2002). This finding is consistent with the literature on self-regulated learning

Figure 1. Self-regulated learning model

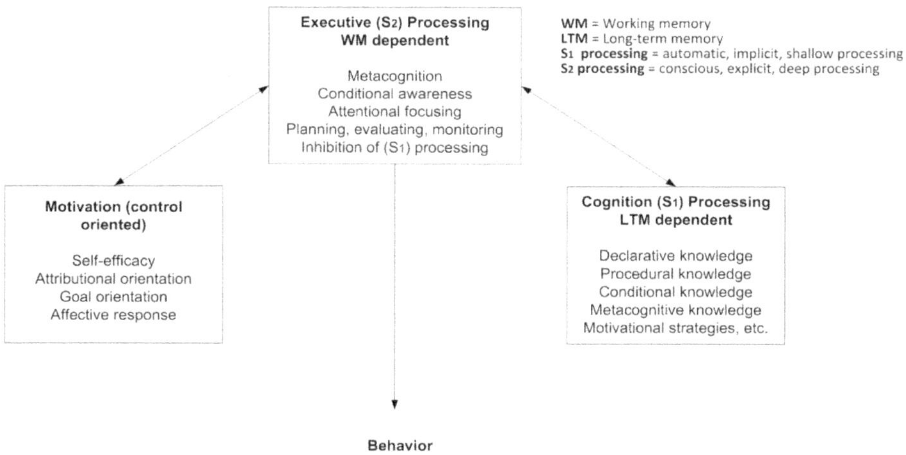

(Boekaerts, 1997; Paris & Paris, 2001; Schunk, 2005; Schunk & Zimmerman, 2008; Zimmerman & Schunk, 2001) in general. Furthermore, our work points to the conclusion that self-regulation remains a significant factor even when academic achievement and aptitude are removed as factors in predicting student performance (Lindner & Harris, 2002).

Measuring Self-Regulated Learning

Over the past decade we have been developing and refining an instrument, the Self-Regulated Learning Inventory (for details, see Lindner & Harris, 1992; 1998; 2002), designed to assess a particular learner's profile and orientation in relation to self-regulation of the learning process. The inventory has gone through several iterations and by now its reliability as an instrument is well established (reliability indices for all three subscales are at.8 or above). In addition, both criterion-related and construct-related evidence point to its validity. Not only can the score on the inventory be used for predictive purposes through analysis of an individual's responses to the inventory, but one is able to use the inventory to obtain a profile of a learner's strengths and weaknesses relative to self-regulation of the learning process.

The three subscales of the inventory are: a) a motivation scale, assessing the learner's motivational profile in terms of goals, attributions, and self-efficacy; b) a learning strategy scale, assessing both the learner's knowledge of learning tactics and tendency to assemble and activate a plan (strategy) when tackling learning tasks; and c) an executive processing scale, assessing the learner's tendency to analyze, monitor, and evaluate both their motivational state and effectiveness of the particular learning strategy they have put into play. The executive also plays an inhibitory role in terms of keeping the learner on task and with respect to suppressing impulsive ideation and/or responses of both the cognitive and affective variety that might interfere with the learning process. A particular score profile on the inventory may indicate, for example, a learner that is highly motivated, and reasonably strategic, but does little in the way of executive processing. Such a learner is likely to be successful when their largely implicitly activated strategy matches up well with a given task, but lacks flexibility and adaptability when the match is less than optimal and requires modification and adaptation to task constraints.

Can Self-Regulated Learning be Taught?

The question arises: Is self-regulation an acquired skill or does it come more or less naturally to some learners and not to others? While we do not have a specific answer to this question based on our own quantitative data, much related literature suggests that self-regulation can be taught and learned, even when students are identified as possessing specific learning disabilities (De La Paz, 1999; Paris & Paris, 2001; VanderStoep & Pintrich, 2003; Schunk, 2005). However, a cautionary note needs to be raised. Self-regulation is a complex skill and complex skills take time and practice to assemble and acquire. As in explicit cognition in general, resource allocation for metaskills (e.g., cognitive monitoring, self-explanation, etc.) must typically be purchased. Executive cognition is heavily dependent on working memory, a limited resource in the general economy of the cognitive system (Baddeley, 2007; Miyake & Shah, 1999), and explicit forms of cognition are particularly working memory intensive. The price here appears to be a constructive motivational orientation and ready access to a variety of flexible learning and problem-solving tactics with specific attention to appropriate contexts of application (conditional knowledge, or knowing when and where to do what). If a learner has not automated and stored a number of ready-made learning tactics that can be retrieved as a strategy, working memory, unless supported externally, will be quickly overwhelmed, leaving insufficient resources for executive processes to function optimally. Even where such resources exist, if the learner fails to recognize the current situation as a condition wherein specific skills and knowledge are applicable, a failure to utilize existing resources is likely to occur. This suggests that some type of scaffolded approach to teaching self-regulation may be most effective. That is, the instructional approach must be designed to temporarily relieve working memory load, and make up for any knowledge deficit in terms of specific tactics on the part of the learner. It is critical, however, that such supports be provided only on a temporary basis and eventually withdrawn as the learner internalizes the process of self-regulation. Again, we emphasize that only after considerable and targeted practice does self-regulation become normative for a given learner.

Online Learning is an Ideal Environment for Teaching Self-Regulated Learning

While it is true that the development and utilization of self-regulated learning skills can contribute to the success of online learners, it is also true that online learning environments can be ideal settings for individuals to obtain self-regulated learning skills and take greater control and responsibility for their learning. For example, Chang (2005) examined the effect of self-regulated learning strategies on learners' perception of motivation within an online course. The course was modified to include a number of self-regulated learning strategies to assist learners to self-observe and self-evaluate their effectiveness and to increase their motivation for learning. The results of the study revealed that the learners' motivation perception benefited from the online instruction in self-regulated learning strategies. Learners became more responsible for their own learning, more intrinsically orientated, and more challengeable. They also tended to value the learning material more and became more confident in their understanding and subsequent class performance.

Looking at self-regulated learning within teacher education, Dettori, Gianetti & Persico (2006) observed that self-regulated learning strategies can be used in both individual and collaborative activities and that access to online social environments can positively influence self-regulated learning skills, such as cognitive and metacognitive reflection. In particular, asynchronous communities of practice (Piña,

Sadowski, Scheidenhelm & Heydenburg, 2008) can be fruitful sources for acquiring self-regulated learning skills (Dettori, Gianetti & Perisco, 2006).

In a recent study, Whipp & Chiarelli (2004) analyzed how graduate students used and adapted traditional self-regulated learning strategies to complete tasks and cope with challenges in an online technology course. They also investigated motivational and environmental influences on the students' use of self-regulated learning strategies. They found that students in the online course utilized a number of common self-regulated learning skills, including using organizers and schedules for goal setting, planning and management, note taking, charts, reducing distractions and help-seeking from the instructor and peers. However, students also utilized a number of strategies that were unique to the online environment, such as coordinating online and offline work, planning for technology problems, offline composing, editing and sorting of online discussion forum postings, frequent checks of online grade books, interaction with online peers, and gauging success by technological performance (Whipp & Chiarelli, 2004).

Dabbagh & Kitsantas (2004) argue that web-based pedagogical tools for communication and collaboration can support the development of self-regulatory skills, which, in turn, will increase student success in online learning environments. Learning management systems such as Blackboard, Moodle and Desire2Learn, bring many of these tools together into a single interface. Piña (2010) provides a list of tools common to most learning management systems that can be utilized for applying self-regulated learning strategies. These include synchronous chat, asynchronous discussion forums, internet-mail systems, whiteboards, online journals & blogs, wikis, grade book, course calendars, announcements, personal notes, and portfolios.

As a consequence of findings like those reviewed above, we have recommended that faculty and instructional designers embed features in instructional materials that encourage learners' self-regulation (Harris, Piña, & Lindner, 2002). The instructor, for example, can include prompts within the course that provide guidance to the learners as to when to use various self-regulated learning strategies. These strategies and techniques for promoting self-regulated learning within an online course will be elaborated on in the next section of this chapter.

STRATEGIES FOR PROMOTING SELF-REGULATED LEARNING

This section of the chapter will primarily discuss several strategies for promoting self-regulated learning in online learning environments. The first part of this section will present several strategies that we have found successful and can readily be incorporated and implemented in an online course. The strategies are organized by the three main components of the Self-Regulated Learning Model discussed previously: executive processing, cognitive processing, and motivation.

The second part of this section will present a scenario that represents an idealized way to promote self-regulated learning in an online learning environment. This scenario involves using an intelligent tutoring component in an online course and describes how such a tool could be used in an online course to promote students' use and development of self-regulated learning strategies. Even though the intelligent tutoring system described in this scenario would take significant time and resources to develop, the scenario illustrates some realistic possibilities using currently available learning technologies.

Strategies That Can be Readily Incorporated into Online Courses

Executive Processing

The Executive Processing component of the Self-regulated Learning Model includes metacognitive functions such as self-monitoring, conditional awareness, attention focusing, etc. Three important executive processing strategies will be discussed in the following section: goal setting and planning, self-monitoring, and self-evaluation.

Goal Setting and Planning

Our experience has shown that encouraging or requiring learners to set specific goals for what they would like to achieve as a result of completing the course and to write specific action plans for achieving those goals is an effective technique that can be easily implemented in an online course. This technique should be incorporated in the introduction section of the course before the learners begin any coursework. A very simple and obvious goal the learners could write down is what grade or score they expect to achieve in the course. Other goals might include specific knowledge, understanding, or skills they expect to obtain from the course. It is important for the instructor to stress that the goals should be measurable and include a date to be accomplished (which is usually the same date as the completion date of the course).

Once the learners have established specific goals, they should be encouraged to write an action plan to achieve the goal(s). The action plans should not only be specific and detailed, but should also be based on conditional awareness of the course. That is, the learner should be encouraged to evaluate the nature of the course and use contextual clues to determine what specific tasks are needed to be completed to achieve their goal. One of the best ways to determine these contextual clues is to evaluate the course syllabus and the instructor's announcements for the course. For example, if the student concludes from the course syllabus that the course grade is primarily determined by the scores on multiple-choice exams and quizzes, the learner's action plan would most likely include reading the course materials and textbook chapters for key vocabulary and definitions. The learner may plan to spend a certain amount of time each week to review self-quizzes or study guides well in advance of the exam so he/she will have time to ask the instructor questions and allow time for the instructor's response.

If, on the other hand, the learner concludes from the course syllabus (and perhaps announcements or emails from the instructor clarifying his/her assumptions) that a course grade is determined primarily by course projects and assignments, the learner's action plan might include some of the following tasks: a) complete the project or the assignment a week before the due date and submit it to the instructor to get feedback concerning to what degree the project meets the criteria, b) establish a study group with other classmates and have them review the project before submitting it, c) review sample assignments provided by the instructor, etc. Other action plans might include such items as: a) deciding to study at a certain place where distractions are likely to be minimal; b) planning out blocks of time each week for studying; c) if an assignment failed to meet all the criteria specified or to achieve the desired score, contacting the instructor to seek clarification, etc.

Once the learners have established their goals for the course, an effective scaffolding strategy for the instructor would be to require the learners to post or submit their goals and action plans to the instructor so he/she can review them and provide feedback and coaching to the learners. At the beginning of the course, the instructor may choose to require the learners to email or post several periodic self-reflections on how well they are following their action plans to achieve their goals and then reduce the number of self-reflections required towards the end of the course. The instructor can provide coaching, if necessary, to help the learners use appropriate strategies to adjust their action plans to achieve their goals. The learners could be required at the end of the course to write a self-reflection paper regarding to what degree they achieved their goals and an evaluation of how successful their action plan was.

Most learning management systems have an option for the learners to keep a journal, or at a minimum, note taking capabilities to post their goals and action plans. Figure 2 illustrates an example goal and action plan that a student posted to an online course for the instructor to review. The goal and action plan shown in Figure 2 was written in WebCT Vista using the Notebook feature; however, the capability for students to post goals and actions plans is available in most learning management systems.

Another technique that we have found effective is to include in the introduction section of the online course a discussion explaining why it is essential for the learners to use self-regulated learning strategies in an online learning environment and why it will make a difference in their academic success. This discussion could include a summary of the information that was presented earlier in this chapter, such as a) the attrition rates for online courses (i.e., generally online courses have higher attrition rates than traditional face-to-face courses), b) online courses generally require more independent learning, and c) students who are academically successful in online courses tend to be self-regulated. The discussion could also explain that several helps have been included in the online course to help the learners use self-regulated learning strategies, but ultimately, it is the learner's decision whether or not to use those strategies suggested during the course.

Figure 2. Example goal and action plan

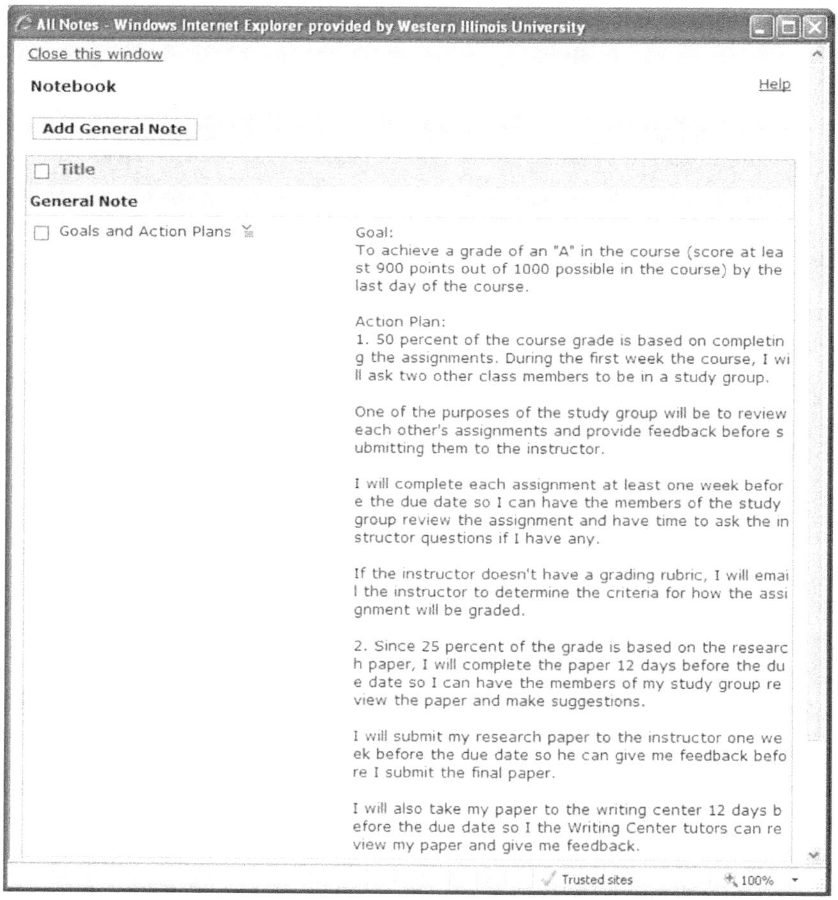

Self-Monitoring

There are several techniques instructors can use to encourage students to use self-monitoring strategies. We have used two types of self-monitoring prompts that have been helpful to learners. The first technique is to embed self-monitoring prompts at strategic points during the course lessons (usually following several frames of course content) in which the learner simply reads the prompt and chooses whether to respond to the prompt or continue on with the lesson. The learner is not required to type a response to the prompt in order to advance to the next page. See Figure 3 for an example of this type of self-monitoring prompt.

The second technique that we have used is to embed self-monitoring prompts in the course lessons that require the learners to type a response to the prompt before they can continue to the next page. See Figure 4 for an example of a self-monitoring prompt embedded in a lesson that requires the learner to type a response before continuing the lesson.

The first technique is rather simple to incorporate in an online lesson that uses a frame-based application, such as PowerPoint or other course lesson development application. The second technique, requiring the learners to actually type in a response to the self-monitoring prompt before they can continue the lesson, usually requires more sophisticated online course development skills than the first technique. We used a web-development authoring system to write the programming code to require the learners to type in a response before they can continue to the next page in the lesson. We also wrote the programming code that allowed us to store the learners' responses in a dynamic database so we could review the students' responses to determine how thoughtful their responses were.

Figure 3. Self-monitoring prompt not requiring a written response

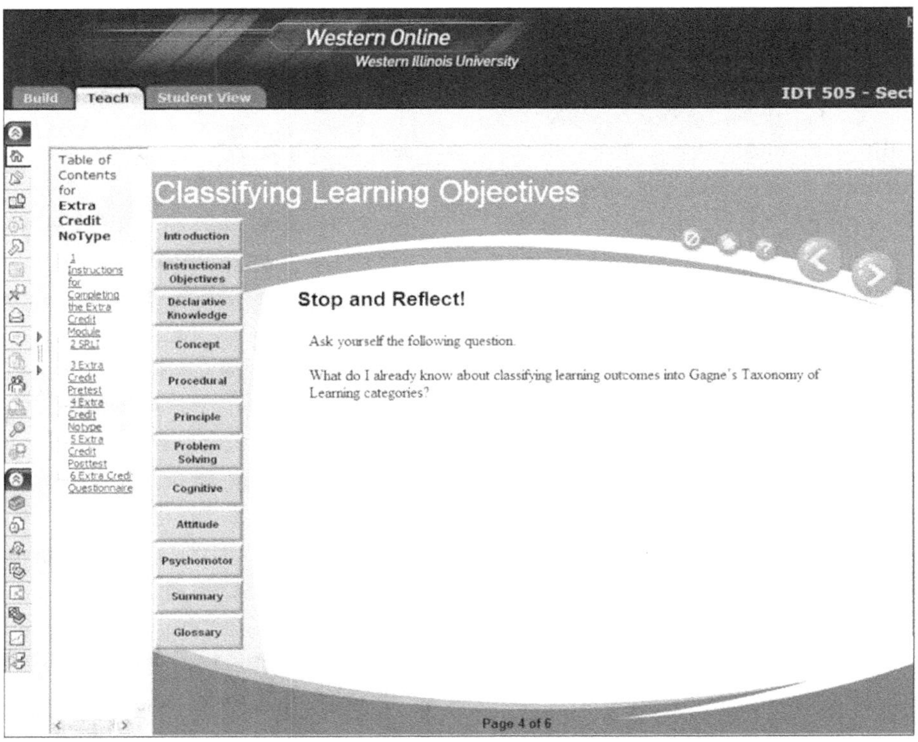

Figure 4. Self-monitoring prompt that requires the learner to type a response

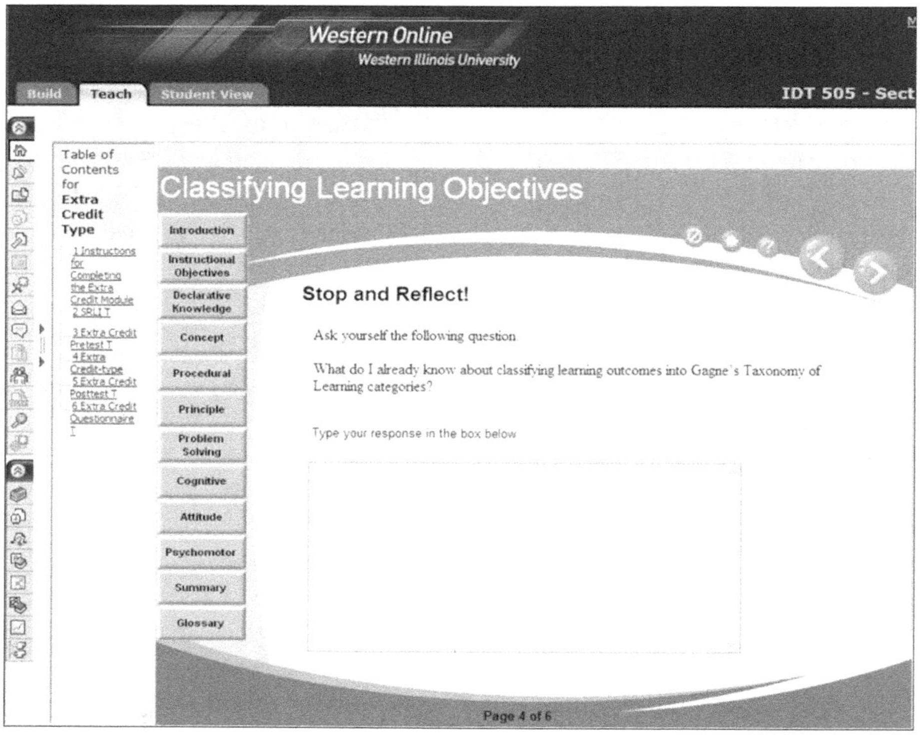

Previously, we conducted a pre-posttest comparison group design study (learners were randomly assigned to one of the two treatment groups) to determine if learner-generated responses to online self-monitoring prompts (in which learners must type in their response to the self-monitoring prompt) results in higher achievement scores than self-monitoring prompts that do not require learners to generate a response (the first technique discussed above). The study showed no significant difference between the two groups (Harris & Linder, 2008). The results from this exploratory study seem to indicate that one technique is not more effective than the other.

We have used a scaffolding approach to determine when and how often to embed self-monitoring prompts in online course lessons. The lessons at the beginning of the course include many more self-monitoring prompts than the lessons nearer the end of the course. Feedback we have received from our students indicate that most people prefer the self-monitoring prompts and feel the prompts help them.

Self-Evaluation

We have found that providing ways for learners to evaluate their quantitative progress in an online course is an effective technique for promoting self-regulated learning. For example, we post the learners' grades (scores) on their graded assignments and course activities in an online gradebook so they can evaluate whether or not they are earning enough points to achieve the goal they have established for the course. The online course materials clearly show the students how to compute their current grade percentage in the course by showing them how to divide the number of points they have earned to date in the course by the total number of points possible for those graded course activities. We try to make it very clear how many points are possible in the course for each assignment or course activity so the student can determine their percentage grade at any time in the course. Many learning management systems include a feature to compute the students' total score to date.

Another technique to promote self-evaluation is to embed prompts (much like the self-monitoring prompts discussed previously) within the course lessons. For example, a self-evaluation prompt such as the following could be embedded at the end of a lesson: "If I were to take a test on this information right now, what grade would I most likely receive?" See Figure 5 for an example of a self-evaluation prompt embedded in a lesson from one of the authors' online courses.

Another technique that helps to promote the use of self-evaluation is to provide self-tests or self-quizzes that learners can complete to help them determine their readiness for taking an exam or quiz. These quizzes are not graded, but are very similar to the exam or quizzes that will be graded. Before learners take the self-test, we encourage them to evaluate how well they think they will perform on the self-test and then compare their estimate with their actual score. This activity gives the learners practice in making more accurate self-evaluations on how well they are prepared for an exam in the future.

Providing grading rubrics for course assignments that the instructor will use to grade the assignments helps to promote learners' use of self-regulated learning strategies. Rubrics provide the learners an opportunity to self-evaluate their work before submitting it to another classmate or the instructor to review. The instructor should explain the importance of reviewing the assignment criteria in the rubric and then review the learner's assignment to ensure the criteria have been clearly met. This process will help the learner practice the skill of evaluating their work based

Figure 5. Self-evaluation prompt that requires the learner to type a response

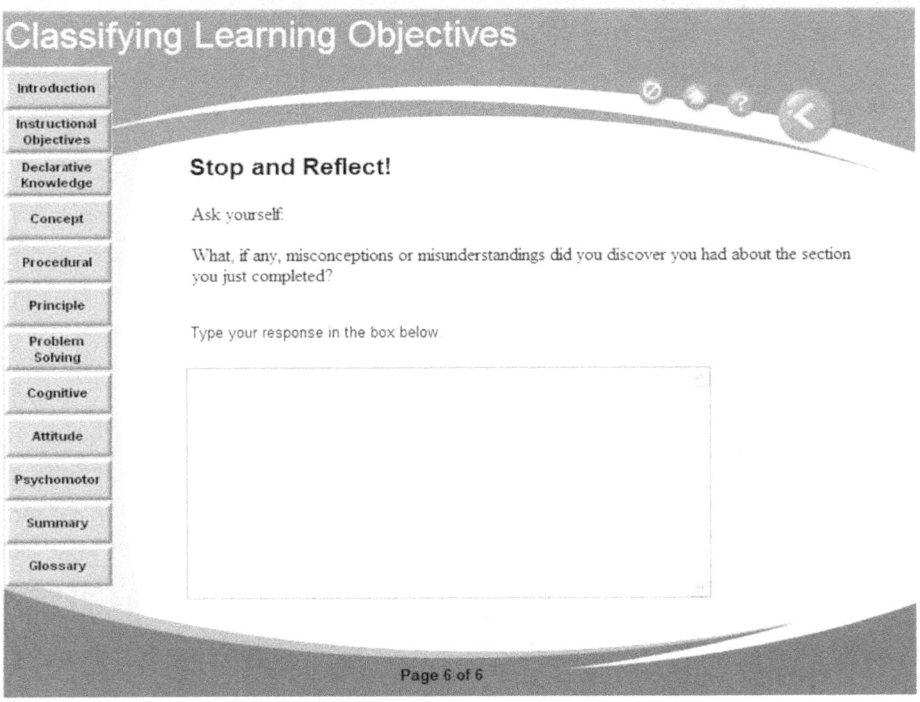

on the instructor's criteria before submitting their assignments.

Cognitive Processing

The cognitive processing component of the Self-regulated Learning Model includes learning strategies, declarative, procedural and conditional knowledge, etc. Specific strategies include managing the learning environment, seeking help from others, using rehearsal, organizational, and elaboration learning strategies, etc. Three cognitive processing strategies will be discussed in the following sections: control the learning environment, organization strategies, and elaboration strategies.

Control the Learning Environment

To promote self-regulation, we encourage our students to control their learning environment. In the course introduction materials, we provide guidelines and/or a checklist for establishing an effective distraction-free study environment. Especially in the beginning of the course, learners could be asked to evaluate their study environment based on a checklist that establishes the characteristics of an effective distraction-free study environment. To ensure that learners complete this activity, we require the students to submit a self-reflection narrative on how well they are doing and a copy of their completed checklist. At periodic times throughout the course (especially at the beginning of the course and less towards the end of the course) we require the students to submit a self-reflection narrative on how well they are doing on controlling their learning environment.

Our online course materials provide suggestions regarding how the students can seek and obtain help depending on the nature of their problems. For example, if a learner is having problems with the learning management systems or other technical problems such as email or the discussion boards, recommendations are provided on how to contact technical support and what do after the helpdesk has closed. Brief scenarios of how previous learners solved their problems are a great resource for the students to be more self-reliant in solving their problems.

Organizational Strategies

We encourage learners to use organizational learning strategies, such as outlining, concept mapping, etc. to promote deeper understanding. The online course lessons include advanced organizers with specific instructions to the learners to develop graphic organizers of the content presented in the lessons. Guidelines, principles, and examples of graphic organizers are provided to help students who lack the knowledge and skills for creating effective graphic organizers. For example, students are encouraged during the beginning course lessons to develop a graphic organizer of the content presented in the lesson. Students are then encouraged to compare the graphic organizer they created with one provided by the instructor to receive immediate feedback. Students are instructed that there is not an absolute right or wrong approach to creating graphic organizers; however, by following the guidelines and principles discussed previously in the lesson, some graphic organizers can be more effective than others in promoting deeper understanding of the content.

Elaboration Strategies

Many of our course assignments and activities require learners to expand on the information presented in the textbook or course materials. These type of activities encourage the learners to be more self-directed and use metacognitive strategies as they reflect on their learning processes. Following is an example of a resource-based assignment from one of our online courses.

This activity gives you a chance to be self-directed and focus your learning on your particular inter-

ests. It is an example of a new paradigm of learning in higher education called resource-based learning. This concept shifts the focus from teaching to learning by requiring students to select their own learning materials from a wide range of real-world information resources. A resource-based approach can help students assume responsibility for their own learning, and provides a practical means of addressing differences in students' educational needs.

For this activity, you should expand (that is, extend what you have learned in this unit and go beyond the materials and resources provided in this unit) on one or more of the ideas, concepts, or topics discussed in this unit. Search for more information available on the idea(s) or concept(s) of interest using information technology resources available.

For example, one of the best information technologies available is the World Wide Web. You may choose to use additional resources such as CD-ROMs, DVDs, computer-based multimedia, contacting experts via E-mail or personal interviews, videotapes, computer programs, etc.

Since this is a course which addresses new and innovative instructional technologies, you should use information technology resources which are non-print based and preferably advanced technologies, such as the Web, DVDs, computer programs, etc.

The requirement for this activity is to write a reflection paper summarizing what you learned from the learning experience. One of the primary purposes of this activity is to facilitate development and usage of metacognitive strategies (i.e., to think about your thinking). Research shows that you will remember and utilize the material you read more effectively if you will reflect on your learning.

Motivation

The motivation component of the Self-regulated Learning Model includes self-efficacy, attributional orientation, goal orientation, affective response, etc. Three motivation strategies will be discussed in the following sections: self-efficacy, self-motivation, and volition.

Self-Efficacy

In the introduction section of our online course materials we include a discussion of the importance of self-efficacy. Self-efficacy is students' confidence about their ability to perform a task. The introduction explains that high self-efficacy students tend to be confident and motivate themselves to acquire learning more than low self-efficacy students (Scott, 1996). The research literature also shows that high self-efficacy students tend to exert more effort than low self-efficacy students when they meet obstacles in learning (Pajares, 2002). One way we determine those students who may be low in self-efficacy is to require the students to answer a question when they submit their goals and actions plans for the course. The question is stated something like the following: "To what degree do you feel confident about your ability to achieve your goal and follow through on your actions plans?" The students respond to the question by selecting one of the following: a) very confident, b) generally confident, c) somewhat confident, d) not much confidence, or e) no confidence at all. We follow-up with those learners who indicated that they have either "not much confidence" and "no confidence at all" to help them increase their self-efficacy. An effective technique we have found to help these students increase their self-efficacy is to follow-up with them after they complete a course assignment or activity by helping them to reflect on what things they did to successfully complete the assignment. We also encourage the students to remind themselves before they begin the next assignment that they successfully completed the previous assignment and to reflect on what learning strategies they used.

Self-Motivation

We provide guidelines and prompts to learners regarding strategies they can use for self-motivation. For example, if learners become discouraged and sense their motivation is dropping in a course, they can remind themselves of past successes and how they overcame similar types of challenges in the past. We sometimes embed prompts in the course lessons or at some point during the completion of a unit that asks learners to assess their motivation level. If a student indicates that his/her motivation is low, he/she can click on a link that provides some suggestions and principles for self-motivation.

Another strategy that works well is to establish a private discussion board for each student. Only the student has access to this discussion board. This discussion board can be used for the student to post his/her goals and action plans for the course. The student can also post reminders of external or internal rewards that will be received if he/she meet the goal. For example, one student posted pictures of an expensive performance car to the discussion board that he planned to purchase when he secured a good paying job (which of course was based on the requirement that he successfully completed his college degree). He frequently reviewed the discussion board to self-motivate himself to successfully complete an online course.

Volition

In the introduction section of the online course we also include a discussion of the importance of volition as a factor in successfully completing the course. Volition refers to a learner's degree of resolve in accomplishing goals. In some courses we include a section that provides an example (a brief case study or scenario) of how a student overcame a major obstacle and was successful in doing well in the course because of his or her resolve to not give up. Another effective strategy is to include brief testimonials from former learners concerning how they overcame obstacles because of their determination to successfully complete the course even when challenges arose.

In addition, another technique is to embed prompts before the learner begins a new unit or module which asks the learner to assess their volition before they begin the new learning module. If the learners feel their will power is low, they can click on a link that provides suggestions to increase their volition, such as reviewing testimonials from previous students, case studies or scenarios of people who increased their volition.

Using Intelligent Tutoring to Promote Self-Regulated Learning

To illustrate how the Model of Self-Regulated Learning and the Self-Regulated Learning Inventory discussed previously could be effectively integrated into an online learning environment, we will borrow and adapt an example from the writing of Winne (1995). This example is based on using a scaffolded approach in which supports are provided only on a temporary basis and eventually withdrawn as the learner internalizes the process of self-regulation. However, in our scenario, unlike Winne's where "Pat" was described as a high functioning self-regulated learner, we will assume that "Pat" is an average (grade point average of "C") psychology student who wouldn't necessarily be considered a strong self-regulated learner. This scenario describes how a hypothetical intelligent tutoring component might be effectively embedded within an online learning environment using a scaffolding approach (see Figure 6 for a general description of the process and phases guiding the design of the online system used to scaffold the learner through a complex learning task) to promote the learner's use of self-regulated learning strategies, with the ultimate aim of helping her become a lifelong learner.

Pat, our hypothetical fourth-year psychology student, is in the midst of taking an online class on formal (mathematical) models and their applications in the social sciences. After the normal online course introductions and background information, Pat is asked to complete the online version of the Self-Regulated Learning Inventory (Lindner & Harris, 1992; 1998; 2002). After Pat completes the inventory, the online system provides a learner profile based on her results. From an analysis of Pat's responses to the inventory items, the learner profile provides feedback to Pat on her strengths and weaknesses as a self-regulated learner, based on the components of the Model of Self-Regulated Learning discussed previously. Subsequently, Pat is provided with general suggestions and recommendations about her approach to learning and, using this feedback, she is allowed to set her own learning goals. For example, if the learner profile indicates that Pat demonstrates characteristics of an instructor-dependent learner, then the system would inform her that learners who tend to show characteristics of an instructor-dependent learner often have a challenging time completing online courses. Recommendations would then be provided to Pat by the system on specific tactics and strategies she should follow to be successful in her online course adventure.

Another powerful function of using the results from the Self-Regulated Learning Inventory is to provide the necessary information to the tutoring system in the online course so that the subsequent instruction is adaptive to Pat's instructional needs, learner profile, and stated learning goals. For example, if Pat's learner profile results were to show that she is weak in self-monitoring or self-evaluation (under the executive processing subscale), or she fails to set specific goals to engage in such cognitive tactics, then the online course instructional scaffolding could be designed to focus its feedback and recommendations on providing instruction and support on to how to use self-monitoring and self-evaluation tactics while completing the course learning activities and assignments.

Figure 6. Self-regulated learning scaffold for online learning environments

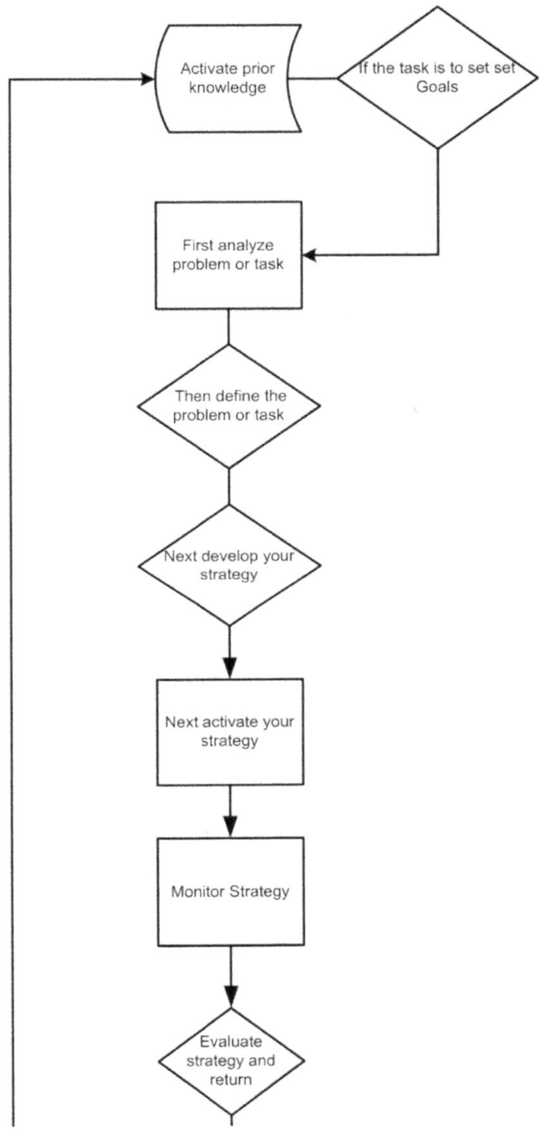

Continuing on…Module Six of Pat's online course is about models of exchange. Pat's assignment is to read, study, and understand the material in Module Six and to prepare a PowerPoint presentation that will be posted to the online course website for next week about how models like this may explain career choice. The topic is complex, and the text is challenging. How can the system help Pat to successfully accomplish this task and acquire strategies to promote self-regulated learning in the process?

Before Pat begins the instruction, she is asked by the system to access and evaluate her relevant prior knowledge on this topic. For example, the system might ask some of following questions. Following each question is a text field for Pat to write her answers.

1. What do you know about this topic already?
2. How would you rate your level of prior knowledge?
3. Review your prior understanding of the topic. If you do have prior knowledge or exposure on this topic, can you generate some specific examples?

Pat responds to the questions, which encourage her to use her prior knowledge to help her understand the new knowledge to be presented. The system then asks her to describe her specific learning goals. As Pat thinks about her goals, she decides to review a previous module on how to use PowerPoint to develop powerful presentations. She clicks on the appropriate menu option or button in order to see a site map of the various modules, contents, and tools available to her. After reviewing the module, she writes out several goals.

The system then asks Pat to analyze and define the learning task and learning environment. For example:

1. What is the nature of the task (memorization, comprehension, performance, etc.)?
2. What issues do I need to consider regarding controlling the learning environment to facilitate the best learning environment possible?

Following Pat's analysis, the system provides Pat with several choices of various learning tactics that can be used to achieve her goal. The system also provides feedback to Pat based on her choices and the variables involved in the instructional situation in which they are embedded.

As Pat begins reading the content in Module Six, she finds the material interesting but finds it somewhat difficult to understand. After Pat completes a section of the module, the system asks questions that encourage Pat to reflect (self-monitor) on her perceived ability to accomplish the learning task. The system asks these particular questions because it is adapting the scaffold to the results of Pat's learning profile, which showed that Pat was weak in self-monitoring and self-evaluation strategies. For example:

1. Are you understanding the material and learning what you need to know to meet your goal, which is (Pat's learning goal is restated here)?
2. Are you prepared for any possible questions others students might ask during your presentation?

Pat believes that by spending a bit more time and effort this week, she will do well. She remembers another student's ordeal from the previous week when she and the other students asked questions in the online chat session after his presentation was posted. She is nervous about that part of her presentation. The other student seemed to know the facts and principles well enough, but he could not work with them very well in responding to the other students' questions. Pat infers that memorization will not build an adequate understanding for surviving the questions that will be asked. "How can I get ready for this?" she mutters. "If I knew what questions they'd ask, I could prepare answers to them beforehand." "So," she continues, "why not invent questions I think the students may ask, then plan answers to those?" "In fact, why not plant those questions in the midst of my initial lecture so people have something to ask for which I'm sure I've prepared?"

This reminds Pat about needing a plan for her presentation. She remembers the professor's seminar two weeks ago and how well it went. She decides to frame her presentation in a similar fashion: selective review of main concepts and models of exchange, a sketch of variables affecting career choice, then presenting and working through one basic model of exchange. The system provides an option for Pat to record notes and thoughts as she is completing the instruction. Pat writes a few thoughts and ideas down in her notes window. As she continues to study the content in the module, Pat plans to link the new material in the text to each part of her talk.

After completing another section of the module, the system detects that Pat has not surveyed the module headings, overview, and summary of the module. Instead, she jumped right into the first frame of instruction without previewing the module first. The system asks Pat if she would like to survey the material in Module Six first before systematically completing all the frames of instruction. Pat responds affirmatively.

The system then provides Pat with an explanation of how to skim the module (surveying its headings and figures, and reading each stop-and-think question, etc.). Pat remembers from earlier modules that stop-and-think questions highlight core concepts and principles. In response to the suggestions of the system, Pat sketches a diagram that links the information in Module Six to an outline for her presentation. As she continues the instruction, she looks for concepts and ideas that can be added to her diagram and outline. After completing a section of the module (or during the section) that is very challenging, the system prompts Pat that now might be a good time to assess her motivation level. For example:

You have just completed a very difficult section of the instruction. Based on your responses to the Self-Regulated Learning Inventory, now might be a good time to assess your emotional state. If you feel like your motivation level is a little low, it might be helpful to remind yourself that you have been a successful student in the past and that you have overcome difficult academic challenges on other occasions.

After reading the comments provided by the system, Pat says to herself, "I just have to pay attention, not get sidetracked, and keep at it. OK, let's just get this next point for now."

In the last section of the module, Pat feels that developing a concept map of the different ideas and topics in this section would help her to identify the right relationships. Pat clicks on the appropriate link provided by the system enabling her to refresh her memory on how to develop a concept map. After reviewing principles and procedures for concept maps, she develops a concept map using an application provided by the system for visual representation of knowledge.

After completing her concept map, a prompt generated by the system appears suggesting that now might be a good time to evaluate her strategy of developing a concept map to determine if this learning tactic was effective.

A little later in this last section, Pat gets stuck when addressing a stop-and-think question about indifference curves. She feels stumped. Recalling the advice of the system from a previous interaction in this module, Pat recalls a study of problem solving from an earlier module that suggested problems become solvable when subjects develop a clear representation of the problem space. She clicks on the link provided allowing her to access the knowledge base related to problem solving. She then reviews the section on representing the problem space.

Following her review, she initiates an adaptive, tactical modification. First, Pat checks whether she understands the question by trying to generate a graph of the information in it. She cannot. She then skims backward in the module looking for concepts named in the question. As she finds them, she translates each onto a mental image of a part in a generic graph for indifference curves.

She had to look at an earlier figure in the module for the generic curve, but that helped. The images she has been building have become rather complicated, so she decides to draw the figure on a piece of paper next to the computer. She annotates each concept on her drawing, checking each concept in the stop-and-think question against the definition she has written on the figure. When the question is fully mapped, Pat predicts what kind of answer the question calls for. She reasons this will help her check the process she is using to address the question. "It works!," she exclaims. Pat writes a note on the piece of paper about the whole process. She decides to keep this kind of log about strategies she invents that help in studying this book.

The system asks Pat several questions after she has completed the module. For example:

1. How would you rate your current level of understanding for this module on a 1 to 5 scale?
   1 2 3 4 5 (1 = no understanding; 5 = completed understanding)

Pat's selects number 3. The system asks Pat is she would like to review the sections she had the most trouble with (based on her past performance of the practice items). The system then asks the second question.

2. Are there any concepts that you missed or are still unclear? Yes No

Pat clicks on the Yes button feeling that there are still a few confusing concepts. The system then asks Pat to specify what is confusing and to identify the unclear concepts. She lists them and then reviews the module to clear up her misunderstandings.

At this point, the system asks Pat to evaluate her learning tactics, strategy, goals, and performance. For example:

1. Which learning tactics were most helpful? Why?
2. What might you have done differently that would improve your learning in the future?
3. If you were to take a quiz on this topic right now, what grade would you expect?
   A B C D F

The system queries Pat if there is anything else she would like to do before finishing Module Six. Pat clicks the No button and exits the module.

What we have described in this scenario is one possible type of application in very general terms. In this case, a type of dynamic, intelligent tutoring was emphasized. The system described in this scenario would take considerable resources, costs, and time to develop. However, similar types of scaffolding could be provided that are more static and would provide similar instruction to all learners. The intent here has only been to provide the reader with a sense of some of the possibilities for promoting self-regulated learning in online learning environments.

CONCLUSION

The primary purpose of this chapter has been to present and discuss several techniques and strategies that we have found successful from our experiences in promoting self-regulated learning in online course environments. We also presented a theoretical model of self-regulated learning that was used as the framework for developing the various techniques and strategies we have used in our online courses. Our experiences, as well as the research literature, suggests that students can acquire self-regulated learning skills and that online courses are an ideal learning environment to nurture these skills.

Generally speaking, online courses entail more independent learning than traditional face-to-face classes. It follows that most online course environments also require that students be highly self-regulated. A major challenge we, as well as others teaching in online learning environments, have experienced is that many students are not able to execute the self-regulated learning skills necessary to successfully achieve their learning goals in online courses. Fortunately, the problem is not intractable.

Our research with, and experiences in, online settings have shown that instructors and course materials can help students learn to be more self-regulated in achieving their learning goals while completing an online course. In fact, an online course can be an ideal learning environment for teaching self-regulated learning skills to learners. We have found that incorporating the techniques and strategies presented in this chapter does indeed promote self-regulated learning in online course environments. Most of the strategies discussed in this chapter can readily be incorporated into most online courses, regardless of the learning management system being used to deliver the course. However, as was stated earlier in this chapter, self-regulation is a complex skill and complex skills take time and practice to assemble and acquire. Online instructors who implement the techniques and strategies presented in this chapter will most likely not see a dramatic improvement or change in students' self-regulation over short time periods; it is only after considerable and targeted practice accompanied by supportive and specific feedback that self-regulation becomes normative for a given learner.

REFERENCES

Allen, I. E., & Seaman, J. (2007). Online Nation: Five Years of Growth in Online Learning. Needham, MA: Sloan-C. Retrieved 12/12/2009 from http://www.sloan-c.org/publications/survey/pdf/online_nation.pdf

Baddeley, A. (2007). *Working memory, thought and action*. New York, NY: Oxford University Press.

Boekaerts, M. (1997). Self-regulated learning: A new concept embraced by researchers, policy makers, educators, teachers, and students. *Learning and Instruction*, *7*(2), 161–186. doi:10.1016/S0959-4752(96)00015-1

Chang, M. (2005). Applying self-regulated learning strategies in a web-based instruction - an investigation of motivation perception. *Computer Assisted Language Learning*, *18*(3), 217–230. doi:10.1080/09588220500178939

Dabbagh, N., & Bannan-Ritland, B. (2005). *Online learning: Concepts, strategies, and applications*. Upper Saddle River, NJ: Pearson.

Dabbagh, N., & Kitsantas, A. (2004). Supporting self-regulation in student-centered web-based learning environments. *International Journal on E-Learning*, *3*(1), 40–47.

De La Paz, S. (1999). Self-regulated strategy instruction in regular education settings: improving outcomes for students with and without learning disabilities. *Learning Disabilities Research & Practice*, *14*(2), 92–106. doi:10.1207/sldrp1402_3

Dettori, G., Gianetti, T., & Persico, D. (2006). SRL in online cooperative learning: Implications for pre-service teacher training. *European Journal of Education*, *41*(3), 397–414. doi:10.1111/j.1465-3435.2006.00273.x

Eastmond, D. V. (1996). *Alone but together: Adult distance study through computer conferencing*. Creskill, NJ: Hampton Press.

Graesser, A. C., McNamara, D. S., & VanLehn, K. (2005). Scaffolding deep comprehension strategies through point & query, autotutor, and iSTART. *Educational Psychologist*, *40*(4), 225–234. doi:10.1207/s15326985ep4004_4

Harris, B. R., & Lindner, R. (2008, October). *Promoting self-regulated learning strategies in online courses*. Paper presented at the annual convention of the Association for Educational Communications & Technology, Orlando, FL.

Harris, B. R., Piña, A. A., & Lindner, R. (2002, October). *Facilitating self-regulation in online courses*. Paper presented at the annual convention of the Association for Educational Communications & Technology, Dallas, TX.

Jonassen, D., Davidson, M., Collins, J., Campbell, B., & Haag, B. (1995). Constructivism and computer-mediated communication in distance education. *American Journal of Distance Education*, *9*(2), 7–26. doi:10.1080/08923649509526885

Kauffman, D. F. (2004). Self-regulated learning in web-based environments: Instructional tools designed to facilitate cognitive strategy use, metacognitive processing, and motivational beliefs. *Journal of Educational Computing Research*, *30*(1&2), 139–161. doi:10.2190/AX2D-Y9VM-V7PX-0TAD

Kunda, Z. (1990). The case for motivated reasoning. *Psychological Bulletin*, *108*(3), 480–498. doi:10.1037/0033-2909.108.3.480

Lindner, R. W., & Harris, B. (1992). Self-regulated learning: Its assessment and instructional implications. *Educational Research Quarterly*, *16*(2), 29–37.

Lindner, R. W., & Harris, B. (1998). Self-regulated learning in education majors. *The Journal of General Education*, *47*(1), 63–78.

Lindner, R. W., & Harris, B. R. (2002, June). *The contribution of self-regulated learning to academic success in college students*. Poster presented at the annual convention of the American Psychological Society, New Orleans, LA.

Loomis, K. D. (2000). Learning styles and asynchronous learning: Comparing the LASSI model to class performance. *Journal of Asynchronous Learning Networks*, *4*(1), 23–31.

Miyake, A., & Shah, P. (Eds.). (1999). *Models of working memory: mechanisms of active maintenance and executive control*. Cambridge, UK: Cambridge University Press.

Moody, J. (2004). Distance education: Why are the attrition rates so high? *The Quarterly Review of Distance Education*, *5*(3), 205–210.

Pajares, F. (2002). Gender and perceived self-efficacy in self-regulated learning. *Theory into Practice*, *41*(2), 116–125. doi:10.1207/s15430421tip4102_8

Paris, S. G., & Paris, A. H. (2001). Classroom applications of research on self-regulated learning. *Educational Psychologist*, *36*(2), 89–101. doi:10.1207/S15326985EP3602_4

Patterson, B., & McFadden, C. (2009). Attrition in online and campus degree programs. *Online Journal of Distance Learning Administration*, *12*(2).

Piña, A. A. (2010). An introduction to learning management systems. In Kats, Y. (Ed.), *Learning management systems: Technologies and software solutions for online teaching*. Hershey, PA: IGI Global Publishing.

Piña, A. A., Sadowski, K. P., Scheidenhelm, C. L., & Heydenburg, P. R. (2008). SLATE: A community of practice for supporting learning and technology in education. *International Journal of Instructional Technology and Distance Learning*, *5*(7).

Pintrich, P. R. (2004). A conceptual framework for assessing motivation and self-regulated learning in college students. *Educational Psychology Review*, *16*(4), 385–407.

Puustinen, M., & Pulkkinen, L (2001). Models of self-regulated learning: A review. *Scandinavian Journal of Educational Research, 45*(3), 269–286. doi:10.1080/00313830120074206

Puzziferro, M. (2008). Online technologies self-efficacy and self-regulated learning as predictors of final grade and satisfaction in college-level online courses. *American Journal of Distance Education, 22*(2), 72–89. doi:10.1080/08923640802039024

Schunk, D. H. (2005). Self-regulated learning: the educational legacy of Paul Pintrich. *Educational Psychologist, 40*(2), 85–94. doi:10.1207/s15326985ep4002_3

Schunk, D. H., & Zimmerman, B. J. (1998). Conclusion and future direction for academic interventions. In Schunk, D. H., & Zimmerman, B. J. (Eds.), *Self-regulated learning: From teaching to self-reflective practice* (pp. 225–235). New York, NY: Guilford Press.

Schunk, D. H., & Zimmerman, B. J. (Eds.). (2008). *Motivation and self-regulated learning: Theory, research, and applications*. New York: Lawrence Erlbaum Associates.

Scott, J. E. (1996). Self-efficacy: a key to literacy learning. *Reading Horizons, 36*(3), 195–213.

Tyler-Smith, K. (2006). Early attrition among first time e-learners: A review of factors that contribute to drop-out, withdrawal and non-completion rates of adult learners undertaking elearning programmes. *Journal of Online Learning and Teaching, 2*(2), 73–85.

VanderStoep, S. W., & Pintrich, P. R. (2003). *Self-regulated learning: from teaching to self-reflective practice*. Upper Saddle River, NJ: Prentice Hall.

Whipp, J., & Chiarelli, S. (2004). Self-regulation in a web-based course: A case study. *Educational Technology Research and Development, 52*(4), 5–22. doi:10.1007/BF02504714

Williams, P. E., & Hellman, C. M. (2004). Differences in self-regulation for online learning between first- and second-generation college students. *Research in Higher Education, 45*(1), 71–82. doi:10.1023/B:RIHE.0000010047.46814.78

Winne, P. H. (1995). Inherent details in self-regulated learning. *Educational Psychologist, 30*(4), 173–187. doi:10.1207/s15326985ep3004_2

Zimmerman, B. J. (1990). Self-regulated learning and academic achievement: An overview. *Educational Psychologist, 25*(1), 3–17. doi:10.1207/s15326985ep2501_2

Zimmerman, B. J., & Martinez-Pons, M. (1986). Development of a structured interview for assessing student use of self-regulated learning. *American Educational Research Journal, 23*(4), 614–628.

Zimmerman, B. J., & Schunk, D. H. (Eds.). (2001). *Self-regulated learning and academic achievement: Theoretical perspectives*. Mahwah, NJ: Lawrence Erlbaum Associates.

KEY TERMS AND DEFINITIONS

Asynchronous Discussion Forum: is an online discussion site where participants can engage in text-based conversation organized into topic-based discussion threads and do not have to be logged in at the same time.

Attributions: the causal factors to which individuals appeal when explaining outcomes in their lives, typically divided into external-internal, controllable-uncontrollable, and stable-unstable.

Automatic Processing: cognitive processing of information that requires little or no conscious awareness.

Cognitive Processing: basic operations of the information processing system of the mind, e.g., storage and retrieval of information from long term memory.

Cognitive Strategies: thoughtfully developed plans for maximizing outcomes in the face of learning tasks and challenges.

Conditional Awareness: conscious awareness that specific knowledge or information can be effectively applied given the conditions or context of a particular situation.

Conditional Knowledge: knowing when and where (under what conditions) to apply (or not apply) specific knowledge in one's possession.

Executive Processing: decision making processes that control attention and the allocation of cognitive resources in problem solving.

Explicit Cognition: conscious, deliberate thinking processes engaged when dealing with novel and/or difficult information or problems

Learning Tactics: specific operations performed in the course of learning (e.g., rehearsing information) that assist in accomplishing learning goals.

Metacognitive Processes: reflexive cognitive processes that are focused on understanding how thinking and learning work; literally thinking about one's thinking.

Self-Efficacy: an individual's subjective sense of the ability to successfully effect desired outcomes.

Self-Regulated Learning: the ability to exercise intentional, proactive control over learning related outcomes through adaptive management of cognitive resources and motivational states.

Synchronous Chat: a way of communicating online by sending text messages to people in the same chatroom in real-time.

Working Memory: the limited work space of conscious awareness used in explicit problem solving or information processing. Working memory includes short term memory, several buffer systems for temporary storage, and the central executive that manages information processing resources.

Chapter 9
Influence of Task Nature on Learner Self-Regulation in Online Activities

Manuela Delfino
Institute for Educational Technology (CNR), Italy

Giuliana Dettori
Institute for Educational Technology (CNR), Italy

Donatella Persico
Institute for Educational Technology (CNR), Italy

ABSTRACT

This chapter analyses Self-Regulated Learning (SRL) in a virtual community interacting through asynchronous textual communication. The community consisted of trainee teachers of a post-graduate blended course in Educational Technology. The online component of this course was based on a socio-constructivist approach. The study aims to compare SRL practice in different types of collaborative activities carried out online. The investigation method is based on interaction analysis, an approach allowing a systematic study of the content of the messages exchanged by the community members. The results of the study consist of quantitative data on SRL-related events that took place during the learning process, allowing the comparison of activities according to the degree and type of self-regulation displayed by the learners. The results of the study suggest that the nature of the task influences the way students self-regulate. The difference, however, does not lie in the total amount of detected SRL indicators but in their type, therefore suggesting that different types of tasks might induce different kinds of SRL actions. These findings can inform the design of online activities by providing suggestions for the choice of tasks, according to SRL-related pedagogical purposes.

DOI: 10.4018/978-1-61692-901-5.ch009

INTRODUCTION

Research in education has been increasingly devoting attention to Self-Regulated Learning (SRL) in the past couple of decades. SRL can be defined as an active process by which learners become aware of their own learning and feel responsible for it, setting goals and monitoring their achievements, controlling their cognitive, motivational and emotional behaviour, evaluating their outcomes and devising strategies to improve them, paying attention to the contextual features of the learning environment in order to take advantage of their affordances (Zimmerman, 1998; Pintrich, 2000). This increased interest has been largely determined by the changes brought about, on one side, by the diffusion of educational approaches that encourage learners' active engagement, and, on the other side, by the increased pervasion of technology in all aspects of life, bringing about both opportunities and needs to constantly keep updated and improve one's competence (Dettori, Giannetti & Persico, 2006).

Understanding how learners become self-regulated is therefore an important issue. Even though some essential abilities improve with age, building SRL competence is neither automatic nor fast (Boekaerts, 1997), and it can take advantage of suitable teaching and practice. For instance, Van den Boom, Paas, Van Merrienboer and Van Gog (2004) argue that the acquisition of SRL competence can be stimulated by embedding aspects of it in instructional strategies. Dabbagh and Kitsantas (2004) claim that web-based learning tools, such as collaboration and communication environments, can support the development of specific self-regulatory skills related to successful work in online environments.

Moreover, there is evidence that SRL skills are context-dependent (Boekaerts, 1999) and are not easy to transfer from one context to the other (Hofer, Yu & Pintrich, 1998). This means, for instance, that people who are able to self-regulate their own individual learning in traditional settings may not be able to self-regulate as effectively in a collaborative context, due to the fact that not only does group activity entail different skills, such as negotiating objectives, strategies and meanings, but it also affects individual behaviour, as its organization must respect the constraints imposed by collective action. If the collaboration takes place online, as in Computer-Supported Collaborative Learning (CSCL), a further variable is added by the technological environment, which offers a number of possibilities and challenges, many of which entailing the need to manage and control the individual activity (Salovaara, 2005). There is evidence in the literature, moreover, that technology influences the relation between self-regulation and learning context, in that computers can be used as metacognitive tools for enhancing learning and hence become catalysts for change (Azevedo, 2005; Lowerison, Sclater, Schmid & Abrami, 2006). These research studies, and many others along the same line, suggest that it is advisable to foster SRL in every learning situation, paying attention to how its practice can be supported in different contexts and by different means.

In this chapter, we analyze SRL in a virtual community whose members interacted through asynchronous textual communication. SRL practice of the participants was examined in four different types of online collaborative activities, aiming to investigate the influence of the task nature on self-regulation in online settings. A better understanding of this influence should improve the design of online activities by informing criteria for task-definition. The study method was based on content analysis of the messages exchanged by the learners during the learning process. This approach provides information on the practice of SRL drawing from the actions performed by the learners, as they emerge from the written interactions, highlighting, at the same time, what type of self-regulated actions are carried out.

The next section presents the method used to study SRL practice in online learning activities, illustrating the set of indicators used for the analysis. The third section describes the context, setting and outcomes of the case study. Finally, the concluding remarks aim to summarize what we can learn from this experience.

A METHOD TO OBSERVE SRL PRACTICE IN CSCL ACTIVITIES

The assessment of learners' practice of SRL is usually made by means of questionnaires and interviews proposed in the course or at the end of a learning experience (Torrano & Gonzales, 2004). Such methods, despite the many advantages they offer (Bryman, 2004; Cresswell, 2003), have the drawback that the data collected may be biased by participants' beliefs and *a posteriori* reflections. Student self-reports, moreover, often result to be inaccurate portrayals of actual SRL behaviour (Jamieson-Noel & Winne, 2003). For these reasons, this study was based on a different approach developed to observe the practice of SRL in the course of online collaborative learning activities (Dettori & Persico, 2008), taking advantage of the fact that online interactions take place in written form and are permanently stored by the Computer-Mediated Communication (CMC) platform. The main asset of such approach is that it does not gather the learners' opinions, but is based on the analysis of the communicative exchanges that have been taking place during a learning activity.

The investigation of learning dynamics by means of Interaction Analysis (IA) is a research methodology which has been increasingly used in the past few years to explore both cognitive and affective aspects of collaborative discourse. It relies on discourse analysis (Gee, 2005) and consists in detecting phrases and expressions that reveal aspects of interest in the written messages exchanged by the learners. The variables investigated may be manifest, that is, objectively recognizable (which makes it possible to automate the analysis process), or latent, that is, implicit in message content (which entails the need for a manual analysis).

In order to be applied, IA requires a set of indicators of the investigation object. Several research studies applying content analysis in CSCL have proposed sets of indicators for different variables, such as participation and interaction effectiveness (Calvani, Fini, Molino & Ranieri, 2010), or social, cognitive and teaching presence (Garrison, 2007). A few studies have focused on single variables which are related to SRL, such as critical thinking (Newman, Webb & Cochrane, 1995), cognitive and metacognitive knowledge (Henri, 1992), social construction of knowledge (Gunawardena, Lowe & Anderson, 1997), without giving a wide-angle view on SRL. The indicators used in this study explicitly address a variety of aspects related to SRL practice. They have been proposed by Dettori and Persico (2008) and consider the intertwining of facets that characterize SRL in CSCL environments.

The considered set of indicators is based on the work of Zimmermann (1998; 2000) and Pintrich (2000) on SRL. Some studies on the potential support to SRL afforded by Technology-Enhanced Learning Environments (Steffens, 2006; Banyard, Underwood & Twiner, 2006; Carneiro, Steffens & Underwood, 2005) were also taken into consideration. In this perspective, SRL appears to be characterized by two independent sets of aspects, that can be called the "process" model and the "component" model of SRL. According to the process model, SRL consists of three phases that are cyclically repeated and influence each other: planning, monitored execution, and evaluation. The component model, on the other hand, distinguishes among the cognitive, metacognitive, motivational and emotional aspects of SRL. The two models can be seen as complementary and can be meaningfully considered both at the individual and at the social level.

The result is a characterization of SRL as a 3-dimensional process, defined by three independent sets of features. The twelve groups of aspects

Table 1. Indicators of SRL actions in CSCL environments

		planning	monitored execution	evaluation
cognitive / metacognitive	individual	- Making personal plans on how to proceed in the learning process: breaking tasks in sub-tasks, establishing deadlines, detecting priorities, etc. - Adjusting plans to overcome failures. Example: *"I will read your proposal again"*	- Enacting plans. - Working consistently on the assigned task. - Monitoring plans fulfilment. - Making syntheses of individual work and objectives reached. Example: *"I'm working on the first part. I'll post it by tonight, as planned"*	- Assessing own learning. - Reflecting on individual learning achieved. - Spotting difficulties and causes of failures. - Comparing one's work with that of peers. - Assessing/expressing awareness of individual time management. Example: *"I'm quite happy about my work, although I know I took longer than I should have"*
	social	- Making proposals on how to proceed in the learning process. - Discussing and negotiating on planning aspects. - Working out together plan changes necessary to overcome failures. Example: *"Let us devote a couple of days to the readings, and then try to summarize them"*	- Quoting peers' contributions, asking questions, reacting to peers' messages. - Mediating among peers. - Checking understanding. - Summarizing the ideas suggested by all group members. - Encouraging peers to act. Example: *"I agree with what you wrote because..."*	- Assessing group learning. - Commenting group achievements. - Reflecting on group learning - Encouraging peers to express their opinions on the work done. Example: *"We have done a very good job, don't you think so?"*
motivational / emotional	individual	- Exploring one's expectations about the current learning activity. - Anticipating possible emotional aspects. Example: *"I expect to learn a lot from this course!"*	- Expressing one's emotions and motivations. - Looking for appropriate support when needed. - Disclosing oneself to peers. Examples: *"I'm really excited by this new activity...".* *"I must admit that I feel uneasy with..."*	- Comparing one's current motivation and emotions with the original ones. - Understanding the reasons of possible changes to plans. - Commenting on emotional aspects developed during the learning process. Example: *"At the beginning of this course I was a bit worried not to be able to handle the technology, but now I find it easier than I thought"*
	social	- Discussing expectations and motivations about the current learning activity or learning in general. - Sharing motivations for own commitment. - Encouraging peers to get involved in planning. Example: *"OK, all of us have been taking it easy until now. What about starting to tackle the task more seriously?"*	- Encouraging peers to express their emotions and motivations. - Encouraging peers and providing them with emotional support. - Taking care of group functioning by informing peers of one's intentions. Examples: *"Don't you agree that we should try to respect our plans more closely?"*	- Expressing appreciation for peers' efforts, contributions and results. - Spotting group's malfunctioning and analyzing its causes. Example: *"Thanks for working so hard! You did a good job!"*

raising from such combination are shown in Table 1, together with examples of possible phrases that would be regarded as clues of each indicator. Following Garrison, Anderson and Archer (1999), cognitive aspects are grouped with metacognitive ones, since it is often difficult to clearly mark the separation between them, especially in a context, like CSCL, that usually fosters metacognitive activities along with cognitive ones. For similar reasons, motivational aspects are grouped with emotional ones.

Self-regulation is revealed by the fact that learners carry out certain kinds of actions and communicate them to other learners. All the variables involved are latent, in that self-regulative instances cannot be strictly associated with the use of particular expressions or constructs. As a consequence, the analysis has to be conducted on the semantic level.

It is important to note that the occurrence of individual indicators in a conversation aiming to support the execution of a collaborative task can not be taken for granted. Learners would mostly find the expression of personal plans, actions, thoughts and emotions out of place, unless they are explicitly encouraged to do so by the tutors and group mates or implicitly by a relaxed and friendly atmosphere in the learning environment. Encouraging personal expressions should not be considered only a way to gather data for interaction analysis, but is actually a way to help learners manifest their social presence. This is considered a necessary condition for successful online learning, as important as cognitive and teaching presence (Garrison, 2007). Hence, in well designed and thoughtfully implemented online courses it is not surprising to find individual indicators of motivation and emotion. Nevertheless, even in the best conditions, we should be aware that such indicators represent only a (possibly small) part of the self-regulated actions actually carried out by the learners.

A CASE STUDY

The activity analysed in this chapter was the online component of a blended course that was designed and run according to a CSCL approach. This entails that participants are asked to collaborate and discuss with peers at a distance in order to reach a common purpose, and by this means gain the knowledge and the skills which are the course's objectives (Koschmann, 1996; Koschmann, Hall & Miyake, 2002).

Context

The considered experience was carried out within a course designed and run for the teacher training school of the University of Genoa (Italy) in the academic year 2004/2005. The subject taught was Educational Technology. Aim of the course was therefore to acquaint the trainee teachers with a variety of ICT tools and methods so as to become able to improve teaching and learning through them.

The participants were 95 trainee teachers with different backgrounds, together with 7 tutors. For most of the students (89%) this was the first exposure to CMC in formal learning activities. Six of the 7 tutors were experts in online learning tutoring while one was novice to this activity. Participating in the online activities was mandatory for the trainees to meet the requirements.

The course lasted 3 months. It adopted a blended approach consisting in the integration of 5 face-to-face meetings with 12 weeks of online activity based on the use of a CMC environment (a customized configuration of Centrinity FirstClass®). Face-to-face sessions were devoted to introduce the subject from a theoretical point of view and to stimulate and launch an effective participation in the online activities.

Online work was mainly collaborative. The student cohort was segmented into virtual workgroups, each supported by a tutor. The groups were re-structured two times during the course, according to the requirements of the various tasks. Learning activities involved web-navigation, readings, collaborative production of documents, peer reviews and analysis of online learning resources. Communication among the participants was mostly asynchronous.

Five tasks were sequentially proposed, in 5 different discussion spaces created on purpose. The tasks were: *Familiarization*, *Peer review* of online resources, *Role-play* on WebQuests, collaborative *Case-study* of school-based learning communities, and *Concluding meta-reflection*.

Only 4 of these modules were considered in the current study; the module on the peer review of online resources was left out because it entailed a low level of online interactions, with students grouped in pairs and mostly working individually.

Task 1: Familiarization

The initial online module, lasting 3 weeks, was devoted to familiarization with the platform and with the new learning mode, as well as to socialization within the community. In this task, the metaphor of navigation was proposed as a unifying theme apt to offer the opportunity to carry out some activity of limited cognitive demand (Delfino & Manca, 2007). The course was therefore described as a sea-journey. Seven discussion areas were created in the familiarization area, each of which took the name of a kind of boat (caravel, cruise liner, fishing-boat, motorboat, sailing boat, steamboat and submarine). Each participant was supposed to choose one of them to "board", which implied joining the discussion group with the corresponding boat name. This operation split the participants into seven working groups. Within each of such groups, the participants had to explain the reason for their choice, and to decide, by negotiating with their group-mates, a name, a motto and a symbol for their boat. The rationale for this activity, and for the long time allocated to it, resides in the importance attributed by the course organizers to letting the students get a good acquaintance with the communication platform and the dynamics of online interaction. This aimed to lay the bases of online collaboration, making the students feel at ease with negotiating some decision with peers at a distance, hence facilitating the subsequent activities on content knowledge.

Task 2: Role-Play on WebQuests

During Task 2, the whole cohort of students was split into twelve different subgroups, the members of which were decided by the tutors by mixing students with different backgrounds, levels of interest, motivations, and behaviours shown in the previous activities.

This task, lasting 3 weeks, was devoted to the analysis of online educational resources. In particular, it was aimed at making trainees acquainted with WebQuests, working directly on one of such educational activities (a WebQuest about educational WebQuests, based on the model proposed by Dodge, 2009). The activity consisted in a role-play scenario, where students had to (1) take on the role of strongly characterized teachers (the technology enthusiast, the technology detractor, the bureaucrat, the headmaster, etc.); (2) discuss strengths and weaknesses of three WebQuests from these different perspectives and produce a summary of the observations made; (3) choose one of the three WebQuests analysed and prepare a written project suggesting how to improve it as though they were actually going to run that project with some colleagues in school.

Task 3: Case-Study of School-Based Learning Communities

The same groups of students engaged in Task 2 carried out Task 3, a case study on school-based learning communities, lasting 3 weeks. Here, the trainees were asked to discuss assets and flaws of real-life school projects, based on documentation provided by the tutors. In particular, they were requested to (a) read three case-studies, (b) individually produce a detailed analysis of each case and share it with peers, and (c) cooperatively write a paper synthesizing the main elements raised.

Task 4: Concluding Meta-Reflection

In Task 4, carried out over the last week of the online course, the metaphorical theme of the familiarization phase was resumed. Hence, the participants were requested to choose a boat (i.e., a sub-conference within the interaction area devoted to this task), which could be the same chosen in

Table 2. Task comparison according to the SRL parameters

	planning	monitored execution	evaluation	cognitive / metacognitive	emotional / motivational	individual	social
Task 1	partially scaffolded	partially scaffolded			scaffolded by the metaphoric framework	contribution to brainstorming	comparison of ideas and joint choice
Task 2	scaffolded	partially scaffolded		explicit cognitive aim		individual points of view	joint synthesis
Task 3	scaffolded			explicit cognitive aim		individual points of view	joint synthesis
Task 4			explicitly required	explicit metacognitive aim	scaffolded by the metaphoric framework	evaluation of own learning	sharing individual evaluations

Task 1 or a different one, and explain the reasons for their choice. This gave rise to an individual reflection based on some questions posed by the tutors about the learning experience (e.g., at the end of your journey, what kind of souvenirs are you bringing home?). After all participants had provided their personal answers, a group discussion was carried out, focusing on aspects such as competences acquired, difficulties met, usefulness of new contents, effectiveness of the learning methods, impressions on CMC and opinions about its usability in the school setting.

Features of the Examined Tasks

In order to better understand the regulative dynamics that took place during the course, a discussion of the features of the four tasks that appear to be relevant to SRL dynamics can be helpful (Table 2 proposes a comparison of the tasks based on these features).

Continuity between the first and the fourth tasks was established by the metaphorical setting. This acted at the beginning of the course as an emotional and motivational framework for the unusual collaborative experience that the participants were asked to undertake. The same setting, at the end of the course, meant to recall a familiar place in which to share reflections on the learning experience.

Tasks 2 and 3 were the two core learning modules of the course. Continuity between them was maintained by keeping the same group composition (i.e., students and tutors), which allowed the participants to take advantage in Task 3 of the reciprocal knowledge and the social interaction dynamics developed in Task 2. On the other hand, the nature of the cognitive tasks and the learning strategies proposed were remarkably different in the two cases. These two tasks lasted 3 weeks each and alternated collaborative and individual activities.

Task 1 partially scaffolded planning because the task description contained detailed instructions about how to proceed (choosing a boat and negotiating a name, a motto and a symbol with the boat-mates) while no indication was provided about how to manage the activity, what intermediate deadlines to set, what to do first and what to do later, how long to devote to each subtask. These decisions were left to the learners as a useful exercise, in order to help them experience that interaction-based online activities need to be planned and constantly monitored towards the goal achievement, no matter how simple is the task assigned. In this activity, the participants were not explicitly requested to self-evaluate their own achievements. The emotional /motivational component of SRL was scaffolded by the metaphoric framework proposed by the tutors. The whole activity was essentially based on social negotiation, given that all the subtasks (except for the boat choice) were to be carried out by proposing, discussing, making decisions together with peers.

Task 2, on the other hand, scaffolded both planning and execution because the groups were formed by the tutors, the task was clearly outlined and the intermediate deadlines were suggested and recalled by the tutors. The reasons for such a thorough scaffolding was that this activity was the first in which students were confronted with the course contents and collaborative learning based on comparison of points of view, negotiation of meanings and joint outcome production. In addition, given that students were not used to this approach, course designers considered it advisable to support planning and execution as much as possible in this task. For example, it is well known that group formation can be very time-consuming when left to inexperienced students, and often it does not even turn out to be profitable because students tend to team up with friends or acquaintances and rarely dare to join people they do not know, therefore avoiding a fruitful mixture of backgrounds. While the cognitive aims of this task were made explicit in its description, no mention was made of the emotional and motivational components of SRL that needed to be handled while carrying it out, neither was self-evaluation.

Task 3 was mostly scaffolded in its planning phase, but not so much in its monitoring. The task was well specified, the groups were already formed because they were the same as in the previous activity, but much of the monitoring was delegated to the students: they decided the intermediate deadlines for the task, they controlled the timing, one member of the team assumed the role of facilitator/ coordinator and another merged all the conclusions in one document. As in the previous activity, neither the need for self-evaluation nor the emotional and motivational aspects were explicitly addressed in the task description.

In Tasks 1, 2 and 3 evaluation was a key element, but participants were not directly asked to elicit their personal evaluation of the ongoing activities. This fact did not prevent them, as we shall see in the following sections, to express an evaluation of the learning processes. This is probably related to the fact that collaborative activities carried out at a distance naturally call for ongoing considerations and expression of opinions regarding the results obtained.

Task 4 was very different from the previous ones, being mostly based on individual meta-reflection. No scaffolding was provided to planning and monitoring: no groups were formed, no intermediate deadlines were proposed, and a lot of freedom was granted about how to proceed (e.g., students could even choose different means of expression from the written form, such as drawings or short-movies attached to their postings). However, the aim of the activity was explicitly stated as the evaluation of cognitive, emotional and motivational aspects of the course.

Method of the Study

Interaction analysis was based on manual coding of the corpus of 1949 messages exchanged by 95 participants, using the set of indicators presented in the previous section (Table 1). Two coders independently looked for occurrences of SRL-related events in students' messages. While one of them had been involved in the online course as designer and tutor, the other was external to the course but expert in SRL.

At the beginning, the coders undertook a coding training session. The inter-rater reliability was computed with Holsti's method on a common subset of 154 messages and resulted 0.83. Disagreements were resolved through discussion and complete consensus was reached. The remaining messages were split in two parts and each coder only processed one half of the total, exchanging views with the other coder in case of doubt.

At the end of this process, it resulted that the students' messages containing at least one SRL indicator were 897 (46.02% on the total of mes-

Table 3. Percentages of indicators detected in the interactions of Task 1

Task 1: Familiarization	**planning**	**monitored execution**	**evaluation**
cognitive/metacognitive individual	3.98	2.52	1.89
cognitive/metacognitive social	19.50	34.80	1.68
motivational/emotional individual	1.26	8.18	2.10
motivational/emotional social	5.03	12.79	6.29
TOTAL	29.77	58.29	11.96

sages), that the total number of SRL indicators was 1247 and that the average number of indicators per SRL-related message was 1.39.

Study Outcomes by Task

This section reports the main results of the study and shows, for each task, how the indicators were distributed along the process and the component models. Although data concerning SRL indicators at individual and at social level are also reported, their direct comparison will not be addressed in this chapter. As pointed out in the section on the research method individual self-regulation could be largely underestimated when the source of data are only the messages exchanged between learners.

Table 3 reports the percentage of indicators found by the coders for each category within the students' messages in Task 1. This table tells us that in this task the highest concentration of indicators concerned the phases of monitored execution and the cognitive/ metacognitive aspects at social level. This depends on the fact that the students accomplished this task by quoting each other often, checking understanding, summarizing the ideas expressed by the group and encouraging the group to act. Similarly, the reasonably high values of social cognitive planning indicators mean that in this task the level of planning of social activities taking place was high. The high values of social motivational monitoring indicators show a prevalence of actions aiming to encourage peers to express their emotions and to provide emotional support. All in all, it is not surprising to find in this initial task, where the students had to socialize with each other and get acquainted with the learning environment, a high share of indicators of emotional aspects. Nevertheless, the cognitive and metacognitive aspects prevail, because all the actions that the students carried out to accomplish the task (choose a boat, propose a name, a motto or a symbol) were considered as cognitive ones by the raters because they entail the execution of the assigned task. If we consider the distributions of the SRL-related actions carried out in the three phases, we may note that execution actions are more than the sum of the other two. This may be due to the fact that, in Italy, students are rarely encouraged to make plans for, and evaluate, their own learning and they are therefore not used to do it. We will see in Tables 4-6 that this difference decreases in the next tasks, possibly due to the encouragements to perform planning and evaluation actions they received over the course.

Table 4 reports the percentage of indicators found in the analysis of the students' messages in Task 2. The highest concentration of indicators is again in the social cognitive execution category, due to a large amount of reciprocal quotations, mediations, summarizing and drawing conclusions. In this task, however, the second highest concentrations refer to the categories individual cognitive planning, individual cognitive evaluation and social motivational evaluation. Within

Table 4. Percentages of indicators detected in the interactions in Task 2

Task 2: Role play on WebQuests	**planning**	**monitored execution**	**evaluation**
cognitive/metacognitive individual	12.94	1.49	10.45
cognitive/metacognitive social	9.45	29.85	2.99
motivational/emotional individual	2.49	8.46	0.00
motivational/emotional social	0.50	6.97	14.43
TOTAL	25.38	46.77	27.87

the individual cognitive planning category falls the choice of the role (the coders regarded it as a planning action of individual nature), while individual cognitive evaluation indicators included expressions aiming to assess one's own learning, spotting difficulties, reflections on individual learning achievements, etc. Finally, social motivational evaluation indicators concerned expressions of appreciation for peers' efforts and attempts to identify the causes of possible group malfunctioning. It should not be forgotten that this task, being a role play, required the students to carry out a collaborative effort adopting the points of view of the chosen roles; it was therefore natural that they expressed appreciation to those in the group who had the role to moderate the discussion and summarize the conclusions in a document. All in all, there is a predominance of cognitive/metacognitive indicators over the motivational/emotional ones, as in the previous task, and even to a slightly greater extent.

Table 5 shows the distribution of SRL indicators in Task 3, the case study on school-based learning communities. In this task, social cognitive monitoring and social motivational evaluation indicators show higher percentages; the first is due to quotation of peers as well as to mediation and summarizing activities, which show attention to coordinate one's contributions with those of group mates, and to control the overall development of the discussion, monitoring its progression towards the expected outcomes; the second is mainly determined by affective factors such as showing appreciation for peers contributions after assessing group activity and spotting group malfunctioning. In addition, there is a relatively high concentration of social cognitive planning indicators, revealing planning of group activities, such as, for example,

Table 5. Percentages of indicators detected in the interactions of Task 3

Task 3: Case study	**planning**	**monitored execution**	**evaluation**
cognitive/metacognitive individual	5.14	2.29	5.71
cognitive/metacognitive social	16.00	21.71	5.14
motivational/emotional individual	0.57	9.71	1.14
motivational/emotional social	3.43	6.86	22.29
TOTAL	25.14	40.57	34.28

Table 6. Percentages of indicators detected in the interactions of Task 4

Task 4: Conclusions	**planning**	**monitored execution**	**evaluation**
cognitive/metacognitive individual	4.82	10.15	21.57
cognitive/metacognitive social	1.78	9.14	5.84
motivational/emotional individual	0.51	7.87	15.99
motivational/emotional social	0.76	3.30	18.27
TOTAL	7.87	30.46	61.67

making proposals on how to proceed. These results are not surprising, because this task required the analysis of best practice experiences (i.e., the case studies) and the identification of their pros and cons: while approaching this task, many students expressed in their messages the difficulties they were facing (not having a "role" to play, as in the previous activity, they needed to express their true opinions, which made them feel somehow shy), and widely supported each other in the decisions on how to proceed.

Task 4 explicitly required trainees to reflect on the course experience and the learning achievements. Its nature is clearly reflected in the outcomes shown in Table 6. A high percentage of individual cognitive evaluation indicators reveals a good amount of individual evaluation of learning achievements, as entailed by the task assigned. Only few students extended their evaluation to group accomplishments, probably because this was not explicitly requested. Many evaluated both cognitive/metacognitive and motivational/emotional aspects. A reasonably high percentage of individual cognitive execution reveals plan enactment. Individual cognitive planning is low, due to the nature of the task assigned, but its value is higher than one would expect in a conclusive activity. This depends on the fact that a number of trainees explicitly expressed their own plans for applying the competence acquired in their future profession. The motivational evaluation indicators, at both individual and social level, are very high because, being this task the last of the course and very focused on drawing a balance on the work done, the trainees express their gratitude to their course mates for the good collaboration, as well as their satisfaction for what they have learned during the course, not only in terms of content knowledge but also of professional competence.

DISCUSSION

After analyzing each of the four tasks individually, we can now take a general view of the outcomes and compare the results across tasks in order to draw some conclusions about the dynamics triggered by each of them (Tables 7 and 8).

An aspect that mostly strikes attention is that the percentage of monitoring indicators, and in particular the monitoring of cognitive aspects in social activities, is the highest for all the collaborative tasks (i.e., all except the last), in which the meta-reflection had to be worked out individually. In other words, whenever the task was collaborative, most of the students' efforts seem to concentrate on monitoring the activity, with specific focus on the cognitive and social aspects (see Tables 3 to 6). This was done by properly reacting to peers' messages, mediating, checking understanding, summarizing what had been said, encouraging others to contribute.

Table 7. An overview of the SRL percentage indicators detected in the 4 tasks summarized from the point of view of the "process model"

	planning			monitored execution			evaluation		
	individual	*social*	*TOTAL*	*individual*	*social*	*TOTAL*	*individual*	*social*	*TOTAL*
Task 1	5.24	24.53	29.77	10.70	47.59	58.29	3.99	7.97	11.96
Task 2	15.43	9.95	25.38	9.95	36.82	46.77	10.45	17.42	27.87
Task 3	5.71	19.43	25.14	12.00	28.57	40.57	6.85	27.43	34.28
Task 4	5.33	2.54	7.87	18.02	12.44	30.46	37.56	24.11	61.67

Table 8. An overview of the SRL percentage indicators detected in the 4 tasks summarized from the point of view of the "component model"

	cognitive/metacognitive			motivational/emotional		
	individual	*social*	*TOTAL*	*individual*	*social*	*TOTAL*
Task 1	8.39	55.98	64.37	11.54	24.11	35.65
Task 2	24.88	42.29	67.17	10.95	21.90	32.85
Task 3	13.14	42.85	55.99	11.42	32.58	44.00
Task 4	36.54	16.76	53.30	24.37	22.33	46.70

Concerning the differences among the percentages of SRL indicators in the four tasks, a few appear particularly noteworthy:

1. the percentage of planning indicators (and to a lower extent, monitoring indicators) is much lower in the fourth task, whereas the percentage of evaluation indicators is higher. The comparison between the observed and the expected values of the planning, monitored execution and evaluation components in Task 4 shows that this difference is statistically very significant (Chi-square = 20.797, DF = 2, $p < 0.0001$, two-tailed), and the reason for this might be the fact that Task 4 was mostly individual and even if individual planning and monitoring might have taken place, the students might not have made it explicit through their messages, as already pointed out before. However, since the nature of the task assigned did not require planning but was rather focused on evaluation, trainees have probably felt necessary to make their evaluation efforts explicit;
2. the social monitoring indicators decrease over time (Chi-square = 21.8882, DF = 3, $p < 0.0001$, two-tailed), which suggests that students feel less and less the need to check on each others' behaviour as they get used to work together online, and establish trust relationships with their group mates;
3. the percentage of evaluation indicators increases (Chi square = 20.257, DF = 3, $p = 0.0002$, two-tailed). Whether this is effect of time or of task, it cannot be said, but it is probably a combination of both. In fact, it is possible that students increasingly understand, over the course duration, the need to assess their own as well as group learning, and that the peak reached in the last activity is consolidated by the fact that evaluation is explicitly solicited by the task;
4. the non-negligible presence of planning indicators, at both individual and social level,

in the final task, is mostly due to messages where trainees expressed plans to use in their future profession what they had learned in the course. The presence of these plans, in a task requiring to carry out an evaluation of the learning experience, at the conclusion of the course, recalls the cyclical model of SRL phases (Zimmerman, 1998), according to which evaluation supports new planning, and can therefore be taken as an indicator of the self-regulation of the trainees involved.

As for the distribution between cognitive/metacognitive versus motivational/emotional aspects, Table 8 shows some interesting data.

Firstly, the indicators of the cognitive/metacognitive aspect are more frequent than those of the motivational/emotional, regardless of the task. Apparently the cognitive component increases while the motivational decreases, but neither trend is statistically significant (cognitive: Chi square = 1.029, DF = 3, p = 0.7942, two-tailed; motivational: Chi square = 1.778, DF = 3, p = 0.61982, two-tailed).

Secondly, in collaborative activities, it is confirmed that indicators of individual SRL are less frequent than those of social SRL, which can be explained, as already mentioned, by the fact that the need to share the social aspects of SRL is perceived more strongly by the learners.

Finally, and somewhat surprisingly, the first and the last tasks do not feature the highest percentage of motivational/emotional indicators. As a matter of fact, one could expect these indicators to be higher where the expression of emotions and motivation is scaffolded by the metaphoric framework. Task 4, explicitly soliciting the students to express their emotions, scores the highest percentage of these indicators, but the difference is not statistically significant (Chi-square = 1.952, DF = 1, p = 0.1624, two-tailed). Task 1, on the contrary, scores even less than Task 3, where no specific scaffolds were provided for the emotional component of SRL: also in this case the distribution is not statistically different from the distribution in the whole data set (Chi-square = 0.746, DF = 1, p = 0.3876, two-tailed). This lack of effect of the metaphoric invitation to express one's motivations/emotions might be due to the novelty of the learning environments and way of working, especially since the expression of one's motivations and emotions is not really fostered in the Italian school system but, in fact, often considered an undue disturbance in classroom learning. In conclusion, the background metaphor probably helped to establish a relaxed atmosphere but was not sufficient to trigger learners' engagement on the motivational/emotional level, especially since at the beginning of the course the participants had still to learn to manifest their social presence. The situation improved over the course duration, thanks to the increased acquaintance of the learners' with each other and with the online way of working, as well as to the encouragement implicitly received from the tutors, who were the first to express appreciation for the participants' engagement and achievement.

It is interesting to note that the highest value of the motivational/emotional indicators at the social level are obtained in Task 3, in which working with the same group in the previous task had already helped establish a good level of acquaintance and reciprocal knowledge. Not surprisingly, again, motivational/emotional engagement at the individual level has its peak in Task 4, since at this point the participants felt at ease with the environment and knew most of their group mates well. We suppose that the evaluative task assigned had a positive influence in this respect, because evaluating the work done led the students to view the course as a real opportunity for professional growth and hence to appreciate it as worth the effort. This hypothesis is confirmed by the expressions of gratitude often included in the posts of this task. These data suggest that not only have motivation and emotion a positive effect on learning, but also that the cognitive pleasure deriving from becoming aware of one's achievements and learning may in turn give rise to supportive and positive emotions (Moos & Azevedo, 2008; Efklides & Volet, 2005).

Last but not least, the ratio between the social and the individual components is generally in favour of the former (Table 9) when the nature of the task is collaborative (Chi square = 26.684, DF = 1, $p < 0.0001$, two-tailed). These data should be regarded with great caution for the reasons mentioned above: the fact that individual indicators were not found does not necessarily mean that individual SRL did not take place. Nevertheless qualitative message analysis showed that in Task 4 the relatively high percentage of indicators concerning individual SRL is due to the many students who interpreted the task strictly in terms of individual self-assessment, without extending meta-reflection and evaluation to the group work. They summarized what they felt they had learnt, but did not try to assess the group achievements. On the other hand, almost 40% of the trainees extended the evaluation of the work done by including some reflections about group learning, showing an awareness of the importance of the group in the learning process they had undertaken and therefore displaying a good amount of self-regulation at the social level.

CONCLUSION

This paper tackles the issue of understanding whether, and to what extent, the nature of the tasks proposed in an online course affected the way learners practiced SRL during the activities carried out to accomplish the tasks.

Table 9. An overview of the SRL percentage of social vs. individual indicators detected in the 4 tasks

	social	individual
Task 1	80.08	19.92
Task 2	64.18	35.82
Task 3	75.43	24.57
Task 4	39.09	60.91

When interpreting the results obtained, the pros and cons of gathering data on SRL-related actions in online courses by analyzing learners' written interactions should be taken into consideration. Notably, among the advantages, there is the fact that interaction analysis is not biased by the opinions of the subjects involved in the process, while the main drawback is the fact that not all the relevant information arises from the analysis of the exchanged messages, because learners do not always explicitly communicate their actions or express all feelings and thoughts. Bearing this in mind, when the balance between individual and collaborative learning strategies is in favour of the former, other methods should be used to assess the adoption of self-regulation strategies.

Given the above results, what can be said about how to design online educational activities when fostering SRL is among the aims? Firstly, it is important to pay attention to both the phases and the components of SRL, and explicitly encourage their practice, possibly addressing them one by one. Secondly, scaffolding and fading techniques can be used, letting learners gradually take control of their learning process while tutors hand over decision-making about planning, monitored execution and evaluation to the learners. Thirdly, explicit metacognitive tasks are essential to foster SRL phases, especially evaluation.

The way different tasks influence SRL development is not straightforward: our study suggests that the way tasks are formulated and scaffolded, more than their type, determines differences in which SRL phases and components are most practiced by students. It stands to reason that the more students are given space and encouragement to choose for themselves, the more they are in the position to take control of their learning process. Similarly, when explicitly requested to carry out one type of activity, like metacognitive reflection, they usually do it.

The role of the tutor appears to be essential in handling the learning process and the development of SRL. Given that mere knowledge acquisition is not sufficient to live and work in the knowledge society, SRL should be among the aims of most training actions and the tutors should be aware of this important objective, therefore paying attention not only to the contents to be learned, but also to the ways the learning process can be controlled from both the cognitive and the emotional points of view.

REFERENCES

Azevedo, R. (2005). Computers as Metacognitive Tools for Enhancing Learning. *Educational Psychologist*, *40*(4), 193–197. doi:10.1207/s15326985ep4004_1

Banyard, P., Underwood, J., & Twiner, A. (2006). Do Enhanced Communication Technologies Inhibit or Facilitate Self-regulated Learning? *European Journal of Education*, *41*(3/4), 473–489. doi:10.1111/j.1465-3435.2006.00277.x

Boekaerts, M. (1997). Self-regulated learning: a new concept embraced by researchers, policy makers, educators, teachers and students. *Learning and Instruction*, *7*(2), 161–186. doi:10.1016/S0959-4752(96)00015-1

Boekaerts, M. (1999). Self-regulated learning: where we are today. *International Journal of Educational Research*, *31*, 445–457. doi:10.1016/S0883-0355(99)00014-2

Bryman, A. (2004). *Social Research Methods*. Oxford, UK: Oxford University Press.

Calvani, A., Fini, A., Molino, M., & Ranieri, M. (2010). Visualizing and monitoring effective interactions in online collaborative groups. *British Journal of Educational Technology*, *41*(2), 213–226. doi:10.1111/j.1467-8535.2008.00911.x

Carneiro, R., Steffens, K., & Underwood, J. (Eds.). (2005). Self-regulated Learning in Technology Enhanced Learning Environments, *Proceedings of the TACONET Conference*. Shaker Verlag, Aachen.

Cresswell, J. W. (2003). *Research Design, Qualitative, Quantitative and Mixed Method Approaches*. London, UK: Sage Publications.

Dabbagh, N., & Kitsantas, A. (2004). Supporting self-regulation in student-centered web-based learning environments. *International Journal on E-Learning*, *2*(4), 40–47.

Delfino, M., & Manca, S. (2007). The expression of social presence through the use of figurative language in a web-based learning environment. *Computers in Human Behavior*, *23*(5), 2190–2211. doi:10.1016/j.chb.2006.03.001

Dettori, G., Giannetti, T., & Persico, D. (2006). SRL in online cooperative learning: implications for pre-service teacher training. *European Journal of Education*, *41*(3/4), 397–414. doi:10.1111/j.1465-3435.2006.00273.x

Dettori, G., & Persico, D. (2008). Detecting Self-Regulated Learning in Online Communities by Means of Interaction Analysis. *IEEE Transactions on Learning Technologies*, *1*(1), 11–19. doi:10.1109/TLT.2008.7

Dodge, B. (2009). A WebQuest About WebQuests, Middle School / High School Version. Ed Tech Department, San Diego State University. URL: http://webquest.sdsu.edu/webquestwebquest.html (last retrieved March, 12th, 2010).

Efklides, A., & Volet, S. (Eds.). (2005). Special issue on Feelings and Emotions in the Learning Process. *Learning and Instruction*, *15*(5), 377–516. doi:10.1016/j.learninstruc.2005.07.006

Garrison, D. R. (2007). Online community of inquiry review: Social, cognitive and teaching presence issues. *Journal of Asynchronous Learning Networks*, *11*(1), 61–72.

Garrison, D. R., Anderson, T., & Archer, W. (1999). Critical inquiry in a text-based environment: computer conferencing in higher education. *The Internet and Higher Education*, *2*(2-3), 87–105. doi:10.1016/S1096-7516(00)00016-6

Gee, J. P. (2005). *An Introduction to Discourse Analysis: Theory and Method*. London, UK: Routledge.

Gunawardena, C. N., Lowe, C. A., & Anderson, T. (1997). Analysis of a global online debate and the development of an interaction analysis model for examining social construction of knowledge in computer conferencing. *Journal of Educational Computing Research*, *17*(4), 397–431. doi:10.2190/7MQV-X9UJ-C7Q3-NRAG

Henri, F. (1992). Computer conferencing and content analysis. In Kaye, A. R. (Ed.), *Collaborative learning through computer conferencing* (pp. 117–136). Berlin, DE: Springer-Verlag.

Hofer, B., Yu, S. L., & Pintrich, P. (1998). Teaching college students to be self-regulated learners. In Schunk, D. H., & Zimmerman, B. J. (Eds.), *Self-regulated Learning. From Teaching to Self-reflective Practice* (pp. 57–85). New York, NY: The Guildford Press.

Jamieson-Noel, D., & Winne, P. H. (2003). Comparing Self-Reports to Traces of Studying Behaviour as Representations of Students' Studying and Achievement. *Zeitschrift fur Padagogische Psychologie*, *17*(3-4), 159–171. doi:10.1024//1010-0652.17.34.159

Koschmann, T. (Ed.). (1996). *CSCL: Theory and practice of an emerging paradigm*. Mahwah, NJ: Lawrence Erlbaum.

Koschmann, T., Hall, R., & Miyake, N. (Eds.). (2002). *CSCL 2: Carrying forward the conversation*. Mahwah, NJ: Lawrence Erlbaum.

Lowerison, G., Sclater, J., Schmid, R., & Abrami, P. C. (2006). Student perceived effectiveness of computer technology use in postsecondary classrooms. *Computers & Education*, *47*(4), 465–489. doi:10.1016/j.compedu.2004.10.014

Moos, D. C., & Azevedo, R. (2008). Exploring the fluctuation of motivation and use of self-regulatory processes during learning with hypermedia. *Instructional Science*, *36*(3), 203–231. doi:10.1007/s11251-007-9028-3

Newman, D. R., Webb, B., & Cochrane, C. (1995). A content analysis method to measure critical thinking in face-to-face and computer supported group learning. *Interpersonal Computing and Technology*, *3*(2), 56–77.

Pintrich, P. R. (2000). The role of goal orientation in Self-regulated learning. In Boekaerts, M., Pintrich, P. R., & Zeidner, M. (Eds.), *Handbook of self-regulation* (pp. 451–502). San Diego, CA: Academic Press. doi:10.1016/B978-012109890-2/50043-3

Salovaara, H. (2005). An exploration of students' strategy use in inquiry-based computer-supported collaborative learning. *Journal of Computer Assisted Learning*, *21*(1), 39–52. doi:10.1111/j.1365-2729.2005.00112.x

Steffens, K. (2006). Self-Regulated Learning in Technology-Enhanced Learning Environments: lessons of a European peer review. *European Journal of Education*, *41*(3/4), 353–380. doi:10.1111/j.1465-3435.2006.00271.x

Torrano, F., & Gonzales, M. C. (2004). Self-regulated Learning: Current and Future Directions. *Electronic Journal of Research in Educational Psychology*, *2*(1), 1–34.

Van den Boom, G., Paas, F., Van Merrienboer, J. J. G., & Van Gog, T. (2004). Reflection prompts and tutor feedback in a web-based learning environment: effects on students' self-regulated learning competence. *Computers in Human Behavior*, *20*(4), 551–567. doi:10.1016/j.chb.2003.10.001

Zimmerman, B. J. (1998). Developing Self-fulfilling cycles of academic regulation: an analysis of exemplary instructional models. In Schunk, D. H., & Zimmerman, B. J. (Eds.), *Self-regulated learning. From teaching to Self-reflective practice* (pp. 1–19). New York, NY: The Guildford Press.

Zimmerman, B. J. (2000). Attaining self-regulation: a social cognitive perspective. In Boekaerts, M., Pintrich, P., & Zeidner, M. (Eds.), *Handbook of self-regulation* (pp. 13–39). New York, NY: Academic Press. doi:10.1016/B978-012109890-2/50031-7

KEY TERMS AND DEFINITIONS

Computer-Mediated Communication: Communication process between humans through ICT.

Computer-Supported Collaborative Learning: Field of study aimed at understanding how people learn together through the use of computers.

Educational Technology: Theory and practice of systematic design of learning processes and resources.

Interaction Analysis: Research methodology that relies on discourse analysis and consists in detecting phrases and expressions that reveal aspects of interest in the communication flow.

Self-Regulated Learning: Learning process controlled by the learner from the cognitive, metacognitive, emotional and motivational points of view.

Teacher Training: Process aimed at making teachers more competent for their work.

Chapter 10
Theoretical and Practical Issues in Designing a Blended e-Learning Course of English as a Foreign Language

Rita Calabrese
University of Salerno, Italy

Filomena Faiella
University of Salerno, Italy

ABSTRACT

The aim of this chapter is to provide an outline of the main theoretical issues in the field of Self-Regulated Learning which have inspired the design and implementation of a blended learning course of English as a Foreign Language (EFL) at the University of Salerno. In particular, the first part of the chapter focuses on some key concepts concerning meaningful learning, self-regulated learning, as well as e-learning in academic settings, as basic components to achieve cognitive academic language proficiency (CALP). The second part of the chapter is devoted to the description of the sequencing and progression of our syllabus design in line with the principles/guidelines for "good teaching practices for using Technology Mediated Instruction (TMI)".

INTRODUCTION

The only real voyage of discovery consists not in seeking new landscapes but in having new eyes.
- Marcel Proust

DOI: 10.4018/978-1-61692-901-5.ch010

Online education, either delivered as part of blended educational models (part online, part face-to-face) or as full distance learning, has become increasingly widespread in different learning domains including academic contexts (Barone & Calabrese, 2005). In this chapter, we will report on an educational experience concerning an EFL (English as a Foreign Language) course that is intended as a contribution to investigating the role of self-regulated learning in a computer-assisted language learning context. We will first provide a brief review of studies which have investigated the cognitive factors involved in the acquisition of a foreign language in relation to the development of meaningful learning and self-regulated learning, in order to outline the theoretical framework underpinning the whole paper. We will then deal with the overall features of a blended course, describing in more detail the online EFL component. This experience is part of a broader e-learning program designed by a team of technologists of the *eLearning_Lab* of the Faculty of Education, University of Salerno (www.eformazione.unisa.it) and delivered by a group of professors from various academic areas (Italian Literature, Art and Design, Music, Philosophy and so on).

THEORETICAL ISSUES

Aspects of Adult Second Language Acquisition

Among the vast amount of studies on first and second language acquisition, two important paradigms have proved to be particularly influential within Second Language Acquisition (SLA) theory. The first is based on cognitive theories derived from psychology and linguistics, while the second is influenced by socio-cultural theories. The cognitive paradigm mainly focuses on the mental processes occurring during language development and acquisition, whereas socio-cultural theorists see language acquisition as contextualized within a social and a cultural contexts. Both theoretical positions can help examine the process of learning and teaching online (Lamy & Hampel, 2007, p. 19) and provide useful hints for distance education designers.

SLA theory gained great impetus from Krashen's (1985) theoretical assumptions built around the central idea of "comprehensible input" for the development of a second language. Thus, the major function of the SL classroom is to provide learners with input for acquisition by setting up meaningful and communicative activities. As a matter of fact, the development of pragmatic competence can actually be achieved through exposure to real language in particular contexts of use. The communicative issue in SLA leads to a further aspect of SLA theory which is characterized by the so called "social turn" influenced by the rediscovery of Vygotsky's constructivist view of learning (1978). In language research, Vygotsky's concept of the "zone of proximal development" proved to be particularly influential in the domain of L2 teaching and learning and gave rise to important tenets in the field of Computer Mediated Communication (CMC) as well: the idea of scaffolding (Faiella, 2005) as the educator/instructor's supporting action that is adapted to the learner's needs (Faiella, 2005), and the more recent concepts of "collaborative dialogue" and "instructional conversation". As a matter of fact, the combination of the input-interaction-output model and the "social turn" view has produced an integrated model that can be applied to both face-to-face communication and virtual interaction (Lamy, Hampel, 2007, p. 20); the only difference is that the latter is accomplished through what appears on the screen and other technological devices, e.g., mouse and keyboard (Clarke, 2008, p. 14).

In order to understand Second Language Acquisition (SLA) processes in instructed conditions within CMC environments, it is important to determine whether SLA processes in adult learners are essentially the same as or different from those involved in child first language acquisition and, if different, how so (Doughty, 2003, p. 275). Given the evident differences in outcomes, a logical inference is that child language acquisition and adult SLA involve different types of processing for language learning which give rise to three different theoretical positions. The Fundamental Difference Hypothesis (Bley-Vroman, 1990) proposes that child language learning is automatic and implicit whereas adult SLA is characterized by explicit and general problem-solving strategies. The Competition Hypothesis (Felix, 1985) claims that implicit Universal Grammar (UG, i.e., language principles underlying every language) and explicit problem solving processes initially compete in adult SLA, with the latter eventually dominating the former. Finally, recent studies on adult SLA which have been carried out adopting the hemodynamic method of FMRI (Functional Magnetic Resonance Imaging) suggest that different brain regions subserve language processing in L1 and L2 during the exposure to language input (Sorace, 2005, p. 73).

The common explanation for these child-adult differences is that there are maturational constraints on language acquisition (Doughty, 2003, p. 275). Children in primary language acquisition face the difficulty of processing and "shaping" the structure of their native language by relying upon the language input they hear as the only cues for segmentation. When acquiring a second language, adult learners generally apply their native-language processing strategies to L2 structure by focusing on specific elements of language which belong to L1 rather than to L2, hence the necessity of exposing L2 learners to continuous audio-visual input.

The major difficulty for adult L2 learners is that L2 declarative or explicit knowledge cannot be matched to the needs of overall processing mechanisms. Nonetheless, L2 formal instruction can help learners in organizing the processing space by enhancing mechanisms that depend upon perceptual acuity. One of the main goals of L2 instruction should therefore be the systematization of learners' processing space in order to enable them to notice the cues located in the input during implicit learning rather than to promote meta-linguistic reflection in the first stages of language exposure.

An important recent claim is that implicit processing is more powerful than explicit thinking for learning complex systems involving multiple tasks (Sorace, 2005, p. 296), implying that instructed SLA processing should enhance implicit learning through appropriate activities and only gradually provide explicit knowledge. Findings from a series of experiments have indicated that contextualized grammar knowledge was acquired better implicitly from exposure to instances of language than by trying to explicitly induce grammar rules (Sorace, 2005, p. 294). Thus, where complex knowledge is learned in context, implicit learning is more successful. It is therefore important to determine how complex learning is processed in particular conditions delivering varied types of information such as repeating either visual (images, video) or auditory (mp3 recordings, podcast) language patterns combined with performative tasks such as writing, completing, filling in, etc..

However, findings show that the main feature of implicit learning is fragmentary knowledge, i.e., subjects are not able to recombine fragments from the input in order to verbalize the underlying rules and consequently "declarative knowledge is a by-product of practice during implicit learning" (Sorace, 2005, p. 295). Accordingly, it appears reasonable that SLA necessarily involves more than one mode of processing, that is, explicit learning takes place alongside implicit learning and such assumption informs the design of effective applications in instructed SLA.

Individual Differences in SLA

Research in the fields of psychology and applied linguistics has highlighted the relationship between maturational constraints and cognitive/learning styles underlying the overall motivational processes of self-regulated learning. It is therefore necessary to review some of the key concepts involved in this area.

Cognitive and learning styles: The study of cognitive and learning styles, borrowed from the discipline of psychology, has greatly influenced SLA researchers. First of all, the issue of a certain predisposition to deal with learning situations or to process information has proved to be very fruitful in SLA research in terms of language proficiency and attainment. However, it is necessary to draw a distinction between learning styles and cognitive styles that is sometimes unclear in the literature. "Learning style is a gestalt combining internal and external operations derived from the individual's neurobiology, personality and development and reflected in learner behaviour" (Keefe & Ferrell, 1990, p. 56). According to this definition, learning style expresses a typical preference for a way to approach learning in general. Cognitive style, on the other hand, refers to information-processing preferences.

It is also possible to go beyond the cognitive domain and apply the concept of style to other fields to include areas such as sensory preference and personality. Regarding the sensory domain, Reid (1995) singled out auditory, visual, kinesthetic and tactile preferences. In relation to personality, Oxford and Anderson (1995) took a wider perspective including six interrelated aspects of learning styles: cognitive (concerning preferred patterns of mental processing), executive (concerning the ability to manage his/her own learning processes), affective (concerning attitudes that influence an individual in a specific learning situation), social (concerning the extent of involvement with other people while learning), physiological (concerning the learner's sensory tendencies) and behavioural (concerning learner's attempt to satisfy his/her learning preferences). Thus, the term "learning style" is used in the literature to cover a wide range of learning patterns and orientations at various psychological and behavioural levels. In this respect, learning styles are closely related to learning strategies in that learning style refers to a cross-situational use of a class of learning strategies.

Self-regulation: The concept of language learning strategy reflects the learner's active contribution to enhance the effectiveness of his/her own learning, which plays an important role in L2 acquisition. In this respect, it is closely related to the less ambiguous term of "*self-regulatory learning*" which was adopted by researchers focusing on the essence of strategic learning (Dörnyei & Skehan, 2003, p. 611). The notion of "self-regulation of academic learning refers to the degree to which individuals are active participants in their own learning: it is a more dynamic concept rather than learning strategy. The self-regulated learner can be portrayed as applying a set of varied skills during studying activities" (Winne, 1995, p. 173), he/she is aware of his/her motivation and what he/she knows and what the differences between these kinds of information imply for approaching a task. Using the new paradigm, researchers have attempted to combine learner-initiated cognitive, meta-cognitive, and motivational processes and strategies. From a self-regulatory point of view, language learners can improve the effectiveness of their learning not only by applying creative operations that suit their learning styles, but also by increasing motivation to learn. In this view, self-regulation and motivation are bound together to enhance learner achievement.

Motivation: This concerns the direction and measure of human behaviour, i.e., the choice of a particular action and the effort to pursue it. There is, of course, a broad range of reasons that can influence human actions and in particular, learners to study a foreign language. In order to better understand the intricate motivational network of a virtual classroom, it is necessary to adopt a comprehensive model which covers a wide range of academic and social motives. Dörnyei (2000) argues that such a model can help explain the relationship between many factors such as general reasons concerning L2-related attitudes, learner-specific motives, reasons related to the micro-context of the language classroom, the teacher's motivational influence, the motivational characteristics of the curriculum and the teaching materials, the learner's self-regulatory activity, the role of time since motivation is a dynamic process rather than a state. In fact, from a process-oriented perspective, it is important to take into account different phases of motivation: choice motivation related to the foreign language chosen as a study subject, execution motivation concerning activities to achieve the learning goals, motivational retrospection in which learners analyse and evaluate their actional processes.

Language Learning in Virtual Environments

Some researchers (Lantolf & Appel, 1994; Lantolf, 2000) have applied Vygotsky's ideas to second language learning, interpreting the process of language acquisition as an overall mediated process occurring in three different domains of mediation:

1. Social mediation or social interaction
2. Self-mediation or private speech
3. Artefact mediation operated by language, tasks and technology (Lamy & Hampel, 2007, p. 26).

Human learning is mediated through interaction with others, using language and other mediational tools, with new technologies having developed new modes of communication that try to reproduce some characteristics of face-to-face communication.

New media offer a wide range of ways of communicating (including spoken and written language, images, video) and this means that learners and teachers cannot simply replicate modes of face-to-face communication, rather they acquire the main ability to cope with these means once, and accordingly, they self-regulate their learning/teaching processes. It is therefore required that teachers and course designers are aware of the affordances computer tools offer by taking into account how learners will use them to construct meaning. As a matter of fact, some researchers (New London Group, 1996; Kress, 2003) talk about the development of "multiliteracies" including the skills of using hardware and software as well as making meaning from the hypertextual combination of multiple linguistic texts, audio, video and symbolic graphics codes such as "smileys" (e.g., ☺, ☹) and other graphic representations. In written CMC language learning the "multimedia character of new electronic media facilitates reading and writing processes that are more democratic learner-centred, holistic and natural" (Warschauer, 1999, p. 11).

However, according to some researchers, environments based on written communication can produce both facilitative and inhibitory effects in terms of learner experience. In fact, anxiety is generally provoked by the "mismatch between foreign language learners' mature thoughts and their immature foreign or second language proficiency which leads to frustration" (Gregersen & Horowitz, 2002, p. 562). On the other hand, written environments can also provide scaffolding by reproducing features of oral language and therefore train students for face-to-face communication. In particular, in face-to-face communication participants can see facial expressions, hear the tone of voice and listen to the words used. In e-learning the only use of written communication can make it more difficult to convey precise meaning and sometimes this fact may lead to misunderstandings. The use of emoticons, a code based on punctuation and other symbols, can help convey participants' intentions and emotions. There is of course considerable variation in the way to combine the elements of traditional and e-learning as well as the learning skills associated with them (Clarke, 2008, p. 3). For example, websites may have been especially designed as part of an education program or for general purposes consequently requiring different search or study skills. Table 1 shows a comparison of traditional and e-learning skills as well as sub- or related skills combined with more specific language skill.

Table 1. Comparing traditional and e-learning skills in SLA

Skill	Traditional	e-Learning
Reading	skimming/scanning	skimming/scanning/browsing are particularly important to locate relevant websites using the Worl Wide Web
Writing	summarising the key points of a text; reporting information; note taking and exercises	referencing information; communicating (e.g. e-mails); keeping records; note taking and exercises
Listening	Understanding the key points of a conversation between two or more speakers or a recorded message	Understanding the key points of a conversation between two or more speakers or a recorded message
Interaction	Ability to understand and respond to a native or non-native speaker/interlocutor in written/oral communications	Ability to understand and respond to a native or non-native speaker/interlocutor in written communications
Searching	Ability to search physical libraries by reviewing the contents page, looking up key words, checking the publication date and the author	Ability to search the world wide web, to analyze information, to assess content and compare alternative sources
Planning	Determined by the teacher along with timetables and study guides	In part determined by the teacher and in part by the learner
Time management	Critical to control the teaching/learning process	Critical to control the teaching/learning process
Self-assessment	Steady	Steady
Creating content	Associated with arts and crafts	Enhanced and facilitated by the availability of technological equipment

Meaningful Learning, Autonomy and Self-Regulated Learning: An Overview

The design of a Virtual Learning Environment (VLE), i.e. the choice of contents as well as the organization of materials and tasks of a learning course, generally relies on meaningful learning and autonomy as main theoretical foundations. These concepts, integrated into the multidimensional construct of Self-Regulated Learning, may contribute to constructing educational environments that are inspired by the constructivist paradigm as well as sensitive to the cognitive aspects of academic learning.

Ausubel's notion (1968) of meaningful learning implies that the learner is in a certain disposition to link new information (concepts and propositions) with existing concepts in his/her cognitive structure and that new material to be learned is potentially significant to him/her.

As theorized by Ausubel, the idea of meaningful learning emphasizes the active role of the learner who relates "new subsuming concepts and principles to be learned" in a "non-arbitrary and substantive" fashion to what he/she already knows, to the learner's existing knowledge, to relevant anchoring concepts already available in cognitive structure, through a major transaction of interaction between new knowledge and his/her cognitive structure. The result of this process is acquisition, retention/reduction and retrieval of meanings of new symbolic expressions. Ausubel assumes that the attribution of meanings is a complex individual cognitive experience which is influenced and determined by many factors that closely relate to the learner, to the instructional design as well as teaching techniques.

The concept of meaningful learning has been developed within the research field that investigates the integration of technology and instruction. Several researchers in this field (Ashburn & Floden, 2006; Jonassen, Howland, Marra & Crismond, 2007) have defined meaningful learning as "deep and enduring understanding of complex idea" and argued that meaningful learning occurs when individuals are engaged in social activities, i.e., in projects and tasks based on the interactions with each other in natural and complex learning environments, using authentic and contextualized tools that promote construction, collaboration and reflection on what learners are studying.

Jonassen (1994) points out that computers, as cognitive learning tools rather than "conveyors of information", afford the most meaningful thinking because they force students to reflect on their knowledge in a new and meaningful way.

In general, studies on Technology Mediated Instruction within the social-constructivist paradigm invite us to be aware of the role of the learner, letting him/her be actively involved in the learning process, since he/she "constructs" conceptual systems in his/her mind (Piaget, 1971). The learner acquires new knowledge in interacting with others, through contextualized practices and activities which have to be rooted in specific historical, cultural and social settings which in turn determine the conditions as well as the purposes, goals, means and tools of such activities. In fact, the social-constructivist paradigm states that learning is neither a transmissive nor submissive process, rather it views learning as a complex process through which the learner assigns meaning to things in the real world. "Meaning exists neither in us, nor in the world, but in the dynamic relation of living in the world" (Wenger, 1998, p. 54), it is itself produced by an active and productive process of negotiation, in terms of controversy, interpretation, amendment, or confirmation. The negotiation of meaning, then, allows its articulation, expression and representation through factors such as conversation, reflective practice, language, relation, comparison, non-verbal interaction. Knowledge, learning and culture therefore derive from the negotiation of meaning.

The social-constructivist perspective suggests a mediated instructional model based on scaffolding instruction that fosters meaning-making. An important inference of constructivism is that students, as "active learners", play a central role in controlling learning, and must accept greater responsibility for generating learning and constructing their own understanding. This does not mean that constructivism advocates the abolition of teacher control, rather a radical change is needed in teaching/learning management: teachers create a learner-centred environment for learners and assume the role of facilitators who help learners to build understanding and support their active participation in their own learning process. A self-regulated learning perspective widens this idea by pointing out that learners' involvement should encompass the meta-cognitive, motivational and behavioural aspects; (Zimmerman, 1990). Its main objective is the development of awareness of cognitive processes and control over learning.

Autonomy has been defined as the learner's ability to take charge of (Holec, 1981), or take responsibility for (Little, 1991), or control over (Benson, 2001) his/her own learning. From this point of view, the concept of autonomy refers to an ability that can be acquired, enhanced and developed in individuals mainly through instruction and learning experiences involving students in the processing of their knowledge.

The concept of autonomy in the field of Second Language Acquisition has been used to indicate the learner who studies alone, or the right of the learner to determine his/her own learning goals. Autonomy is not self-instruction and does not imply learning in isolation without a teacher, rather it implies "a holistic view of the learner that requires us to engage with the cognitive, meta-cognitive, affective and social dimensions of language learning and to worry about how they interact with one another" (Little, 2003). Autonomy can manifest itself in different ways and "take various forms for different individuals, and even for the same individual in different contexts or at different times" (Benson, 2001, p. 37).

The ability of a learner to understand and control his/her own learning processes is defined as Self-Regulated Learning (SRL). Self-regulating students "set goals for their learning and then attempt to monitor, regulate, and control their cognition, motivation, and behaviour, guided and constrained by their goals and the contextual features in the environment" (Pintrich, 2000, p. 453). Self-regulation is generally accepted as an important construct in student success within environments as online courses that enhance learner choices.

It is worth noting that the point of view of self-regulated learning is extremely important in order to examine how a blended learning course can help students become more autonomous in their meaningful learning of English as a Foreign Language.

In order to increase the understanding of the relationship between e-learning and motivational processes, it is necessary to gain a greater understanding of the learning materials that are developed to increase motivation.

Students need to understand the current state of their knowledge and build on it, improve it, and make decisions.

The use of e-learning offers an added value to face-to-face training (Vento, D'Esposito, Faiella, 2008), especially in relation to the possibility for students to study according to the variables of time, ways and places appropriate to their learning styles. The need to set up an online environment that is "learner-oriented" and designed to increase the potential of asynchronous and synchronous communication tools, as well as improve the effectiveness of guidance and counselling training during lessons, should be emphasized.

APPLICATIONS

Designing an e-Learning Course for Language Learning

In the first part of the chapter we have focused on the developmental and individual differences that can influence the individual regulation efforts within virtual learning environments. In the second part, we will relate the cognitive and motivational processes underlying SRL to the constituent parts of the EFL e-learning course.

Virtual Learning Environment

The EFL blended learning course was carried out in the experimental set up of a teaching laboratory called *eLearning_Lab* (www.eformazione.unisa.it). The main aim of the course was to respond to the increasing educational and training demands of the students as well as creating more flexible study paths based on innovative teaching and learning methods.

The EFL course was designed in a blended format (Ligorio, Cacciamani & Cesareni, 2006) since this has proved to be particularly effective in terms of flexibility and individualization of the educational proposals.

"Resource-based learning offers learners the opportunity to exercise control over learning plans, the selection of learning materials and the evaluation of learning" (Benson, 2001, p. 113).

The online English component of the course was delivered through the Moodle platform as a parallel instructional pathway complementing the traditional face-to-face learning environment of classroom lessons by means of learning materials and resources available 24 hours a day. Moodle is an Open Source learning management system developed to help teachers to create effective online courses. It manages the teaching activity through specific tools devised to prepare lessons in text format and hypertext links to web pages, write glossaries, create multiple-choice test items, true-false tests, short essays and evaluative tests. It allows the development of learning environments in which communication can be achieved through different services integrated into the system. The integrated platform, when properly set, meets the principles of constructivism based on the dialogue and negotiation of meaning, social relationships and active participation to meaningful practices. Therefore, it can support an active approach to the process of knowledge building and stimulate the natural abilities to learn through the creation of genuine virtual learning environments in which learners can actively cooperate in a mutual support (scaffolding). Moodle also leads students to focus on the processes of (self-) learning and (self-) evaluation without distracting with the complexity of use, and encourages communication and interaction among students, teachers, web-administrators, through various synchronous and asynchronous communication tools. Learners, in fact, should not feel alone with resources and materials, at the mercy of technical difficulties. The centrality of the student does not exclude a fundamental role for the teacher who is there not only to ensure the quality of content but also as a point of reference for each student.

The learning environment set up on Moodle for our course provides a number of tools, ideas and alternatives which have been carefully evaluated and pedagogically calibrated to guarantee a constant monitoring of the actions undertaken by participants and accordingly to improve, correct and adapt the educational intervention to the training needs of the students.

In order to enhance goal orientation, the student can find a brief description of the course syllabus on the right side of the interface with specifications concerning progression and sequence of the learning experiences, course objectives, expected outcomes and topics to be covered. The central part of the user interface is divided into ten sections and each of them includes a general overview of the language contents, homework, reading assignments and recommended readings. There is no restriction on the sequencing from the first to the tenth module: each student can plan the selection of materials and activities which are relevant to his/her personal needs and goals. The only restriction is placed on the self-assessment test each student has to do as a self-reflection operation on his/her prior language knowledge and competence.

Research Framework

Earlier on, we have described the theoretical framework within which the blended e-learning model we adopted in designing the EFL course can be placed. In this section, we will deal with the application of SLA theories to the practice of an online EFL course by presenting the main features which have informed its planning and design.

Vygotsky's "scaffolded" model of instruction seemed to provide a valuable and appropriate basis upon which we could build the EFL course.

Following Skehan's (1998, p. 132) indicators for task-based learning in SLA which are worth considering in a Computer Assisted Language Learning (CALL) system (Chapelle, 2001, p. 46), we have selected five guidelines concerning cognitive conditions that can influence the linguistic content of a course along with the more general criteria adopted in designing online courses. By matching the five preliminary conditions for SLA task-based instruction with the stages of the Instructional System Design (ISD), we obtained the overall architectural structure of the language course. Figure 1 gives an overview of all these elements combined together (arrows indicate correspondences between the two models).

On a practical level, we adopted a blended approach to cope with both the linguistic demands of face-to-face communication and the constructivist principles of language learning. In particular, the learning activities carried out during self-study sessions were adopted as warm-up or grammar reflection stages before the in-class face-to-face sessions.

Figure 1. Overview of task-based instruction indicators and stages of ISD

Cognitive conditions for SLA	Stages of the Instructional System Design
1. Selection of a range of target structures following the sequence of natural development	I. Analysis: students' pre-requisites and future professional profile, expected outcomes, affordances (technological tools, web resources, degree of students' involvement in performing online tasks)
2. Choice of tasks which meet utility conditions, i.e. degree of likelihood that a particular structure will be used by learners as they perform a task	II. Design: in terms of Bloom' s taxonomy of educational objectives; learning modules, learning objects, virtual visits, assessment tests
3. Sequencing of tasks to achieve balanced goal development according to the three dimensions of language performance, i.e. fluency, accuracy and complexity (Chapelle, 2001, p. 46)	III. Development: of text files or web pages
4. Maximizing chances of focus on form while performing meaningful tasks essentially based on communication tasks	IV. Implementation: in the Moodle integrated platform
5. Use of accountability cycles to let learners become responsible of their own learning, aware of the language they are acquiring	V. Evaluation/assessment: formative and summative tests

In the following pages, we will illustrate how this model was implemented within the University context aiming at:

1. Enabling learners to achieve the pre-requisites required by the language course objectives (elementary pathway).
2. Increasing/supporting learners during self-study sessions at home or university self-access centre (pre-intermediate pathway).
3. Empowering learners' both "traditional" study and e-learning skills (autonomy in searching the web, consulting web resources specifically designed to improve their foreign language knowledge and increase motivation to learn).

Course Format

Participants: The EFL e-learning course was delivered at the Faculty of Education (University of Salerno) during the autumn term 2008. The participants were 83 students aged 21-22 attending a regular face-to-face course of English for Primary School teachers, but only 50 of them actually attended both the regular and the online course.

Method: The first step in designing the scaffolded structure of the course was to choose the A2 and B1 levels from the Common European Framework (CEF) of Reference (Council of Europe, 2001) as respectively the starting and expected levels of language proficiency. Students were therefore provided with *e-documentation* for self-assessment and placement test, a detailed description of the course syllabus, a list of the planned activities and guidelines for the final tests and assessment criteria in order to assist students in achieving course objectives. For this reason, they were first given an online version of a self-assessment test (http://www.tolearnenglish.com/test-de-niveau-anglais-grammaire.php) in order to determine the learner's starting level of language competence along with the table of CEF descriptors for the levels involved (A2, B1/B2).

Tools and Materials: The description of the course syllabus also included a wide range of study resources (Online Dictionary with pronunciation (http://www.thefreedictionary.com/), Bilingual Dictionary (http://www.wordreference.com/enit/word), BBC Learning English with video recordings devoted to pronunciation (http://www.bbc.co.uk/worldservice/learningenglish/grammar/pron/sounds.shtml) and learning materials carefully selected from a number of accredited websites devoted to EFL acquisition (Oxford University Press Study Links (http://www.oup.com/elt/global/products/englishfile/englishfile2/c_games/), Australian Network Living English Stories (http://australianetwork.com/livingenglish/stories/le_ep01.htm). The availability of a variety of resources has to be seen as the pedagogically/theoretically motivated necessity to intersperse instructional methodologies using different learning styles: logical/deductive with text-based material, verbal-visual with audio-video materials and visual-kinesthetic with interactive exercises.

Syllabus: Then, two interrelated learning pathways were designed, an elementary pathway and a pre-intermediate pathway. The term pathway generally refers to a collection of recommended materials selected and organized on the basis of levels of language competence, learning needs and styles. Therefore, the concept of learning pathway can be seen as a personalized educational plan characterized by built-in choice of contents, self-assessment techniques, flexibility in terms of time.

The basic characteristics of the two learning pathways were the same. Both pathways were planned as 9-week courses and structured around "Thematic Modules" focusing on everyday speech and communicative topics (*Introducing yourself and people you know; Personality and Physical Appearance; Leisure and Free Time, etc.)* "conveyed" and introduced by specific episodes selected from a TV movie series. Each module was subdivided into "grammar units" composed of grammar notes with written exercises, listening activities such as "video-dictation", text completion and note taking while watching the selected video. It is worth noting that the pre-intermediate course was intended as complementary to the in-class lessons, whereas the elementary one was conceived as "preparatory" to achieve the expected language objectives of the course corresponding to the pre-intermediate level indicators.

Activities: The related activities (video-dictation, listening exercises, revision tasks) were placed in meaningful and real contexts in order to expose students to the different communicative functions of language without explicitly providing grammar knowledge. The primary concern was to choose materials and learning tasks which could meet the expectations of the group of students involved in the study and their potential professional profile as future primary school teachers. Assuming that exposure to real contexts of language use enhances comprehensible input and implicit thinking for learning, students were exposed to a given communicative situation performed by the characters in the video. This activity could then respond to both the "criterion of authenticity" (with the correspondence of language learning and real-world task) and "learner fit" taking into account learner's characteristics and learning styles which increase the effectiveness of Computer Assisted Language Learning tasks (Chapelle, 2001, p. 8). Each of the 42 episodes was provided with the transcript on the right of the screen (Figure 2) and a grammar focus area containing explanations and examples taken from the video. The language material could also be used to make brief assignments proposed at the end of the section. Finally, a complete list of the episodes was also provided allowing fast and easy browsing to search for a specific topic.

During the face-to-face session in the classroom, warming-up activities started with a review of the communicative patterns found in the episodes through learner-learner or learner-teacher interactions. This activity had two main aims: 1. to give students the opportunity to negotiate meaning, 2. to draw students' attention to specific syntactic patterns in certain contexts of use.

Assessment and Evaluation - At the beginning of the course students were provided with a downloadable portfolio (Figure 3) in the form of

Figure 2. Overall structure of TV movies

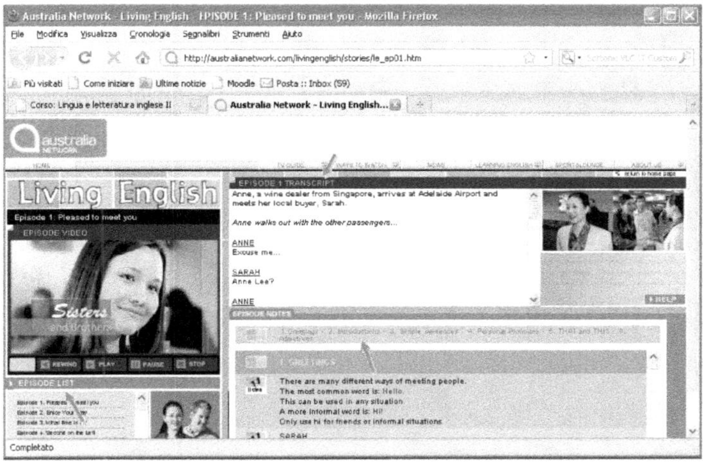

a diary to be completed during self-study sessions where students could record important information such as date, type of activity, topic, grammar areas involved, and achieved outcomes with comments on their learning experience, study skills implied, drawbacks, backwash.

Students were asked to carefully complete this kind of documentation since it was used as the starting point of the oral exam.

Summative assessment took place in the classroom in both written and oral form as this still remains the central focus of the national examination system. Given the characteristics of the course informed to a constant (self)-monitoring of students' progression, we assumed as our primary aim the standards recently fixed in relation to foreign language assessment:

1. **Monitoring** to ensure students' progress.
2. **Assessment** including formal and informal assessment, ongoing evaluation, target-setting, regular self-assessment.
3. **Recording** marks from tests and formal assessments.
4. **Translating** assessment results into frequent descriptive feedback for students, providing them with specific insights as to how to improve.

Translating, in particular, proved to be crucial, since it related the learning experience of the self-study sessions carried out in the virtual environment to the educational reality of the university classroom.

DISCUSSION

A careful evaluation of the students' self-assessment test and portfolio gave evidence of the expected outcomes for the whole group attending the course (50 out of 83 students successfully passed the English exam set out according to the Preliminary English Test format). In other words, the high percentage of students (60,2% out of the total number of students attending the face-to-face course) who provided evidence of having self-organized and self-evaluated their own learning activity, besides passing the exam, shows that they had effectively set up the necessary actions to become self-regulated learners (Pintrich, 2000, p. 455).

The first section of the course provided students with the tools to (self-)evaluate the level of their prior language knowledge and skills. The students had at disposal a self-assessment grid to compare their actual skills with the CEF descriptors of the

Figure 3. Portfolio for Self-Study

LANGUAGE PORTFOLIO

NAME ____________ SURNAME ______________ ID. ____________

DEPARTMENT ______ ____________ curriculum ________ prof. ______________

HOURS ___ NUMBER OF CREDITS __________

DECLARATION OF RESPONSIBILITY ACCORDINIG TO THE LAW N. 76 DPR 445 2000:

SIGNATURE: ______________

PART A (VIDEO VISION): _______ HOURS

Date	Hours	Materials (Title / Unit)	Topics/Comments

PART B (Phonology and Listening): _____ HOURS

Date	Hours	Materials (Title / Unit)	Topics /Comments

PART C (ONLINE RESOURCES): _____ HOURS

Date	Hours	Materials (Title / Unit)	Topics/Comments

Sezione D (GRAMMAR ACTIVITIES): _____ HOURS

Date	Hours	Materials (Title / Unit)	Topics/Comments

B1 level (Council of Europe, 2001). They were asked to give specific examples of how they actually use the English language in real-life situations and then compare these examples with the CEF standards specified for each language skill (listening, reading, spoken and written interaction). Furthermore, the experience of comparing the learning goals with their ability to evaluate their own language competence would encourage students to reflect on the path of learning they need to plan and become active participants in their learning process.

The online course was seen as an alternative way to make the teaching/learning process a highly motivating experience through which students could become aware of their own learning and progress in language skill mastery.

Starting from the premise that students need advanced self-regulated learning skills to succeed in online environments, as well as to learn a second language, we specifically designed the courseware to help students take control of their learning and support students to plan, monitor, control and evaluate their EFL learning. The focus was also placed upon the learner's interaction with the learning resources and support tools specifically designed for planning, monitoring and (self-) evaluating the learning process. In this respect, the most effective means of self-monitoring proved to be the personal portfolio compiled by each student as the report on his/her tailor-made learning pathway.

CONCLUSION

In recent years, foreign language teaching has undergone a profound methodological transformation. First the slate-board and book have been complemented with a wide range of audiovisual aids, and then with a series of tools based on the Internet and real time communication. Teachers that live daily with their students can not exempt themselves from experiencing new ways of meeting their students' educational needs.

When designing our course, we considered a scaffold ed structure for fostering self-regulated learning in order to promote learner autonomy and meaningful learning. In this view, the student does not passively undergo the training intervention made by the teacher, but personally contributes to building his/her learning. In order to reach this aim, we have provided the online course with tools that facilitate and stimulate planning, monitoring and self-assessment of students' learning.

By self-assessing their language skills, students obtain the necessary information concerning the primary language abilities of understanding, interacting and producing verbal messages in both written and oral forms. Then, the possibility to present the contents of the discipline in e-documents linked together in a non-linear but associative architectural structure allows the students to *navigate* through various kinds of information (text, graphics, sounds, audio, images, videos) following routes which combine the learners' autonomous choices and teachers' instructional design decisions. Thanks to the frequent links, students can *navigate* from one node to another as active players able to structure their own learning pathways with contents in a dynamic educational framework. The Portfolio, moreover, is a tool designed and conceived with the express purpose of directing students' learning efforts towards specific goals, helping them to plan study time and monitor their own learning process. This document accompanies students throughout their self-study sessions as a valuable means for raising their language learning awareness by recording the accomplished tasks, activities, and objectives.

Relying on these conceptual tools, the overall structure of our course was aimed at enhancing self-regulation by setting up a series of operations ranging from self-evaluation to specification of study items, language skills and objectives, all contributing to defining the profile of the self-regulated language learner.

Finally, guidelines to the final tests and the specification of the evaluation criteria were designed as means to help students to make sense of the assigned tasks and relate them with the final goal, plan their own learning path consistently with the expected result and, most importantly, facilitate the overall process of self-regulation in a behavioural dimension.

At the end of the course, the class participation and learners' achievements pointed out that Technology Mediated Instruction can offer additional delivery tools and innovative teaching/learning strategies which should complement those currently experienced in the academy and actually support language learning. When used appropriately, web resources can facilitate language learning in a highly motivating dimension and allow the teacher/facilitator to search for continually updated instructional tools and undertake research on new teaching methodologies, in an ongoing process of professional training and long-life learning.

ACKNOWLEDGMENT

Rita Calabrese is the author of Introduction, Aspects of adult Second Language Acquisition, Language Learning in virtual environments, Research framework, Course Format. Filomena Faiella is the author of Abstract, Meaningful learning, autonomy and self-regulated learning: an overview, Virtual Learning Environment, Discussion and Conclusion.

REFERENCES

Ashburn, E. A., & Floden, R. E. (2006). *Meaningful learning using technology: what educators need to know and do*. New York, NY: Teachers College Press.

Ausubel, D. P. (1968). *Educational Psychology. A Cognitive View*. New York, NY: Holt, Rinehart & Winston.

Barone, L., & Calabrese, R. (2005). Didattica nella rete. Aspetti positivi e negativi dell'uso di Internet nel campo della didattica, con uno sguardo particolare alla lingua inglese (Teaching through the Net. Advantages and disadvantages of the Internet in EFL teaching). *Rassegna Italiana di Linguistica Applicata, 2*(3), 33–52.

Benson, P. (2001). *Teaching and Researching Autonomy in Language Learning*. London, UK: Longman.

Bley-Vroman, R. (1990). The logical problem of foreign language learning. *Linguistic Analysis, 201*(2), 3–49.

Chapelle, C. A. (2001). *Computer Applications in Second Language Acquisition. Foundations for Teaching, Testing and Research*. Cambridge, UK: Cambridge University Press.

Clarke, A. (2008). *e-Learning Skills*. New York: Palgrave MacMillan.

Council of Europe. (2001). *Common European Framework of Reference for Languages: Learning, Teaching, Assessment*. Cambridge, UK: CUP.

Dörnyei, Z. (2000). Motivation in action: towards a process-oriented conceptualization of student motivation. *The British Journal of Educational Psychology, 70*(4), 519–538. doi:10.1348/000709900158281

Dörnyei, Z., & Skehan, P. (2003). Individual Differences in L2 Learning. In Doughty, C., & Long, M. H. (Eds.), *The Handbook of Second Language Acquisition*. Oxford, UK: Blackwell. doi:10.1002/9780470756492.ch18

Doughty, C. (2003). Instructed SLA. In Doughty, C., & Long, M. H. (Eds.), *The Handbook of Second Language Acquisition*. Oxford: Blackwell. doi:10.1002/9780470756492

Faiella, F. (2005). Metodologie di scaffolding per il blended learning (Scaffolding Methodologies for blended learning). *Form@re, 39(2)*. Retrieved from http://formare.erickson.it/archivio/novembre_05/2_FAIELLA.html.

Felix, S. (1985). More evidence on competing cognitive systems. *Language Research, 1*(1), 47–72.

Gregersen, T., & Horowitz, E. K. (2002). Language Learning and Perfectionism: Anxious and Non-anxious Language Learners' Reactions to Their Own Oral Performance. *Modern Language Journal, 86*(4), 562–570. doi:10.1111/1540-4781.00161

Holec, H. (1981). *Autonomy in Foreign Language Learning*. Oxford, UK: Pergamon.

Jonassen, D. H. (1994). Technology as Cognitive Tools: Learners as Designers. *ITForum*. Retrieved from http://itech1.coe.uga.edu/itforum/paper1/paper1.html

Jonassen, D. H., Howland, J., Marra, R. M., & Crismond, D. (2007). *Meaningful learning with technology*. Columbus, OH: Merrill/Prentice Hall.

Keefe, J. W., & Ferrell, B. G. (1990). Developing a defensible learning style paradigm. *Educational Leadership*, *48*(2), 57–61.

Krashen, S. (1985). *The Input Hypothesis: Issues and Implications*. Harlow, UK: Longman.

Kress, G. (2003). *Literacy in the New Media Age*. London, UK: Routledge. doi:10.4324/9780203164754

Lamy, M. N., & Hampel, R. (2007). *Online Communication in Language Learning and Teaching*. New York, NY: Palgrave MacMillan. doi:10.1057/9780230592681

Lantolf, J., & Appel, G. (1994). *Vygotskian Approaches to second Language Research*. Norwood, NJ: Ablex.

Lantolf, J. P. (2000). Second Language Learning as a Mediated Process. *Language Teaching*, *33*(2), 79–96. doi:10.1017/S0261444800015329

Ligorio, M. B., Cacciamani, S., & Cesareni, D. (2006). *Blended learning. Dalla scuola dell'obbligo alla formazione adulta (From compulsory to adult education)*. Roma, IT: Carocci.

Little, D. (1991). *Learner Autonomy 1: Definitions, Issues and Problems*. Dublin, EI: Authentik.

Little, D. (2003). Learner autonomy and second/foreign language learning. *Subject Centre for Languages, Linguistics and Area Studies Good Practice Guide*. Retrieved from http://www.llas.ac.uk/resources/gpg/1409

Matthews, R. (1997). *Guidelines for Good Practice: Technology Mediated Instruction*. Sacramento, CA: The Academic Senate for California Community Colleges.

New London Group. (1996). A Pedagogy of Multiliteracies: Designing Social Futures. *Harvard Educational Review*, *66*(1).

Oxford, R. L., & Anderson, N. J. (1995). A cross-cultural view of learning styles. *Language Teaching*, *28*(4), 201–215. doi:10.1017/S0261444800000446

Piaget, J. (1971). *Biology and Knowledge*. Chicago, IL: Chicago University Press.

Pintrich, P. R. (2000). The role of goal orientation in self-regulated learning. In Boekaerts, M., Pintrich, P., & Zeidner, M. (Eds.), *Handbook of self-regulation* (pp. 451–502). Orlando, FL: Academic Press. doi:10.1016/B978-012109890-2/50043-3

Reid, J. M. (1995). *Learning Styles in ESL/EFL Classroom*. Boston, MA: Heinle and Heinle.

Skehan, P. (1998). *A cognitive Approach to Language Learning*. London, UK: Edward Arnold.

Sorace, A. (2005). Selective optionality in language development. In Cornips, L., & Corrigan, K. P. (Eds.), *Syntax and Variation. Reconciling the Biological and the Social. Amsterdam, NL/New York*. NY: J. Benjamins Publishing.

Vento, M., D'Esposito, M.R., Faiella, F. (2008). *Percorsi di formazione a distanza "e-learning". L'esperienza dell'ateneo salernitano (Rooting for distance learning "e-learning". The experience of the University of Salerno)*. Lecce: Pensa Editore.

Vygotsky, L. S. (1978). *Mind in Society: the Development of Higher Psychological Processes*. Cambridge, MA: Harvard University Press.

Warschauer, M. (1999). *Electronic Literacies: Language, Culture, and Power in Online Education*. Mahwah, NJ: Lawrence Erlbaum Associates.

Wenger, E. (1998). *Communities of practice: learning, meaning, and identity*. Cambridge, UK: Cambridge University Press.

Winne, P. H. (1995). Inherent details in self-regulated learning. *Educational Psychologist*, *30*(4), 173–187. doi:10.1207/s15326985ep3004_2

Zimmerman, B. J. (1990). Self-regulated learning and academic achievement: An overview. *Educational Psychologist*, *25*(1), 3–17. doi:10.1207/s15326985ep2501_2

KEY TERMS AND DEFINITIONS

eLearning Lab: teaching laboratory of Faculty of Education (University of Salerno, Italy) that was created with the aim of responding to the increasing educational and training demands of the students as well as create more flexible study paths using innovative methods of teaching and learning.

Explicit/Implicit Knowledge: Explicit knowledge refers to information acquired and stored as such in the mind ("*knowing that*"). Implicit knowledge is "*knowing how*" to go about doing something. These terms are often associated to the concepts of *declarative knowledge*, which is defined as the factual information stored in memory and known to be static in nature and *procedural knowledge* as the knowledge of how to perform, or how to operate.

Moodle: a Learning Management System (LMS). It manages the teaching activity through specific tools devised to prepare lessons, glossaries, wiki and tests. Communication is achieved through different services (chat, forum, messages, blog) integrated into the system.

Motive: the hidden reason for doing something.

Multiliteracy: the mastery of different abilities in the overall communications environment.

Portfolio: a collection of self-study session recordings and documents also including placement and formative tests.

Rote Learning: Ausubel considered rote learning as opposed to meaningful learning. He claimed that the rote level of learning does not allow learners to anchor new knowledge into concepts that are already available in the cognitive structure.

Scaffolding: in the educational field, the word "scaffolding" is used metaphorically by researchers, trainers and teachers to denote the support and assistance provided by an adult or more knowledgeable peer to a learner for conduct a task too complex for his/her skill levels.

Chapter 11
Evaluating Web Content for Self-Directed Language Learning

Yoko Hirata
Hokkai-Gakuen University, Japan

ABSTRACT

Recently, information and communication technology (ICT) in Japanese tertiary institutions have begun to play an increasingly important role in teaching and learning of foreign languages. This technology helps students have access to various kinds of language learning materials and resources through the websites any time and anywhere. Online or web-based language courses provide Japanese students with the variety and flexibility to work at their own level and pace through this technology. However, a major issue to be considered when implementing these courses is the fact that traditionally Japanese students are not culturally self-directed or autonomous language learners. The purpose of this study was to examine how Japanese students perceived two different approaches of self-directed language learning projects through the evaluation of English language websites. The findings suggested that the students' perceptions of the research-based project using websites were positive and they were able to regulate their own learning process.

INTRODUCTION

In recent years, online or web-based courses have been recognized to be one of the effective methodologies in foreign language education (Felix, 1999; Kung & Chuo, 2002; McBride, 2002). This is for the purpose of maximizing the efficiency and quality of these approaches and improving students' overall language proficiency levels. In Japanese educational settings, online education is regarded as effective in dealing with the diversity of language learners (Jung & Suzuki, 2006). Unlike classroom learning materials, however, materials on the websites often require students to engage in solitary activities (Egbert, 2005; Walraven, Brand-Gruwel, & Boshuizen, 2009). Although

DOI: 10.4018/978-1-61692-901-5.ch011

a variety of websites, either free or commercial, have been used in the classroom, there appears to be very little research on how to help students make the most of them (Coates, James & Baldwin, 2005). Since using various web-based teaching tools and methodologies in the language classroom is a new development in Japan (Jung & Suzuki, 2006), the implementation of this technology in combination with self-directed approaches should be thoroughly explored. This chapter aims to explore self-directed English learning by reporting on a websites evaluation project which uses self-directed language learning methodologies to examine the Japanese university students' thoughts of the web evaluation processes.

SELF-REGULATED LANGUAGE LEARNING

It has been widely acknowledged that students' more active and direct involvement with the learning process leads to a clearer understanding of the language (Aston, 1993; Dickinson, 1995). According to a research finding that focuses on students' attitudes and beliefs about their own learning capacities, incorporating autonomous and self-directed learning is useful (Usuki, 2007). There are several research studies which offer various kinds of approaches and they suggest that language learning is best facilitated by the development of students' responsibility to learn by themselves (Dickinson, 1987; Sturtridge, 1997). It has been argued that the development of autonomy and the improvement of the various language skills are closely connected with each other (Little, 2007). Nguyen (2008) claims that students can take initiative and assume responsibility for their own learning if they have some control over the learning process. In addition, studies have indicated that autonomous and self-directed learning approaches do not necessarily work well simply when students are given plenty of opportunities to explore their various practical options outside the classroom (Galloway & O'Brien, 1998). In spite of this recent growing interest, the effectiveness of self-directed online learning approaches has not yet been fully examined. It is still unknown to what extent students' self-directed tasks or activities successfully promote their independence and autonomy in a certain educational context.

Self-directed or self-regulated language learning is impeded by Asian cultural factors (Nguyen, 2008). When implementing autonomous and self-directed learning in Japanese educational settings, a major issue to be considered is the fact that traditionally students are not culturally self-directed or self-regulated learners. Like many Asian students, Japanese students are still widely perceived as passive learners and, therefore, accustomed to the teacher-centered learning environment and the controlled teaching methodology (Wei, 2008). Therefore, as Usuki (2007) points out, learner autonomy is not promoted to any meaningful extent in Japan. The instructor's role is to be in charge of the class, have a greater initiative and "transmit knowledge to the students" (Kennedy, 1991, p. 63). Students are not likely to "see learning as exploration, but instead wait for the instructor to lead them" (Galloway & O'Brien, 1998, p. 5). Another issue to be considered is the fact that the Japanese educational setting does not value independence nor assign creative or imaginative tasks (Usuki, 2007). The focus in the Japanese secondary school is on rote-memorization for examinations and communicative language activities, although the latter has not been fully emphasized (Shucart, Mishina, Takahashi & Enokizono, 2008). Therefore, Japanese students in general tend to display a lack of engagement in any language learning activities (Usuki, 2007). In the online learning environment, in particular, these students often have problems choosing by themselves websites or resources which are appropriate for their own needs and preferences. This results in the situation where the instructor generally has control over the materials that the students use (Friedman, 2009). In addition, students' cultural reticence to self-disclose (Galloway & O'Brien, 1998) and an unwillingness to give their opinions as individuals (Koike & Tanaka, 1995) make it difficult to examine their thoughts and comments. This is a major problem for Japanese students to become self-directed language learners. It has remained a matter of concern on how to encourage students to engage effectively in web-based environments which require students' "self-initiation and their ability to self-regulate their own learning" (Nguyen, 2008, p. 68). More emphasis should be placed on developing effective methods for promoting students' real motivation and commitment to study in online courses.

DATA-DRIVEN LANGUAGE LEARNING

In order to promote students' responsibility of their learning and their self-directedness, Wenden (1991) emphasizes the importance of raising students' awareness of their own learning through the process of planning, monitoring, and evaluating. Johns (1991) also points out that Data-Driven Language Learning, based on observing, hypothesizing, and experimenting, is important for students to raise their awareness of improving their language learning skills and strategies. This effective method, as Johns (1991) claims, is based on the notion that "language learner is also, essentially, a research worker" (p. 2). It makes students access various language data and draw their attention to language patterns and chunks of words as 'researchers'. This approach is particularly effective in allowing students to pay close attention to the target word in rich authentic contexts, such as scientific, conversational, journalistic, and academic texts and, as a result, they are able to make reasonable conclusions about the structural rules of the target word without any help from the instructor.

For the purpose of retrieving and displaying lexical combinations from any kind of data, a computer program called 'concordancer' is used. This program performs the basic function of searching and extracting all the occurrences of a certain key word or phrase in a database. In order to find lexical patterns which are associated with the key word easily, Key Word In Context (KWIC) mode has been used. Adopting this approach, students are expected to gain an extensive knowledge of words and expressions in successful ways based on reliable facts about frequency and typicality of words (Dodd, 1997). This approach is particularly important for Japanese students, who have little opportunity to be exposed to authentic English in their everyday lives, to improve their language proficiency levels. In addition, this approach has been regarded to help students to become more self-reliant and confident in understanding the relationship between the meaning of a word and the context in which it is included more effectively.

THE WEB AS LANGUAGE DATA

In spite of an increasing emphasis on Data-Driven Learning, the use of language data has only been limited to certain endowed educational settings with large authentic language databases. In order to reduce these major barriers for language learning and ensure the basic fairness of the educational environment, the benefits of introducing web-derived data into the classroom for language learning have recently been determined (Fletcher, 2007; Friedman, 2009; Kehoe & Renouf, 2002). Hirata and Hirata (2007) point out that web-derived data has various possibilities as language learning resources, which stimulate students' intellectual curiosity and deepen their knowledge of language. Friedman (2009) also claims that, by using web-derived data as a substitute for existing authentic language resources, students are encouraged to choose texts on the Web, for themselves, of personal or professional relevance. This approach is regarded to strengthen their intrinsic motivation (Skehan, 1991) as it provides students with an opportunity to encounter various expressions and language use in a natural, non-artificial context (Friedman, 2009). The approach also helps them feel personally involved in the language process (Ellis, 1994, quoted in Friedman, 2009). The potential benefits of web-extracted data for students to organize and assess their own language study should be explored.

PURPOSE OF THE STUDY

The purpose of the present study was to examine how Japanese students perceived two different approaches of self-directed language learning projects, research-based and non-research-based, through the evaluation of English language websites. Both types of the projects were based on 'planning', 'monitoring', and 'evaluating', which are Wenden's (1991) concepts of learner training for self-directedness. As previously noted, this concept includes Awareness Raising approaches. However, only the research-based project included Data-Driven Language Learning approaches which were based on 'observing', 'hypothesizing', and 'experimenting' (Johns, 1991). Therefore, the present study was also aimed at determining how these two different approaches affect students' appreciation of the projects and how students can be encouraged to learn in a self-directed way by making the most of the websites. For the purpose of investigating the influence of different procedures on the outcome of the self-directed language learning projects and the students' perception, the two different projects with and without Data-Driven Language Learning approaches were examined. The study sought to answer the following two questions:

1. What are the benefits and problems of students' evaluation of the websites, based on Awareness Raising approaches, for the purpose of promoting their self-regulated language learning?
2. How can Data-Driven Language Learning activities affect the students' evaluation of the websites and encourage them to motivate themselves to manage their own tasks and help them have confidence in completing these tasks?

The answers to these two questions will also help instructors understand how to help students acquire self-initiation and skills to self-regulate their own language learning in the online learning environment.

THE PROJECTS

Settings and Participants

The two projects described in this chapter, research-based websites evaluation project and non-research-based websites evaluation project, were carried out in two different semester-long undergraduate English language courses, Course A and Course B, at a university in Japan. Both courses were required for students to take as compulsory English subjects. They were addressed to Japanese students who wanted to improve English language skills fully by making the most of the various websites. These courses were blended learning English courses in that both regular teacher-directed instructions and web-based instructions co-existed in the classroom. These courses were mainly designed to foster students' English skills and for students to engage in various exercises by using a textbook. The textbook included practices of recognizing proper usage of words and phrases through multiple-choice questions, true or false drill exercises, and fill-in-the-blank comprehension tests. The English language websites, which students were required to use for their study, were a variety of ESL/EFL (English as a second language/English as a foreign language) self-access and independent language learning websites. The courses were also designed to help students use information and communication technology (ICT) as much as possible in language learning and to acquire English lexical skills as well. The class in both courses was scheduled for one and a half hours each week over a 15-week course in a computer room.

The total number of the students in Course A was thirty five and all of them participated in the research-based project during the course. The total number of the students in Course B was twenty and all of them participated in the non-research-based project during the course. The students' profiles of the research-based evaluation project were generally the same as those of the non-research-based evaluation project. All of these students were full time students between the age of eighteen and twenty. Almost all the students were thoroughly accustomed to the Japanese traditional lecture-type language learning approach where the instructor supplied textbook-based teacher-centered instructions (Cooker & Torpey, 2004). Therefore, some of the students tended to depend more on the instructor in the classroom. Many of them had learned English at least for six years in secondary school and had attained at least a lower-intermediate level of proficiency. Although the students had general experience in using computers, they were fairly new to the Web. They had no real experience in self-directed language learning in or beyond the conventional classroom, either.

Research-Based Evaluation Project

First Phase: Data-Driven Language Learning Activities

This research-based evaluation project was divided into two phases: The first phase being Data-Driven Language Learning Activities and the second phase being Websites Evaluation. These two different phases were closely related with regard to the content and the resources the students used. At the beginning of the project, focus was placed on the use of the navigational functions of web browsers, various software applications, and basic computer literacy.

In the Data-Driven Language Learning Activities, the instructor gave students a guidance of how to analyze the language data and evaluate the relevant websites. This guidance included the procedure of the first phase and examples of various websites, which were supposed to be useful for the students. The examples of these websites were given to the students as practical options to choose from.

Students then created language data based on web resources. These resources included English texts, listening lessons' scripts, and practices which were readily available online. The data also included drill exercises and gap filling listening comprehension tests. After the students browsed web pages of various genres and categories of English language websites, they were required to open up a text document and to collect the language data by dragging the web page into the text document. Once the language data was constructed from the websites, at the 'observing' stage students were required to examine various English expressions, and, at the 'hypothesizing' stage, they tried to provide possible explanations of what they had found. After discovering the lexical patterns that were associated with a specific word or phrase, at the 'experimenting' stage, the students compared their findings with the corresponding entries and their example sentences in online dictionaries. This was for the purpose of increasing students' awareness of various language uses in the English texts and listening scripts, and improving their language skills. The students also compared the web-derived data with the data taken from the textbook they had been using in the classroom. Understanding lexical patterns which are frequently repeated in English texts and listening scripts is one way of raising awareness of useful chunks of words (Willis, 2000). A computer program used as a concordancer at this phase, called *Lex*, was a user-friendly in-house computer program

Figure 1. A students' search results by Lex

```
[fridnds__D] er: Well, that's not going to work. You have piano lessons after school and then
[fridnds__D] piano lessons after school and then you have to babysit for the neighbors until
[fridnds__D] t all . . . for an entire month, so you have to catch up on that. And, don't you
[fridnds__D] ave to catch up on that. And, don't you have an essay due in your English class
[fridnds__D] help me with that.] Oh no. And then, we have to clean out the garage. You said y
[fridnds__D] fixing anything around the house. We do have a few things that you could repair
[fridnds__D] ey, are you free this Saturday night? I have two tickets to the football game, a
[fridnds__D]  been really nice talking to you, but I have to run. Hey. Here are the tickets.
[fridnds__D] ean you're always saying that you don't have any friends . . . and that your lov
[fridnds__D] because you study too much. Well, we'll have a blast. Come on! Relax. [Well . .
[fridnds__D] e rinsed off all the soap, they usually have two or three large baths where you
[fridnds__D]  page of the recipe, but don't worry. I have plenty of experience around the hou
[fridnds__D]  know, these things go fast. I mean you have to listen to my advice as your olde
[fridnds__D] I need a calculator for geometry, and I have heard you can download free softwar
[fridnds__D]    Dave: Nah. I'm gonna hit the sack. I have a test tomorrow morning, and I wann
[fridnds__D] Woman: Well . . . you ought to at least have a physical before you begin. I mean
[fridnds__D] of 18 and 55 years old, and the results have been compiled in the following age
[fridnds__D] ly jogging and tennis footware. We also have to come up with a more appealing sl
[fridnds__D] ting to go until ten thirty, but now we have someone from the international cent
[fridnds__D] Okay.] And, immediately after lunch, we have reserved a campus shuttle to give s
[fridnds__D] Okay, that makes sense.] And unless you have a can opener or the can has a pull-
[fridnds__D] r or the can has a pull-tab lid, you'll have to use a rock or something to open
[fridnds__D]  like a couple of candy bars. Then, you have to decide on the type of shelter yo
[fridnds__D] n option. The reality is that you might have to evacuate to a shelter, possibly
[fridnds__D] l, lightweight tent in the event that I have to survive on the street or in a pa
[fridnds__D] those either. So, what do you do if you have small kids? They'd probably go stir
[fridnds__D]  preparation for them is needed. If you have to evacuate to a shelter to wait ou
[fridnds__D] ill be bored out of their minds, so you have to pack small card games, paper, or
[fridnds__D] t, let's go over the to-do list. Do you have everything ready? I mean, did you p
[fridnds__D] he written test anyway? [Well . . .]You have to give way to any pedestrians cros
[fridnds__D] hange the cat litter daily, or he might have an accident on the carpet. [Oh, gre
[fridnds__D]  it couldn't be worse. [Oh]. You see, I have to pick my wife up from the airport
[fridnds__D] st game, but a spark from the fire must have blown out and burned the manuscript
[fridnds__D]  trouble now. [I know.] You're going to have a hard time digging yourself out of
[fridnds__D] extensive, we can't repair it, we might have to put a crown on your tooth. Or as
[fridnds__D] Donalds!]. Yeah, but, you know, I don't have the resources at the moment to do t
[fridnds__D] ng at will pay for some classes after I have been with the company for six month
[fridnds__D] Kelly: Times are changing, Mr. Adams. I have to figure in expenses for a benefit
[fridnds__D] t change my point of view. I guess I'll have to settle for TV dinners and my dog
[fridnds__D] t. I like women who are open-minded and have something interesting to say. Hey,
```

for retrieving and displaying lexical patterns. This program performs the essential function of searching and extracting all the occurrences of a certain word or phrase in language data. Since this is a in-house program, its access was restricted to the students participated in this project. With this program, students could independently consult the target words and expressions which were associated with the key word. Figure 1 shows a student's search results by *Lex*.

The data in Figure 1 is an excerpt of the search results taken from the conversation between two adults, by focusing on the word *have* as a key word. This student also constructed the conversation data between an adult and a child and compared the results with the conversation data between two adults. After analyzing the data, the student identified what type of context is associated with the word *have* and noticed what kinds of words were used with the word *have*. The key word *have* is displayed with approximately six words on either side. The letters on the left-hand column provide the origin of each line. An excerpt of the student's analysis is as follows:

The word have plays a very important role in both types of the conversation. For example, the word have, as an auxiliary verb, which is used to say that someone should do something and have which is used to form percept tenses. Basically the usages of the word have in both types of conversation are the same. Most of these examples include have which is used to say that someone owns something. However, in the conversations between an adult and a child, the variety of nouns which co-occurred with have as an object were limited, such as money and the names of toys. In addition, in this type of conversation, certain conversational styles were identified and they included informal and commonly used expressions between adults and children. The results taken from Lex clearly indicated several patterns and styles, such as utterances which persuade someone to do something. Understanding these patterns and styles is very important for my study.

Second Phase: Websites Evaluation

Students at this phase were instructed to examine some ESL/EFL websites and to evaluate the quality and the appropriate use of these resources. At the beginning a list of websites was available to help students make appropriate choices about what they wanted to work on. Later on at the planning stage, however, students were required to organize their plan to choose the appropriate websites for their own study. The evaluation was based on the measures for evaluating self-access materials (Cooker, 2008). The criteria used in this study was divided into three major sections: 'Navigability', 'Achievable Challenge', and 'Attractiveness'. In the 'Navigability', students were required to examine if everything on the websites was clearly indicated and well-organized. In the 'Achievable Challenge', students were asked to determine if authentic English was provided and the information had been updated. In the 'Attractiveness', students evaluated websites considering if the website attracted their interest. They also checked if answer keys and hints, as well as a large amount of selection of language resources, were available. These processes based on the criteria were categorized as Wenden's 'monitoring' (1991). The author's pilot study suggested that the criteria was an effective measure for evaluating language learning websites for students (Hirata & Hirata, 2009). The criteria also facilitated website analysis and provided students with a standard for evaluating websites. At the evaluation stage, the students completed the project by filling out the evaluation form and making comments concerning the advantages and disadvantages of the project.

Non-Research-Based Websites Evaluation Project

In the non-research-based websites evaluation project, students were not required to do Data-Driven Language Learning activities. The students simply examined ESL/EFL websites and assessed the quality and the appropriate use of these resources for their English study. The procedure for this project was the same as the one explained above in the second phase of the research-based evaluation project. After filling out the same evaluation form which was used in the research-based project, the students finished the project by filling out the evaluation form and making comments concerning the advantages and disadvantages of the project.

Data Analysis

After the course had been completed, a 17-item questionnaire was given to the students for the purpose of collecting their opinions, attitudes and perceptions of the evaluation, including the benefits and problems of the learning process. The questions sought information about attitudes toward the evaluation of the websites. The rating scale used in the questionnaire was a 10-point Likert Scale with 1 representing "strongly disagree" and 10 representing "strongly agree". In order for students to fully understand the questions, the questionnaires were written in Japanese. For the purpose of attaining a mean response for each question, the responses were totaled and averaged. Standard deviation was then obtained for the purpose of examining statistically significant differences between students' responses. The data is presented in this paper as mean ±SD. The questionnaire was also analyzed by using Spearman's correlation to determine correlations between responses and significant factors underlying their responses. Correlation is significant at the.01 level (2-tailed). These results of the present study were compared with those of the previous study which didn't include Language Data Analysis (Hirata & Hirata, 2009).

FINDINGS

The results of the questionnaire revealed the students' different perceptions of this project. The

Table 1. Correlation between overall evaluation and understanding of the features

	Overall evaluation of the project	Understanding of the websites' features
Overall evaluation of the project	1.00	-----
Understanding of the websites' features	.573**	1.00

*Notes: Correlation Matrix (N=35), **p <.01.*

findings indicated that almost all the students of both courses used websites daily and the majority of them were required to use a computer for various assignments in other courses in and outside the classroom. The Average (±SD) of this response was 8.65 (±2.00). There were only a few students who had used English instructional websites for their own English study. The Average (±SD) of this response was 3.15 (±2.10). For most of the respondents, it was difficult to read English websites and to find information they needed for the evaluation.

As for the students who felt the project was valuable and found the evaluation useful, there was a large degree of different opinions. Those who highly rated this project had a moderate correlation understanding the websites features as shown in Table 1 (r =.573, p <.01). This indicates that those who highly rated the project appreciate the importance of evaluating the websites.

Although many students found the project challenging, many respondents highly valued the criteria of the project. The Average (±SD) of this response was 6.65 (±1.40). This result suggests that the criteria used in the project was easy for everyone to adopt. There was no significant correlation of students who highly rated this project and the criteria evaluation as an effective measure. Even those who had never browsed English websites were able to understand the various features of search engines and the use of the websites. The Average (±SD) of this response was 7.30 (±1.30). In addition, those who used English websites frequently had a weak correlation rating the criteria as an effective measure as shown in Table 2 (r =.386, p <.01).

It is interesting to see that there was a significant correlation of students who highly rated this project and rated the effectiveness of this project for their future development and the creation of the websites by themselves as shown in Table 3 (r=.404, p<.01). This project encouraged students, who valued the project, to have more incentive to explore the websites and to utilize the web resources for their own long-term benefits.

Qualitative data analysis also suggested what kinds of benefits and problems the students perceived during the evaluation process. Below are comments from the students in the research-based project.

When the instructor told me about the tasks I was required to do, I was not interested in them. While I was doing these tasks, it looked like I was getting nowhere. However, when I finished both tasks, I understood the importance of these two different

Table 2. Correlation between the use of English websites and effective criteria

	Frequent use of English websites	Criteria as an effective measure
Frequent use of English websites	1.00	-----
Criteria as an effective measure	.386**	1.00

*Notes: Correlation Matrix (N=35), **p <.01.*

Table 3. Correlation between overall evaluation and future development of the websites

	Overall evaluation of the project	**Students' future development of the websites**
Overall evaluation of the project	1.00	-----
Students' future development of the websites	.404**	1.00

*Notes: Correlation Matrix (N=35), **p <.01.*

approaches. It was kind of strange that I hadn't realized it until I completed both of these tasks.

The most important thing that we should do to have good English skills is to have a list of good educational websites based on my interests and to keep working on them whenever I have time. We should not rely on the learning approaches we have acquired during our secondary school years. This course made me aware of what I needed to do if I wanted to develop my language skills.

The comments from the students in the non-research-based project were as follows.

The websites that the instructor gave me were quite useful. I didn't know that such good websites existed out there. It seemed that finding such good websites was challenging for me. A Google search for 'English listening' didn't work to find good results. Both tasks were laborious, but when I completed them I felt that they were beneficial to me.

I hadn't used the educational websites until I took this course. I was shocked to know that there were a wide range of useful websites out there which I could use to improve my English skills. They were all well designed to target students who were not good at learning English and who just needed to improve English skills. Browsing various websites in our everyday lives is a must for anyone who wants to improve their language proficiency levels.

Table 4 shows the averages (± SD) of the students' responses to the ease of evaluating of the websites. With regard to the usability of the websites (Question1) and the number of language exercises attached to the websites (Question 8), almost all of the students in the research-based evaluation project felt that the evaluation was not difficult. The averages (± SD) of these responses were 7.65 (± 1.30) and 6.85 (± 2.10) respectively. In response to Question 6 and 7, students indicated that evaluating the availability of updated materials and the attractiveness of the content of the websites was relatively easy for the students. In addition, the availability of the answers and feedback was a feature that the students could easily evaluate. The averages (± SD) of these responses were 6.45 (± 1.60) and 6.65 (± 1.87) respectively. There were three criteria which students in the research-based project thought difficult to evaluate: (1) the layout and usability of the websites, availability of the relevant websites, and the quality and consistency of the materials. This is the major difference between the responses of the research-based evaluation project and those of the non-research-based evaluation project. For students in the research-based evaluation project, the only potentially difficult aspect of the evaluation may involve the evaluation of the authenticity of the English.

Table 5 shows that there was a moderate correlation between the students' future development of the websites and their understanding of the websites' features (r =.527, p <.01). This result indicates that the more students understand the details of the websites' features, the more confident

Table 4. Averages of the students' responses to the ease of evaluating

	Research-Based Evaluation Mean (SD)	Non-Research-Based Evaluation Mean (SD)
1. Layout of the websites and usability of the icon.	7.65 (1.30)	6.70 (1.95)
2. Availability of the relevant websites.	5.40 (2.80)	5.05 (2.31)
3. Consistency of the materials and the quality of audio files.	6.45 (1.80)	6.35 (1.95)
4. Availability of language levels for the materials.	5.75 (2.70)	6.25 (2.47)
5. Meaningful English in authentic use.	3.70 (2.50)	4.20 (2.70)
6. Availability of materials recently updated.	6.65 (2.30)	4.95 (2.56)
7. Attractiveness of the content of the websites.	6.60 (1.90)	6.65 (1.87)
8. Appropriate number of language exercises.	6.85 (2.10)	6.85 (2.10)
9. Availability of keys and feedbacks.	6.45 (1.60)	6.65 (1.87)

the students become to get involved with the development of the websites in the future.

The comments from the students in the research-based project which are relevant to this point were as follows:

Thinking back about my own learning experience in my secondary school, what I learned in the classroom was quite limited and biased toward the memorization of grammar and vocabulary. The project I engaged in, in this classroom, was a powerful way to reflect on my own learning experience and on the contents I have acquired in the last six years.

The project helped me realize that it was necessary for us to consider whether the content of the learning materials, either on the textbooks or on the websites, was appropriate for our needs. Analyzing the expressions by using the computer system was not an easy process for me to complete, but it was worth understanding what I have learned through classroom textbooks, such as TOEFL or TOEIC. There was a big difference between the words, expressions and idioms I had learned for the preparation of taking standardized tests and those I will encounters in authentic situations.

Two different tasks provided me with a clear guideline for what I should do to improve my language skills. I am desperate for developing my English skills. Although the only studying English experience I had was when I was studying for university entrance exams, these two tasks I have completed helped me realize that I should use the materials on the websites more on a daily basis, otherwise I can't understand everything about authentic English language use.

The completely different tasks that I was required to do during this course were worthwhile. I could understand that these tasks were closely related

Table 5. Correlation between future development and understanding of the websites

	Students' future development of the websites	Understanding of the websites features
Students' future development of the websites	1.00	-----
Understanding of the websites features	.527**	1.00

*Notes: Correlation Matrix (N=35), **p <.01.*

and both of them had synergistic effects. The tasks I engaged in also helped me realize that the contents of the websites were very large and it was not easy for me to handle them. Of course, I need to learn techniques and skills to browse the websites, but analyzing the language data derived from the websites decreased handling difficulty. If we could take this kind of approach all the time in the classroom, I don't think we should learn various techniques and skills to navigate the websites.

Concerning various skills and strategies which were necessary to navigate and locate the important information on the websites, many participants stated that, they needed to increase their awareness of words and expressions they had studied in the classroom. This suggests that the project raised the students' awareness of how they should improve their language skills more effectively. Many of them stressed the importance of comprehensive knowledge and information concerning the difference between the words and expressions they encountered in their everyday lives and those which they were required to learn in the classroom.

IMPLICATIONS

The findings of the present qualitative and quantitative data suggested that the overall students' perceptions of both research-based and non-research-based projects using websites were positive and the students were able to regulate their own learning process for their own study. Many students fully utilized various language learning resources on different websites which provided them with massive exposure to authentic English and its cultural contexts. With the help of the evaluation checklist which covered the three areas such as 'Navigability', 'Achievable Challenge' and 'Attractiveness', the findings have shown that the students in both types of goal-focused projects found it relatively easy to evaluate the websites. This is consistent with the findings of Walraven, Brand-Gruwel & Boshuizen (2009) and Hirata and Hirata (2009). Although there appears to be no big difference between the results of the research-based evaluation project and the non-research-based evaluation project, the findings described earlier indicate that there were some major benefits of using two-fold language activities to implement the self-regulated project in web-based language learning. First of all, the results of the questionnaire suggested that the students who participated in the research-based websites evaluation project greatly benefited from accomplishing their work by completing two different tasks in online learning. This is particularly clear from the comments from the students in Course A. As shown in the previous section, the research-based project encouraged students to reflect on their own learning experience and to raise their awareness of the language use they had learned. This indicates that, compared with the non-research-based project, the research-based project had a greater influence on students' understanding of how they should improve their own English skills. The students expressed their distinct opinions about both the advantages and disadvantages of the project. Although statistically the number of participants in this study was relatively small, the results also suggest that the process of 'observing', 'hypothesizing', and 'experimenting' can be highly beneficial to students. This finding is in accordance with the results of the students' strategies for self-regulated approaches (Mclnerney, 2008).

The results have shown that for students in the research-based evaluation project, evaluating the layout of the websites and the usability of the icon, the availability of the relevant websites, and the consistency of the materials and the quality of audio files was easier, compared with students in the non-research-based project. This finding suggests that students who have finished the Data-Driven Learning activities are more conscious of examining these features of the websites because of the analytical process of the activities. With regards to some question items concerning the ease of the evaluation of the authenticity of the websites, the averages of students' responses in the research-based project were lower than those in the non-research-based project. Although the students in the research-based evaluation project had some difficulties in evaluating the authenticity of the English on the websites, in combination with the students' comments, it can be assumed that the students had a more critical attitude toward what they had evaluated through the project. Many students expressed their appreciation of the process to reflect on their own online experience. Students' comments suggest that due to the completion of language awareness activities, the students became more concerned about the authenticity of the words used on the websites. It can be asserted that this integration of 'observing', 'hypothesizing', and 'experimenting' process encouraged students to engage in critical thinking and to have a sense of ownership of their learning (Ridley, 1997). As data retrieved from the research-based evaluation project suggest, both Wenden's (1991) and John's (1991) concepts are, therefore, effective to help students become sensitized to their language learning. These are a significant component for the future design of successful web-based learning for independent study in Japanese educational settings. The overall result indicates that, as Usuki (2007) claims, the students were not "merely passive learners" (p. 22). The students who had had no online independent language learning experience received the substantial benefits of understanding how they should "explore knowledge themselves and find their own answers" (Littlewood, 2000, p. 34).

The findings have shown that there were some issues the instructor should take into consideration. First of all, the students' comments indicated that some students had difficulties in engaging in the two-fold project without the instructor's direct support until they realized the real benefits of the project for their own study. This negative reaction by the students at the beginning of the project could have resulted from the fact that they didn't have any experience in evaluating any websites by themselves. Students' limited experiences of navigating the English websites could have created discouragement to try something new (Nguyen, 2008). In contrast, with the help of the instructor, many students could be trained to complete the project and determine what works and what does not work for their improvement of the language skills. Since reserved students were observed to be actively engaged in the project if the instructor was directly involved in their learning process even at the beginning, more focus should be placed on considering students' specific cultural background and educational context. In this regard, the instructor should play an important role for maximizing the specific student's learning potentials. A deep understanding of what students really need and how they act in particular educational settings is indispensable for the instructor. This will help students increase awareness of their learning process and to gain the benefits of understanding their own contribution to their own learning process (Usuki, 2007). This approach will also promote students' self-regulation and autonomy in their language learning and make significant contributions to the development of web-based learning.

CONCLUSION

This study examines the website evaluation methods, in combination with language data analysis, for developing students' self-directed learning based on their perceptions and attitudes toward evaluating websites for their own study. Although more research is necessary and various different factors should be explored, the results of this study provided valuable insights into how students can develop their self-regulated learning in the language classroom. The findings suggested that students' own planning, monitoring, and evaluating process helped students acknowledge the benefits they could derive from their learning environments and contribute to the development of a positive and responsible attitude with regards to learning. The results also indicated that the instructor plays an important role in motivating students to manage their own tasks and encouraging them to have confidence in completing their tasks during the continuing learning process. In order for students to make maximum use of their own learning strategies to achieve their own goals, more emphasis should be placed on the development of teaching systems which caters for individual needs and preferences as well as helping students make choices to suit their learning styles. Further research based on a larger scale study should be conducted to draw more valid conclusions regarding various possibilities for the evaluation to measure student progress in rapidly changing learning environments.

ACKNOWLEDGMENT

This project is supported by a research grant provided by Hokkai-Gakuen.

REFERENCES

Aston, G. (1993). The learner's contribution to the self-access center. *ELT Journal*, *47*(3), 219–227. doi:10.1093/elt/47.3.219

Coates, H., James, R., & Baldwin, G. (2005). A critical examination of the effects of learning management systems on university teaching and learning. *Tertiary Education and Management*, *11*(1), 19–36.

Cooker, L. (2008). Self-Access Materials. In Tomlinson, B. (Ed.), *English Language Learning Materials* (pp. 100–132). London: Continuum.

Cooker, L., & Torpey, M. (2004). From the classroom to the self-access centre: a chronicle of learner-centred curriculum development. *Language Teaching*, *28*(6), 11–16.

Dickinson, L. (1987). *Self-instruction in language learning*. Cambridge, UK: Cambridge University Press.

Dickinson, L. (1995). Autonomy and motivation. A literature review. *System*, *23*(2), 165–174. doi:10.1016/0346-251X(95)00005-5

Dodd, B. (1997). Exploiting a corpus of written German for advanced language learning. In Wichmann, A., Fligelstone, S., Knowles, G., & McEnery, T. (Eds.), *Teaching and Language Corpora* (pp. 131–145). London, UK: Longman.

Egbert, J. (2005). *CALL essentials: Principles and practice in CALL classrooms*. Alexandra, MD: TESOL.

Ellis, R. (1994). *The Study of Second Language Acquisition*. Oxford, UK: Oxford University Press.

Felix, U. (1999). Web-based language learning: a window to the authentic world. In R. Debski & M. Levy (Eds.), *WORLDCALL: Global Perspectives on Computer-Assisted Language Learning* (pp.85-98). Amsterdam, NL: Swets & Zeitlinger.

Fletcher, W. H. (2007). Concordancing the web: promise and problems, tools and techniques. In M. Hundt, N. Nesselhauf & C. Biewer (Eds.), *Corpus Linguistics and the Web* (pp. 25-45). Amsterdam, NL: Rodopi.

Friedman, G. L. (2009). Learner-created lexical databases using web-based source material. *ELT Journal*, *63*(2), 126–136. doi:10.1093/elt/ccn022

Galloway, I., & O'Brien, D. (1998). Learning online: choosing the best computer-mediated communication activities. *Language Teaching*, *22*(2), 7–9.

Hirata, Y., & Hirata, Y. (2007). Independent research project with web-derived corpora for language learning. *JALTCALL Journal*, *3*(3), 33–48.

Hirata, Y., & Hirata, Y. (2009). Students' Evaluation of Websites in Hybrid Language Learning. In F.L. Wang, J. Fong, L. Zhang, V.K.S. Lee (Eds.), *Hybrid learning and education* (pp. 186-196), Lecture Notes in Computer Science 5685, Berlin, DE: Springer.

Johns, T. (1991). Should you be persuaded - two samples of data-driven learning materials. In Johns, T., & King, P. (Eds.), *Classroom Concordancing* (pp. 1–13). Birmingham, UK: Birmingham University English Language Research Journal.

Jung, I., & Suzuki, K. (2006). Blended learning in Japan and its application in liberal arts education. In Bonk, C. J., & Graham, C. R. (Eds.), *The Handbook of Blended Learning* (pp. 267–280). San Francisco, CA: Pfeiffer.

Kehoe, A., & Renouf, A. (2002). WebCorp: applying the web to linguistics and linguistics to the web. Retrieved September 29, 2007, from http://www2002.org/CDROM/poster/67/

Kennedy, J. (1991). Perspectives on cultural and individual determinants of teaching style. *RELC Journal*, *22*(2), 61–78. doi:10.1177/003368829102200205

Koike, I., & Tanaka, H. (1995). English in foreign language policy in Japan: toward the twenty-first century. *World Englishes*, *14*(1), 13–25. doi:10.1111/j.1467-971X.1995.tb00336.x

Kung, S.-C., & Chuo, T.-W. (2002). Students' perceptions of English learning through ESL/EFL websites. *TESL-EJ, 6*(1). Retrieved May 5, 2009, from http://www-writing.berkeley.edu/TESl-EJ/ej21/a2.html

Little, D. (2007). Language learner autonomy: some fundamental considerations revisited. *Innovation in language learning and teaching, 1*(1), 14-29.

Littlewood, W. (2000). Do Asian students really want to listen and obey? *ELT Journal*, *54*(1), 31–36. doi:10.1093/elt/54.1.31

McBride, N. (2002). Web-enhanced approaches to the teaching of linguistic variation in French. *ReCALL*, *14*(1), 96–108. doi:10.1017/S0958344002000812

Mclnerney, D. M. (2008). The motivational roles of cultural differences and cultural identity in self-regulated learning. In Schunk, D. H., & Zimmerman, B. J. (Eds.), *Motivation and Self-Regulated Learning. Theory, Research, and Applications* (pp. 369–400). New York, NY: Routledge.

Nguyen, T. C. L. (2008). Learner Autonomy and EFL Proficiency: A Vietnamese Perspective. *Asian Journal of English Language Teaching*, *18*, 67–87.

Ridley, J. (1997). *Learner autonomy 6: developing learners' thinking skills*. Dublin, EI: Authentik.

Shucart, S. A., Mishina, T., Takahashi, M., & Enokizono, T. (2008). The CALL lab as a facilitator for autonomous learning. In Zhang, F., & Barber, F. B. (Eds.), *Handbook of research on computer-enhanced language acquisition and learning* (pp. 483–495). New York, NY: Information Science Reference.

Skehan, P. (1991). Individual differences in second-language learning. *Studies in Second Language Acquisition*, *13*(2), 275–298. doi:10.1017/S0272263100009979

Sturtridge, G. (1997). Teaching and language learning in self-access centres: changing roles? In Benson, P., & Voller, P. (Eds.), *Autonomy & independence in language learning* (pp. 154–165). London, UK: Longman.

Usuki, M. (2007). *Autonomy in Language Learning: Japanese Students' Exploratory Analysis*. Nagoya, JP: Sankeisha.

Walraven, A., Brand-Gruwel, S., & Boshuizen, H. P. A. (2009). How students evaluate information and sources when searching the World Wide Web for information. *Computers & Education*, *25*(1), 234–246. doi:10.1016/j.compedu.2008.08.003

Wei, M. (2008). Increasing Oral participation in ESL/EFL conversation Classrooms. '. *Asian Journal of English Language Teaching*, *18*, 67–87.

Wenden, A. L. (1991). *Learner strategies for learner autonomy. Planning and implementing learner training for language learners*. London, UK: Prentice Hall.

Willis, J. (2000). A holistic approach to task-based course design. *Language Teaching*, *24*(2), 7–11.

KEY TERMS AND DEFINITIONS

Concordancer: A computer program which counts and lists the occurrences of a keyword/key term, showing examples of its use from text data.

Data-Driven Learning: A term coined by Tim Johns (1991) that refers to the learner examining large amounts of language data to find out certain lexical patterns or rules of language use.

Chapter 12
Using Video as a Retrospective Tool to Understand Self-Regulated Learning in Mathematical Problem Solving

I-Pei Tung
McGill University, Canada

Kevin Chin
McGill University, Canada

ABSTRACT

This chapter presents a novel approach that combines self-regulated learning (SRL) with Activity Systems Theory (AST). While SRL focuses primarily on individual cognitive and social aspects, it does not address sociocultural factors that inherently play a role in learning processes. The combination of SRL and AST is effective due to the central role that feedback plays in both theories. The viability of this approach is tested with data collected from Canadian secondary school-level students engaged in mathematical problem solving (MPS) using video as a retrospective feedback tool. Analysis using this theoretical framework based on SRL and AST provides a richer understanding of how video can contribute to learning within technology-enhanced learning environments (TELEs). Based on these findings, suggestions for implementation are provided for educators who would like to effectively use video in classroom situations.

INTRODUCTION

The face of learning is changing substantially. Classroom learning is surrounded by informal learning opportunities with new tools, lectures are complemented with project-based learning, and frontiers between instruction and learning are fading as students collaborate in forming knowledge communities. It is clear that these developments are driven by technological advancements which

DOI: 10.4018/978-1-61692-901-5.ch012

require psychological and pedagogical models of learning that take into account both: (a) the diversity of situations in which learning takes place, and (b) the specific needs of learners, instructors, and school administrators.

Within such contexts, self-regulated learning (SRL) is a useful perspective to draw upon as an important theoretical framework that describes significant aspects of successful learning in terms of individual goals, motivations, volitions, and emotions (Butler & Winne, 1995; Pintrich, 2000a; Winne & Hadwin, 1998; Zimmerman, 2000). Learners employ self-set learning goals and use environmental features to regulate their cognition, motivation, and behaviour. One weakness in this framework lies in its oversight of contemporary educational perspectives that view learning as a sociocultural process involving the individual, as well as peers, educators, and the environment. In order to better reflect current understandings of teaching and learning contexts, this chapter presents a combination of the psychological perspective of SRL and the sociocultural perspective of Activity Systems Theory (AST) so as to provide a more comprehensive understanding of how learning might take place within technology-enhanced learning environments (TELEs).

Following a brief overview of SRL, this chapter provides an explanation of how its omission of sociocultural learning components can be addressed by drawing upon relevant elements from AST. The viability of this new theoretical framework will be illustrated through its application to data collected from an empirical study with Canadian secondary school-level students using video as a retrospective tool in mathematics problem solving (MPS). Based on this study and its findings, a number of implementation guidelines are also provided for educators who are interested in using video to promote SRL in classroom environments.

BACKGROUND

The concept of self-regulation is a relatively new area of educational research. Two decades ago, Glaser (1987) characterized self-regulation as an "executive" skill for monitoring problem-solving and other socio-cognitive processes. Early studies suggested that self-regulation skills were important in: (a) helping learners use and transfer their knowledge to new situations, and (b) selecting, combining, and coordinating cognitive resources while carrying out specific learning tasks. Since then, many studies have identified important relationships between motivation and learning, suggesting that self-regulated learners achieve better learning outcomes (Pintrich, 2000a, 2000b; Zimmerman & Martinez-Pons, 1988, 1990). Given the significance of such benefits, researchers have become interested in fostering the abilities of students to become self-regulated learners (Pintrich & Zusho, 2002).

SRL has been defined as "an active, constructive process whereby learners set goals for their learning and attempt to monitor, regulate, and control their cognition, motivation, and behaviour, guided and constrained by their goals and the contextual features in the environment" (Pintrich, 2000a, p. 453). It has also been seen as involving "cognitive, affective, motivational and behavioural components that provide the individual with the capacity to adjust his or her actions and goals to achieve the desired results in light of changing environmental conditions" (Zeidner, Boekaerts, & Pintrich, 2000, p. 751). Reaching a consensus definition remains a challenge, as different models of self-regulation emphasize slightly different aspects of the overall process (Pintrich, 2000a). Corno's (2001) work has emphasized volitional aspects of self-regulation, Winne's (1996) approach has focused on cognitive aspects of self-regulation, while McCaslin and Hickey (2001) have highlighted sociocultural aspects of self-regulation. In spite of their differences, nearly all models of SRL share the following four basic assumptions (Pintrich, 2000a):

1. Active construction. Learners actively construct their own meanings, goals, and strategies based on information from external physical and social environments as well as internal psychological environments.
2. Potential for control. Learners can monitor, control, and regulate (to various degrees) aspects of their cognition, motivation, behaviour, and environment. All models of SRL recognize that biological, developmental, contextual, and individual differences can interfere with individual efforts at regulation.
3. Recognizing goals/criteria/standards. There are standards (e.g., goals, reference values) against which comparisons are made to assess whether a process should continue, or whether any type of change is necessary. The general example for learning assumes that learners set goals, monitor their progress, as well as adapt and regulate their cognition, motivation, and behaviours to reach these goals.
4. Social and contextual aspects of learning. Competence and learning are influenced by the cultural, demographic, personality, and contextual characteristics of learners, as well as by the self-regulation of their cognition, motivation and behaviour that mediate relations between persons, contexts and achievements.

Based on this set of assumptions and for the purposes of this chapter, SRL is defined as an active and constructive process in which learners set goals for learning, and then endeavour to monitor, regulate and control their cognitions, motivations and behaviours, while guided and constrained by goals and contextual features of their learning and achievement (Butler & Winne, 1995; Zimmerman 1989, 2000).

THE COMBINATION OF SRL AND AST

Current views of SRL focus primarily on individual thinking processes, and overlook the critical role of social interaction, leaving social components of learning relatively untouched. This is somewhat surprising, as the roots of SRL can be traced back to work by Vygotsky and Bandura, both of whom stressed the importance of an individual's interactions with others, i.e., people who engage in verbal interactions with this individual, and the environment, i.e., tools and artefacts that exist in the external world. A handful of researchers have begun integrating social dimensions of learning into models of SRL, seeking to better articulate the socially constructed nature of learning and learners (Winne & Hadwin 1998; Zimmerman, 2000; Zimmerman & Schunk, 2001). In support of this important position, this chapter outlines a new approach that combines SRL, a psychological theory that focuses primarily on the individual learner, with AST, a sociocultural theory that focuses primarily on the social and individual dimensions of learning (Engeström, Miettinen, & Punamäki, 1999).

Researchers have long recognized that learning is a social activity, and that classrooms are social systems. AST views human learning from a macro level, and is an elaborated version of Vygotskian assumptions that are used to understand complex human actions from a systemic perspective. Building on Leont'ev's (1978) work on joint and collective activity, Engeström et al. (1999) proposed a collective activity system that situates local human activity within broader social-cultural-historical structures. Within such a framework, learning activities are viewed as the following constructs: (a) subjects, (b) tools, (c) objects, (d) rules, (e) divisions of labour, and (f) communities (Engeström et al., 1999). The conceptualization of AST by Engeström et al. (1999) shows that historically formed mediating constructs transform the nature of the initial three components of individuals, environments, and populations, into subjects, objects/motives, and communities, respectively (Engeström, 1987).

The actions of human agents (subjects) occur within the context of purposeful goal-directed activities toward the solution of problems or purposes (objects/motives), which are mediated by tools and artefacts (instruments) in collaboration with others (communities). AST also stresses the mediational role of communities and their conventions (rules) and of social structures (divisions of labour), which refer to horizontal actions and interactions among community members and to vertical divisions of power and status respectively. Thus, the concept of 'activity systems' addresses the specific needs of active agents in socially mediated activities; their essence is to take situations or conditions and transform them "in an effort to create something qualitatively new" (Lantolf & Thorne, 2006, p. 210). The concepts of mediation and transformation are necessarily connected. AST identifies the various components of activity systems, characterizes them, and describes how they interact. For instance, subjects use various psychological and material tools to achieve desired goals or objectives. These subjects belong to specific communities, and are constrained by systems of rules and operate within more or less complex divisions of labour.

In summary, the combination of SRL and AST is quite natural as they address the combined phenomenon of human action as having both internal psychological and external sociocultural dimensions (Vygotsky, 1978; Wertsch, 1998). While the former tends to focus on internal psychological aspects of learning in academic contexts, the latter tends to focus on external sociocultural dimensions of human activities. In both cases, it is clear that learners participate in systems of activities in which they use material and conceptual tools to interact with each other and their surroundings. While meanings or concepts are important (Vygotsky, 1978), the fundamental unit of analysis is activity. Activity was not a key concept for Vygotsky, as he did not consider the analysis of the whole structure of purposeful, goal-oriented actions a necessary precondition for understanding the psychological processes that are embedded in these actions. Although Vygotsky (1978) was not especially intrigued by external activities unless they constituted a context of semiotic mediation, he was predominantly interested in signs, meanings, intersubjectivity and the appropriation of cultural objects. For the purpose of the current study, selecting activity as the basic unit of analysis allows for an investigation into the interactions that naturally occur in classrooms, i.e., everyday actions lived by students, teachers and other participants. Students and teachers construct meanings in obvious and concrete manners. They do things. They use tools. They interact with each other and with computers. They converse, tell and write stories, and solve math problems.

Importance of Internal and External Feedback

Learners embedded in classroom activity systems use feedback from internal cognitive environments as well as external physical and social environments to regulate their cognitions, motivations and behaviours in order to achieve learning goals. These feedback loops are seen as operating both internally and externally. More specifically, internal feedback is produced by the learner's own monitoring processes when she or he evaluates personal performance, whereas external feedback is provided by other people or events. Both types of feedback can inform the learner about domain knowledge errors, but can also enhance learners' effective engagement in tasks (Winne, 1996). For instance, when a teacher initiates a conversation with a student, the ensuing exchange goes back and forth, becoming a 1st stimulus-feedback- 2nd stimulus unit with a feedback loop. Consequently, the process and product of this feedback loop become internalized within the student. In other words, the external feedback generated from others or events can lead to internal feedback on the part of the learner. Using feedback as a central concept in learning, it is critical to combine cognitive and motivational/affective processes addressed by SRL with regulatory processes addressed by AST, i.e., rules, community, division of labour, object, subject, and tools. In doing so, researchers will be able to more effectively identify and investigate internal and external processes that contribute to learning in TELEs.

Role of Feedback within AST and SRL

Goal-oriented learning activity systems rely on feedback that focuses on: (a) learners, and (b) the activity systems in which they are embedded. Learners in classroom activity systems use internal and external feedback loops that supply information from both the outside sociocultural world and internal cognitive world, in order to regulate cognitions, motivations, and actions as a way to achieve learning goals. Information supplied by these feedback loops can include verbal as well as social inputs, e.g., language, gestures, which are then used to regulate the various components of classroom activity systems, i.e., systems of rules, communities, divisions of labour, objects, subjects, and tools. Learners use various psychological and material tools to achieve their desired goals or objectives. They also belong to specific communities that provide constraints in the form of systems of rules, and operate within more or less complex divisions of labour. Within an SRL framework, feedback from reflecting on one's own learning plays an essential role in moving upwards to a higher level of development (Winne & Hadwin, 1998). According to Winne and Butler (1994) "feedback is information with which a learner can confirm, add to, overwrite, tune, or restructure information in memory, whether that information is domain knowledge, meta-cognitive knowledge, beliefs about self and tasks, or cognitive tactics and strategies" (p. 5740).

Internal and external feedback can occur at four different levels (Hattie & Timperley, 2007). First, feedback can be focused on a task or product, regardless of whether the work is correct or incorrect. At this level, feedback includes directions to acquire more, different, or correct information. Second, feedback can be aimed at the process used to either create a product or complete a task, and more directly focused on the processing of information. Third, student feedback can be focused on the self-regulation level, which includes helping to create greater skill in self-evaluation, or confidence to engage further with a task. This type of feedback has more influence on student self-efficacy, self-regulatory proficiencies, and self-beliefs. For example, students may be encouraged or informed by an instructor on how to better and more effortlessly continue with a task. Fourth, feedback can be personal in the sense that it is directed to the 'self,' which is too often overlooked in regard to task performance.

From an AST perspective, these four levels of feedback might take form as interactions between subjects, tools, and objects within activity systems. In other words, through focusing on feedback loops, relevant subjects, tools, and objects can be identified and then described in regard to how they interact in self-regulatory, goal-directed learning activities. For example, learners receive instruction and feedback for improvement of learning outcomes, and the connection between instruction and feedback helps to better understand how students develop SRL in their classroom performance. From an AST perspective, feedback from teachers, peers, and parents takes the form of socially constituted signs, which learners internalize and transform into psychological tools that can then be used to regulate their learning.

One example of such signs might be a teacher's comments written in the margins of student essays.

It is clear that all learners participate in activity systems in which they use material and conceptual tools to interact with each other and their surroundings. In such situations, the fundamental unit of analysis is activity, such as a learning event.

Video as Feedback Tool

Video has been used in classroom for assessment. Research has highlighted the educational potential for documenting students' performance and integrating them into a digital portfolio to enhance student learning (Tung, 2004; Tung, 2007). Thus, video can be used to help students, even parents and other people interested in educational processes, to understand the dynamics of how students learn. Video mediates the objectives of different participants in classroom activity systems. Video has been – and continues to be – an invaluable tool for providing feedback to learners in a variety of domains (Hodges, Chua, & Franks, 2003; Raab, Masters, & Maxwell, 2005; Vickers, Livingston, Umerls, & Holden, 1999) because of its visual and aural representation of rich, complex situations (Brophy, 2004). For example, research on learning sports have involved participants who watched their own performance on video, and received visual feedback that was used to self-correct behaviours. Video has also been used in teacher training (Dysthe, 2002; Newhouse, Lane, & Brown, 2007; Pea, 2008), where educators have watched themselves in the classroom, using visual and aural feedback to improve their professional practices (Derry, 2007; Sherin, 2004).

It is not surprising that researchers have also taken advantage of these technological benefits when studying cognition. In this study, video is used to provide students with visual and aural feedback on their academic performance, with the intent to self-correct as athletes and teachers have done. With respect to using video as a tool for enhancing SRL, Baggetun and Wasson (2006) adopt a sociocultural perspective to highlight the role of tools as social mediators of learning in students' self-regulation (Lave, 1988; Lave & Wenger, 1990; Vygotsky, 1978). This context is particularly noteworthy as it facilitates situations where SRL is both embedded in and mediated by a community and its cultural artefacts. Of particular importance to this study is the potential of video in addressing the four levels of feedback previously mentioned (Hattie & Timperley, 2007).

TESTING THE THEORETICAL VIABILITY OF COMBINING SRL WITH AST

This novel combination of SRL and AST is meant to highlight the strengths and diminish the weaknesses of both approaches. The work presented in this chapter has been guided by the following question: How can mathematical problem solving be understood when using a new theoretical framework based on SRL and AST to analyze students' SRL processes as they use video as a retrospective tool?

Context, Participants, and Procedures

Data for this study were collected at a Canadian all boys secondary-level school located in an urban setting. Five secondary three (Grade 8) students who participated in this study were asked to solve five mathematical problems related to the Pythagorean theorem (which students had studied two months earlier). A facilitator was also present to prompt students when necessary. This process took place in two phases. In Phase 1, students were videotaped individually engaging in MPS, using a calculator or any other tool they deemed necessary, using a think-aloud process. These videos were subsequently used in the next phase of the study. Phase 2 took place approximately one week later, where students were asked to try again and solve any incorrect problems from the previous week. During these individual sessions, participants were: (a) shown video clips of their first attempts to solve the problems, and (b) again videotaped as they reviewed and commented on their performance, using a think-aloud process. Discourse data extracted from the videos—which contains digital records of student performance activities and student retrospective process activities—were analyzed to gain a better understanding of student learning in regard to MPS.

The data collected from this study consisted of approximately 100 minutes of audio and video recordings from five participants, who provided extensive verbalizations while they engaged in MPS.

A semantic framework developed by Bracewell, Breuleux, & Le Maistre (2003) was used to code the transcripts. Bracewell et al.'s (2003) framework was created on the bases of Frederiksen's (1975, 1986) theory of propositional structures and their relations, allowing for the determination of student goals, evaluations, conditions on goals and evaluations, and relations among them that they apply in attempting to solve problems. Goals, evaluations, and relations are defined by semantic criteria, such as propositional components and relations (Bracewell & Breuleux, 1994; Bracewell, 1999; Frederiksen, 1975). The reliability of identifying goals, evaluations, and relations depends heavily upon a coder's knowledge about propositional structures and the way that these structures appear as grammatical and lexical forms in language. This semantic structure reveals functional behaviour by identifying explicit and intentional goals, results, and evaluations that constitute the task. The characteristics of this semantic structure fit well with an SRL process, particularly in the current study, where problem-solving tasks were used to measure how students self-regulate.

Contextual Description Using an AST Framework

For the purposes of brevity, only one case is presented in this chapter as it is considered exemplary in terms of how a student can successfully use video as a feedback tool to self-regulate his problem-solving processes. Andrew (the student) is considered an excellent model as he was able to use video of himself engaged in MPS as a feedback tool to solve a problem that he was initially unable to answer. Not only was he able to determine the correct response to the problem, but viewing himself on video also helped him spontaneously demonstrate important cognitive skills.

Andrew determined that there was a need or motive to fulfil, and therefore constructed an *object*, i.e., to solve five mathematics problems. The object usually relates to an action that has a clear beginning and end. Instead, the object often takes the form of finding a problem, defining a problem, and then determining what the possible action(s) are. From an SRL perspective, this follows the sequence of forethought, planning, monitoring, controlling, and reflecting. The mediating instruments (i.e., tools), were video, software (The Geometer's Sketchpad®), worksheets, and laptops connected to a wireless network. Using these mediating instruments, the subject moved toward accomplishing the object, i.e., problem-solving. However, the subject did not accomplish the object in isolation. The student worked with a facilitator (community member) who had explicit and implicit interactions with the student. The student knew he could approach the facilitator for any inquiry during this problem-solving process (rules of interaction) and established what kind of tools were available (division of labour) in order to solve the problems (outcome).

The study took place in two phases. In Phase 1, Andrew was videotaped solving mathematics problems and using a think-aloud protocol. He was very comfortable in front of the camera while solving the problems because of prior experience with being videotaped in this classroom. A week later during Phase 2, while he was watching himself problem-solving, he used the laptop to rewind and forward his own video, and sometimes pointed with his finger to the screen and said "Oh yes", "No, that was wrong" or "Now, I got it". This type of interaction demonstrated that he was taking advantage of the affordances of video. In essence, this process of interacting with video allowed Andrew to self-direct his own learning. From an AST perspective, we can identify two levels of learning: (a) the macro, which involves viewing the whole problem-solving phenomenon, and (b) the micro, which includes the subject, tools and objects.

Stages in Andrew's Problem Solving Process

In analyzing Andrew's case, the problem-solving process can be seen as being composed of four stages. In Stage 1, Andrew was assigned several mathematical problems to solve within a 15 minute time period all the while using a think-aloud protocol. A video camera was placed in front of him to document the problem-solving process, and he was able to later watch himself solve the problems on a computer. In Stage 2, Andrew was asked to review the video on a computer to track his own process. In Stage 3, Andrew focused on the last problem, defined as an individual task, in which he performed the arithmetic operation to determine his own solution process. At this moment, he started to point out how it went wrong with the last problem solving process. He said to himself: "Oh, that's wrong." In Stage 4, Andrew described what he was doing in the video. Andrew's utterances provided data that would be later used for analysis, and defined as a type of verbal interaction with a category of codes broadly adapted from SRL theory. In each category, codes were created to verify Andrew's utterances. In this last step, the goal was to define the sequences of his problem-solving processes. His solutions and integration of knowledge took place in the following sequential manner: evaluation-plan-monitor-reflect.

Data Analysis

The question that Andrew tried to solve in Step 1 is shown in Figure 1.

Here is an excerpt from Andrew's transcript in Step 2:

Andrew: No, that's wrong. So, I'm gonna take my compass, and I found out that middle is always 0, not 5. I thought it was 5, but it's not, so what I'm gonna do—I'm just gonna scratch that out—what I'm gonna do is, I'm gonna draw a circle five points out from the middle. I'll draw a circle around it, and then see what happens. [Adjusts compass] Okay, a bit less. There. [Draws circle] Aha! There. Okay. I'll put this back.

The identification of these components (goals, evaluations, conditions, types of relations, actions) permitted the construction of a graphical representation of the discourse and actions made by students that were analogous to problem-behaviour graphs (Newell & Simon, 1972). This graphical representation is presented in Figure 2,

Figure 1. The problem to be solved in the described experience

On a coordinate plane, the distance between A (x, y) and the origin is 5. If both x and y are integers, then which of the following numbers CANNOT be a possible value of y?
(A) 0
(B) 1
(C) 3
(D) 4
(E) 5

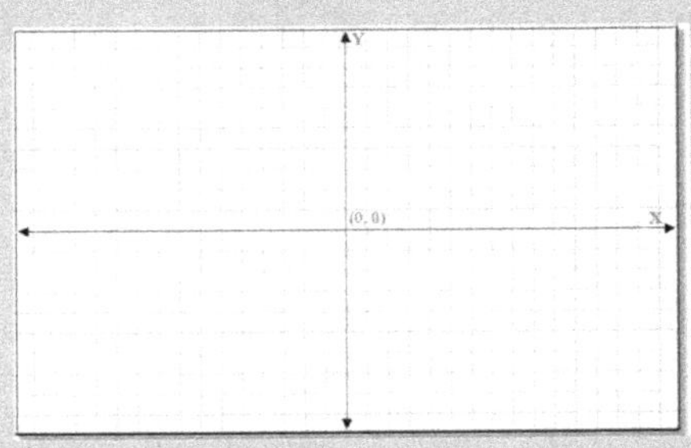

- Do you have trouble solving this problem? If yes, what is it?
- Do you need anything (instruments such as a ruler, a compass, The Geometer's Sketchpad®)?
- Please tell me how you got the answer.
- Do you know how to teach others?

Figure 2. An example of coding sequence

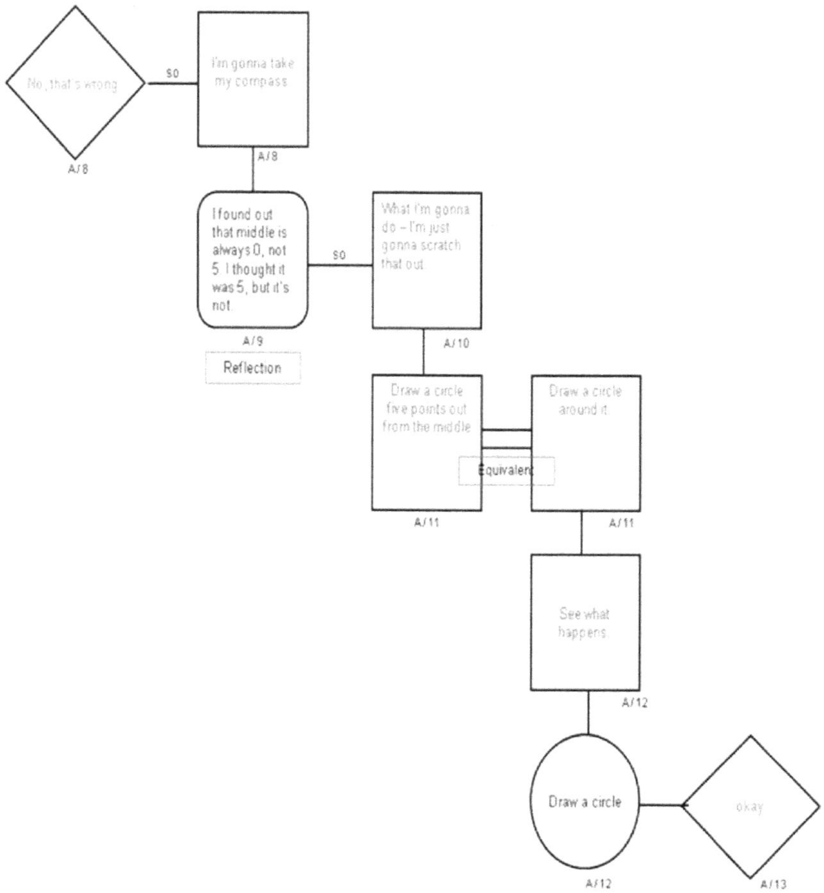

where in general, temporal sequence is indicated vertically and other relations (e.g., subgoaling, conditionals) are indicated horizontally.

The coded sequence shows a number of self-regulation processes:

- Evaluation of previous performance – "no that's wrong" (A/8)
- Reflection on prior problem representation, and its use as conditional information for planning – "I found out the middle is always 0….I'm just gonna scratch that out" (A/9, A/10)
- Planning that leads to deliberate action – "What I'm gonna do…See what happens" (A/10-A/12)

This type of coding thus holds promise of being sufficiently reliable and detailed to allow for a more comprehensive analysis of self-regulation strategies.

Subsequent activity does not lead to a satisfactory solution of the problem, because Andrew is unable to recognize that the distance from A to the origin is 5, with the actual distance of 5 units that appear on the grid. At this point, Andrew was watching himself on video, trying to problem-solve, and the facilitator prompted him to continue the think-aloud process. Andrew reviewed his own performance from the video, and indicated that the action he did in the past was incorrect. He announced a new plan, and the action he was going to take was interacting with the video tool and re-conceptualizing his understanding. Then, as a reflective question, he asked himself what to do. He activated the next step by drawing a circle. The next expected step was a monitoring state. The facilitator invited Andrew to verify his thinking about the origin of the problem, and how it related to the Pythagorean theorem that he had previously learned. This was a very strong hint to help Andrew think about the problem he was solving. In the excerpt above, Andrew used the video to reflect on what he was doing and what he was planning to do, which indicated the phases of SRL as reflecting-planning-controlling. The facilitator was interacting in the state of forethought to bring Andrew's attention back to his experience with the Pythagorean theorem. The role of the facilitator was relatively limited during this learning event. Similar to a teacher who is student-centred and using inquiry-based instruction, the facilitator used prompts to scaffold Andrew and help him negotiate his own conceptualization in his zone of proximal development. In this particular case, Andrew independently developed his own self-regulatory processes, with little assistance from the facilitator.

The presentation of these data and results are seen as an exemplary case that demonstrates the utility of combining SRL with AST in order to gain a better understanding of how self-regulating processes can be enhanced using video in a classroom-based setting while engaged in MPS. The data analysed illustrate how Andrew instantiated the four basic assumptions regarding SRL: (a) constructing meaning based on information from external and internal feedback, (b) monitoring, controlling, and regulating cognition, motivation, behaviour and environment, (c) recognizing standards to use in assessing progress, and (d) using self-regulatory activities to mediate between personal and contextual characteristics and actual competence and/or performances. In addition, learning was viewed simultaneously using an AST framework, which allowed for an understanding of how self-regulating processes can be enhanced when complemented by acknowledging social factors such as community members, rules of interaction, division of labour, and outcomes. The subject, Andrew, was seen as working towards objectives, using instruments, and interacting with others. The actions of human agents (*subjects*) occur in the context of purposeful goal-directed activities toward the solution of problems or purposes (*objects/motives*), which are mediated by tools and artefacts (*instrument*) in collaboration with others (*communities*). Feedback was constantly being used by Andrew, originating from himself based primarily on video viewing, as well as from a facilitator.

Solutions and Recommendations: Guidelines for Educators

Based on the application of a novel approach combining SRL and AST which recognizes both individual and social dimensions of learning, a set of general guidelines can be suggested to assist educators in the design, development, and implementation of integrating video in the classroom. Working with students and video has led to a number of useful guidelines that may assist educators in successfully using this medium in the classroom to foster SRL. This section elaborates upon principles suggested by Ley and Young (2001), and provides some useful starting points in carrying out effective activities that maximize the use of SRL, video, and student learning.

Principle 1: "Guide learners to prepare and structure an effective learning environment" (Ley & Young, 2001, p. 94). Using video in the classroom can be challenging. In order to prepare students to engage in this activity, it is important to acclimatize students to being on video, using a think-aloud protocol, developing concise language skills, and if possible, inviting peers to watch videos in small groups (Tochon, 1999). For example, students need time to practice think-aloud protocols until they are comfortable speaking aloud and describing their actions. Students also need time and opportunities to develop skills in offering each other feedback. By engaging in such preparatory activities, relevant communication skills will be developed that will assist in fostering SRL processes. With sufficient practice and familiarization, it is reasonable to expect that students can use video to work in pairs or small groups in problem solving.

Principle 2: "Organize instruction and activities to facilitate cognitive and metacognitive processes" (Ley & Young, 2001, p. 94). Consider how guidance at macro, meso, and micro-levels of instruction can be provided to students.

At the macro level, i.e., the curricular level, it is important when designing video activities to schedule sufficient time between viewing and think-aloud sessions to allow students to consolidate their new knowledge, skills, and attitudes. SRL skills demand time to mature, and will not develop in a 5-10 minute period. At a meso level, i.e., the learning event, it is important to be explicit with participants regarding the goals and objectives of the video activities to help direct their efforts. At the micro level, i.e., specific learning task, it is important to provide a set of specific questions to foster self-regulated behaviours in students as they view videos. Before they view themselves on video, prepare students by asking them: (a) What do you want to focus on in this session? and (b) Why do you think this is important? After watching the video, ask students: (c) What did you do well? and (d) What did not go so well? Finally, ask students to take some time and consider: (e) What could you improve upon in the future? and (f) What factors influenced you to take these actions? Based on the results of this study, it is theorized that using video to facilitate their problem-solving process would have been improved had they been supported by these questions.

Principle 3: "Use instructional goals and feedback to present the learner with monitoring opportunities" (Ley & Young, 2001, p. 94). Students can be instructed to take advantage of reviewing their video multiple times, and have them pay attention to significant moments (Pea, 2006; Sherin & van Es, 2005). After an initial session of watching themselves, a subsequent session may involve addressing different objectives from the instructor's perspective. Provide students with a sheet with specific questions that facilitates taking a third person perspective by having them self-assess using the following questions: (a) What is the topic of the task? (b) What kind of activity are you engaged in? (c) What are you seeing in the video? and (d) What do you think about the performance in the video?

Principle 4: "Provide learners with continuous evaluation information and occasions to self-evaluate" (Ley & Young, 2001, p. 95). Opportunities to self-evaluate can be effectively brought together by continuous recording of problem-solving behaviours. Current software and hardware advances permit video to provide a lasting record of such instances, which can be edited to present a particular set of information for students to use in self-assessment (Calandra, Gurvitch, & Lund, 2008; Le Fevre, 2004; Sherin, 2004).

The following scenario is offered in order to illustrate how these four principles might be applied by educators who are interested in improving student learning processes through the use of video in the classroom. Robert, a secondary-level teacher, has some knowledge of educational psychology and wants to improve SRL processes in his students. He has read about the potential for using video in the classroom, and spent time developing a system to introduce his class to using video as a retrospective feedback tool in a very gradual process. At the beginning of the school year, Robert focuses on guiding his students to prepare for using video through structuring an effective learning environment (Principle 1). He sets up a system where students are videotaped during everyday MPS. They are asked to watch themselves, without making any comments on what they are seeing. After a few weeks, Robert has students practicing think-aloud on video, and then has them watch themselves again. In order to develop effective communication skills, Robert has students work in pairs, practicing how to provide feedback to each other on their work. Next, when students seem prepared to move on to the next stage, Robert develops a targeted MPS activity that organizes instruction and activities to facilitate cognitive and metacognitive processes (Principle 2). He designs an activity that takes place over an extended period of time, understanding that students need time to develop new knowledge and skills related to SRL. The project is clearly explained to students, articulating the need to improve thinking skills through using a set of guiding questions to self-assess. As students progress with their MPS, individual video sessions are reviewed many times, complemented by a set of more sophisticated questions to help foster self-assessment skills (Principle 3). Finally, students can be involved in editing their own video clips to provide a series of opportunities to self-assess and monitor based on individual learning needs (Principle 4).

IMPLICATIONS FOR RESEARCH AND APPLICATION

This chapter outlines how a theoretical framework based on SRL and AST can be used to analyze students' SRL processes as they use video as a retrospective tool in MPS. Based on these ideas, it is clear that there are many implications for future work in regard to research and application. One particularly intriguing area is that of using video as an assessment tool. Cascallar, Boekaerts and Costigan (2006) discuss how the advent of the "new modes of assessment" movement has given rise to the call for more ecologically valid instruments. While assessment can involve the use of standardized instruments, as well as observational techniques, self-report questionnaires, interviews, think aloud protocols, and diaries, they can be seen as far removed from the operational behaviours or productions that indicate partial facets of what is being assessed. Video has demonstrable benefits as a self-assessment tool, as it is excellent in capturing "rich and thick representations of practice" that include language and action, e.g., aural and visual learning evidence (Le Fevre, 2004, p. 239). The possibility of video being used to fill the need for assessment tools needs to be further explored, with an emphasis placed on how its ability to capture the complexity and richness of contextual information can aid in determining whether key knowledge, skills, and attitudes are being developed.

CONCLUSION

This chapter provides a description of a novel combination of SRL and AST, and illustrates the viability of this approach as an analytic framework using empirical data collected with secondary school students using video while engaging in MPS. In response to the question, "How can mathematical problem solving be understood when using a new theoretical framework based on SRL and AST to analyze students' SRL processes as they use video as a retrospective tool?", findings suggest that this approach is useful in providing a better understanding of how learning takes place in classroom settings within TELEs. Within this new framework of SRL and AST, learning is viewed at an individual level, as well as a sociocultural level where tools and community members contribute to learning. A critical component in this approach is that of feedback, and the use of video is seen as a promising tool that acts as a catalyst in allowing individual and peer-monitoring to take place.

Students are surrounded by technology and tools, such as desktop and laptop computers with built-in videocameras and microphones. Personal video recorders are becoming more accessible over time, and smartphones are increasingly video capable. Wireless connectivity extends the reach of technology to access resources and tools that are located beyond the bricks and mortar of the classroom. Without appropriate guidance from educational research in the form of meaningful and useful theories, the adoption of technology for instructional purposes will result in ill-conceived activities that do not have any impact except for providing temporary respite from traditional didactic teaching. It is hoped that continued research on fundamental concepts in educational psychology will assist educators to maximize the affordances of technology to best meet the ever-changing needs of students.

ACKNOWLEDGMENT

We greatly appreciate the comments and suggestions from anonymous reviewers and editors. The first author would also like to express her gratitude to the Fonds Québécois de la Recherche sur la Société et la Culture for its support of her doctoral research. The content of this book chapter is part of first author's unpublished doctoral dissertation. Dr. Robert J. Bracewell has been very helpful in providing feedback on earlier conceptualizations of ideas, as well as with guidance concerning semantic analysis.

REFERENCES

Baggetun, R., & Wasson, B. (2006). Self-regulated learning and open writing. *European Journal of Education, 41*(3/4), 453–472..doi:10.1111/j.1465-3435.2006.00276.x

Bracewell, R., Breuleux, A., & LeMaistre, T. (2003). Semantic coding model. Unpublished manuscript, McGill University, Montreal, Canada.

Bracewell, R. J. (1999). Objects of Study in Situated Literacy: The role of representations in moving from data to explanation. *Written Communication, 16*(1), 76–92. doi:10.1177/0741088399016001004

Bracewell, R. J., & Witte, S. P. (2003). Workplace tasks, ensembles, and activity: Linkages between text production and situation of use in the workplace. *Written Communication, 20*(4), 511–559.. doi:10.1177/0741088303260691

Brophy, J. (2004). Introduction. In Brophy, J. (Ed.), *Using video in teacher education* (pp. ix–xxiv). New York, NY: Elsevier.

Butler, D. L., & Winne, P. H. (1995). Feedback and self-regulated learning: A theoretical synthesis. *Review of Educational Research, 65*(3), 245–281. doi:.doi:10.3102/00346543065003245

Calandra, B., Gurvitch, R., & Lund, J. (2008). An exploratory study of digital video editing as a tool for teacher preparation. *Journal of Technology and Teacher Education, 16*(2), 137–153.

Cascallar, E., Boekaerts, M., & Costigan, T. (2006). Assessment in the evaluation of self-regulation. *Educational Psychology Review, 18*(3), 297–306. doi:10.1007/s10648-006-9023-2

Corno, L. (2001). Volitional aspects of self-regulated learning. In Zimmerman, B. J., & Schunk, D. H. (Eds.), *Self-regulated learning and academic achievement: Theoretical perspectives* (pp. 191–225). Hillsdale, NJ: Erlbaum.

Derry, S. J. (2007). Video research in classroom and teacher learning: Standardize that! In Goldman, R., Pea, R., Barron, B., & Derry, S. J. (Eds.), *Video research in the learning sciences* (pp. 305–320). Mahwah, NJ: Erlbaum.

Dysthe, O. (2002). *Theoretical background for portfolios as learning and assessment tools in teacher education*. Paper presented in Nordic Educational Research Association Congress, Tallin, Estonia.

Engeström, Y. (1987). *Learning by expanding: An activity-theoretical approach to developmental research*. Helsinki, FI: Orienta-Konsultit.

Engeström, Y., Miettinen, R., & Punamäki, R.-L. (Eds.). (1999). *Perspectives on activity theory*. New York, NY: Cambridge University Press.

Frederiksen, C. H. (1975). Representing logical and semantic structure of knowledge acquired from discourse. *Cognitive Psychology*, *7*(3), 371–458. doi:10.1016/0010-0285(75)90016-X

Glaser, R. (1987). Learning theory and theories of knowledge. In De Corte, E., Lodewijks, H., Parmentier, R., & Span, P. (Eds.), *Learning and instruction. European research in an international context* (pp. 397–414). Oxford, UK: Pergamon Press, University Press.

Hattie, J., & Timperley, H. (2007). The power of feedback. *Review of Educational Research*, *77*(1), 81–112..doi:10.3102/003465430298487

Hodges, N. J., Chua, R., & Franks, I. M. (2003). The role of video in facilitating perception and action of a novel coordination movement. *Journal of Motor Behavior*, *35*, 247–260. doi:10.1080/00222890309602138

Lantolf, J., & Thorne, S. L. (2006). *Sociocultural theory and the genesis of second language development*. Oxford, UK: Oxford University Press.

Lave, J. (1988). *Cognition in practice: Mind, mathematics, and culture in everyday life*. Cambridge, UK: Cambridge University Press. doi:10.1017/CBO9780511609268

Lave, J., & Wenger, E. (1990). *Situated learning: Legitimate peripheral participation*. Cambridge, UK: Cambridge University Press.

Le Fevre, D. M. (2004). Designing for teacher learning: Video-based curriculum design. In Brophy, J. (Ed.), *Using video in teacher education* (pp. 235–258). New York, NY: Elsevier.

Leont'ev, A. N. (1978). *Activity, consciousness, and personality*. Englewood Cliffs, NJ: Prentice-Hall.

Ley, K., & Young, D. B. (2001). Instructional principles for self-regulation. *Educational Technology Research and Development*, *49*(2), 93–103. doi:10.1007/BF02504930

McCaslin, M., & Hickey, D. T. (2001). Educational psychology, social constructivism, and educational practice: A case of emergent identity. *Educational Psychologist*, *36*(2), 133–140. doi:10.1207/S15326985EP3602_8

Newell, A., & Simon, H. A. (1972). *Human problem solving*. New York, NY: Prentice-Hall.

Newhouse, C. P., Lane, J., & Brown, C. (2007). Reflecting on teaching practices using digital video representation in teacher education. *Australian Journal of Teacher Education, 32*(3) Retrieved Dec. 28, 2009, from http://ajte.education.ecu.edu.au/issues/PDF/323/Newhouse.pdf.

Pea, R. D. (2006). Video-as-data and digital video manipulation techniques for transforming learning sciences research, education and other cultural practices. In J. Weiss, J. Nolan & P. Trifonas (Eds.), *International handbook of virtual learning environments* (pp. 1321-1393). Dordrecht, NL: Kluwer Academic Publishing.

Pintrich, P. R. (2000a). The role of goal orientation in self-regulated learning. In Boekaerts, M., Pintrich, P. R., & Zeidner, M. (Eds.), *Handbook of self-regulation* (pp. 451–502). San Diego, CA: Academic Press. doi:10.1016/B978-012109890-2/50043-3

Pintrich, P. R. (2000b). Multiple goals, multiple pathways: The role of goal orientation in learning and achievement. *Journal of Educational Psychology*, *92*(3), 544–555. doi:10.1037/0022-0663.92.3.544

Pintrich, P. R., & Zusho, A. (2002). The development of academic self-regulation: The role of cognitive and motivational factors. In Wigfield, A., & Eccles, J. S. (Eds.), *Development of achievement motivation* (pp. 249–284). San Diego, CA: Academic Press. doi:10.1016/B978-012750053-9/50012-7

Raab, M., Masters, R., & Maxwell, J. (2005). Improving the 'how' and 'what' decisions of elite table tennis players. *Human Movement Science*, *24*(3), 326–344. doi:10.1016/j.humov.2005.06.004

Sherin, M. G. (2004). New perspectives on the role of video in teacher education. In Brophy, J. (Ed.), *Using video in teacher education* (pp. 235–258). New York, NY: Elsevier.

Sherin, M. G., & van Es, E. A. (2005). Using video to support teachers' ability to notice classroom interactions. *Journal of Technology and Teacher Education*, *13*(3), 475–491.

Tochon, F. V. (1999). *Video study groups: For education, professional development, and change*. Madison: Atwood Publishing.

Tung, I.-P. (2004). *Documenting the use of digital portfolios in a primary school classroom*. Unpublished master thesis, McGill University, Montreal, Canada.

Tung, I.-P. (2007). Acting with technology in a multicultural learning community: Participant reactions to digital portfolio assessment in K-6. In Hirashima, T., Hoppe, U., & Young, S. (Eds.), *Supporting learning flow through integrative technologies* (pp. 51–58). Amsterdam: IOS Press.

Vickers, J. N., Livingston, L., Umeris, S., & Holden, D. (1999). Decision training: The effects of complex instruction, variable practice and reduced delayed feedback on the acquisition and transfer of a complex motor skill. *Journal of Sports Sciences*, *17*(5), 357–367. doi:10.1080/026404199365876

Vygotsky, L. S. (1978). *Mind in society*. Cambridge, MA: Harvard University Press.

Wertsch, J. V. (1998). *Mind as action*. Cambridge, MA: Harvard University Press.

Winne, P. H. (1996). A metacognitive view of individual differences in self-regulated learning. *Learning and Individual Differences*, *8*(4), 327–353. doi:10.1016/S1041-6080(96)90022-9

Winne, P. H., & Butler, D. L. (1994). Student cognitive processing and learning. In Husen, T., & Postelthwaite, T. N. (Eds.), *The international encyclopedia of education* (pp. 5738–5745). Oxford, UK: Pergamon.

Winne, P. H., & Hadwin, A. F. (1998). Studying as self-regulated learning. In Hacker, D. J., Dunlosky, J., & Graesser, A. (Eds.), *Metacognition in educational theory and practice* (pp. 277–304). Hillsdale, NJ: Erlbaum.

Zeidner, M., Boekaerts, M., & Pintrich, P. R. (2000). Self-regulation: Direction and challenges for future research. In Boekaerts, B., Pintrich, P., & Zeidner, N. (Eds.), *Handbook of self-regulation* (pp. 749–768). San Diego, CA: Academic Press. doi:10.1016/B978-012109890-2/50052-4

Zimmerman, B. J. (1989). A social cognitive view of self-regulated academic learning. *Journal of Educational Psychology*, *81*(3), 329–339.. doi:10.1037/0022-0663.81.3.329

Zimmerman, B. J. (2000). Attaining self-regulation: A social cognitive perspective. In Boekaerts, B., Pintrich, P., & Zeidner, N. (Eds.), *Handbook of self regulation* (pp. 13–39). San Diego, CA: Academic Press. doi:10.1016/B978-012109890-2/50031-7

Zimmerman, B. J., & Martinez-Pons, M. (1988). Construct validation of a strategy model of student self-regulated learning. *Journal of Educational Psychology*, *80*(3), 284–290. doi:10.1037/0022-0663.80.3.284

Zimmerman, B. J., & Martinez-Pons, M. (1990). Student differences in self-regulated learning: Relating grade, sex, and giftedness to self-efficacy and strategy use. *Journal of Educational Psychology*, *82*(1), 51–59. doi:10.1037/0022-0663.82.1.51

Zimmerman, B. J., & Schunk, D. H. (Eds.). (2001). *Self-regulated learning and academic achievement: Theory, research, and practice*. New York, NY: Longman.

KEY TERMS AND DEFINITIONS

Activity System Theory: A collective activity system that situates local human activity within broader social-cultural-historical structures. Within such a framework, learning activities are viewed as the following constructs: (a) subjects, (b) tools, (c) objects, (d) rules, (e) divisions of labour, and (f) communities.

Mathematical Problem Solving: The process of solving a mathematic problem.

Retrospective Rool: A tool all users to view their performance in the past, which is used in a reviewing process.

Self-Regulated Learning: An active, constructive process whereby learners set goals for their learning and attempt to monitor, and control their behaviours, guided and constrained by their goals and the contextual features in the environment.

Chapter 13
Activating a Self-Regulated Process:
The Case of a Remedial Activity within an ICT Environment

M. Alessandra Mariotti
University of Siena, Italy

Laura Maffei
University of Siena, Italy

ABSTRACT

This contribution is based on a research study which aims at investigating the benefits coming from the use of the Aplusix ICT environment in a remedial intervention in Algebra. The authors start from elaborating a theoretical reference frame for Self-Regulated Learning in order to make it suitable to reformulate and investigate the specific pedagogical problem of a remedial activity in Algebra (first year of the upper secondary school, 9th grade). Then, the authors present the design of a teaching intervention that has been carried out at school, centred around the use of Aplusix. Finally, they discuss some results from the analysis of the data collected during the experiment. The study's results show clear evidence of the evolution of students' awareness and self control, i.e. they have become self-regulated learners.

INTRODUCTION

This contribution is based on a research study which aims at investigating the benefits coming from the use of the Aplusix ICT tool (Nicaud, Bouhineau, Chaachoua & Trgalova, 2004; 2006) in a remedial intervention in Algebra. Assuming the intervention of the teacher on students' failures to be non-effective, a different pedagogical approach has been designed to help students to overcome

DOI: 10.4018/978-1-61692-901-5.ch013

their difficulties (Maffei & Mariotti, 2006; 2007). The leading principle involves creating a learning environment centred on the use of the ICT environment in which students engage directly in the remedial activity. As students interact with the new element, Aplusix, a new net of relationships emerges in which the role of the teacher changes dramatically. In particular, the traditional teacher-student relationship with respect to the didactical problem of overcoming encountered difficulties, is substituted by a new ICT-student relationship with respect to the *specific* difficulties encountered by each student.

In this chapter, after a general description of our specific didactical problem, we elaborate on the *Self-Regulated Learning (SRL)* theoretical perspective in order to reformulate the pedagogical problem of the remedial intervention. Then, we present the design of a teaching experiment aimed at investigating the role of the Aplusix ICT tool in such a remedial intervention. Finally, we discuss some results coming from the analysis of the data collected during the experiment in the perspective of SRL.

Even if we present a very limited contribution to the general issue of developing SRL, we believe that our analysis acquires its value beyond the limits of the specific case of the Algebra domain. The obtained results seem to open new perspectives concerning the use of a particular ICT environment with respect to both the specific objective of calculus skills and the more general objective concerning the development of meta-cognitive attitudes, i.e. consciousness and control of one's own activity. In other words, the specific learning environment set up on the basis of the student-tool interaction, seems to determine significant changes in pupils' attitude towards their own errors and impasses, showing its effectiveness in fostering self-regulated learning processes.

SELF-REGULATED LEARNING: THE CASE OF A REMEDIAL INTERVENTION IN ALGEBRA

The Didactical Problem

Italian curricula consider students' acquisition of algebraic skills in performing calculation a main goal of the first year of the upper secondary school, e.g. 9th grade. After being introduced to the main rules for expanding and factorizing, students are expected to memorize the formulas of the main products (second, third power of a binomial, difference of squares…), and to apply them to solve symbolic manipulation tasks. A didactical problem arises: *How to help pupils overcome the difficulties that they meet in gaining basic competences related to memorization and application of algebraic formulas?*

Approaching this didactical issue requires considering two interlaced problematics. On one hand, we must address the intrinsic difficulty of memorizing and applying formulas; on the other hand, we must address the general problem of remedial, that is, the meta-cognitive (Brown, 2002) problem of recovering from longstanding failures and restoring a meaningful relationship between the student and the mathematical knowledge.

The data coming from experiments with the Aplusix environment (Maffei, 2004; Maffei & Mariotti, 2006) led us to make some hypotheses concerning the benefits deriving from the use of such ICT tools to support pupils in overcoming different types of difficulties, both at the cognitive and at the meta-cognitive level. Specifically, the hypotheses concerning how the use of Aplusix in a remedial intervention could foster students' awareness and management of their own learning difficulties make sense within the perspective of Self-Regulated Learning. Here we take the more general perspective of SRL for describing and interpreting some of the results obtained in our teaching experiments. In particular we will discuss results that show students' improvement in algebraic manipulation and the development of self-regulated processes.

Remedial Intervention in a Meta-Cognitive Perspective: a Self-Regulated Process

Facing the didactical problem of remedial, we started from a basic assumption drawn from the work of Zan (2002), who writes:

Even if the teacher recognises the student's error and intervenes, it is up to the student to modify his behaviour: but if the student is to significantly change his behaviour he first has to be convinced that the change has to be made, that the existing behaviour lead to failure. (p. 94)

As a consequence, any remedial intervention will be successful provided that students become responsible of their own errors and difficulties, as well as responsible of correcting them. As long as the responsibility for the error, in detecting and in correcting it, remains under the teacher's control, remedial may remain ineffective. On the contrary, when the students become aware of their own responsibility, that is, when they become aware that their own behaviour does not fit with their own aims, it is possible to undertake an effective remedial intervention. All this can be interpreted as a particular case of developing SRL, intended as a practice that encompasses cognitive strategies, meta-cognition and motivation in one coherent construct, stressing how the 'self' is the agent that establishes learning goals and strategies and how this engagement influences the quality of the learning process.

As a basic assumption in facing the didactical problem of remedial, we consider the need for students to change their attitude towards their difficulties, their errors, and more in general towards learning. Since difficulties and errors are evidence of unsuccessful learning, an effective remedial intervention needs to act at changing the way of learning. As some authors argue (see, for example, Pintrich, 2000; Zimmerman, 2001), students who become self-regulated learners are more effective learners, and both their performance and their ability to find resources when facing difficulties improve. Moreover as Zimmermann (Zimmerman, 1998; 2001) claims, stimulating students to become self-regulated learners makes them less dependent on the teacher in all the classroom activities. Pintrich and Zusho (2002) in their research show how any student, even a weak student, can become a self-regulated learner.

In our study, elaborating on the mentioned assumptions, we emphasize the benefits that weak students can gain once they become self-regulated. Moreover, all the authors mentioned above (see for example Zimmerman & Schunk, 1998; Pintrich, 2000) stress that planning a 'suitable' learning environment is crucial in order to promote SRL. They suggest to design activities that improve self-monitoring and make the learning process explicit by means of reflective activities aimed at developing meta-cognition.

With respect to our research study, both the specificity of the remedial situation and the leading principle of SRL on which it is based, require a finer analysis in order to identify the main characteristics to be considered in designing an appropriate learning environment to trigger students' awareness of their own resources, self monitoring and practicing self-regulation in general.

The Key Role of Feedback

Necessity of remedial may be diagnosed through the appearance of behaviours that are inadequate for the solution of specific tasks. However it is only when the student becomes aware of the inadequacy of her/his behaviour with respect to her/his own goals that she/he may accept the challenge of trying to correct it. The possibility of reaching such awareness depends on the type of feedback generated by student's behaviour, thus it is through feedback that the student can become aware of her/his inadequate behaviour and the remedial action may be initiated.

In the perspective of SRL, different authors (Balzer, Doherty & O'Connor, 1989; Butler & Winne, 1995) explore the role of feedback in acquiring self-regulation. Balzer, Doherty and O'Connor (1989) argue that not only any specific output about the task (*external feedback*) gives a hint on how to perform it, but it may also enhance learners' self-consciousness and consequently stimulate the learner to renew her engagement in the task's solution. Briefly, the function of external feedback consists in generating this development of an internal process that the authors call *internal feedback*. This external/internal exchange is described as follows: when after an external feedback an internal feedback is activated, the acquired self-control may alter knowledge and beliefs that might influence subsequent self-regulation. As a consequence, when the subject is newly confronted with the same external feedback, she/he may have a different reaction. From the perspective of SRL in which our didactical problem has been framed, the interplay of this external/internal exchange may contribute to describing the development of the self-regulated learning process both at the cognitive and the meta-cognitive level.

Elaborating on these assumptions we designed our remedial intervention. Firstly, making students regain autonomy and responsibility in the management of their own errors became the main goal to be achieved in order to make the remedial intervention successful. Thus, we set up a learning environment where errors might be perceived as failures with respect to students' own goals, so that students could be encouraged to autonomously modify those behaviours that led them to failure.

In order to promote the 'self' in the learning process, we decided to exploit Aplusix' potential, given by the feedback component, which makes it a tool able to engage and re-engage the student in the task and to trigger the external/internal exchange.

In the following section we focus on the main characteristics of Aplusix, especially on its potentialities in terms of providing feedback. Then we describe the rationale of the remedial intervention reformulated in the SRL frame, that is, how the Aplusix facilities were exploited and amplified to provide adequate feedback for specific activities, and in doing so, setting up a learning environment promoting SRL practices, apt to make remedial effective.

THE APLUSIX ICT ENVIRONMENT

Aplusix is a CAS (Computer Algebra System) in which students can perform both arithmetical and algebraic calculations (Nicaud, Bouhineau & Chaachoua, 2004; 2006). The main feature of the software is the possibility of activating a feedback based on the equivalence of algebraic expressions. In fact, two different contexts of use are available to the user: the *Training mode* and the *Test mode*. The Training mode is characterized by the presence of feedback, while in Test mode no feedback is provided.

More in detail, in Aplusix it is possible to create algebraic expressions and operate on them. Each expression appears on the screen enclosed in a box, expressions obtained in successive steps of elaboration appear in successive boxes (Figure. 1), feedback is given by means of three different signs between two successive boxes containing the algebraic expressions; each sign refers to the relationship of algebraic equivalence between the expressions contained in the two successive boxes. These signs appear automatically on the screen at each typing in the active box. Black lines point out that an expression is equivalent to the previous one; red crossed lines mean that the expressions not equivalent; blue crossed lines indicate that equivalence or not equivalence cannot be stated, for instance if the expression contained in the last box is not well formed (e.g. the user opened a parenthesis but forgot to close it) (Figure 1). Since what appears on the screen between the boxes of successive steps are effectively 'signs' which can be seen as feedback, we called them *feedback-signs* (Maffei, Sabena & Mariotti, 2009, p. 67). No explicit reference to the algebraic interpretation of the feedback signs is given by the environment. Nonetheless, an immediate interpretation related to common sense in considering red crossed lines as indications of 'error' can be expected. Moreover, in case one tries to go ahead when the CAS gives a sign of non-equivalence, that is, red crossed lines, a *message* is displayed telling the user that it is impossible to go ahead because the current box is not equivalent to the previous one. This message provides a new feedback with an explicit reference to the algebraic interpretation of the relationship between the expressions contained in the boxes.

When the user performs activities in Test mode, as mentioned above, no feedback is provided, however Aplusix offers the possibility to revise one's own work, and if necessary, to correct the solutions given to the tasks, by means of the modality *Observation activity*. This allows the user (the student, the teacher or the researcher) to revise the solution given to a particular task step by step. When the replay of the solution/s is activated, the user receives from the system the feedback both in case she/he has performed the task within the training mode and in the test mode. Two different options of this modality are possible: *observation* and *observation/correction*. In the first case the user can only observe the process without interacting with the machine, in the second case the user can also intervene, for instance correcting the mistakes that are highlighted.

Another interesting component of Aplusix is the *Detached Step* tool *(DS)*. While solving a task, the user can open a new independent working space (the Detached Step), where new calculations can be carried out independently of the solution sequence. The feedback provided in the Detached Step is consistent with the mode set in Aplusix, that is, when Aplusix is set in Training mode feedback is provided (Figure 1), whereas when it is set in Test mode no feedback is given.

THE EXPERIMENTATION

The Objectives and the Participants

According to the original design, concerning the study of the role of ICT in a remedial activity, attention was primarily focused on the functioning of Aplusix tools with respect to the cognitive processes involved in formula-memorization. The former project, involved two classes and was included in the official remedial session organized by the school. A control on the didactic effectiveness was designed but it concerned students' achievements in terms of disciplinary goals. The activity started with an initial test in paper and pencil; the same test was going to be repeated at the end of the teaching intervention, with the aim of evaluating the improvement of pupils' performances.

As long as the experiment was being carried out, results emerged concerning meta-cognitive processes related to becoming aware of one's own difficulties and to managing one's own resources to improve calculation performances. The potentialities offered by Aplusix seem to be very promising, especially thanks to the kind of control on pupils' actions provided by the presence of a particular feedback. The following section will present a brief analysis of this control function according to the specific theoretical framework of SRL.

Exploiting Aplusix' Tools from an SRL Perspective

We analyze the Aplusix environment and the feedback facilities from the point of view of SRL based on the classic model of Zimmerman (Zimmermann, 1998; 2001). We discuss the sense in

Figure 1. Different types of feedback provided in Aplusix in Training mode

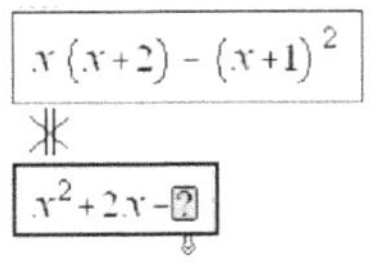	The user has started to edit (second box), trying to obtain an expression equivalent to the expression enclosed in the first box. The Aplusix system has put blue crossed-lines between the two boxes, indicating that, at that instant, the expression the user is editing is 'not well-formed'.
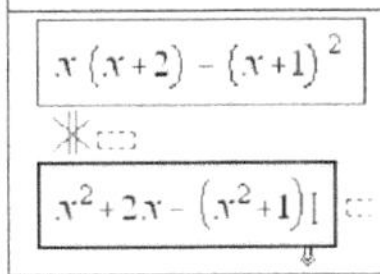	The user has continued to edit the expression in the second box. The Aplusix system has put red crossed-lines between the two boxes indicating that, at that instant, the expression the user is editing is 'not equivalent' to the previous expression.
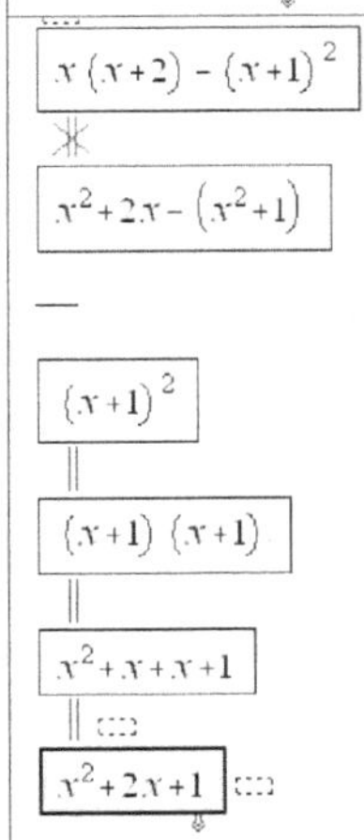	After having activated the Detached Step (a black segment shows the activation of this tool), a new working space, independent from the solution space is available. In the new working space the user has carried out the manipulation of a sub-expression. The Aplusix system has put black lines between consequent boxes indicating that the expression in a box is 'equivalent' to the expression in the previous box.
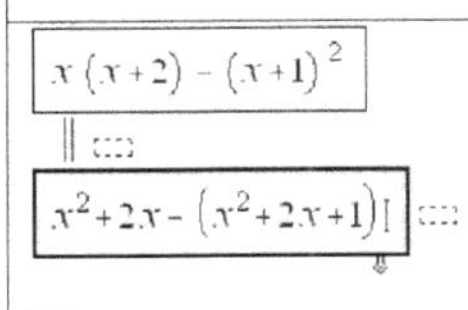	The user has copied the expression obtained in the Detached Step in the solution sequence substituting it to a sub-expression he/she had previously edited (cfr. Fig. 1.c.). As a consequence, the Aplusix system has put black lines between the expression the user has modified and the previous one.
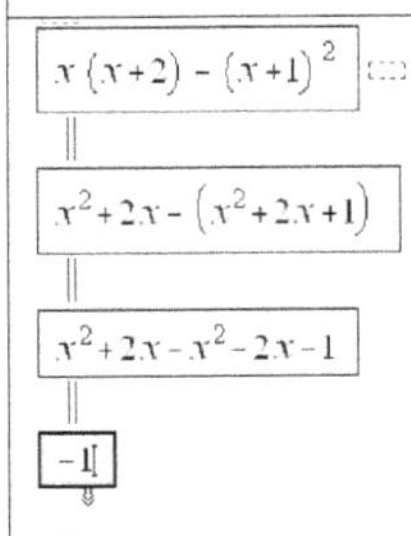	Finally, the user has completed the manipulation of the expression creating two new boxes. The Aplusix system has put black lines between successive boxes.

which Aplusix' facilities, and feedback in particular, fit those features outlined by Zimmermann as 'desirable' when SRL practice and development are among the learning objectives. The model identifies three types of features in promoting SRL: 'features supporting planning', 'features supporting task execution and activity monitoring', and 'features supporting self-assessment'.

Let us now consider the different kinds of feedback provided by Aplusix in Training mode. All of them can be classified as 'features supporting task execution and activity monitoring'. In the Training mode, each step of the calculation is monitored by the machine that checks whether algebraic equivalence has been respected or not. Feedback on the equivalence, as described above, offers a constant support for the correct execution of the task providing information about the status of the equivalence between the expressions enclosed in two consecutive boxes. On the other hand, the explicit statement about non-equivalence, and the interdict for the solver to proceed when she/he tries to go on neglecting the indirect information, provide not only a feedback about the correctness of the procedure, but also a strong input to search for what is wrong in order to overcome the impasse. In this respect, this kind of feedback message plays a crucial role in triggering students' personal engagement in identifying the origin of the impasse.

As far as the Detached Step is concerned, it can be considered a 'feature supporting planning'. Its activation makes sense when the solver intends to address a specific problem, namely a sub-problem with respect to the task in progress. As we will see, the students are going to use this working space whenever they identify a specific goal that may consist in obtaining the formula they need to apply in solving the task, or in carrying out an alternative strategy in order to solve the task, i.e. not using the formula but the distributive law, or in isolating a particularly difficult sub-expression to be treated simultaneously with other manipulations of the expression which it is part of. The use of DS is expected to support students in isolating the element that is causing difficulty in the solution of the task, e.g. separating it from other elements involved and preventing them from interfering with it.

Finally, the Observation/correction modality can be classified among the 'features supporting self-assessment'. This feature of Aplusix gives the possibility to revise the solution performed in the Test mode: the feedback on the equivalence highlights the errors and the user can come back to the task and correct them.

In the following section we describe how the design of the remedial intervention exploited the potential offered by Aplusix in order to promote self-regulated processes. More specifically, in the design of the remedial learning environment we decided to amplify 'supporting and monitoring features' present in Aplusix setting up an 'ad hoc' *Help-Message (HM)* with the interdict of proceeding. The HM suggests opening a DS and it provides a hint for solving the impasse. This new form of feedback may be seen as 'supporting planning' mainly because of the activation of the DS.

The Rationale of the Remedial Intervention

Since the didactical issue we dealt with concerned difficulties in symbolic calculations and was related to memorization and application of algebraic formulas, the hypotheses we put forward concerned both the cognitive and the meta-cognitive sphere. In particular we aimed to help students memorize formulas but we also aimed to help students develop their confidence on the possibility of activating rescue strategies of calculation based on the algebraic meaning of a formula.

The first objective became fostering the use of an alternative strategy when the required formula is not available to memory; specifically promoting the application of the distributive law as the fundamental equivalence relationship that may be used to develop symbolic calculations. Any algebraic formula states the equivalence between two expressions, and such equivalence condenses in a single relationship the iteration of the application of a chain of equivalences stated by the distributive law (Kieran, 1992; Kieran & Drijvers, 2006; Mariotti & Cerulli, 2003). In other words, the iteration of the process starting from one of the two expressions and producing the other by means of algebraic manipulations is supposed to be compacted into the memorized formula. The procedure therefore 'disappears' behind the equal sign between the initial and the final expression.

This should support students' awareness of the fact that their performances do not depend only on their memory and that a rescue strategy is always available since any formula can be reconstructed by using the main properties and specifically the distributive law.

Moreover, the repetition of the same calculation procedure, again and again, should have an effect at the meta-cognitive level motivating the use of the formulas as a personal need to speed up the calculation. According to the objective described above which consisted in promoting such a 'rescue strategy', we designed a specific learning environment based on Aplusix where calculation tasks are expected to be solved interacting with the machine by means of a specific system of feedbacks. Since Aplusix' default feedback appears as a pop-up message (*'The step is not equivalent to the previous one'*) when Aplusix is in a non-equivalence state and the user tries to proceed, we created a specific *Help Message*. The message says:

The step is not equivalent to the previous one. Hint: remember that 1) $a^2=a \cdot a$; 2) $a^3=a \cdot a \cdot a$ or $a^3=a^2 \cdot a$ and apply the distributive law. You can use the Detached Step to develop the calculation.

This new message of feedback is assumed to amplify the effect of the original feedback provided by Aplusix. In fact, it can be considered both a feature 'supporting task execution' and a feature 'supporting planning', mainly because of the suggestion to activate the DS.

Similarly, approaching the second objective of supporting the memorization process, the default message of error was re-elaborated to include a help message offering the possibility of opening a *Help Window* (Maffei, 2004; Maffei & Mariotti, 2007). The new help message says:

The step is not equivalent to the previous one. Are you sure you have not forgotten any term? If you are sure of that and you cannot find the error, you can open the Help Window.

The Help Window (HW) consists of a list of formulas. Two different HWs are available depending on the type of task that is to be performed; when the task requires to expand an expression, the HW displays the expanded parts of the main products; when the task asks to factorize an expression the HW displays the list of the factorized parts of the main products (Figure 2).

Once a student decides to open the HW by activating an icon at the bottom of the worksheet, this new window remains available for 60 seconds

Figure 2. On the left, the Help Window in case of expanding tasks; on the right, the Help Window in case of factorizing tasks. In both cases the message on the top says: "This windows will remain visible for 60 seconds. Choose the expression that you need among the following ones"

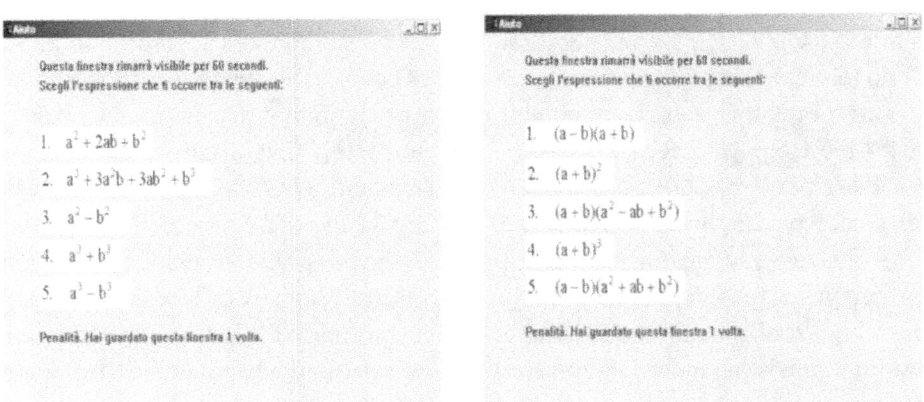

at most. When the HW is open, writing on the worksheet is interdicted. A penalty is counted every time a student uses the HW. The short time given to choose a formula in the list and the penalty in consulting the window should motivate the students to memorize the formulas as soon as possible.

The choices made in planning the remedial intervention can be explained by our intention of designing a particular kind of learning environment. In this one a task is proposed and specific feedback is set up, aimed at triggering an interplay (Balzer, Doherty & O'Connor, 1989) between external feedback – that is, a given input about the task – and internal feedback – that is, an internal process concerning the task and its solution. Such interplay is expected to promote SRL development.

Moreover, we can explain such development of SRL through a *SRL-cycle* (Figure 3). The first phase of the cycle is triggered by the feedback provided by the Aplusix ICT on the equivalence between two consequent steps. This external feedback generates an immediate interpretation that allows the student to keep control over the solution process, orienting her/his decisions on what to do. In particular, the appearance of the crossed red lines should make the student stop and reflect. Students find this feedback puzzling because, beyond the immediate interpretation that something is wrong, the student wonders why it is wrong and how to proceed (internal feedback). The internal feedback raises attention and curiosity, and a first question comes to their minds 'Why?'. The second question is 'What can I do to make these red crossed lines disappear?'. At this point, the second component of the feedback - the Help Message or the Help Window - provides a new external feedback that not only gives some specific hints to overcome the impasse (for instance, suggesting to exploit the definition of power or recognizing the correct formula to apply) but it also suggests activating an external resource consisting in the DS. In other words, a second phase of the cycle occurs: the combination of the default feedback and the designed "helps" constitutes a new complex external feedback leading to an internal feedback involving both the cognitive and the meta-cognitive level; it generates an internal process re-launching the solution process that can progress either directly, using the new information available to reach the correct answer, or indirectly, leading the students to define a new task in the working space offered by the DS - where the default control of Aplusix is still available. In this way a cycle, initiated by the appearance of the red lines, closes, and a new one starts. A sequence of cycles may occur, until the students is able to go back to the original task and complete it (Figure 3).

The Remedial Sequence

On the basis of the hypotheses made, two different phases of intervention are planned, centred on the use of Aplusix. Each phase is characterized by four lab-sessions in which students are requested to solve a list of tasks in the *Training mode*. As explained above, the specific *Help Message* accompanies the default message of error.

The first phase aims at consolidating the basic understanding concerning symbolic manipulation of 'expansion' as the successive application of the distributive law, and at making pupils conscious of the fact that memorization, although not indispensable, is useful and possible. The tasks are common tasks requiring the expansion of different powers of polynomials. The sequence of the tasks is organized according to increasing levels of complexity: from second to higher powers and from binomial to trinomial. Consistently with its goals, this first phase is characterized by making available a *Help Message* which, besides suggesting the use of a *Detached Step,* reminds the solver of the possibility of using the distributive law. As said, this suggestion is meant to induce the students to appropriate what we called a rescue strategy to be activated whenever they feel at risk of failure.

Figure 3. The solution of a task in different phase constituting a SRL-cycle

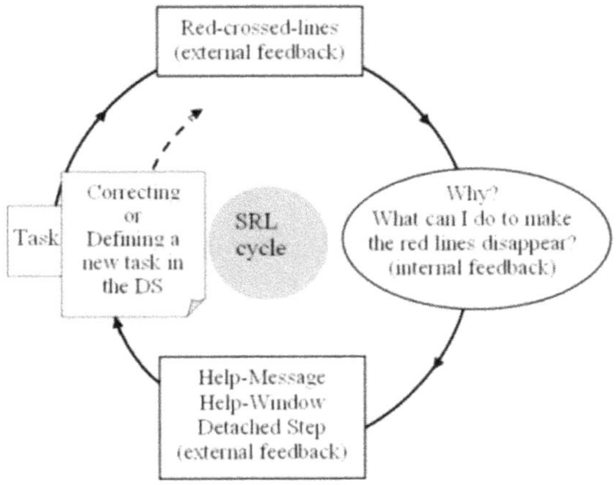

The second phase, aims at fostering memorization of the main formulas. The tasks propose either expansion or factorization. The *Help Message* invites the students to reflect in order to correct their mistakes and, as a last possibility, it offers to open a *Help Window*, which shows a list of possible formulas from which the students are invited to select what they need.

After each session, as homework, students are requested to write a report on what they think they have learned, both in terms of formulas and in terms of solution strategies, and comment on their use of Aplusix. These reports are designed as meta-cognitive tools, because they stimulate the student to reflect and make explicit what has been done during the lab-sessions and the specific role of the ICT, both in terms of 'what' it is possible to do with Aplusix facilities (direct potentiality of the environment) and 'how' it is possible to exploit such facilities (indirect potentiality). We highly value this reflective practice because of its potential in terms of making students aware of their own resources and the ways of managing them.

At the end of the two phases, a test is passed in the *Test mode*, and at the end, in the last session of this phase, students are asked to revise their work using the *Observation modality*. In this revision session the machine control will indicate the errors, while the students are asked to correct them on their notebook. We do not choose the Observation/correction modality since our aim is to evaluate the level reached by pupils in correcting their possible errors only counting on the output given by Aplusix.

During all the phases, students work individually; the teacher is present in the room but her interventions are limited and mainly concern technical questions about the functioning of the software. In our view, this choice should stress the importance of personal engagement in any remedial process. The role of the teacher in this specific context was not under investigation.

At the end of the experimentation the participants where interviewed to collect their final comments about the experiment carried out with Aplusix.

During the teaching experiment different kinds of data were collected. Besides the records of all the Aplusix sessions for each student, we collected all the reports and the transcripts of the interviews carried out at the end of classroom intervention.

SOME RESULTS

As already said, the results of the remedial intervention can be interpreted from two different perspectives. On one hand, according to the aim of its design, the intervention had a direct positive impact on the learning process, leading students to overcome specific difficulties in the symbolic calculation. On the other hand, sort of unexpectedly, the intervention had an indirect positive impact on self-regulating the learning process, since it showed effectiveness in stimulating awareness of one's own difficulties as well the activation of autonomous and responsible decisions about how to overcome them.

In this contribution we focus on this second type of results and provide a qualitative analysis illustrating the functioning of the remedial cycle and in particular the role of the specific ICT–based feedback in supporting SRL development. The presentation is organized according the functioning of the different components of the environment, that we identified as elements activating SRL, specifically: the feedback on equivalence, as well as the functioning of both the Help Messages and of the Observation activity.

The first result concerns the synergy between the red-crossed-lines feedback and the use of the DS in developing reflective strategies and autonomous learning paths towards the memorization of a formula. The second result concerns students' appropriation of the DS as a resource that can be exploited in case of difficulties, and that constitutes a basic achievement in the development of SRL practices. Moreover, we found that using written reports stimulates students' reflection on their own strategies and ultimately helps them develop a personal approach to memorizing formulas.

The Synergy Between the Red Crossed Lines Feedback and the Detached Step

The analysis of the solution processes shows the use of the feedback provided by the software as a constant means of control experienced during the calculation. As expected, the appearance of the red crossed lines triggers a sequence of attempts of corrections that most of the times seems not to be well organized.

We present an excerpt from the final interview in which Ylenia is talking about her impressions on the use of Aplusix and she explains that the red lines tell her that something has to be corrected:

when I see the red lines where I made a mistake, I know I have to correct (Ylenia).

This comment exemplifies how the constant feedback induces the student to correct her own errors.

Another significant issue is formulated by Alessio. This excerpt also comes from the final interview:

Aplusix helps to understand the errors because it tells that you have committed an error but it does not tell the exact point and it makes you find it on your own. At the beginning, if one is really bad, he would also need to know exactly where the error is, then little by little one becomes familiar with the way Aplusix helps you to correct (Alessio).

As it appears clearly, when the red crossed lines appear, students are not aware of what exactly is wrong in their behaviour, nevertheless they realize a failure with respect to a general goal concerning the interaction with a machine: everything must run smoothly. Thus students engage themselves in a task that is primarily the task of making red lines disappear. The first and immediate reaction is proceeding through trial-and-error. Sometimes this behaviour is successful, but most of the times the students face an impasse: whatever they do, the red lines do not disappear. This experience of impasse is crucial because it lets the student live the failure of a random behaviour and it stimulates her/him to overcome it by looking for a reflective (thoughtful) strategy. At this point, the Help Message provides an adequate stimulus allowing the re-launch of the task together with the search of a possible solution strategy (see Figure 3). The following protocol is exemplar of what happens during the first phase of the remedial sequence in relation to the feedback of Aplusix and the Help Message.

Protocol # 1(Alessio)

Alessio sees the red crossed lines (Figure 4, on the left), and carries out a number of trials to fix the third and the fourth monomial, (he changes the signs and the literal parts) but without any success. Only at this point does Alessio decide to activate the help message and subsequently follow the given hint.

He opens the DS (see Figure 4, on the right) and within this new working space he applies the distributive law. When the calculation is completed, he copies and pastes the results in the solution space.

Protocol # 2 (Alessio)

Going on in the activity Alessio faces an impasse similar to the previous one (Figure 5).

Similarly, he decides to follow the suggestion of the Help Message and comments on his decision as follows:

Yes, I'll do (I will open the Detached Step), but then it is better if I calculate $(a+b)^3$ straight out, because after that there will be other (expressions) like that.

As announced, Alessio opens a Detached Step (Figure 5, on the right), and instead of calculating the expansion proposed by the task, he sets the general goal of expanding the third power of a binomial: he inserts a first expansion of the expression $(a+b)^3$, and then he calculates it applying the distributive law. Once he has obtained the correct expansion, the student moves again to the solution space where he accomplishes the correct substitution. This behaviour and the comment accompanying the opening of the DS, show a move from a strategy based on pure adaptation to the hint to an autonomous decision of setting a specific learning goal: the derivation and application of formulas. It is interesting to remark that there is no explicit reference to memorization although the comment implicitly refers to the need of a formula available to be applied.

This second example of the use of the DS shows that Alessio established a connection between the use of the distributive law and the derivation of a formula, but it overall shows that he became aware of the possibility of a rescue strategy as well as of the usefulness of a formula.

The autonomous decision of deriving the general expansion for the cube of a binomial and the explicit statement about the usefulness of getting

Figure 4. The left hand screenshot shows Alessio's calculation and the appearance of the red lines. In the right hand screenshot, Alessio moves to the Detached Step and calculates by using the distributive law

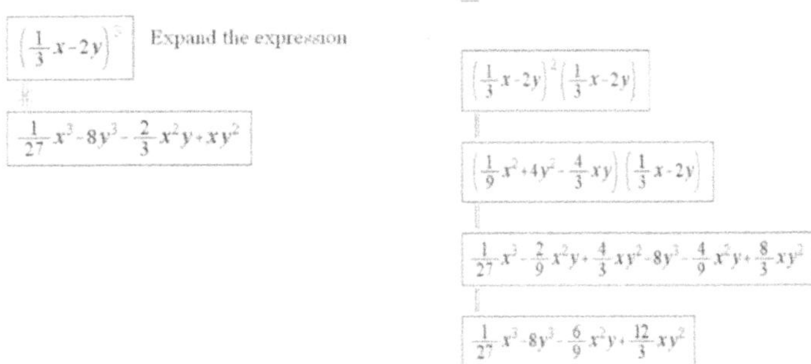

Figure 5. In the left hand screenshot, Alessio realizes that something is wrong and decides to move in the DS. In the right hand screenshot, Alessio calculates the general formula by using the distributive law

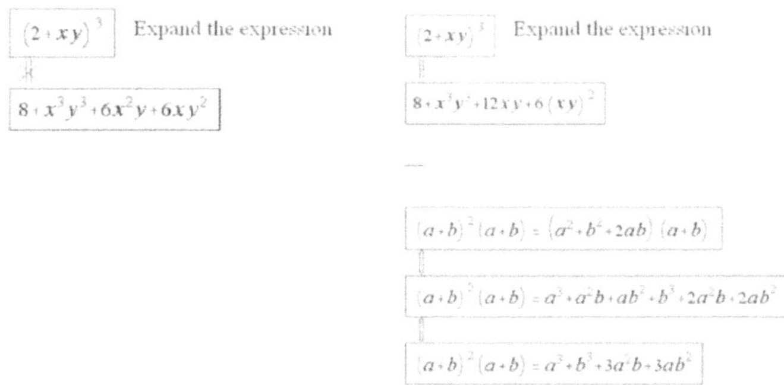

a formula is, in our view, a good example of the evolution of SRL practices related to the availability of the working space offered by the DS.

Mattia shows a similar behaviour. Firstly, he follows the hint and uses the DS to accomplish the calculation using the distributive law; then, in the next task, Mattia calculates directly, but Aplusix provides a feedback of error (Figure 6, on the left). Let us see how he reacts (Figure 6, on the right). We also provide the transcript of a short interaction with the teacher.

Protocol # 3 (Mattia)

Mattia: Nothing has been changed, there is only a minus, why it is not correct?

Teacher: You are right, there is only a minus instead of a plus with respect to the previous expression.

Mattia: And now? I have to follow the help. Should I do everything again as in the previous exercise? I don't want to waste time.

Teacher: Let's try Mattia. It's the last task for today.

(Mattia opens a DS)

Mattia: Yes, because before I was stupid… Now I do that with a and b and then, you will see, now I will find the right formula.

As announced, the student opens a DS where he derives the general formula that immediately after he will apply in the solution space (Figure 6, on the right).

Figure 6. The left hand screenshot shows Mattia facing an impasse with a square of a trinomial. In the right hand screenshot, Mattia uses the Detached Step to derive a new formula

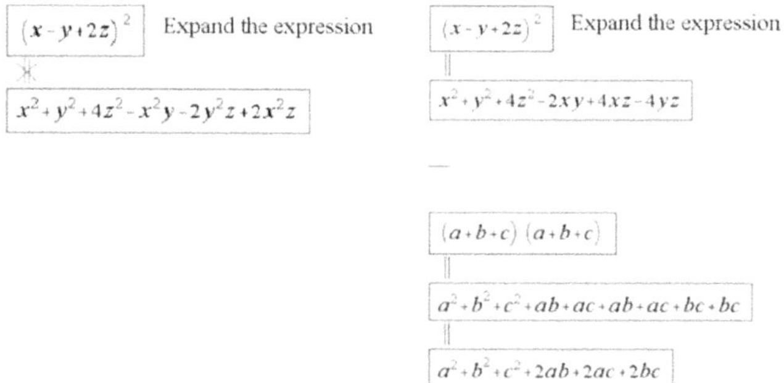

To conclude and summarize the content of this section, we present an excerpt from the final interview with Alessio. After the question of the interviewer, Alessio is explaining how he used the DS specifically with respect to a goal.

Protocol #4 (Alessio)

Interviewer: What was your goal using the Detached Step?

Alessio: It has been useful for the cube of a binomial, since at the beginning I did not remember the formula, thus I wrote the formula in order to know it better. Most of the times I used the step for these reasons, well… but sometimes also to accomplish some calculations.

Interviewer: What kind of calculations?

Alessio: Well, because if I noticed that I had made an error in applying a formula, then I wouldn't have erased all, I was used to use the help, I mean...in the Detached Step I made the whole exercise with the distributive law. Then, before pasting the result above I saw what was wrong by comparing the wrong solution with the correct solution. I usually made mistakes with signs and so I was used to make all the multiplications, in that way I made fewer errors; by the way, I knew that if you know the formulas you also have to know to apply them. Sometimes I made the multiplication because I was not sure about formulas. But now I know formulas, I know all the formulas.

The use of the DS described by Alessio is consistent with what we saw in the previous examples: DS is perceived of great use for accomplishing calculations when memory fails (as Alessio says: *"at the beginning I did not remember the formula"*).

Alessio is also aware of the role that working in that space had in terms of constructing and learning the formula (*"I wrote the formula in order to know it better"*, Alessio says). Moreover he declares that he found it useful to combine the feedback on the equivalence provided both by the DS and by the solution in order to reflect on his own errors. When he realized to be wrong he took advantage of the possibility of comparing the wrong solution with the correct one developed in the DS environment. Using this strategy, Alessio seems to have also elaborated a clear idea of his own specific difficulties and "classified" his errors.

In the following section we are going to show other possible utilizations of the DS. They appear as evolved strategies where the potential offered by the control of equivalence in this working space is exploited and elaborated by students into refined strategies.

Using the Detached Step and Becoming Aware of One's Own Resources

As expected, the working space offered by the Detached Step induced the students to develop more refined calculation strategies, based on the identification of their specific difficulties. The following example shows how a student, Adriano, is able to exploit the control of Aplusix within the DS in order to accomplish the calculation that he considered at risk. As appears in the trace of the solution (Figure 7), Adriano seems to have reached a good level of awareness about his difficulties. The moment he foresees the difficulty he stops the calculation and before letting the machine react with red lines, he decides to move to the DS.

Protocol #5 (Adriano)

When Adriano has to face at the calculation of the first triple product, he writes the minus sign and moves to the DS (Figure 7, on the left). He seems to foresee the possibility of the appearance of red lines and stops before it happens. Once in the DS, he sets the calculation and completes it taking advantage of the control of equivalence. When he is certain of the correctness of the result, he feels confident to paste the expression obtained in the solution space. He repeats the same procedure for the second triple product (Figure 7, in the

Figure 7. In the left hand screenshot, Adriano calculates a triple product in the Detached Step. The central screenshot shows the same strategy. The right hand screenshot shows how Adriano completes the task correctly

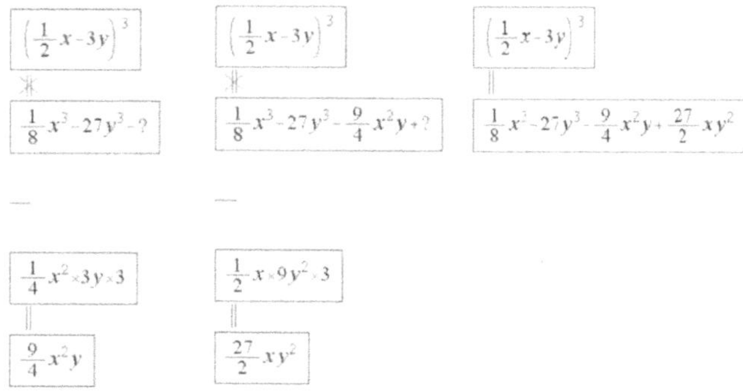

middle) and completes the task correctly (Figure 7, on the right).

This constitutes a good example of the kind of practices that may be developed within the designed remedial learning environment. Specifically, in this example we can observe how the student reaches a positive attitude towards his own errors: besides self-consciousness that makes him stop when he perceives the risk of error, he develops a fluent capacity of taking advantage of the resources provided by the environment.

We now show a different way to use the DS component. In the following protocol Renée finds it sufficient to have a free space where a delicate part of the calculation can be treated.

Protocol # 6 (Renée)

Renée explicitly recognises the formula that is needed – she writes in the comment *"it is the cube of a binomial"* (Figure 8)- but at the moment of calculating the triple products she decides to isolate these calculations and moves to the DS.

Definitely taking a sub-expression out of an expression allows one to better concentrate on it, but there may be another reason for doing so, as shown by the case of Renée. The use of DS may constitute an external support for both retrieving formulas and performing the mental calculation.

Generally speaking, the use of the DS seems to foster the development of personal strategies of calculation, supporting our findings about the role of DS in guiding the development of self-consciousness and an efficient management of resources, as shown in the following short excerpt drawn from a written report. Talking about the strategies that he developed during the remedial sessions (see Protocol 5), Adriano writes: "*Then in these last sessions I found a terrific way of using the Detached Step. Now I am sure that using Aplusix I will not be wrong any more, not even when there are the fractions.*"

Help Window and Detached Step Fostering SRL

As shown in the previous examples (see Alessio and Mattia), during the first phase of the sequence

Figure 8. Renée 's use of the Detached Step

Figure 9. The solution produced by Adriano

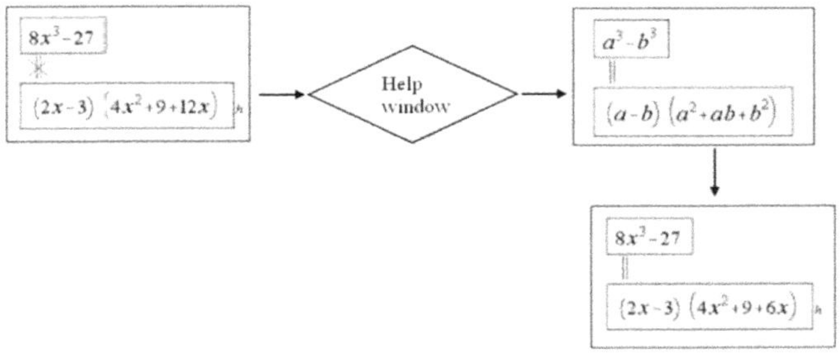

the issue of memorizing has already emerged, but students still manifest difficulties in the retrieval of the correct formula and its application. In the second phase of the remedial sequence, where the tasks and the support were specifically designed for supporting the memorization of formulas, the learning environment has functioned not only to foster the memorization process but also to make students aware and responsible of their own learning process.

We now present a protocol from the second phase where the student uses the strategy to exploit the DS after having seen the formula he needs in the Help Window.

Protocol #7 (Adriano)

Adriano realises his difficulty and opens the Help Window.

Adriano is not able to correctly retrieve the formula of the difference of cubes (Figure 9, on the left).

He thinks that the formula contains the double product of the factors and hypothesizes an error with the signs; he makes some changes on the signs of the expression. Then he decides to open the Help Window. Although he only sees the factorized part of the formula in the list of formulas contained in the HW, it seems that he recognizes it and in fact, after having closed the window, he immediately writes the whole formula in the DS (Figure 9, top right). Afterwards, he moves to the solution space where he produces the correct factorization (Figure 9 bottom right). In our view, this provides evidence of the fact that Adriano exploits the DS as additional support to memorization. He does not seem confident in his memory yet, but he seems at ease in controlling the resources offered by the environment.

It is possible to observe different uses of HW, in both the case of expansion and that of factorization. According to the SRL-cycle described above, after the feedback on non-equivalence the HW is activated and efficiently exploited in combination with the use of DS. The following protocol is exemplar.

Protocol # 8 (Mattia)

When the red lines feedback warns Mattia that his solution is not correct, he makes different attempts at fixing it. At each attempt he changes the signs of the monomials, but the red lines still remain (Figure 10, on the top). At the end, he opens the HW, looks at the list of expressions, and, going back to the solution sequence (Figure 10, on the bottom), makes the proper correction. It is not easy to interpret what happens when Mattia opens the window, what kind of help he gets from the list of expressions. We hypothesize that opening the window and looking at the formula lead Mattia to stop and reflect on what he is doing. As the protocol shows, the product of his reflection is not only the correction but also the commentary

Figure 10. Mattia cannot find the error and opens the Help Window. Afterwards he corrects the error and comments on it. He writes: "In fact, they were not the signs. I was doing a wrong calculation."

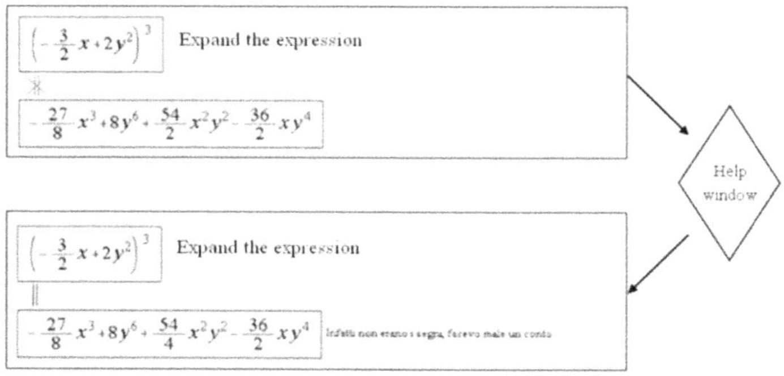

on his own mistake. He writes: "*In fact, they were not the signs. I was doing a wrong calculation.*"

Thus it seems that Mattia has overcome his first evaluation of his difficulties related to the management of signs, gaining consciousness of the possibility of making mistakes in the calculation because of inefficiency in the retrieval of the formulas.

This interpretation is supported by the following statement that Mattia writes in one of his reports:

As regards the formulas, I knew them already but not well (sometimes I obtained them in the Detached Step). Now I really know the formulas, I have them stamped in my mind.

This comment clearly shows what we argued before about the need for a combination of the two different types of feedback in order to promote SRL. In fact, the feedback of the red lines acts as a spark that triggers the necessity to recognize the proper formula in the HW. In the meantime, having no possibility of copying the formula (when the window is still opened) triggers the memorization process. The repeated iteration of the process consisting in shifting from the formula, displayed in the window, to its application in the worksheet seems to positively affect the process of memorization. Specifically, the separation into two consequent phases, retrieving the formula and applying it, is amplified by the fact that the HW gives information only on the formula, while the students remain completely in charge of the second phase, its application.

In summary, the variety of the discussed strategies, also considering those developed by the same student during the experiment, suggests a students' 'creative' way of planning and monitoring their own solution process, and more generally of developing self-regulated approaches in memorizing formulas.

Many cases show evidence of the fact that the students acquired self-consciousness about the level of memorization reached. We find a confirmation of this fact in the reports written by the students, like in the following example. In the last report, at the end of the experiment, answering to the question *"Did you memorize any strategy using Aplusix?"*, Alessio writes:

I used the strategy to look often at the Help Window because in one way or another it was always useful. With the square of a binomial and the difference of squares, I have no problem; I have some hesitations, even if small, with the cube of a binomial and the product of polynomials. The help was very useful, but I think it should be very penalizing in an evaluation test.

Alessio seemed to make a meta-analysis of his own cognitive and meta-cognitive processes. In particular he seems capable of self-evaluation both in terms of what he has not yet achieved (he speaks of small hesitations) and of the contribution coming from the support provided by the HW.

The Final Test and the Self-Correction

At the end of the remedial intervention a test was assigned, followed by a self-correction session. The test was to be carried out with Aplusix set in the Test mode, that is, without any feedback or help activated. The tasks required the expansion and the factorization of main products. In the self-control session students used the Observation modality to revise their work in the Training mode. In this modality, in which the feedback indicated the errors, the students were asked to correct their errors on their notebook. The analysis of the data shows evidence not only of improvements of students' performance but mainly of achievements in terms of development of SRL practices, related to the use of the specific ICT-based learning environment. Consider the following example.

Antonio (Figure 11, on the left) stops at the very beginning, he seems to be conscious of his difficulty in retrieving the formula and he does not even open a new step in the working space. Taking advantage of the resources of Aplusix and pursuing the goal of overcoming the impasse, he activates the DS command. In the new working space he writes the formula of the cube of a binomial. Then, back to the main working space (Figure 11, on the right), Antonio successfully applies the formula and writes the comment 'cube of binomial' in the solution sequence, with the aim of stressing the recognition of the formula. Antonio shows a good level of control on his own resource both in terms of memorization and in terms of exploiting the support provided by the machine.

Even more significant is Antonio's behaviour during the revision session after the final test. The control activated in this phase (Observation modality) makes Antonio realize his error (Figure 12, on the left); according to the requirements of the task the student provides the correct version of the exercise in his notebook (Figure 12, on the right).

After the feedback on the error the student makes an hypothesis on the type of error, actually he writes *"Maybe the signs?"* and then he decides to take a global approach, moving to a *new line* where he reconstructs the needed formula by using the distributive law. In other terms, Antonio reproduces the way of acting as if he worked in the environment created by the DS. Such behaviour can be interpreted as the effect of the development of new practices that originate in student's experience with the support provided by the ICT. Now the student seems able to activate the rescue strategy in a different context, in the paper and pencil environment. We notice how

Figure 11. The left hand screenshot shows Antonio's calculation. In the right hand screenshot, he writes the comment "Cube of binomial"

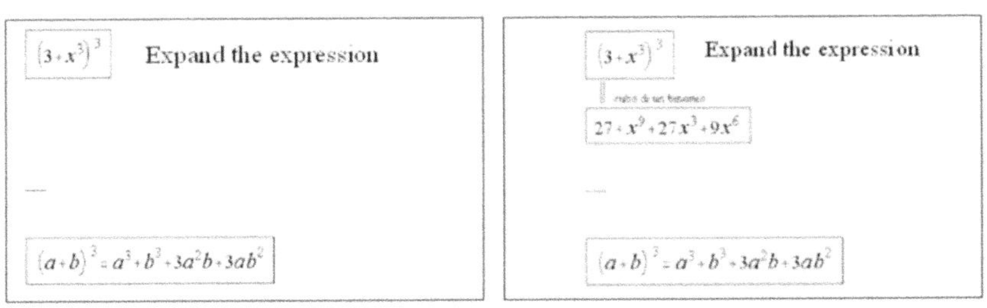

Figure 12. What appears on the screen and Antonio's self-correction. The final comment means "So the correct exercise is (...) I had got the last sign wrong"

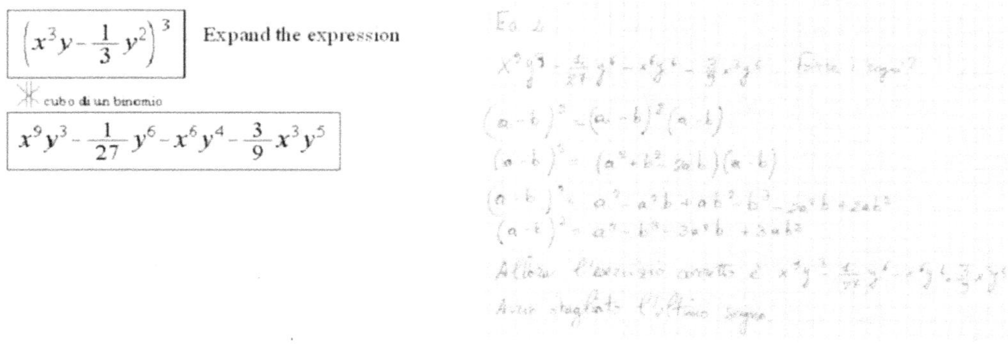

significant features of the original Aplusix context are reproduced, for instance the structure of the writing, that is, lines are organized respecting the equivalence between two successive lines. The experience in Aplusix seems to have fostered the development of general strategies for the retrieval of formulas, and in our view this contributed to develop students' confidence in their own capacities in managing their resources. Moreover, according to our assumptions, a personal student-error relationship seems to be consolidated: Antonio, like many other students, seems to have achieved a good level of self-consciousness and self-control as a consequence of his work within Aplusix.

Generally speaking, the control offered by the ICT seems to lead students to change the way they relate to their own errors. The following excerpt (Ylenia), drawn from one of the final interviews, provides very good evidence of this point: *"Aplusix has some good features: everything seems easier than in paper and pencil. When I see the red lines, I understand that I have made an error (or more than one), I like it very much...on the contrary, when my teacher corrects my test I don't even look at the errors, the most important thing I pay attention to is the good or bad mark I got."*

This comment is particularly eloquent and shows a development in perceiving ones own errors. It is interesting to remark how the student compares the feedback provided by Aplusix (*"when I see the red lines"*) with the common feedback coming from the teacher.

The change of attitude with respect to errors is consistent with our hypothesis about the effectiveness of the remedial activity, but it also witnesses the development of effective practices of SRL that seem promising far beyond the specific case of remedial on symbolic calculation tasks.

CONCLUSION

The remedial intervention discussed above starts from the assumption that a remedial can be successful if and above all the students involved can experience their errors and assume the responsibility for overcoming them. The design of the remedial session was based on general assumptions concerning algebra calculation but also on the assumption that one effective way to develop self-regulation in students is to provide them with opportunities to practice regulating aspects of their own learning and to reflect on that practice.

The design of the experiment was centred around the use of the Aplusix environment. In particular the feedback component was elaborated and exploited to make students face their errors.

Then, the Help Messages and the Help Windows were designed to propose specific strategies to students. What we find is that students not only followed the hints, but they developed them in a personal and creative way. This can be considered a by-product of students' engagement in the specific learning environment. Our findings provide evidence that in the designed learning environment students can self-regulate internal states and behaviour as well as some aspects of the environment, for instance the development of personal strategies exploiting the control offered by Aplusix' feedback on the equivalence and the working space offered by the Detached Step. Moreover, the modes of use of the different types of feedback establish a student-ICT interaction in which the teacher's action is temporarily hidden. In this way, besides personal engagement in the remedial process there was the beginning of a new personal engagement in the learning processes.

These promising findings induce us to hypothesise that the kind of learning environment set up, characterized by the interaction with the Aplusix' features, may promote the development of SRL processes in particular for those students at risk of failure. The main characteristics that showed their effectiveness are linked to the following issues:

- Temporarily suspending teacher's control in favour of automatic control;
- Automatically highlighting the presence of errors and preventing/inhibiting to progress without fixing the error (that is, implicitly re-launching the task);
- Providing a free working space to elaborate personal solution strategies (DS);
- Providing a specific support to memorization (HW);
- Requiring an explicit reflection on one's own solution processes. The writing of reports asks specifically for a reflection on how to use the different facilities available in Aplusix.

All the components of Aplusix analysed in this contribution with respect to their potentialities in triggering SRL processes can be modulated according to the didactical aims. As we pointed out, each component has a specific role in promoting SRL processes.

Other studies have been planned in order to further investigate how each component of Aplusix (and the components created to interact with it) may contribute to making a learning environment become a 'self-regulated learning environment'.

ACKNOWLEDGMENT

Our thanks go to the anonymous students who participated in our teaching experiment. We are also indebted to Anna Baccaglini-Frank whose help went far beyond language revision.

REFERENCES

Balzer, W. K., Doherty, M. E., & O' Connor, R. (1989). Effects of cognitive feedback on performance. *Psychological Bulletin*, *106*(3), 410–433. doi:10.1037/0033-2909.106.3.410

Brown, A. L. (2002). Design Experiments: Theoretical and Methodological Challenges in Creating Complex Intervention in Classroom Settings. *Journal of the Learning Sciences*, *2*(2), 141–178. doi:10.1207/s15327809jls0202_2

Butler, D. L., & Winne, P. H. (1995). Feedback and Self-Regulated Learning: A Theoretical Synthesis. *Review of Educational Research*, *65*(3), 245–281.

Kieran, C. (1992). The learning and teaching of School Algebra. In Grouws, D. A. (Ed.), *Handbook of Research on Mathematics Teaching and Learning* (pp. 390–414). New York, NY: MacMillan.

Kieran, C., & Drijvers, P. (2006). The Co-Emergence of Machine Techniques, Paper-and-Pencil Techniques, and Theoretical Reflection: A Study of CAS use in Secondary School Algebra. *International Journal of Computers for Mathematical Learning*, 205–263. doi:10.1007/s10758-006-0006-7

Maffei, L. (2004). *Un intervento di recupero in algebra. Progettazione e analisi di una sperimentazione basata sull'utilizzo del software Aplusix [A remedial intervention in Algebra. Design and analysis of an experimentation based on the use of the Aplusix software tool*]. Unpublished Master dissertation, University of Pisa, 2004.

Maffei, L., & Mariotti, M. A. (2006). A remedial intervention in algebra. In J. Novotna, H. Moraova, M. Kratka & N. Stehlikova (Eds.), *Proceedings of the 30th International Conference for the Psychology of Mathematics Education* (pp.113-120). Charles Universit, Prague.

Maffei, L., & Mariotti, M. A. (2007). Memorizing algebraic formulas: the support of a microworld. In D. Pitta Pantazi & G. Philippou (Eds.), *Proceedings of the 5th Congress of the European Society for Research in Mathematics Education* (pp. 1460-1469). University of Cyprus. Larnaca.

Maffei, L., Sabena, C., & Mariotti, M. A. (2009). Memorizing algebraic formulas: the support of a microworld. In M. Tzekaki, M. Kaldrimidou & H. Sakonidis (Eds.), *Proceedings of the 33rd International Conference for the Psychology of Mathematics Vol. 4.* (pp.65-72). University of Thessaloniki and University of Macedonia, Thessaloniki.

Mariotti, M. A., & Cerulli, M. (2003). Espressioni numeriche ed espressioni letterali: continuità o rottura? [Numerical expressions and literal expressions: continuity or break?]. *La matematica e la sua didattica, 1*, 43-65.

Nicaud, J-F., Bouhineau, & D., Chaachoua, H. (2004). Mixing microworld and CAS features in building computer systems that help students learn algebra. *International Journal of Computers for Mathematical Learning*, *9*(2), 169–211. doi:10.1023/B:IJCO.0000040890.20374.37

Nicaud, J-F., Bouhineau, D., Chaachoua, H., & Trgalova, J. (2006). Developing interactive learning environments that can be used by all the classes having access to computers. The case of Aplusix for algebra. *Le cahiers Leibniz*, 148.

Pintrich, P. (2000). The Role of Goal Orientation in Self-Regulated Learning. In M. Boekaerts, P. Pintrich & M. Zeidner (Eds.), *Handbook of Self-Regulation*, 452-502. New York, NY: Academic Press.

Pintrich, P. R., & Zusho, A. (2002). The development of academic self-regulation: The role of cognitive and motivational factors. In Wigfield, A., & Eccles, J. S. (Eds.), *Development of achievement motivation* (pp. 249–284). San Diego, USA: Academic Press. doi:10.1016/B978-012750053-9/50012-7

Zan, R. (2002). Episode II: Marco and Anna - Each following their own path from Learning from learners. Plenary Panel, In A.D. Cockburn & E. Nardi (Eds.), *Proceedings of the 26th International Conference for the Psychology of Mathematics* (pp.81-110), University of East Anglia, Norwich, UK.

Zimmerman, B. J. (1998). Developing self-fulfilling cycles of academic regulation: an analysis of exemplary instructional models. In Schunk, D. H., & Zimmerman, B. J. (Eds.), *Self regulated learning, from teaching to self-reflective practice* (pp. 1–19). New York, NY: Guilford Press.

Zimmerman, B. J. (2001). Theories of self-regulated learning and academic achievement: an overview and analysis. In Schunk, D. H., & Zimmerman, B. J. (Eds.), *Self-regulated learning and academic achievement* (pp. 1–37). Mahwah, NJ: Lawrence Erlbaum Associates.

Zimmerman, B. J., & Schunk, D. H. (1998). *Self-regulated learning and academic achievement: Theory, research and practice*. New York, NY: Springer.

KEY TERMS AND DEFINITIONS

Aplusix CAS Environment: educational software based on a CAS, allowing users to perform algebraic calculations and characterized by the precence of a feedback on the equivalence between algebraic expressions.

Computer Algebra System (CAS): piece of software that allows the user to perform algebraic alculations.

Feedback (External /Internal): external feedback is any specific output about the task; internal feedback is an internal process which may be triggered by the occurrence of an external feedback.

Remedial Activity: didactical intervention aiming at making students overcome their learning difficulties.

SRL-Cycle: theoretical construct designed to describe the development of SRL through an exchange of external and internal feedback.

Chapter 14
Assessing Self-Regulation Development through Sharing Feedback in Online Mathematical Problem Solving Discussion

Bracha Kramarski
Bar-Ilan University, Israel

ABSTRACT

This study examined the relative efficacies of two different metacognitive teaching methods – problem solving (M_PS) and sharing knowledge (M_SK). Seventy-two Israeli sixth-grade students engaged in online mathematical problem solving and were each supported using one of the two aforementioned methods. M_PS students used a problem-solving and feedback process based on the IMPROVE model (Kramarski & Mevarech, 2003). In contrast, M_SK participants were instructed to reflect and provide feedback on the solution without an explicit model. This study evaluated each method's impact on the students' mathematical online problem solving. It also examined self-regulated learning (SRL) processes by assessing students' online feedback using a rubric scheme. Findings indicated that M_PS students outperformed the M_SK students in algebraic knowledge and mathematical reasoning, as well as on various measures of sharing cognitive and metacognitive feedback. The M_SK students outperformed the M_PS students on measures of sharing motivational and social feedback.

DOI: 10.4018/978-1-61692-901-5.ch014

INTRODUCTION

Self-Regulated Learning in ICT Education

Although Information Computer Technology (ICT) environments present significant opportunities for fostering learning (e.g., Lajoie and Azevedo, 2006), relatively little focus has been placed on understanding how students of different ages successfully manage these environments. Researchers and educators recommend investigating methods for the effective use of these powerful but frequently underused learning environments (Azevedo, 2005; Azevedo & Jacobson, 2008). This research is particularly important as there have been numerous calls for more implementation of technology in education (e.g., National Council of Teachers of Mathematics (NCTM), 2000; Program of International Students Assessment (PISA), 2003).

Several researchers have suggested that one potential mediator between the potential of ICT and academic performance is the quality of students' self-regulatory learning (SRL) processes (e.g., Azevedo, 2005; Kramarski & Dudai, 2009; Kramarski & Mizrachi, 2006). SRL refers to a cyclical and recursive process that utilizes feedback mechanisms (e.g., Butler & Winne, 1995; Zimmerman, 2000). Students are considered self-regulated learners to the degree that they are cognitively, metacognitively, motivationally and behaviorally active participants in controlling their own learning process (e.g., Pintrich, 2000; Zimmerman, 2000). In terms of cognitive and metacognitive processes, self-regulated students are good strategy users. They plan, set goals, select strategies, organize, monitor, and evaluate at various points during the acquisition process (Pintrich, 2000; Zimmerman, 2000). In terms of motivational processes, self regulation refers to students' willingness to learn and to attain academic self-efficacy.

Researchers have also begun to direct more and more attention to self regulation on the social level, where knowledge is distributed among the group members, each of whom uses their knowledge and skills to contribute to the group endeavor. Learning on the social level is known as a "community of practice" and as "situated learning" (Wenger, 1998). Wenger (1998) maintains that in order for practice to generate coherence within a community, the essential factors of *mutual engagement, joint enterprise and shared repertoire* must be present. Mutual engagement means that members of a community of practice are engaged in a common negotiated activity. Joint enterprise allows a community to extend the boundaries and interpretation of practice beyond those that were created. Shared repertoire means that members negotiate their communal resources (routines, sensibilities, artifacts, vocabulary, styles).

A survey of current research on online communities of practice (e.g., Johnson, 2001) raises the question as to what exactly constitutes a community of practice in ICT education. An online community is a virtual community that can easily be set up across cultures via the World Wide Web (WWW), a group whose members are separated in space and time (i.e., geographic location and time zone). The other key concept behind online communities is the use of networked technologies so that the members may collaborate and communicate with each other (Johnson, 2001).

Online Mathematical Problem Solving Discussion Forums

Online mathematical problem solving discussion forums are an example of virtual communities in practice. A discussion is a group situation in which many people share their knowledge and opinions with the others in the group, and argue in favor of their opinion. Online discussions provide students with a wide variety of ways to interact with each other (Han & Park, 2008). They allow asynchronous exchange and enable one-on-one and one-on-many interactions (Johnson, 2001; Kramarski & Mizrachi, 2006). During online discussions, learners can solve problems and share feedback through collaboration with their peers, both for metacognitive and mathematical understanding. Students must explain their own thinking to other group members and adapt their own thinking to the solutions proposed by other members, which in turn may facilitate more efficient use of metacognitive skills. Through critically examining others' reasoning and participating in resolving disagreements, students learn to monitor their own thinking, which in turn improves their mathematical reasoning concepts (e.g., Artz & Yaloz-Femia, 1999).

Unfortunately, research (e.g., Azevedo & Jacobson, 2008; Kramarski & Mizrachi, 2006) indicates that online discussions do not naturally realize their full potential. It was found that without coaching and scaffolding, the content of asynchronous discussion can become poor and superficial (e.g., Oliver & Herrington, 2000; Kramarski & Dudai, 2009; Kramarski & Mizrachi, 2006). Furthermore, research has demonstrated that students have difficulty adopting SRL processes on both the individual and social levels during online discussions (e.g., Kramarski & Mizrachi, 2006).

Students often do not realize *what* they need to do, *how* to do it, *when* and with *whom* to interact in the forum in the most productive manner.

Following Zimmerman (2008) suggestions, as well as findings showing the importance of metacognitive support in the development of SRL in the social context (e.g., Kramarski & Dudai, 2009; Kramarski & Mizrachi, 2006; Schraw, Crippen & Hartley, 2006), the complexity of SRL processes, and the importance of the measures chosen to assess these processes (see later an elaboration on the issue of assessing SRL), this study focuses on two main questions: (1) What are the effects of different metacognitive support methods used in online discussions on SRL, in the context of mathematical problem solving? (2) How can we assess these processes?

This study focuses on a comparison study of developing SRL with two metacognitive methods through online mathematical problem solving discussion forums: Metacognitive Sharing Knowledge (M_SK) with feedback strategy, and Metacognitive Problem Solving (M_PS) with IMPROVE self-questioning method.

Metacognitive Sharing Knowledge (M_SK) with Feedback Strategy

Research has indicated that feedback is an important metacognitive strategy in sharing knowledge in the process. Feedback can be provided/received from sources both external (i.e., peers, computer) or internal (i.e., self-directed guidance). According to Butler and Winne (1995), feedback plays a multidimensional role in aiding knowledge construction. Feedback may encourage students to engage in reflecting on *why* a solution or explanation is right or wrong, and it may further encourage them to consider alternative solution strategies. Feedback stimulates students to adopt more metacognitive strategies in their learning tasks, and may also encourage them to explain their own problem-solving processes (e.g., Kramarski & Mevarech, 2003). According to Artz & Yaloz-Femia (1999), through critically examining other participants' solutions and receiving/providing feedback, students learn to monitor their own thinking, which in turn strengthens their mathematical reasoning concepts. Some evidence indicates that students learn more when they provide explanations and feedback than when they receive explanations or feedback (Kramarski & Dudai, 2009; Webb, 1989). Research on feedback in computer-based learning environments has shown differential effects of feedback strategies on students' learning. Corrective feedback, pertaining to the correctness of the solution, helps immediate learning, whereas explanatory (elaborated) and metacognitive feedback, pertaining to the solving process, helps ensure deep understanding and the ability to transfer knowledge (Aleven &

Koedinger, 2002; Kramarski & Zeichner, 2001; Moreno, 2004). However, most studies have examined the effects of metacognitive feedback provided by external agents such as computers, whereas little research has been conducted on feedback provided by students in online mathematical discussion forums.

Metacognitive Problem Solving with IMPROVE Self-questioning (M_PS)

One promising metacognitive instructional support seems to be the use of self-questioning (King, 1991, Kramarski & Mevarech, 2003; Schoenfeld, 1992). Self-questioning can be directed to the problem-solving process itself (e.g., comprehending the problem) or to metacognitive strategies such as providing explanation or feedback which are essential in the SRL process (Kramarski & Dudai, 2009). Many researchers have emphasized the importance of extensive practice followed by explicit guidance using the *WWWH* self-questioning strategy (*what, when, why,* and *how*). This strategy helps students select a specific self-regulatory strategy, approach, or response within learning (e.g., King, 1991; Kramarski & Mevarech, 2003; Schoenfeld, 1992; Schraw, Crippen & Hartley, 2006; Veenman, van Hout-Wolters & Afflerbach, 2006).

To support students' involvement in regulatory learning, Mevarech and Kramarski (1997) designed the IMPROVE metacognitive self-questioning method. The metacognitive questioning encourages students to actively engage in self-regulating their learning by using four kinds of questions: comprehension, connection, strategy, and reflection. *Comprehension questions* help students understand the information of the task/problem to be solved (e.g., "What is the problem/task?;" "What is the meaning of…?"). *Connection questions* prompt students to understand tasks' deeper-level relational structures by articulating thoughts and explicit explanations (e.g., "What is the difference/similarity?"; "How do you justify your conclusion?"). *Strategy questions* encourage students to plan and to select the appropriate strategy (e.g., "What is the strategy?;" "Why?"). *Reflection questions* help students monitor and evaluate their problem-solving processes, encouraging students to consider various perspectives and values regarding their selected solutions (e.g., "Does the solution make sense?;" "Can the solution be presented otherwise?").

The IMPROVE method is grounded in the SRL theoretical framework. According to Zimmerman, self-regulation models and the problem solving approach are closely linked because they emphasize similar processes (Zimmerman, 2000). The four metacognitive questions empower learners' self-regulation. The questions direct learners' thoughts and actions throughout the SRL phases of the problem solution process (planning, monitoring and evaluation; Zimmerman, 2000). The method is also grounded in socio-cognitive theories of learning, which extend the view of metacognition to encompass both self-directed dialogue and social aspects, such as practice of tasks and group discussion between peers of comparable expertise, thus making the monitoring and regulation processes overt (Brown & Campione, 1994).

In general, the research reports that metacognitive support with IMPROVE self-questioning demonstrated positive effects on students' learning outcomes and SRL processes in different technology environments (e.g., Kramarski & Dudai, 2009; Kramarski & Hirsh, 2003; Kramarski & Gautman, 2006; Kramarski & Mizrachi, 2006). However, most of the studies were conducted on older students and assessed SRL processes with self-report (aptitude) measures. The current study intended to assess SRL with online (event) measures, as described in the following section.

Ways of Assessing SRL

One of the main issues relating to SRL complex structure addresses the methodological ways of assessing students' SRL processes during a knowledge construction activity.

The research literature (Winne & Perry, 2000; Zimmerman, 2008) presents two main measures for assessing SRL processes: *aptitude* (offline measure) and *event* (online measure). The *aptitude* measure is defined as a "relatively enduring attribute of a person that predicts future behavior" (p. 169, Zimmerman, 2008). These measures are mostly used in questionnaires. An alternative approach assesses SRL as an *event*, which is defined as a temporal entity that is examined in real time during learning. An event may be a more accurate measure of processes related to SRL (Azevedo, in press; Veenman, 2007; Winne & Perry, 2000; Zimmerman, 2008). Winne and Perry (2000) asserted that too little has been achieved in measuring SRL as an event, and they suggested the need for measurements and methods to characterize temporal unfolding patterns of engagement in terms of tactics and strategies that constitute SRL.

The current study examined the process of SRL as an event measure by analyzing online feedback with a rubric scheme. The event measure was assessed under the two metacognitive methods: sharing knowledge (M_SK and self-questioning (M_PS).

Current Study Objectives

Based on our research findings, we designed our study to investigate the effects of two metacognitive methods M_SK and M_PS for sixth grade students' online mathematical problem solving and their SRL processes. The M_SK method encouraged students to provide an elaborated feedback when working online to solve mathematical problems in a forum discussion. This, method required that the student *reflect* on the entire solution process of all participants and suggest *modifications* to the solution, if needed. M_SK students were encouraged to engage in a *reflective* online discourse by eliciting questions, clarifications, and explanations without an explicit model. The M_PS self-questions (based on the IMPROVE method) encouraged students systematically to provide feedback and elaborate explanations for their thinking, to clearly refer to the data in the problem and to the problem-solving process, and to suggest a conclusion when working online to solve mathematical problems in forum discussions. These students were encouraged to use the IMPROVE questions during the problem solving and feedback process (see an extended description of both methods in the Method section).

Based on prior studies regarding feedback effects on the learning methods of older students (e.g., Kramarski & Dudai, 2009; Moreno, 2004), we assumed that the M_PS method would be more effective in enhancing both the online mathematical problem solving and SRL of younger students, than the M_SK method, for the following reasons. First, providing online metacognitive feedback in the M_PS group enables the younger students to act systematically as external regulators at a social level and to share multidimensional perspectives regarding solution processes. The M_PS method may help younger students monitor, evaluate, and modify their solutions (if necessary) and may challenge them to try new ways to solve problems. Consequently, we suggested that younger students exposed to a M_PS method would more easily augment their mathematical problem solving processes, and SRL than students exposed to the M_SK.

The present study investigated the effects of M_SK and M_PS methods on students' (a) online mathematical abilities for problem solving: procedural knowledge, algebraic knowledge, and mathematical reasoning; and (b) SRL processes: cognitive, metacognitive, motivational and social aspects assessed by online-event discussions. We developed a unique rubric-scheme measure to assess event-based SRL processes of sixth grade students. The scheme was based on online feedback students provided according to the tasks they performed.

METHOD

This study investigated 72 Israeli sixth graders who were engaged in two different methods M_PS and M_SK. At the outset of the study, there were no significant differences between the two groups in their prior knowledge of mathematics. Means were 73.30 and 71.23 respectively for M_PS, and M_SK (*SD* = 14.80, 15.70).

Shared Structure and Curriculum

Students in the M_SK and M_PS groups participated in five weekly 4-hour workshops. The goal was to enhance mathematical understanding as defined by the new curriculum standards for young students (problem solving, mathematical reasoning, and communication; NCTM, 2000; PISA, 2003). During training, students completed arithmetic practice sets (e.g., numbers and operations) and discussed algebraic concepts (e.g., symbols, expressions, patterns, and representations).

All workshops in both groups were structured similarly. The teacher presented the lesson's subject and contents to the students and emphasized the importance of communities of practice in ITC. A class discussion was then held about working in a group and how to provide feedback by discussing structure, clarity and conclusions.

Students from both groups completed the same practice sets in small online forums of four participants, once a week in the computer lab (90 min). Practice sets were based on the Web Based Learning environment (WBLe), which provided tasks of varying complexity. The sets were followed by links for additional resources in order to solve the tasks. The sets were based on the conceptual framework of the Programme for International Student Assessment (PISA) (2003) for solving authentic tasks requiring different levels of algebraic abilities. Students were asked to solve two tasks every week, send their solution to the forum, provide feedback for the other forum participants' solutions, receive feedback for their own solution, and adjust their solution according to the peers' suggestions as needed. One task was practiced online during the workshop, and the second task was practiced asynchronously during the week.

At the end of each weekly meeting, every group sent their solutions of both tasks and their feedback exchanges to the teacher, and the teacher then provided online feedback within 24 hours to each student in the group. The teacher encouraged each student to provide the other three students in the forum with feedback. The teacher's feedback included comments regarding both the accuracy of the solution and the problem-solving process. Table 1 summarizes the main components in each training program.

M_SK Group

The aim of providing feedback by *sharing knowledge* was to improve students mathematical problem solving according to the NCTM standards. Small groups discussed mathematical problem solving by referencing solutions, mathematical explanations, and difficulties. The *sharing knowledge* method encouraged students to be critical of their peers' work, with the intention that feedback can effect change in their peers' ability to formulate solutions process.

The teacher explained that by sharing methods, discussing written work, and reflecting on problems and solutions, students could improve their understanding of problem solving.

M_PS Group

Students in the M_PS group received problem solving support for metacognitive problem solving. The support was based on the IMPROVE metacognitive self-questioning model (Kramarski & Mevarech, 2003; Mevarech & Kramarski, 1997).

The *comprehension questions* were designed to prompt students to reflect on the problem

Table 1. Summary of the mathematics training program by metacognitive support methods

Learning Approaches	**Metacognitive Problem Solving M_PS**	**Metacognitive Sharing Knowledge M_SK**
Theoretical teaching and learning framework	New curriculum standards (NCTM, 2000) for early childhood education: Problem solving, mathematical reasoning, and communication; numbers and operations and algebraic ideas; teaching methods for *student-centered* learning.	
Workshop structure	Five weekly 4-hour workshops for a period of five weeks that included 6 main activities: (a) Instructors present the lesson's subject and contents. (b) Students practice authentic online task solutions. (c) Students provide/ receive feedback for their partners' solutions. (d) Students adjust their solution (if needed) according to their peers' suggestions. (e) Students discuss the structure of providing feedback.	
Guidance	(a) *Problem solving* based on IMPROVE self-questioning method: • Comprehension • Connection • Strategy • Reflection.	*Sharing knowledge* based on NCTM standards: • Problem solving • Mathematical reasoning • Communication

before solving it. In addressing comprehension questions, the student was required to focus on the basic features of the problem (e.g., givens, terms). *The connection questions* were designed to prompt students to focus on similarities and differences among problems, and to explain *why*. In addressing the connection questions, students were required to focus on prior knowledge, and to define the structural features of the task and the information provided. The *strategic questions* were designed to prompt students to consider which strategies were appropriate for solving the given problem/task, and the basis and reasons for doing so. In addressing the strategic questions, students described "*what*" strategy they selected, "*how*" the strategy should be implemented, and "*why*" the strategy was the most appropriate one for solving the problem/task. The *reflection questions* were designed to prompt students to control their problem solving. In responding to reflection questions, students monitored and evaluated their understanding and the different methods employed to solve problems or implement learning methods. The metacognitive questions were presented on a printed card and embedded in the text of the task. The students were encouraged to use these questions in solving their tasks when providing explanations and feedback, and to provide written responses to metacognitive questions. The teacher explained that by using the *problem solving method* students might be able to reflect more easily during the solution stages and provide feedback as needed.

Supervision of Workshops

During the study, an assistant researcher visited all of the workshops and observed how teachers were engaged in the process.

MEASURE

Online Mathematical Problem Solving

An authentic mathematical problem solving task (the Parking Lot Task) was administered at posttest to the M_KS and M_PS. The task included six items that assessed students' *procedural* knowledge of simple calculations (item a), *algebraic* knowledge of algebraic expressions and the use of graph representations (items c, d, e) and mathematical *reasoning* referring to providing explanations and conclusions (items b, c, f).

The Parking Lot Task

Shir and Shira need to park their cars in a public parking lot. There are two parking lots nearby. Read the signs at the entrance to the parking lots:

Central Lot – every hour costs 6 shekels; My Parking Lot – entrance costs 20 shekels + 2 shekels per hour.

a. Assume Shir needs to park for 4 hours, and Shira needs to park for 7 hours. Which parking lot should each one select?
b. Explain the factors that guide Shir and Shira in their choices.
c. Find an algebraic expression that represents the total cost for each parking lot. Explain in detail.
d. Draw a graph that represents these algebraic expressions.
e. Ask three questions that can be answered by the graph.
f. List three main guidelines for choosing a parking lot so that it costs the least amount possible. Explain.

For each of the six items, students received 1 point for submitting a correct solution. In addition, students also received a grade of either 0 (incorrect), 1 (partially correct) or 2 (clear argument) for each item of the explanation and conclusion. We converted all categories scores into percentages. Cronbach alpha reliability coefficient was .82.

Feedback in Online Discussion

Feedback provided during the online discussion was assessed in two elements: *general feedback*, and *feedback based* on *SRL processes: cognitive, metacognitive, motivational*, and *social*. General feedback refers to two categories: *Number* of statements and the *structure* of feedback. Cognitive feedback refers to mathematical-solution *accuracy*, *knowledge*, *process* and *explanations*. Metacognitive feedback refers to *planning*, *monitoring*, and *evaluation*, motivational feedback refers to investment of *effort* and *interest*, and social feedback refers to *communication* style.

For each feedback category, each student received a score calculated by the total number of references they provided during the online discussion, divided by the total statements in the forum (i.e., 60). The scores were converted into percentages. Table 2 provides the scheme index and examples of scoring for online discourse feedback categories based on excerpts from the online discussion. Participants' responses were coded by two trained raters with expertise in mathematics and SRL categories. Inter-rater reliability, calculated with Cohen's kappa measure for the same 30% of the responses coded by both raters, yielded high reliability coefficients for the feedback (general-.96; cognitive- .91; metacognitive- .89; motivational and social - .92). Disagreements on coding of discourse feedback were resolved through discussion.

RESULT

Mathematical Measures

Mathematical Online Task Performance

To examine the M_SK and M_PS students' ability to solve online mathematical tasks, we performed a one way of multi-analysis of variant (MANOVA), followed by a one way of analysis of variant (ANOVA) on each skill: procedural, algebraic, and mathematical reasoning. Table 3 presents means and standard deviations of the task by metacognitive method.

The MANOVA yielded significant differences between groups, $F(3, 68) = 7.34, p < .001$, $\eta^2 = 0.18$. Further analysis of ANOVAs indicated significant differences between groups in the test regarding two skills: *algebraic*, $F(1, 70) = 6.78$, $p < .01$, $\eta^2 = 0.18$, and *reasoning*, $F(1, 70) = 14.03, p < .001$, $\eta^2 = 0.24$. No significant differ-

Table 2. Scheme Index for Scoring Online Discussion Feedback

	Description	Example
	General Feedback	
Number (average) of statement in the feedback	The average was calculated by dividing the number of statements in the feedback by the number of forums.	
Feedback structure (score 1-3)	Feedback is provided in a logical, clear, and coherent format.	"Hi, your work is correct, and the solutions are good". (score 1).
	Cognitive Feedback	
Solution accuracy	Refers to the final solution	"42 N.I.S is the correct solution for 7 hours".
Mathematical knowledge	Mathematical terms, and signs, and principles.	" you can substitute 5 in the expression: 6·□ = 20+2·□"
Strategy use	Using representations (e.g., tables, graphs, and calculation); systematic solution process.	" You based your solution on the table but you can also substitute the numbers in the expression".
Explanations	Explanations; using full description, examples, conclusions; clarity.	.."I suggest that if you need parking for more than 5 hours, it is more profitable to park in "My parking lot".
	Metacognitive Feedback	
Planning	Coordinate the selection of operators, goals, prior knowledge activation	"We suggest to substitute 7 in each algebraic expression as presented in the graph".
Monitoring	Monitor progress toward goals	"An explanation is missing from your answer".
Evaluation	Judgment of the whole process, Suggestions for modifications	"You formulated three similar questions". Now, it is easier to understand your method for determining the solution after you added the table".
	Motivation and Social Feedback	
Motivation	Statements of interest, effort investment, encouragement	"I enjoyed your work".
Social communication	Sharing knowledge, asking for help	"Can you help me? I don't know how to draw the graph".

ences between groups emerged in *procedural* skills, F (1, 70) = 1.15, $p > .05$, $\eta^2 = 0.09$. The M_PS students significantly outperformed the M_SK on algebraic skills and mathematical reasoning. The differences were most effective on mathematical reasoning ($d = 0.94$).

Feedback in Online Discussion

Feedback measures were assessed by analyzing the discussion according to *general feedback*, and *feedback based* on *SRL processes: cognitive, metacognitive, motivation* and *social.*

General Feedback

To compare the M_SK and M_PS groups' differences in providing *general feedback* during online discussion, we performed an ANOVA, for each general feedback category. The ANOVA indicated significant differences between groups in the two categories: *number of statements*, F (1, 71) = 4.93, $p < .01$, $\eta^2 = 0.12$; and *feedback structure*, F (1, 70) = 4.15, $p < .05$, $\eta^2 = 0.18$. In both measures the M_PS group outperformed the M_SK group (Table 4).

SRL-Cognitive Feedback

We performed a MANOVA, followed by an ANOVA for each cognitive feedback category. Table 5 presents means and standard deviations of online cognitive mathematical feedback by metacognitive method. The MANOVA yielded significant differences between groups, F (4, 68) = 8.68, $p < .01$, $\eta^2 = 0.21$. Further ANOVA

Table 3. Means, Standard Deviations, and Cohen's d Effect-size for Online Problem-Solving Performance by Metacognitive Method

	Method	
Skill	**M_PS** $n = 36$	*M_SK* $n = 36$
Procedural knowledge		
M	86.14	87.17
SD	15.28	16.32
d	0.07	
Algebraic knowledge		
M	84.52	73.48
SD	18.70	21.32
d	0.55	
Mathematical reasoning		
M	71.02	58.04
SD	13.42	14.12
d	0.94	

Range. 0-100

Table 4. Means, Standard Deviations and Cohen's d Effect-size for Providing Online General Feedback by Metacognitive Method

	Method	
	M_PS $n = 36$	**M_SK** $n = 36$
General Feedback		
Number of statements[a]		
M	5.44	2.86
SD	2.6	3.2
d	0.80	
Feedback structure[b]		
M	1.84	0.92
SD	0.96	1.75
d	0.53	

[a]Scores are percentages and calculated as total references provided for each category divided by the total statements in the forum, multiplied by 100.

[b]Range. 1-3

analysis indicated significant differences between groups in three of the four feedback categories: *solution accuracy*, F (1, 70) = 4.93, $p < .05$, $\eta^2 = 0.13$; *mathematical process*, F (1, 70) = 10.15, $p < .001$, $\eta^2 = 0.28$; and *mathematical explanations*, F (1, 70) = 6.43, $p < .001$, $\eta^2 = 0.26$. No differences were found between the groups on referencing mathematical knowledge, F (1, 70) = 3.03, $p > .05$, $\eta^2 = 0.06$.

The M_PS students significantly outperformed the M_SK students in providing feedback that referred to *mathematical process* (Cohen's d = 0.98) and to *mathematical explanations* (Cohen's d=0.41). However, the M_SK group outperformed the M_PS group in the category of *solution accuracy* (Cohen's d = 0.36).

SRL-Metacognitive Feedback

We performed a MANOVA, followed by an ANOVA for each feedback category (see Table 6).

The MANOVA yielded significant differences between groups, F (4, 68) = 7.68, $p < .01$, $\eta^2 = 0.19$. Further ANOVA analysis indicated significant differences between groups in all three feedback categories: *planning*, F (1, 70) = 3.4, $p < .05$, $\eta^2 = 0.12$; *monitoring*, F (1, 70) = 4.5, $p < .05$, $\eta^2 = 0.14$; and *evaluation*, F (1, 70) = 5.63, $p < .01$, $\eta^2 = 0.16$. The M_PS students significantly outperformed the M_SK students in two categories: *monitoring* (Cohen's d = 0.50), and *evaluation* (Cohen's d = 0.67). However, the M_SK outperformed the M_PS in *planning* strategies (Cohen's d = 0.35).

SRL-Motivation and Social Feedback

We performed a MANOVA, followed by an ANOVA for each feedback category.

The MANOVA yielded significant differences between groups, F (2, 70) = 4.68, $p < .05$, $\eta^2 = 0.12$. Further ANOVA analysis indicated significant differences between groups in both feedback categories: *motivation*, F (1, 70) = 3.6, $p < .05$,

Table 5. Means, Standard Deviations, and Cohen's d Effect-size for Providing SRL-Cognitive Feedback by Metacognitive Method

	Method	
	M_PS *n* = **36**	**M_SK** *n* = **36**
SRL-Cognitive Feedback		
Mathematical solution accuracy		
M	24.15	31.23
SD	19.48	21.34
d	0.36	
Mathematical knowledge		
M	15.08	14.09
SD	15.02	19.2
d	0.05	
Mathematical process		
M	25.09	17.53
SD	12.68	7.69
d	0.98	
Mathematical explanations		
M	24.13	18.25
SD	16.19	14.27
d	0.41	

Range. 0-100

Table 6. Means[a], Standard Deviations and Cohen's d Effect-size for Providing SRL-Metacognitive Feedback by Metacognitive Method

	Method	
	M_PS *n* = **36**	*M_SK* *n* = **36**
SRL-Metacognitive Feedback		
Planning		
M	32.84	39.48
SD	17.02	18.97
d	0.35	
Monitoring		
M	26.32	21.86
SD	9.42	8.91
d	0.50	
Evaluation		
M	25.08	20.63
SD	5.17	6.68
d	0.67	

[a] Range. 0-100

$\eta^2 = 0.11$; and *social communication* $F(1, 70) = 4.5$, $p < .05$, $\eta^2 = 0.13$. The M_SK students significantly outperformed the M_PS students in the two categories (Table 7).

DISCUSSION

The purpose of this study was to investigate online mathematical problem solving discussion and SRL processes among students exposed to one of two metacognitive support methods. We found that students exposed to online discussion based on the IMPROVE self-questioning strategy (M_PS group), significantly outperformed students that were exposed to sharing knowledge (M_SK group) in a reflective discourse in most measures. Better performance was noticed in the M_PS group for algebraic knowledge, and in mathematical reasoning knowledge of problem solving. Furthermore, students' online mutual SRL feedback while attempting to solve an authentic mathematical task indicated a significant difference between the M_SK and M_PS's ability in terms of cognitive, metacognitive, motivational and social aspects. The M_PS students displayed a higher tendency to refer in their feedback discussion to cognitive (mathematical accuracy; mathematical process; and mathematical explanations) and metacognitive aspects (monitoring and evaluation). Furthermore the M_PS group outperformed the M_SK group in the general measures of the forum discussion: The number of statements and structure of the feedback. In contrast, the M_SK

Table 7. Means, Standard Deviations and Cohen's d Effect-size for Providing SRL-Motivation and Social Feedback by Metacognitive Method

	Method	
	M_PS *n* = **36**	*M_SK* *n* = **36**
SRL-Motivation and Social Feedback		
Motivation		
M	7.84	10.84
SD	6.02	8.97
d	0.33	
Social communication		
M	6.02	7.46
SD	5.42	4.71
d	0.31	

students referred more often to planning strategies, motivational and social aspects.

Sharing Metacognitive Feedback in Online Mathematical Discussion

Several possible reasons may explain the beneficial effect of IMPROVE metacognitive support in mathematical problem solving. First, it seems that performing online problem-solving strategies, using metacognitive IMPROVE tools, can help students think about the steps they need to take in their solution to the problem, and can help them articulate their mathematical thoughts. This, in turn helps them to provide a mathematical clear feedback. When students explain and justify their thinking, and challenge the explanations of their peers, they also engage in clarifying their own thinking and recognizing potential conflict points for further discussion (e.g., Kramarski & Dudai, 2009; Kramarski & Mizrachi, 2006; Lampert, 1990).

According to Brinko (1993) effective feedback needs to be considered in relation to *what* (content of the message), *when* (occasion of the feedback), and *how* (mode of the feedback). Each variable will vary depending on *who* is in the process (source or recipient). Indeed, findings indicated that the M_PS group outperformed the M_SK group in the quality of providing feedback in the online discussion; their feedback was provided in a more logical, clear and coherent format. Whereas, the M_SK's feedback was more focused on corrective solution feedback.

Perhaps the IMPROVE self-questions helped the participants to be more specific in using examples to support their explanations by pointing out the good and bad parts of the solution, suggesting alternatives and different perspectives in the solution process.

We confirmed the assumption concerning SRL models that self-regulation needs feedback about strategy use (Butler & Winne, 1995), and our findings substantiate previous research conclusions regarding the differential effects of feedback types on computer-based learning (Kramarski & Dudai, 2009; Kramarski & Zeichner, 2001; Moreno, 2004). For example, Moreno found that exposing novice students to multimedia supported by explanatory feedback helped them attribute meaning to the process and promoted deeper learning than identical materials using corrective feedback alone. We suggest that further research investigate students' performance more comprehensively under different conditions of online embedded feedback, and examine how feedback relates to student variables such as reading and comprehension texts.

SRL Assessment: Event Measures of Metacognitive Feedback Discussion

Our study is grounded in socio-cognitive theories of learning which extend the view of metacognition to encompass both self-directed dialogue and social aspects. Such aspects include group discussion between peers of comparable expertise, thus making the processes of monitoring and regulation overt (Kramarski & Dudai, 2009). Our study addresses one of the main methodological issues related to SRL: *how* students regulate their learning during online mathematical discussion. As mentioned in the introduction, most of the research focuses on SRL as an *aptitude* – based on students' self-reports of perceived strategy use, resource use, and goal selection that depends on the instructional context (e.g., mathematical problem solving). These studies focus almost exclusively on declarative measures of learning. The current study examined the SRL process as an event measured by analyzing a rubric scheme. Our study extends event methodologies and thereby contributes to an emerging set of trace methodologies. These methodologies capture the dynamic and adaptive nature of SRL during learning with interactive learning environments (Azevedo, in press; Veenman, 2007; Winne & Perry, 2000; Zimmerman, 2008).

Our findings support previous conclusions that self-questioning offers metacognitive tools that may help students shift their attention from procedural thinking to a metacognitive processing level, as they consider strategies, establish sub-goals, and evaluate moves (Ge, Chen, & Davis, 2005; Kramarski & Dudai, 2009; Kramarski & Gutman, 2006; Kramarski & Mizrachi, 2006).

Several findings require further consideration. Why did the M_PS students exhibit higher SRL levels on the cognitive and metacognitive measures than the M_SK group? Some possible reasons are as follows. Communities of practice in the form of online discussion forums require sharing what was understood (i.e., "negotiation of meaning", Wenger, 1998) in such a way as to not only impart the relevant knowledge but also to clarify and explain their understanding to themselves and to their peers. Through critically examining others' reasoning, students learn to monitor their thinking, which in turn fosters their SRL processes (Zimmerman, 2000; 2008). Our findings indicate that the M_PS method is better at meeting such demands. The M_PS students more often used high-order discussion by referring to mathematical processes and explanations, and they addressed more metacognitive aspects, such as monitoring and evaluation. In contrast, the discussions in the M_SK exposed the students to open online interactions which allowed them to learn new ideas and solution strategies; however such open discussions might increase difficulties in integrating ideas and solutions that were raised in the group. Indeed, our findings indicate that the participants in the M_SK group implemented more procedural knowledge. They more often referred to basic discussion elements, such as strategy planning, and thus were less involved in sharing meaning.

Our findings are in line with other researchers, who argued that appropriate support may encourage students to consider issues they may not have considered otherwise (Zimmerman, 2000; 2008). Our findings also support the claim that students' practice should not be an abstract concept but instead the result of being engaged in activities in which students negotiate with each other (Wenger, 1998).

Furthermore, why did the M_SK group exhibit higher SRL levels than the M_PS group on the motivational and social feedback? Perhaps the sharing knowledge strategy provided students with a more open learning space which fostered the motivational-social online discussions. However, our findings indicated that such feedback was not enough to enhance students' higher order problem solving skills (algebraic knowledge and mathematical reasoning). Further research should examine more deeply and explicitly the development of motivational-social skills in online mathematical problem solving environments under different metacognitive methods.

Practical Implications and Future Research

The current study makes an important contribution to theoretical and practical implications about fostering SRL in online mathematical problem solving discussions. Specifically, the study suggests the merit of providing self-questioning support using group feedback in young elementary school students. SRL and online environments using different metacognitive support methods comprise a relatively new topic, as there has been only a few studies in mathematics education from a developmental perspective.

Further studies should devise and apply other metacognitive models for students studying in different technology environments. We recognize the need to deepen the use of mixed measures (aptitude and event) to assess SRL processes under different metacognitive instructional support methods. Methods that analyze students' verbalizations such as thinking aloud, conducting observations, and recording log files may shed further light on the differential benefits of various metacognitive methods. Future researchers would do well to apply the current empirical directions (M_PS and M_SK methods) to other subject matters, age groups, perspectives, and populations, for example, comparing different kinds of online student communities such as large and small groups, older and younger students, or children with and without learning disabilities. In particular, similar future outcomes among younger students will support recommendations to build a mathematical and SRL schools culture (NCTM, 2000).

REFERENCES

Aleven, V., & Koedinger, K. R. (2002). An effective metacognitive strategy: Learning by doing and explaining with a computer-based cognitive tutor. *Cognitive Science*, *26*(2), 147–179. doi:10.1207/s15516709cog2602_1

Artz, A., & Yaloz-Femia, S. (1999). Mathematical reasoning during small-group problem solving. In Stiff, L., & Curio, F. (Eds.), *Developing mathematical reasoning in grades K-12: 1999 yearbook of National Council of Teachers of Mathematics* (pp. 115–126). Reston, VA: National Council of Teachers of Mathematics.

Azevedo, R. (2005). Using hypermedia as a metacognitive tool for enhancing student learning? The role of self-regulated learning. *Educational Psychologist*, *40*(4), 199–209. doi:10.1207/s15326985ep4004_2

Azevedo, R. (in press). The role of self-regulation in learning about science with hypermedia. In Robinson, D., & Schraw, G. (Eds.), *Current perspectives on cognition, learning, and instruction*.

Azevedo, R., & Jacobson, M. (2008). Advances in scaffolding learning with hypertext and hypermedia: A summary and critical analysis. *Educational Technology Research and Development*, *56*(1), 93–100. doi:10.1007/s11423-007-9064-3

Brinko, K. T. (1993). The practice of giving feedback to improve teaching: What is effective? *The Journal of Higher Education*, *64*(5), 575–593. doi:10.2307/2959994

Brown, A. L., & Campione, J. C. (1994). Guided discovery in a community of learners. In McGilly, K. (Ed.), *Classroom lessons: Integrating cognitive theory with classroom practice* (pp. 229–270). Cambridge, MA: MIT press.

Butler, D. L., & Winne, P. H. (1995). Feedback and self-regulated learning: A theoretical synthesis. *Review of Educational Research*, *65*, 245–281.

Ge, X., Chen, C.-H., & Davis, K. A. (2005). Scaffolding novice instructional designers' problem-solving processes using question prompts in a web-based learning environment. *The Journal of Educational Research*, *33*(2), 219–248.

Han, I., & Park, I. (2008). The effects of epistemic belief and discussion-facilitating strategy on interaction and satisfaction in online discussion. *Journal of Interactive Learning Research, 19*(4), 649–662.

Johnson, C. M. (2001). A survey of current research on online communities of practice. *The Internet and Higher Education, 4*, 45–60. doi:10.1016/S1096-7516(01)00047-1

King, A. (1991). Effects of training in strategic questioning on children's problem-solving performance. *Journal of Educational Psychology, 83*(3), 307–317. doi:10.1037/0022-0663.83.3.307

Kramarski, B., & Dudai, V. (2009). Group-metacognitive support for online inquiry in mathematics with differential self-questioning. *Journal of Educational Computing Research, 40*(4), 365–392. doi:10.2190/EC.40.4.a

Kramarski, B., & Gutman, M. (2006). How can self-regulated learning be supported in mathematical e-learning environments? *Journal of Computer Assisted Learning, 22*, 24–33. doi:10.1111/j.1365-2729.2006.00157.x

Kramarski, B., & Mevarech, Z. R. (2003). Enhancing mathematical reasoning in the classroom: Effects of cooperative learning and metacognitive training. *American Educational Research Journal, 40*(1), 281–310. doi:10.3102/00028312040001281

Kramarski, B., & Mizrachi, N. (2006). Online discussion and self-regulated learning: Effects of instructional methods on mathematical literacy. *The Journal of Educational Research, 99*(4), 218–230. doi:10.3200/JOER.99.4.218-231

Kramarski, B., & Zeichner, O. (2001). Using technology to enhance mathematical reasoning: Effects of feedback and self-regulation learning. *Educational Media International, 38*(2/3), 77–82.

Lajoie, S. P., & Azevedo, R. (2006). Teaching and learning in technology-rich environments. In Alexander, P. A., & Winne, P. H. (Eds.), *Handbook of educational psychology* (2nd ed., pp. 803–821). Mahwah, NJ: Erlbaum.

Lampert, L. (1990). When the problem is not the question and the solution is not the answer: Mathematical knowing and teaching. *American Educational Research Journal, 27*(1), 29–63.

Mevarech, Z. R., & Kramarski, B. (1997). IMPROVE: A multidimensional method for teaching mathematics in heterogeneous classrooms. *American Educational Research Journal, 34*, 365–394.

Moreno, R. (2004). Decreasing cognitive load for novice students: Effects of explanatory versus corrective feedback in discovery-based multimedia. *Instructional Science, 32*, 99–113. doi:10.1023/B:TRUC.0000021811.66966.1d

National Council of Teachers of Mathematics. (2000). *Principles and standards for school mathematics*. Reston, VA: Author.

Oliver, R., & Herrington, J. (2000). Using situated learning as a design strategy for Web-based learning. In Abbey, B. (Ed.), *Instructional and cognitive impacts of Web-based education* (pp. 178–191). Hershey, PA: Idea Publishing Group.

Pintrich, P. R. (2000). Multiple goals, multiple pathways: The role of goal orientation in learning and achievement. *Journal of Educational Psychology, 92*, 544–555. doi:10.1037/0022-0663.92.3.544

Program for International Student Assessment - PISA. (2003). *Literacy skills for the world of tomorrow: Further results from PISA 2000*. Paris: Author.

Schoenfeld, A. H. (1992). Learning to think mathematically: Problem solving, metacognition, and sense making in mathematics. In Grouws, D. A. (Ed.), *Handbook of research on mathematics teaching and learning* (pp. 165–197). New York, NY: MacMillan.

Schraw, G., Crippen, K. J., & Hartley, K. (2006). Promoting self-regulation in science education: Metacognition as part of a broader perspective on learning. *Research in Science Education*, *36*, 111–139. doi:10.1007/s11165-005-3917-8

Veenman, M. V. J. (2007). The assessment and instruction of self-regulation in computer-based environments: A discussion. *Metacognition and Learning*, *2*, 177–183. doi:10.1007/s11409-007-9017-6

Veenman, M. V. J., Elshout, J. J., & Meijer, J. (1997). The Generality VS Domain-Specificity of Metacognition skills in Novice Learning Across Domains. *Learning and Instruction*, *7*(2), 187–209. doi:10.1016/S0959-4752(96)00025-4

Veenman, M. V. J., Van Hout-Wolters, B. H. A. M., & Afflerbach, P. (2006). Metacognition and learning: Conceptual and methodological considerations. *Metacognition and Learning*, *1*, 3–14. doi:10.1007/s11409-006-6893-0

Webb, N. M. (1989). Peer interaction and learning in small groups. *International Journal of Educational Research*, *13*, 21–39. doi:10.1016/0883-0355(89)90014-1

Wenger, E. (1998). *Communities of practice: Learning, meaning and identity*. Cambridge, UK: Cambridge University Press.

Winne, P. H., & Perry, N. E. (2000). Measuring self-regulated learning. In Boekaerts, M., Pintrich, P., & Zeidner, M. (Eds.), *Handbook of self-regulation* (pp. 532–566). San Diego, CA: Academic Press. doi:10.1016/B978-012109890-2/50045-7

Zimmerman, B. J. (2000). Attainment of self-regulation: A social cognitive perspective. In Boekaerts, M., Pintrich, P., & Zeidner, M. (Eds.), *Handbook of self-regulation*. San Diego, CA: Academic Press. doi:10.1016/B978-012109890-2/50031-7

Zimmerman, B. J. (2008). Investigating self-regulated and motivation: Historical background, methodological development, and future prospects. *American Educational Research Journal*, *45*(1), 166–183. doi:10.3102/0002831207312909

KEY TERMS AND DEFINITIONS

Feedback: A metacognitive strategy in sharing knowledge and meanings in the SRL process.

Metacognitive Methods: Learning support based on metacognitive approaches

Online Mathematical Problem Solving Discussion: Forums in which many persons advocate for their own individual position in the context of mathematical problem solving.

Self-Regulated Learning: Being active in controlling own learning process.

SRL Online Measures: Assessing self-regulated learning processes as an event.

Chapter 15

The Role of Self-Regulated Learning in Enhancing Conceptual Understanding of Rate of Chemical Reactions

Eunice Eyitayo Olakanmi
The Open University, UK

Canan Blake
The Open University, UK

Eileen Scanlon
The Open University, UK

ABSTRACT

The authors have investigated the effects of self-regulated learning (SRL) prompts on the academic performance of 30 year 9 students (12-13 year olds) learning science in a computer-based simulation environment by randomly assigning participants to either a SRL prompted or non-SRL prompted group. Mixed methods approaches were adopted for data collection and data analysis. Students in the SRL prompted group were given activity sheets which contained SRL prompts, whereas students in the non-SRL prompted group received no SRL-prompts in their activity sheets but some general prompts regarding how to follow the activity sheet. The incorporation of SRL prompted instructions into a computer-based simulation environment that teaches the rates of chemical reactions facilitated the shift in learners' academic performance more than the non-SRL-prompted condition did. This shift was associated with the presence of the SRL behavioural prompts in the activity sheets. This study is a starting point in understanding the impact of the application of SRL-prompted instructions to the teaching of topics in a computer-based learning environment with a view to improving students' academic attainment.

DOI: 10.4018/978-1-61692-901-5.ch015

INTRODUCTION

Self-regulated learning (SRL) is an active and constructive process whereby learners set goals for their learning and then attempt to monitor, regulate, and control their cognition, motivation, and behaviour as well as the contextual features of the learning environment (Zimmerman, 1989). SRL helps learners to choose what to learn, determine how long they want to learn, determine how to learn, access relevant instructional materials effectively, as well as assessing their level of comprehension of learning materials (Zimmerman, 1989). The three main phases of self-regulation, namely planning, monitoring and evaluation, are described as being consistent with the regulatory processes that students engage in during the process of learning (Manlove, Lazonder & De Jong, 2007). Planning is a very important strategy that students need to deploy when learning with computer-based simulation learning environments. It is not surprising, therefore, that recent studies in the field of science learning revealed that science learners are required to manage and evaluate their own efforts in order to attain the specified learning goals, when using technological resources, such as a simulation learning environment, for inquiry and problem-based learning (Manlove, Lazonder & De Jong, 2007). Manlove, Lazonder, & De Jong, (2007) used embedded static scaffolds to facilitate self-regulatory processes (e.g., planning, monitoring, and evaluating) during inquiry learning. They based their assumption that these would be effective on the hope that learners will make use of them during learning; they will know when to use them and thereby monitor their learning processes.

Other researchers, such as Azevedo, Cromley, Winters, Moos, & Greene, (2005); Demetriadis, Papadopoulos, Stamelos, & Fischer (2008) have examined the role of students' ability in regulating their metacognitive activities when engaging with a computer learning environment. According to Zimmerman (2000), self-regulating students will set goals and sub-goals the first time they are introduced to the learning task which in turn help them to decide on specific outcomes of the learning or performance. Once self-regulating students begin to carry out their strategic plans, they begin to monitor their comprehension and task performance. Effective strategies for monitoring include self-questioning and elaboration which includes note taking (Demetriadis, Papadopoulos, Stamelos, & Fischer, 2008). Evaluation of learning processes involves reflection on the quality of the students' planning or how well they execute their plans. Learners who self-regulate will try to evaluate their learning based on the goals they set for themselves at the beginning of the task which should include adequate prediction as well as very clear inferences. Furthermore, research has shown that the challenges that self-regulated learners encounter in a computer-based learning environment differ from those in the conventional classroom (Zimmerman 2000). This might be associated with the large amount of information available as well as the attractive but irrelevant materials such as pictures and animations contained in the computer-based learning environment. These challenges of overload and irrelevant information may result in learners' inability to control and regulate their learning activities effectively (Narciss, Proske, & Koerndle, 2007). For instance, they often have badly constructed plans or they do not have plans at all. In situations where most students determine what to do as they move on with the learning, they make ad-hoc plans rather than taking a systematic approach. Zimmerman (2000) described this method of self-regulation as generally being ineffective because it fails to provide the necessary goal structure and strategic plans for students to progress consistently, monitor and evaluate their learning effectively.

In order to overcome these challenges, various instructional interventions that could help learners to regulate their cognitive and metacognitive activities are necessary when engaged with a simulation learning environment. Prompting students with SRL behaviours might help to overcome these planning, monitoring, and evaluating problems when learning in a computer-based simulation learning environment. The studies reviewed above reveal that there are several outstanding issues related to SRL when using a simulation learning environment which have not been addressed by educational researchers. For example we need to understand how 12-13 year old science learners (KS3), regulate their own learning in a computer-based simulation learning environment and how this helps to improve their understanding of scientific concepts. Therefore, this study is concerned with the question of how students regulate their learning in a computer-based simulation learning environment related to the rate of chemical reactions, in an attempt to improve their academic performance. Specific questions addressed are: Do different instructional conditions (a computer-based simulation learning environment with and without self-regulatory prompts) affect learners' conceptual understanding leading to greater academic performance? How do different instructional conditions affect learners' ability to self-regulate their learning? In this study we examine SRL from the social cognitive perspective in which the introduction of SRL instructional prompts into a social environment (a computer-based simulation learning environment) is assumed to influence the students' self-regulatory processes and their academic progress. The next section of this chapter presents the theoretical frameworks and models guiding SRL in an attempt to explain what it entails for a learner to be successful with regard to the setting of learning targets in a computer supported learning context.

THEORETICAL FRAMEWORK AND MODEL GUIDING SELF-REGULATED LEARNING IN SIMULATION LEARNING ENVIRONMENT

In the 1980s, theoretical frameworks and models guiding self-regulated learning (SRL) were proposed in an attempt to explain what it entails for a learner to be successful with regard to the setting of learning targets in a given context (Zimmerman 1989). While most theorists concur that learners' thoughts, feelings and actions, that are planned and cyclically adapted to the attainment of personal goals, have interrelated cognitive, affective, motivational and behavioural dimensions (Zeidner, Boekaerts & Pintrich, 2000); dissenting opinions among theorists have been noted with regard to which dimensions ought to be emphasised; and consequently; what strategies and processes they tend to encourage learners to adopt in order to enhance their academic performance. Among the theories and models of self-regulated learning strategies already developed are the operant models, information processing and social cognitive models. According to Mace, Belfiore and Hutchinson, (2001), operant models of SRL are founded on the principle that self-regulated learning and behaviour emanate from the strategic manipulation by external stimuli and the consequences spontaneously follow an action. Operant models involve self-application of reinforcement strategies which allows students to set target behavioural goals that will lead to higher academic attainment, systematically observe, record, and evaluate progress, and adapt rewards towards reaching learning goals (Mace, Belfiore, & Hutchinson, 2001). Information processing models of SRL emphasise the use of metacognitive strategies such as self-monitoring and self-evaluation to carry out complex academic tasks (Winne 2001). Information-processing theories focus on covert rather than overt processes. Moreover, information processing theories consider motivational influences on whether a learner will use a particular learning strategy. In summary, information-processing theories do not consider social or environmental factors that may affect metacognition and academic achievement. Social cognitive models of SRL are distinguished from other models in that they investigate interrelationship among self-regulated learning strategies, beliefs, feelings, and social and physical environment (Bandura, 1986; Zimmerman, 2000). For the purpose of this study, we will be considering Bandura's social cognitive theory of SRL as a comprehensive theoretical framework to conceptualize the effect of SRL strategies on learners' performance in a simulation learning environment. Bandura's social cognitive theory suggests that SRL is context dependent, that is, the unique features of a learning environment may influence whether or not a learner enacts SRL strategies. Therefore, choosing this theory as a guiding framework for this study allows us to examine the interaction between learners' personal characteristics (e.g. cognitive, motivation), elements of the computer-based simulation learning environment (e. g. prompting with SRL behaviours) and mediating self-regulatory processes that learners adopt (e.g. planning, monitoring activities).

Bandura's (1986) social cognitive theory highlights how personal, behavioural, and environmental factors affect learners' thoughts when faced with instructional choices (see Figure 1). This he refers to as learners' ability to manage their behavioural responses in a learning environment. This present study adopted Bandura's social cognitive theory as the theoretical framework for investigating the learners' self-regulated learning behaviour in a computer-based simulation learning environment. Some of the personal characteristics of students learning in a computer-based simulation learning environment include an affective factor (e.g. 'how do I feel about this task?') and self-efficacy (e.g. 'can I do the task?'). The behavioural factors influencing students learning in a computer-based simulation learning environment are the use of proper metacognitive learning strategies (e.g. planning task, monitoring their learning, effort regulation and help seeking behaviour). Research has shown that the use of metacognitive strategies has positive effects on students' academic performance (Zimmerman 1989; Pintrich, 2000).

Figure 1.

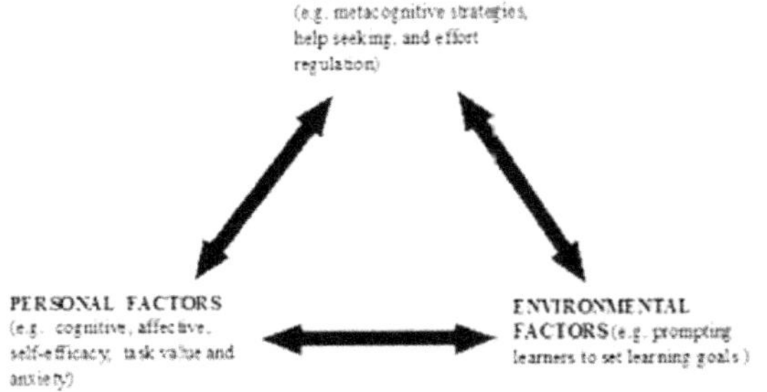

Environmental influences entail helping students to monitor their learning, setting learning goals, and behavioural modelling. Pintrich and Schunk (2002) found that students were capable of learning complex skills through observing modelled performances. Therefore, modelling the effects of SRL prompts on the students' academic performance and how they self-regulate their learning when studying about the rates of chemical reaction in a computer-based simulation learning environment will be considered as an important source of environmental influence in this study.

In Bandura's social cognitive theory, the interaction between the person and the behaviour involves the influences of a person's thoughts and actions. The interaction between the person and the environment involves human beliefs and cognitive competencies that are developed and modified by social influences and structures within the environment. The interaction between the environment and behaviour involves a person's behaviour determining the aspects of their environment and in turn their behaviour is modified by that environment. Zimmerman (1989) is of the opinion that learners are not just being controlled by external factors but rather they possess self-directed capabilities to influence their own behavioural responses in a learning environment. This implies that learners have the ability to control their activities by applying cognitive, meta-cognitive, and behavioural learning strategies when given learning tasks. In addition, Schunk (2001) explains that students' efforts to self-regulate during learning are not determined merely by personal processes such as cognition or affective issues; but rather; these processes are assumed to be influenced by environmental and behavioural events in a reciprocal manner. Bandura (1986) also shared a similar view that self-regulated learning occurs to the degree that a student can use personal processes to strategically regulate his or her behaviour and the immediate learning environment. Based on an adaptation of Bandura's theory to this particular context, we hypothesize that students using a computer-based simulation learning environment are required to analyze the learning situation, set meaningful learning goals, and determine which strategies are effective as well as evaluating their emerging understanding of the topic they are studying. Students also need to monitor their understanding and modify their plans, goals, strategies and effort in relation to task conditions (e.g. cognitive, and motivational) that are contextualised in a particular learning situation (e.g. learning factors concerned with the study of rates of chemical reactions in a computer-based simulation learning environment). Due to these demands, we expect computer-based simulation learning environments to be less effective if learners do not regulate their learning (see e.g. Azevedo, Cromley, Winters, Moos & Greene, 2005). Therefore, our present study examines self-regulated learning from the social cognitive perspective in which the introduction of SRL instructional prompts into a learning environment (a computer-based simulation learning environment) is assumed to influence the students' self-regulatory processes and their academic progress.

Research has also shown that once learners have acquired skills in using self-regulatory strategies, this will help them to promote their own learning and the perception of greater competence, which in turn sustains their motivation to attain new goals (Zimmerman & Martinez-Pons, 1986; Pintrich, 2000; Schunk & Zimmerman, 2006). According to Zimmerman and Martinez-Pons (1986), learners who were high in their overall use of self-regulated strategies sought help more frequently from peers, teachers and parents and learned more than students who did not seek help. Learners' reported use of self-regulated learning strategies (e.g. goal setting, monitoring, help seeking), have been shown to be highly correlated with various academic performance indicators such as exam scores, grades and essays/reports (Zimmerman & Martinez-Pons 1988). These authors found that students' reports of using self-regulated strategies correlated with teachers' judgments of students' self-regulation behaviour during the class. However, their study took place in a conventional classroom setting and not in a simulation learning environment where there are arguably more challenges for students to face in order to be successful in their given tasks. In order to overcome these challenges, various instructional interventions that could help learners to regulate their cognitive and metacognitive activities are necessary when engaged with a simulation learning environment. Hence, this present study aims to investigate the effects of introducing self-regulated learning prompts in a computer-based simulation learning environment on the accomplishment of task performance.

Not surprisingly, several of the variables such as goal setting, monitoring, help seeking associated with self-regulated learning have been shown to have significant impact on the learners' performance in a computer-based learning environment. Previous studies on self-regulated learning have demonstrated that lack of effective metacognitive skills lead to the ineffective use of instructional strategies and poor academic performance when learning in a computer based learning environment (Garhart & Hannafin, 1986; Azevedo, 2005; Narciss, Proske & Koerndle, 2007). Garhart and Hannafin (1986) indicated that learners were not aware of when they needed additional instructional support when learning in a computer based environment. Therefore, learners' inability to metacognitively monitor their learning may lead to their inability to make use of effective instructional support and this could make learners to be ineffective in regulating their learning. Narciss, Proske & Koerndle, (2007) examined how to promote meta-cognitive activities in a computer based learning environment. The results of the study demonstrated a high variability in the total study time adopted by students; working with texts, learning tasks, elaborating tools for surfing, scanning, and trial-and-error-like exercise, and monitoring tools for fostering learners' active information processing in a learning environment; with some spending only a few minutes, while others spent 7 hour with the learning environment. Further analysis of their findings showed that on average, students spent 70% of their study time on text material, 15% with learning tasks, and 13% with elaborating tools, whereas monitoring tools were hardly used (< 1%), with only a few students using them. Moreover, it was discovered that the more time students worked on learning tasks, the higher their performance with the students working less than 3 hour processing a significantly lower percentage of texts and tasks and achieving significantly poorer scores. Meanwhile, the research did not explain why lots of students decided not to use elaborating and monitoring tools for their learning and even for the few students that used them, the reason was not clearly stated. Could individual use of self-regulated learning strategies determine the way each learner learns in a simulation learning environment? Are the students motivated to use monitoring tools? Therefore, due to the challenges above, we argue that successful learning with computer-based simulation learning environments requires learners to self-regulate their learning. Given the importance of self-regulated learning model in learning, next section explains self-regulated learning with computer-based simulation.

Self-Regulated Learning with Computer-Based Simulation

For the purpose of this study, a computer simulation will be regarded as a representation of activities that users learn about through interaction with the computer software (Alessi & Trollip, 2001). Simulations have been used in science education since the early 1970s, and research is still on-going on how best to use simulations in the teaching of science. Blake and Scanlon (2007) evaluated three examples of simulation software developed by the Open University. From their evaluations, they were able to develop a set of features that could make learning with simulation by the distance learners to be effective. Student support, multiple representations, and tailorability were the suggested features that could be considered to enable the most effective use of simulations. Their conclusion was that the success of simulation as effective learning tools in science education is dependent on how simulations are used; this present study therefore seeks to look at the effect of the usage of self-regulated learning prompts on learners' performance in a simulation learning environment.

Self-regulated models offer a comprehensive framework with which to examine how students learn in computer-based learning environment. Several researchers (Shapiro & Niederhauser, 2004; Azevedo, 2005; Lajoie & Azevedo, 2006) have begun to examine the role of students' ability to regulate their personal and behavioural aspects when learning in computer-enhanced environments. The three main phases of self-regulation, planning, monitoring and evaluation are described to be consistent with the regulative processes that students engage in during inquiry learning (Njoo & de Jong, 1993). According to Zimmerman (2000), self-regulating students will set goals and sub-goals the first time they are introduced to the learning task which in turn help them to decide on specific outcome of the learning or performance. Once self-regulating students begin to carry out their strategic plans, they begin to monitor their comprehension and task performance. Effective strategies for monitoring include self-questioning and elaboration which include note taking (Chi, de Leeuw, Chiu & LaVancher, 1994). Evaluation of learning processes involves any reflection on the quality of the students' planning or how well they execute their plans. Self-regulating students will try to evaluate their learning based on the goals they set for themselves at the beginning of the task which should include adequate prediction as well as very clear inferences.

Research has shown that when students engage in computer-based simulation learning environments, they perform very few of the self-regulatory activities discussed earlier on (Narciss, Proske & Koerndle, 2007, Azevedo, 2005). This present study was designed to explore the possibility that prompting students with self-regulated learning behaviours might help to overcome these planning, monitoring and evaluating problems when learning in a computer-based simulation environment.

This brief review of the literature reveals that there are several outstanding issues related to self-regulated learning when using simulation learning environments which have not been addressed by educational researchers. To date, there seems to have been little research into how science learners regulate their own learning in a computer-based simulation learning environment with a view to improving their conceptual understanding of scientific concepts. Therefore, this study is concerned with the question of how students regulate their learning in a computer-based simulation learning environment related to rates of chemical reaction, in an attempt to improve their academic performance. We will be focussing on the introduction of self-regulated learning instructional prompts (e.g. related to goal setting, time management) into computer-based simulation learning environment and any effect of this on students' performance.

METHODOLOGY

Sampling

In this section, we present our classroom research on SRL and computer-based simulation learning environment. This research explored the high school students' use of a computer-based simulation learning environment to learn about rates of chemical reactions. The sample of the study consisted of a class of 30 science students (16 boys and 14 girls) in year 9, in a high school in the UK. All the 30 participants were randomly assigned to two groups: computer-based simulation learning environments using SRL-prompted instructions and non-SRL-prompted instructions. Meanwhile, all the participants were drawn from the same class taught by the same teacher so that no possible outcome of this study could be attributed to any variations in teaching styles. Permission was given to carry out the study in this particular class because they were studying chemical reaction rates at a time that was suitable for our data collection. Pre-tests were administered to all participants in order to establish their knowledge about chemical reaction rates and their Self-regulated Learning (SRL) skill levels. This was designed to determine whether there were real differences in the experimental (SRL-prompted group) and control groups (non-SRL-prompted group) at the beginning of the activity. Moreover, preliminary enquiry from the participating teachers also corroborated that all the participants were acquainted with using computer-based simulations for learning science and the topic had not been previously covered for the students.

Instruments of Data Collection

Self-Regulatory Strategies Questionnaire (SRSQ), pre-test and post-test for reaction rates knowledge test (RRKT), Students' Activity Sheets (SAS) and the observation of the participants as they learnt with the simulation program were used for data collection. The SRSQ, a sub-set of Motivated Strategies for Learning Questionnaire (MSLQ) was adapted by Young (1997) from Pintrich, & Smith, (1993). The MSLQ has received broad acceptance and use by researchers. Pintrich, Smith, Garcia and McKeachie, (1991) and Pintrich and Smith (1993) have demonstrated that the MSLQ is a reliable and valid measure of self-regulated learning. The total reliability of the motivation scales is 0.79 and the values of Cronbach's alpha for each motivational subscale are acceptable, ranging between 0.57 and 0.84. The total reliability of the learning strategies scale is 0.89 and the values of Cronbach's alpha for each of the learning strategies subscales are also acceptable, ranging between 0.62 and 0.83. The questionnaire has a seven-point Likert scale ranging from 1 (Not at all true of me) to 7 (Very true of me) that was used to determine how true each statement is to participants. The questionnaire items which had been previously used for college students were accepted to be suitable for these science students. This present study employed SRSQ because it addressed the learners' use of cognitive and self-regulatory strategies and not the learners' motivational beliefs. The SRSQ consisted of 31 items detailing the cognitive learning strategy scales of: metacognitive self-regulation, time and study environment, effort regulation, peer learning and help seeking.

The RRKT consisted of a 14 item paper-based test (developed according to tests used in the GCSE chemistry in the UK) on the rates of chemical reactions. These included short-answer questions, matching, and multiple-choice tasks. Two types of students' activity sheets were designed for the purpose of this research to collect data on students' interaction with the computer-based simulation learning environment. Activity sheets for the SRL-prompted group consisted of eight SRL instructional learning prompts such as; "set three specific learning goals you would want to achieve after learning about rates of chemical reactions", "if you need help, please ask", and "please raise your hand if you need any help at any point during the activity". Meanwhile, the activity sheets for the non-SRL group contained no SRL prompts but instead some general prompts such as "don't forget to read the aims of this activity", "please make use of the timer provided in the tools section", and "please record your observations on activation energy in this table".

Students in both SRL- and non-SRL-prompted groups responded to either SRL or general prompts by reporting their predictions, observations and inferences on the given tasks by filling in the gaps and the tables in the activity sheets. The students' activity sheets were collected after the lesson and the contents were analysed. Furthermore, the use of participant observation in this study provided valid basis for accurate description of what learners were doing instead of what they remembered or thought that they were doing. This approach was adopted in order to relate learners' behaviour to the conditions dictated by the task, decrease difficulties such as bias in questionnaire completion and learners' constraints in elucidating cognitive processes that they use during task performance. According to Turner (1995), this approach is associated with measuring the process of task performance in students. In addition, informal interviews were conducted with three students from each of SRL-prompted and non-SRL-prompted groups. This was undertaken to validate the findings from the observation.

Simulation Learning Environment and Learning Task

This study involved two versions of a computer-based simulation learning environment (Figure 2) that taught rate of chemical reactions. The simulation was produced by Sunflower© for teaching in high schools in the UK. The two versions of the

Figure 2. Rates of chemical reactions simulation

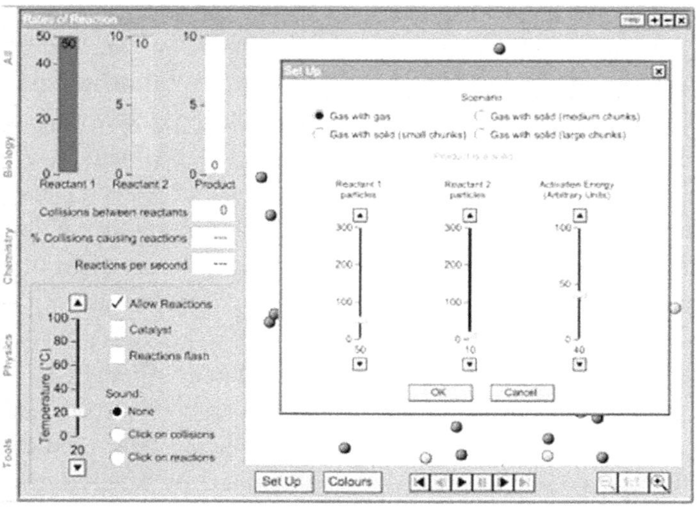

computer-based simulation learning environment represented two varying conditions in which SRL-prompted instructions and non-SRL-prompted instructions were provided to the students for each computer-based simulation learning environment.

An experimental approach in which participants were randomly assigned to computer-based simulation learning environments using SRL-prompted instructions and non-SRL-prompted instructions had been adopted because it minimized the risk of extraneous variables that might have confounded the outcome of this study (Cohen & Manion, 1989).

The SRL-prompted instructions were expected to help students develop an understanding of their own strategies and procedures in learning about rate of chemical reactions in a computer-based simulation learning environment. The learning activity sheet was structured in such a way that students were initially asked to write things they already know about rates of chemical reactions in order to activate their prior knowledge, observe the task through their interaction with the simulation and record their observation in the form of a laboratory report in the learning activity sheets.

Design

The goals of our research are to conduct classroom research which addresses the following questions pertaining to the use of computer-based simulation learning environment to learn scientific concepts: (1) Does a computer-based simulation learning environment employed for teaching rate of chemical reactions lead to overall attainment of higher test scores? (2) Do different instructional conditions (a computer-based simulation learning environment with and without self-regulatory prompts) affect learners' conceptual understanding leading to greater academic performance? (3) How do different instructional conditions affect learners' ability to self-regulate their learning?

In order to address these questions, we combined experimental designs with classroom observation as well as semi-structured interviews. This was because it enabled us to explore both the outcome measures (i.e., shifts in the students' academic performance from pre-test to post-test).

Pre- and post- reaction rate knowledge tests (RRKT) were used to determine the level of prior knowledge of students about the rate of chemical reactions and evaluate the effect of

SRL-prompted instructions on the students' academic performance respectively. The conduct of pre- and post-reaction rate knowledge tests facilitated controls for time-related threats to validity (Blaxter, Hughes, & Tigh, 2006). Both pre- and post- reaction rate knowledge tests (RRKT) are similar in that they are identical in the level of difficulty and complexity to the computer-based simulation problems. For the qualitative aspect of the study, students' activity sheets were scored on the basis of note taking, effort regulation, and task completion. Moreover, participants in each group were observed as they learnt in their respective learning contexts with the computer-based simulation learning environment teaching rate of reactions. One of the authors undertook a participant observer-role in which she was involved in the classroom at helping students if they request for help and made field notes during and after the lesson. Patton (1987) stated that field notes are a description of what was observed during the field work, therefore, the participant observation allowed us to identify the students' behaviour and their interaction patterns with the computer-based simulation learning environment as against using a non-participant observation approach. The observations however, were coded according to the emerging categories of how students made use of SRL skills when learning using the rate of chemical reactions simulation environment.

Our methodological paradigm involves pretesting the participants, randomly assigning them to either an experimental (computer-based simulation learning environment using SRL-prompted instructions) or control group (computer-based simulation learning environments using non-SRL-prompted instructions), content analysis of classroom observation and students' activity sheets, followed by post-testing participants from both groups. Using these methods assisted us to obtain a richer account of the effect of self-regulated learning prompts on learners' performance in a simulation learning environment. Bird and Hammersley (1996) are of the opinion that the use of several methods to explore an issue greatly increases the chances of accuracy. Therefore we employed both qualitative and quantitative methodological tools for this study in order to lead to a more valid, reliable and diverse construction of realities.

Data Analysis Techniques

The data collected were analysed using both quantitative and qualitative methods. Various t-test analyses were used to examine the data collected. Students' scores in Self-Regulated Learning Strategies Questionnaire (SRSQ) administered before and after learning with computer-based simulation learning environment were computed and analysed using independent-samples t-test. This test ensured that there were no differences between the two groups in terms of their SRL behaviours at the beginning of the activity.

A t-test for repeated measures was used for the comparison of the means of pre- and post -reaction rates knowledge test (RRKT) scores for all students. This was employed to determine the effect of learning with computer-based simulation learning environment. In addition, an independent-samples t-test procedure compared the shift in the means of test scores for SRL-prompted and non- SRL-prompted groups. The adoption of t-tests to evaluate the differences in means between the SRL-prompted and non-SRL-prompted groups was based on the fact that this study is using a relatively small sample. Theoretically, it has been established that the t-test can be used for small samples sizes provided the variables are normally distributed within each group and the variation of scores in the two groups is not reliably different as confirmed by sample K-S test. The equality of variances assumption was verified with the Levene's test. The results of all the t-tests met Levene's test condition.

Qualitative analysis approach included content analysis of the students' activity sheets and classroom observations. The content analysis involved using the students' activity sheet to categorize how they made use of self-regulated learning behaviours when learning with the computer-based simulation learning environment. The categorisation was based on the entries made by students in the blank spaces that were available in the two types of the activity sheets as well as whether they were able to complete the given task within the allocated time or not. The use of content analysis in this study was based on signs or observable indicators regarding SRL skills that student employed when performing tasks in the computer-based simulation learning environment. For example, some of the SRL indicators that were used to measure the students' performance were whether the students write more information in the available spaces on their activity sheets or not, whether they were able to manage their time effectively by completing the given task on time or not. Therefore, content analysis enabled us to describe and draw appropriate inferences about the effect of self-regulated learning prompts on the students' performance as measured by their scores on the activity sheets. In order to establish whether or not the incorporation of the SRL prompts in the experimental group's learning materials actually have an effect on the learning outcome; an independent sample t-test was conducted to compare the mean scores of the learners in each group (SRL prompts and non SRL prompts). Also, the classroom observations were analysed through working with and organizing the data, breaking the data into manageable units, synthesizing the data in order to search for certain patterns, deciding on vital aspects and dissemination of the findings (Bogdan & Biklen, 1998). Therefore, the existence, meanings, and the relationships of the words or concepts that were related to self-regulated learning were explored and noted down during the process of analysis.

FINDINGS

Computer–Based Simulation Learning Environment and Academic Performance

The results of the t-test for repeated measures which compares the means of pre-test scores and post-test scores is presented in Table 1 for all students irrespective of their group (SRL-prompted and non-SRL-prompted). This had been adopted with a view to determining whether all students, regardless of instructional conditions, improved significantly from pre-test scores to post-test scores. The analysis revealed that the hypothesis was supported by $t(29) = -8.545$, $p<0.05$. Significance value ($p = 0.000$) of the t-test for repeated measures indicates that there is a significant difference between the students' pre-test scores and post-test scores regardless of instructional conditions.

The results presented in Table 1 showed that all students improved in their test scores after learning from the computer-based simulation learning environment. This lends credence to the hypothesis that all students irrespective of whether the simulation environment they were using for learning is supported by SRL-prompted or non-SRL-prompted instructions would gain some conceptual understanding when learning in a scientific simulation environment. This result agrees with the outcome from the work of Azevedo, Cromley, Winters, Moos and Greene (2005), which focussed on the use of a hypermedia learning environment to learn about the circulatory

Table 1. Means of the pre and post Reaction Rates Knowledge Test (RRKT) scores for all participants

	Mean (M)	Standard deviation (SD)
Pre-RRKT	6.37	2.930
Post-RRKT	10.67	2.551

system in biology. Their research findings showed that students' learning about a challenging science topic with hypermedia irrespective of whether they are provided with support or not tends to gain a declarative knowledge from pre-tests to post-tests. The finding by Azevedo, Cromley, Winters, Moos and Greene, (2005) agrees with the result of this present study in the sense that all students who participated in this study gained some conceptual understanding of the rate of chemical reactions as measured by the overall knowledge test scores.

Differences between Computer-Based Simulation Learning Environment with SRL-Prompts and Non-SRL Prompts

The results of the t-test for independent groups show that there is a statistically significant difference between the shift in the means of the RRKT test scores of the SRL-prompted group and non-SRL-prompted group (Table 2).

The independent-samples t-test analysis revealed that the hypothesis was supported by a significance value ($p = 0.026$), thus, re-affirming that there is a significant difference between the shift in the means of the RRKT test scores of the SRL-prompted group and non-SRL-prompted group.

The significant difference ($p = 0.00$) in the scores of students in both learning contexts implies that most students in the SRL-prompted group gave correct answers on their activity sheets while in non-SRL-prompted group, most students did not give correct answers. This might be associated with the presence of SRL-prompts on the SRL-prompted group's activity sheets.

The outcome of our investigation into whether there are differences in the students' academic performances when learning with computer-based simulation learning environments with and without SRL prompts suggests that there is a statistically significant difference between the shift in the means of the RRKT test scores of the SRL-prompted group and non-SRL-prompted group.

These findings are supported by previous research outcomes which suggested that students who are provided with SRL-prompted instructions and supported with the technology enhanced learning environment, displayed significant learning gains in different domains and scientific tasks (Njoo & De Jong, 1993; Chi, de Leeuw, Chiu & LaVancher, 1994; Azevedo, Guthrie & Seibert, 2004; Azevedo, Cromley & Seibert, 2004; Azevedo, Cromley, Winters, Moos & Greene, 2005). These outcomes contribute to the literature on the usage of simulation learning environment in the teaching of chemical concepts by illustrating that SRL-prompted instructions aimed at facilitating students' ability to regulate their learning processes is associated with improved test score attainment during learning with simulation environment.

Furthermore, students in the simulation learning environment with SRL-prompts were found to have attained higher marks in their activity sheets' scores than students in the non-SRL-prompted group (Table 3). The significant difference in the scores of students in both learning contexts implies that most students in the SRL-prompted group gave correct answers on their activity sheets while in non-SRL-prompted group, most

Table 2. Means and standard deviation of shift in the RRKT test scores of SRL-prompted and non-SRL-prompted groups

	Mean (M) Difference (pre-test to post test)	Standard Deviation (SD)
SRL-prompted group	5.40	3.089
Non-SRL prompted group	3.20	1.897

Table 3. Means of the students' activity sheets' scores

Learning contexts	Mean (M)	Standard Deviation (SD)
SRL-prompted group (n =15)	12.47	2.532
Non-SRL prompted group (n = 15)	7.33	3.309

students did not give correct answers. This might be associated with the presence of SRL-prompts on the SRL-prompted group's activity sheets. In summary, students in the SRL-prompted context were found to have outperformed students in the non-SRL-prompted group in both reaction rates knowledge test (RRKT) and students' activity sheets (SAS) scores.

Self-Regulation in a Computer-Based Simulation Learning Environment with SRL-Prompts and Non-SRL Prompts

This section provides insight into how different instructional conditions influence students' ability to self-regulate their learning behaviour. The t-test statistical analysis of the shift in SRSQ scores confirmed that the SRL-prompted group had made use of SRL-skills; such as metacognitive strategies, effort management, help seeking and time management; to a greater degree than the non-SRL-prompted group (Table 4).

We examined our field notes in order to examine whether there were any qualitative differences in the way learners learn from the simulation in their different instructional groups. Prompting the SRL students was intended to elicit different responses in the way they self-regulate their learning with the learning materials and determine whether this actually produces any difference in their academic performance as measured by the test scores. Our observations of how learners behaved while learning were later grouped into three categories (see Table 5).

The first category is what we will describe as students' understanding of the learning context. After reading out the general introduction to the students in both conditions, it was clear that all the students in the SRL-prompted group just picked up their pens to set sub-goals, as well as activating their prior knowledge of the topic, rate of chemical reactions. We noted that they actually read the prompts and responded through their actions. On the contrary, students in the non-SRL-prompted group did not really settle down on the learning material in time. Moreover, some of the non-SRL-prompted students did not follow the instructions on the number of reactants that they should have on their screen, the colours to use for the reactants and product. For example, students in the non-SRL group were having various colours of their choosing rather than following instructions.

The second category of behaviours observed can be described as the controlling level of task difficulty and demands (Table 5). Students in the SRL-prompted group were able to control their learning context; they allocated time they would like to spend on each goal. We discovered that they made use of prompts on time and they still

Table 4. Means and standard deviation of shift in the Self Regulating Strategies Questionnaire (SRSQ) scores of SRL-prompted and non-SRL-prompted groups

	Mean (M) Difference (pre-SRSQ to post SRSQ)	Standard Deviation (SD)
SRL-prompted group	20.73	16.167
Non-SRL prompted group	2.20	23.094

Table 5. Categorisation scheme for learners' behaviour in both SRL-prompted and non- SRL-prompted groups

Categories	SRL-prompted group	non-SRL prompted group
Understanding of the learning context: Planning, skimming, note taking, and control of context.	Activated their prior knowledge, planned the task and managed their time effectively.	They did not settle down with the task on time and they didn't follow the general instructions on how to carry out the activity.
Task difficulty and demand: Help seeking behaviour and task difficulty	They controlled their learning context by allocating time they would like to spend on each goal. They asked for help when necessary.	They dealt with the difficulty task by playing round the simulation, didn't request for help, and most of them were unable to complete the task on time.
Emotional level: Interest in the learning environment.	They showed more interest in learning and very positive about it.	They expressed some negative words of frustration and confusion.

went ahead to confirm with the instructor, how much time they had left. This showed that they were able to manage their time and put in effort into the learning. Students in the SRL-prompted group were reminded (in the activity sheet) to seek for help at anytime while learning, therefore, they were able to handle some of the difficulties they encountered while learning with the simulation by seeking help from the tutor or the researcher. In contrast, majority of the students in the non-SRL-prompted group just dealt with the task difficulty by playing round the simulation learning environment without requesting for help. Therefore, majority of the students in the non-SRL-prompted group did not manage their time effectively and did not complete their tasks on-time.

Another category that emerged from our observations was students' emotional-level as depicted by their interests in the learning environment (Table 5). We noticed that students in the SRL-prompted group answered their questions and showed more interest in the learning. Moreover, they were very positive about the learning as a whole. A lot of students in the non-SRL-prompted group expressed some form of negative words of frustration and confusion. These feelings and lack of strategies to use when learning with the simulation might have contributed to their low performance in the learning outcome.

The outcomes of the informal interviews we had with the two students from both groups are in agreement with our observational findings. For example a student in the SRL-prompted group when asked how he went about with learning from the computer-based simulation learning environment commented that he really enjoyed learning during the simulation lessons and task activities. He regarded the task activity to be better than writing a long report. He responded to all the prompts in the text boxes and found out that it was a good thing that he planned the whole activities right from the beginning of the lesson. He discovered that SRL-prompts enabled him to be time conscious, making sure that he completed the whole activities on time. He did report that he requested for help when he didn't know what to do when he was told to get the timer out from the tool bar. He commented that it was good that the teacher asked him to request for help which aided him in managing the allocated time efficiently.

For the learners in the non-SRL-prompted group, their comments were contrary to the one expressed above. A student in the non-SRL-prompted group said that she did not really enjoy the simulation lesson and task activities. She reported that she did not understand what to do on time. She found herself trying to find her way round the whole simulation and before she knew it, the whole lesson had almost finished. On reporting her observations from the simulations in the activity sheets, she found this to be very boring. She wished her experience working with the simulations environment would have been different. She pointed out that she did not know how she could effect this change in her experience which she would have liked to do. Overall, she found this exercise to be a bit of a fun as she enjoyed watching the multi-coloured molecules colliding on the computer screen.

From these two responses, it can be deduced that prompting students with SRL-behaviour helps them to focus on the activity, plan how to carry out the giving task, monitor their own learning, seek for help, and manage effectively the allocated time for the given task.

The outcomes of our analysis above confirm that students in the SRL-prompted group were more likely to concentrate on important information, generate more metacognitive activities, and construct a richer rational understanding of themes and purposes within the learning environment compared to students in the same learning environment supported with non-SRL-prompted instructions. Moreover, content analysis of the students' activity sheets suggests that in the SRL-prompted group, students employed the SRL-prompted instructions, to facilitate their own regulatory behaviour when learning in a simulation environment. This finding is supported by the results of the classroom observation and informal interviews reported earlier on. Finally, Table 4 provides support for the hypothesis that student in the SRL-prompted simulation learning environment made use of key SRL processes during learning as a consequence of the intervention occasioned by the introduction of SRL-prompted instructions. These results are in conformity with other studies that explored the deployment of SRL processes used by students interacting with technological learning environments (Azevedo & Cromley, 2004; Azevedo, Guthrie & Seibert, 2004; Azevedo, Cromley & Seibert, 2004; Lajoie & Azevedo, 2006). The findings by these SRL researchers suggest that students who are not integrated into the learning environments are at the risk of being unable to use SRL strategies effectively. Therefore, lower scores in the activity sheets obtained by students in the non-SRL-prompted group attest to this claim.

It is pertinent to note that the success of the incorporation of SRL-instructional-prompts into technological resources such as simulation learning environments for enhancing learning and self-regulation could be ascribed to the fact that the designed SRL-prompts did not impose the burden of additional information processing that may interfere with the students' aim of concentrating on the to-be-learned information. Furthermore, because the designed SRL-prompted instructions were pedagogically integrated into the learning resources, it assisted the students to work towards achieving their target goals within the allocated time. Contrary to our expectations, the SRL-prompted instructions group was able to achieve instructional efficacy despite a brief exposure to prompts. The work of Young (1997) which adopted specific SRL-prompted instructions such as planning and monitoring in a learner-controlled and a program-controlled computer-based instruction (CBI) is in agreement with the findings of this study. Meanwhile, the work of Graesser, Wiley, Goldman, O'Reilly, Jeon & McDaniel, (2007) has explored the impact of a Web tutor on college students' critical stance and learning while exploring Web pages on science. They emphasised that its effects were subtle or nonexistent on multiple measures, except for the articulation of critical stance principles in the essay. This may have been a consequence of the short intervention time frame during which the project had been carried out. On the other hand, we attribute the success of our short-time frame incorporation of SRL-prompted instructions into a scientific simulation learning environment to the fact that it addresses specific goals or activities, rather than a broader intervention which might have required a longer period of intervention in bringing about self-regulatory behaviours.

Implications for Theoretical and Conceptual Models of SRL

This study addresses the concerns about current theoretical and methodological issues among SRL researchers with respect to studies concentrating on the understanding of the inter-relationship between SRL behaviours and learning environments employed during the cyclical and iterative phases of planning, monitoring, time management and reflection (Njoo & De Jong, 1993; Azevedo, 2005). Buttler and Winne (1995) pointed out that a comprehensive model of SRL-prompted instructions incorporated in a technological learning environment would assist greatly in promoting theory-development and research that investigates the complexities of controlling the introduced scaffolds in a technological learning environment. Moreover, one of the major aims of our present study in using SRL-prompts in a computer-based simulation learning environment is that it might help the students to take control of their own learning through critical thinking which may lead to better academic performance. This present study however, provides a platform for researchers to understand how the introduction of SRL-prompted instructions is applicable to the teaching of different topics in a computer-based learning environment.

This study shows that the introduction of SRL-prompted instructions in a science learning environment has effectively supported the learning process. Providing students with SRL-prompts in a computer-based simulation learning environment has kept students engaged as this was noted from our observational analysis because it assisted them to know what to do at the appropriate time during the given task.

Furthermore, it could be helpful for teachers using technology–enhanced learning environments to assist students in the appropriate selection of supports for task activities. For example, a teacher might prompt students to use a particular strategy for the given learning task. Also, designers of educational software need to be aware of the need for prompting students with self–regulated learning behaviour when learning in technology enhanced learning environments; they could do this by placing the prompts next to the task activities and make clear how their use could benefit the learning outcomes.

FUTURE DIRECTIONS IN RESEARCH ON SRL IN TECHNOLOGY ENHANCED LEARNING ENVIRONMENT

There remain numerous theoretical, conceptual, methodological, and educational issues in the area of SRL on learning in a technology enhanced learning environment. Theoretically and conceptually, there is a need for educational researchers to adopt existing theoretical frameworks or models of learning and instruction to drive their research hypotheses regarding learning in any technology enhanced learning environment. We would also encourage the continuation of research through the adaptation of SRL and extend current models (e.g. Bandura 1986) by integrating them with other context-relevant SRL models such as Zimmerman and Schunk's (2001) in order to investigate the nature of SRL behaviours in technology enhanced learning environments.

Our on-going research is exploring other SRL constructs such as co-regulated learning with a view to clarifying the nature of co-regulated learning strategies adopted by 12-13 year old learners employing computer supported collaborative learning (CSCL) environment for learning scientific concepts. We plan to adopt the Dettori and Persico's (2008) approach to conceptualise the practice and development of Co-regulated learning (CRL) and Self-Regulated Learning (SRL) strategies among key stage-three students using CSCL environments. This kind of research may be used to inform the design of CSCL environments that allow learners to learn and build their learning strategies together in order to improve their academic performance as well as their motivation of learning science. We will be employing different methods to capture and analyse behavioural, motivational, and affective states as well as the learners' academic achievement during learning with CSCL.

Methodologically, educational researchers need to conduct more mixed-methods research which will involve multiple measures of learning to capture the various components of science learning and SRL (Azevedo & Jacobson, 2008). There is also a need to conduct research that will extend beyond individual use of simulations of less than an hour to studies that require repeated and extended periods of student engagement with the technology enhanced learning environments in order to examine the self- and co-regulatory processes. Our on-going research is aiming to address this issue by doing longitudinal research with a view to testing several self- and co-regulatory developmental levels. Furthermore, we are of the opinion that an in-depth interview, rather than an informal one, with substantial number of participants would help us to understand students' expectations, feelings and give details of how they make use of CRL/SRL behaviours while learning in a computer-based simulation learning environment. Another area worth exploring for further study is the determination of the appropriate time to introduce SRL supports to the students when learning in technology enhanced learning environments.

CONCLUSIONS

Our research provides a description of the self-regulated learning processes in a computer-based simulation learning environment. This study explored the effect of SRL-prompts on students' academic attainment in the simulation learning environment that teaches the rate of chemical reactions. One of the main methodological issues related to SRL that we address in our research is how students regulate their learning during knowledge construction in a computer-based simulation learning environment. We used our designed activity sheets to capture the nature of SRL during learning of science topic (rate of chemical reaction) with computer-based simulation.

We used a mixed approach by combining experimental design with classroom observation as well as informal interviews to produce both learning outcome measures and process. Our data haves been critical in allowing us to begin to understand the role of SRL in learning about science topics with a computer-based simulation. The results obtained from our research demonstrate that the introduction of SRL-prompted instructions into a science simulation learning environment could be employed to improve students' understanding of difficult science topics. Supporting 12-13 year old science students learning in a simulation environment with SRL-prompted instructions, designed to teach about the rate of chemical reactions facilitated the use of key SRL processes and lead to statistically significant improvement in the attainment of higher test scores. The classroom observation studies carried out during this research provided the supporting evidence that students who had access to SRL-prompted instructions made use of key SRL processes and mechanisms that have been found to have ultimately resulted in significant shifts in the test scores and attainment of sophisticated conceptual understanding on other declarative knowledge measures such as prior knowledge activation. Lastly, our findings suggest that designing technology enhanced learning environments to include various metacognitive tools would go a long way in fostering students' conceptual learning of science topics.

REFERENCES

Alessi, S. M., & Trollip, S. R. (2001). *Multimedia for Learning*. Boston, MA: Allyn and Bacon.

Azevedo, R. (2005). Using hypermedia as a metacognitive tool for enhancing student learning? The role of self-regulated learning. *Educational Psychologist, 40*(4), 199–209. doi:10.1207/s15326985ep4004_2

Azevedo, R., & Cromley, J. (2004). Does training on self-regulated learning facilitate students' learning with hypermedia? *Journal of Educational Psychology, 96*(3), 523–535. doi:10.1037/0022-0663.96.3.523

Azevedo, R., Cromley, J. G., & Seibert, D. (2004b). Does adaptive scaffolding facilitate students' ability to regulate their learning with hypermedia? *Contemporary Educational Psychology, 33*(3), 29, 344–370.

Azevedo, R., Cromley, J. G., Winters, F. I., Moos, D. C., & Greene, J. A. (2005). Adaptive human scaffolding adolescents' self-regulated learning with hypermedia. *Instructional Science: An International Journal of Learning and Cognition, 33*(5-6), 381–412.

Azevedo, R., Guthrie, J. T., & Seibert, D. (2004a). The role of self-regulated learning in fostering students' conceptual understanding of complex systems with hypermedia. *Journal of Educational Computing Research, 30*(1), 87–111. doi:10.2190/DVWX-GM1T-6THQ-5WC7

Azevedo, R., & Jacobson, M. (2008). Advances in scaffolding learning with hyper-text and hypermedia: A summary and critical analysis. *Educational Technology Research and Development, 56*(1), 93–100. doi:10.1007/s11423-007-9064-3

Bandura, A. (1986). *Social foundations of thought and action*. Englewood Cliffs, NJ: Prentice Hall.

Bird, M., & Hammersley, M. (1996). *Educational Research in Action. Milton Keynes*. UK: Open University Press.

Blake, C., & Scanlon, E. (2007). Reconsidering simulations in science education at a distance: features of effective use. *Journal of Computer Assisted Learning, 23*(6), 491–502. doi:10.1111/j.1365-2729.2007.00239.x

Blaxter, L., Hughes, C., & Tigh, M. (2006). *How to research. Milton Keynes*. UK: Open University Press.

Bogdan, R. C., & Biklen, S. K. (1998). *Qualitative research in education: An introduction to theory and methods* (3rd ed.). Needham Heights, MA: Allyn & Bacon.

Butler, D. L., & Winne, P. H. (1995). Feedback and self-regulated learning: A theoretical synthesis. *Review of Educational Research, 65*(3), 245–281.

Chi, M. T. H., de Leeuw, N., Chiu, M.-H., & LaVancher, C. (1994). Eliciting self-explanations improves understanding. *Cognitive Science, 18*(3), 439–477.

Cohen, L., & Manion, L. (1989). *Research Methods in Education*. London, UK: Routledge.

Demetriadis, S. N., Papadopoulos, P. M., Stamelos, I. G., & Fischer, F. (2008). The effect of scaffolding students' context-generating cognitive activity in technology-enhanced case-based learning. *Computers & Education, 51*(2), 939–954. doi:10.1016/j.compedu.2007.09.012

Dettori, G., & Persico, D. (2008). Detecting Self-Regulated Learning in online communities by means of Interaction Analysis. *IEEE Transactions on Learning Technologies, 1*(1), 11–19. doi:10.1109/TLT.2008.7

Garhart, C., & Hannafin, M. (1986). The accuracy of cognitive monitoring during computer assisted instruction system. *Journal of Computer Based Instruction, 13*(3), 88–93.

Graesser, A. C., Wiley, J., Goldman, S. R., O'Reilly, T., Jeon, M., & McDaniel, B. (2007). SEEK Web Tutor: Fostering a critical stance while exploring the causes of volcanic eruption. *Metacognition and Learning*, *2*(2-3), 89–105. doi:10.1007/s11409-007-9013-x

Lajoie, S. P., & Azevedo, R. (2006). Teaching and learning in technology-rich environments. In P. Alexander & P. Winne (Eds.), *Handbook of educational psychology* (2 ed., pp. 803-821). Mahwah: NJ: Erlbaum.

Mace, F. C., Belfiore, P. J., & Hutchinson, J. M. (2001). Operant Theory & Research on Self-Regulation. In B. Zimmerman & D. Schunk (Eds.), *Self-regulated learning and academic achievement: Theoretical perspectives* (2 ed.). Mahwah NJ: Erlbaum.

Manlove, S., Lazonder, A., & De Jong, T. (2007). Software scaffolds to promote regulation during scientific inquiry learning. *Metacognition and Learning*, *2*(2-3), 141–155. doi:10.1007/s11409-007-9012-y

Narciss, S., Proske, A., & Koerndle, H. (2007). Promoting self-regulated learning in web-based learning environments. *Computers in Human Behavior*, *23*(3), 1126–1144. doi:10.1016/j.chb.2006.10.006

Njoo, M., & De Jong, T. (1993). Exploratory learning with a computer simulation for control theory: Learning processes and instructional support. *Journal of Research in Science Teaching*, *30*(8), 821–844. doi:10.1002/tea.3660300803

Patton, Q. M. (1987). *How to use Qualitative Methods in Evaluation*. Newsbury Park, UK: Sage Publications Inc.

Pintrich, P. R.(200). The role of goal orientation in self-regulated learning. In Boekaerts, M., Pintrich, P. R., & Zeidner, M. (Eds.), *Handbook of self-regulation* (pp. 451–502). New York, NY: Academic Press.

Pintrich, P. R., & Schunk, D. H. (2002). *Motivation in education: Theory, research, and applications* (2 ed.). Englewood Cliffs NJ: Prentice Hall.

Pintrich, P. R., & Smith, D. A. (1993). Reliability and Predictive Validity of the Motivated Strategies for Learning Questionnaire (MSLQ). *Educational Measurement: Issues and Practice*, *53*(3), 801–814.

Pintrich, P. R., Smith, D. A. F., Garcia, T., & McKeachie, W. J. (1991). Reliability and predictive validity of the Motivated Strategies for Learning Questionnaires (MSLQ). *Educational and Psychological Measurement*, *53*(3), 801–813. doi:10.1177/0013164493053003024

Schunk, D. H. (2001). Social cognitive theory and self- regulated learning. In B. J. Zimmerman & D. H. Schunk (Eds.), *Self-regulated learning and academic achievement* (2 ed., pp. 125-151). Mahwah, NJ: Erlbaum.

Schunk, D. H., & Zimmerman, B. J. (2006). Competence and control beliefs: Distinguishing the means and the ends. In P. Alexander & P. Winne (Eds.), *Handbook of educational psychology* (2 ed., pp. 349-367). San Diego, CA: Academic.

Shapiro, A., & Niederhauser, D. (2004). Learning from hypertext: Research issues and findings. In Jonassen, D. H. (Ed.), *Handbook of research for education communications and technology*. Mahwah, NJ: Erlbaum.

Turner, J. C. (1995). The influence of classroom contexts on young children's motivation for literacy. *Reading Research Quarterly*, *30*(3), 410–441. doi:10.2307/747624

Winne, P. H. (2001). Self-regulated learning viewed from models of information processing. In Zimmerman, B., & Schunk, D. (Eds.), *Self-regulated learning and academic achievement: Theoretical perspectives* (pp. 277–304). Mahwah, NJ: Erlbaum.

Young, J. D. (1997). The effect of self-regulated learning strategies on performance in learner controlled computer-based instruction. *Educational Technology Research and Development*, *44*(22), 17–27.

Zeidner, M., Boekaerts, M., & Pintrich, P. R. (2000). Self-regulation: Directions and challenges for future research. In Boekaerts, M., Pintrich, P. R., & Zeidner, M. (Eds.), *Handbook of self-regulation*. New York, NY: Academic Press. doi:10.1016/B978-012109890-2/50052-4

Zimmerman, B. J. (1989). Models of self-regulated learning and academic achievement. In Zimmerman, B. J., & Schunk, D. H. (Eds.), *Self-regulated learning and academic achievement: Theory, research and practice* (pp. 1–25). New York, NY: Springer.

Zimmerman, B. J. (2000). Attaining self-regulation: A social cognitive perspective. In Boekaerts, M., Pintrich, P. R., & Zeidner, M. (Eds.), *Handbook of self-regulation*. San Diego, CA: Academic Press. doi:10.1016/B978-012109890-2/50031-7

Zimmerman, B. J., & Martinez-Pons, M. (1986). Development of a structured interview for assessing students' use of self-regulated learning strategies. *American Educational Research Journal*, *23*(4), 614–628.

KEY TERMS AND DEFINITIONS

Computer-Based Simulation: This is a computer program that attempts to simulate an abstract model of a particular system. In the context of this chapter, computer-based simulation is an environment that represents a chemical system composed of molecules represented by balls that move about, collide and react. The environment allows the students to change parameters of the system such as temperature, concentration etc, and observe the impact of the change on the rate of chemical reactions within the system.

Prompt: Act of giving a reminder or cue. Prompts given to the students in this context enabled them to self-regulate their learning activities.

Self-Regulated Learning: This is learning that is guided by metacognition (knowledge about one's thinking), strategic action (planning, monitoring, and evaluating personal progress against a standard), and motivation to learn.

Chapter 16
Enriching Quality of Self-Regulated Learning through Technology-Enhanced Learning Environments:
A Malaysian Case Study

Vighnarajah
Universiti Putra Malaysia, Malaysia

Su Luan Wong
Universiti Putra Malaysia, Malaysia

Kamariah Abu Bakar
Universiti Putra Malaysia, Malaysia

ABSTRACT

Current development in the Malaysian educational policies observed heightened interest in the integration of self-regulation of the learning process through engagement in technology-enhanced learning environments. This study attempts to provide empirical evidence to the effectiveness of the iELC discussion platform in enhancing practice of self-regulation among Malaysian secondary school students. This involved participation of 102 Physics students from four regular national secondary schools. Practice of self-regulation was measured using the Motivated Strategies for Learning Questionnaire (MSLQ) and was analyzed using the two-way between-groups analysis of variance (ANOVA) on a.05 level of significance. Findings provided evident arguments that engagement in this technology-enhanced learning environment warrants for self-regulation in the learning process.

DOI: 10.4018/978-1-61692-901-5.ch016

INTRODUCTION

In the past couple of decades, the educational system in Malaysia aimed only at producing students with good academic results (Smart School: The Story So Far, 2003). The authors of the cited report indicted the educational system for encouraging "mindless memorization and regurgitation of facts and figures, which they [students] do not know how to apply" (p. 1).

However, recent developments of the educational system in Malaysia induced by current global changes and socio-political interest (Osman, Halim & Meerah, 2006) persuaded for the long needed transformation of educational policies. Among the much anticipated effect of these educational policies were the development of self-regulated learning skills and engagement in technology-enhanced learning environments. Literature has established that self-regulation is an important skill for learners (Vighnarajah, Wong & Abu Bakar, 2009; Chang, 2005; Driscoll, 2005; Chen, 2002). Furthermore, Ng (2005) and Brooks, Nolan and Gallager (2001) emphasize that in traditional learning environments monotonous memorizations of information are typical with lack or poor self-regulation skills.

Pintrich (2000, p. 453) defines self-regulated learning as "an active, constructive process whereby learners set goals for their learning and then attempt to monitor, regulate, and control their cognition, motivation, and behavior, guided and constrained by their goals and the contextual features of the environment". Santrock (2001) argues that students who are able to manage effective practice of self-regulated learning are more competent in generating and monitoring their thoughts, feelings and behavior in achieving an objective. Similarly, Ruohotie (2002) and Zimmerman (2002) both share views on practice of self-regulated learning as a self-directive process that channels and amplifies students' cognitive abilities into developing successful academic achievement.

Realizing the numerous benefits of self-regulation, more studies on self-regulated learning have been extended into other areas of interest including technology-enhanced learning environments. This measure is timely and of particular importance since the prevalence of the Internet technology has made it more possible for learners to access abundance of information, and to connect with other learners in a 'borderless' learning environment. Such studies are also evident in the Malaysian educational landscape.

Ng (2005) investigated practice of self-regulated learning in IT-integrated classrooms in secondary Smart Schools. The Smart School Project was an audacious attempt by the Malaysian government to initiate and cultivate learning in technology-enhanced learning environments. In such learning environments, learners were given access to electronic resource centers, computers in science labs, computers in classrooms, computer labs, and multimedia labs. The study involved 409 students from six randomly selected Smart Schools. Students' practice of self-regulated learning was measured using the Motivated Strategies for Learning Questionnaire (MSLQ) (Pintrich, Smith, Garcia & McKeachie, 1991). Results of Pearson correlation coefficient analysis indicates that levels of IT integration ($r = .49$, $p < .01$) were both positively and significantly related to effective practices of self-regulated learning. This finding clearly indicates that students' engagement in technology-enhanced learning environments determines the extent to which they are able to effectively practice self-regulation in their learning.

Meanwhile, Vighnarajah, Luan & Abu Bakar (2009) conducted studies to identify practice of self-regulated learning in online community discussion in regular national secondary schools. Practice of self-regulated learning was again collected using the MSLQ instrument adapted to the specific needs of the study. Among significant findings transpired from two rounds of semi-structured interviews were that participation in this technology-enhanced learning environment improved students' understanding on how to practice self-regulation more effectively in the learning process. Moreover, the students pointed out that they had opportunities to practice self-regulation through interaction with other students engaged in the online community discussion. During this participation, the students identified self-regulated learning components, particularly, rehearsal, elaboration, critical thinking, peer learning and help seeking to be of much importance. These components of self-regulated learning were emphasized in the MSLQ instrument (Pintrich, Smith, Garcia & McKeachie, 1991).

METHODOLOGY

Purpose

The purpose of this study was to investigate the effectiveness of participation in the activities conducted in the *i*ELC discussion platform in enhancing practice of self-regulated learning. Based on the research design and the statistical analysis, the study focused on these research questions:

1. Is there significant pretest effect on the practice of self-regulation in the learning process?
2. Is there significant treatment effect on the practice of self-regulation in the learning process?
3. Is there significant interaction effect between the two independent variables (pretest and treatment) on the practice of self-regulation in the learning process?

Context

The *i*ELC is acronym for Interactive E-Learning Community. The *i*ELC discussion platform was developed using the open source Moodle software. Moodle is a software package used for producing Internet-based courses and websites; its name is acronym for Modular Object-Oriented Dynamic Learning Environment. According to Perens (1997), Moodle is based on Open Source Software (OSS) under the GNU Public License (GPL), hence it is not restricted to licensing costs, and may be used for free. One of the many benefits of the Moodle software is its ability to operate on PHP scripts, MySQL database, and Apache web server; suggesting its competence to run on most operating systems like Windows, Unix, Linux, and Mac-OS (de Zwart, 2003).

The *i*ELC discussion platform supports online learning communities; it was proposed to promote and enhance effective practice of self-regulation in the learning process. However, in the described case study, participation in the activities conducted in the *i*ELC discussion platform was conducted in blended learning approach where classroom discussion precedes the *i*ELC asynchronous forum discussions. The Physics subject was used as the subject of discussion in the *i*ELC forum. We describe below how the students exercised self-regulation in their learning process.

1. Students were instigated to engage in the learning process by means of a brief classroom discussion. This measure was necessary for the instructor to disclose fundamental Physics formulas, definitions and concepts. For instance, students had to understand that momentum is a product of mass and velocity.
2. Students were then suggested to participate in the *i*ELC asynchronous forum discussion to improve their understanding of the application of momentum. For instance, students were encouraged to rehearse and elaborate on the various applications of momentum with other students participating in the asynchronous forum. The instructor posted relevant questions in the forum when students hesitated in initiating the discussion. In this phase, the instructor also monitored

Table 1. Subject exercises to promote self-regulation during participation in the activities conducted in the iELC discussion platform

Strategy	Subject exercises to promote self-regulation
Intrinsic goal orientation	To answer and post questions in forum discussion. This exercise encouraged students to practice intrinsic goal orientation with desire to construct knowledge and understanding.
Extrinsic goal orientation	Participation in the *i*ELC discussion forum increases the students' desire to prove their abilities to other members of the online community
Control of learning beliefs	To answer and post questions in forum discussion. This exercise encouraged students to take appositive perception, and hence, take control of their individual learning behaviors and outcome.
Self-efficacy for learning and performance	Participation in the *i*ELC forum discussion allows the student to assess his/ her ability to contribute useful and critical posts to the forum discussion threads
Rehearsal	Stating formulae and definitions in forum discussion to assist in the activation of information
Elaboration	Elaborating on formulae, definitions and concepts in forum discussion to encourage connection of prior knowledge to the newly acquired knowledge
Organization	Identifying constructive clusters of information in the forum discussion thread
Critical Thinking	To critique on forum posts to apply knowledge in meaningful ways
Metacognitive self-regulation	Skimming the forum discussion threads for facts and opinions. Generating questions after reading the forum discussion threads
Time and study environment	Students need to manage the time and study environment between schools and participating in the *i*ELC discussion platform
Effort regulation	To participate in online group discussions by posting constructive and critical posts to minimize the common perception that Physics is a difficult and boring subject
Peer learning	To engage in online group discussion to clarify doubts on Physics problems that may not have been possible to achieve individually
Help seeking	To seek assistance to solve Physics problems by posting relevant questions in the forum discussion and/ or to contact other members of the online community through use of the chat and dialogue tools

and facilitated the learning process carried out in the forum discussion.

3. The instructor then selected forum posts to be discussed in classroom.

In this study, constructivism was the driving theory for exercising self-regulation in the learning process during participation in the activities conducted in the *i*ELC discussion platform. Ng (2005) argues that embracing the constructivist approach promotes students to actively exercise self-regulation in their learning process. Moreover, McMahon (1997) argues that the strength of the constructivist approach is the construction of knowledge through interaction with the learning environment. Hence, adoption of the constructivist theory works excellently as participation in the activities conducted in the *i*ELC discussion platform amplifies students' interaction with the learning content, the instructor, and more importantly with the other students. Table 1 underlines how self-regulation was practiced in the learning process during participation in the activities conducted in the *i*ELC discussion platform.

Research Design

The study adopted the pretest-posttest and posttest only nonequivalent control group design. This research design was formed by merging the pretest-posttest nonequivalent control group design and the posttest only nonequivalent control group design. According to Trochim (2006), new types of research designs can be formed by

Table 2. The pretest-posttest and posttest only nonequivalent control group design

(E) School Group 1: O_1 X O_2
(C) School Group 2: O_3 O_4
(E) School Group 3: X O_5
(C) School Group 4: O_6
Symbols
X : Treatment
$O_{1,3}$: Pretest
$O_{2,4,5,6}$: Posttest
E : Experimental
C : Control

merging characteristics of established research designs. This research design was favored because it minimized possibilities of data contamination due to pre-testing procedures and reactive effects of testing, in addition to other forms of internal and external validity threats. Table 2 is a graphical representation of this quasi-experimental research design.

In total, the study involved participation from four schools selected in a two-stage cluster sampling technique. The first stage looked at randomly selecting two different zones in Kuala Lumpur to represent the selection of the experimental and control groups. The second stage looked at randomly selecting two schools from each zone to construct the four schools to participate in the pretest-posttest and posttest only nonequivalent control group design. Then, one science stream classroom was randomly selected from each school. Both the experimental and control schools possessed similar ICT infrastructures, with access to Internet facilities and at the very least a 2:1 student computer ratio. Schools without such ICT infrastructures and facilities were not considered. This measure aimed also to ensure that participation in the activities conducted in the *i*ELC discussion platform were not impeded by technical difficulties such as delay in posting and receiving forum posts and downloading on learning materials.

Participants

This study involved a total of 102 Form Four students divided into the experimental groups (n=50) and the control groups (n=52). The Form Four Physics subject is equivalent to Grade 11. In the experimental group, there were 23 male and 27 female students. In the control group, there were 30 male and 22 female students.

Internet access and computer ownership was not a problem as almost all the students in both the experimental and control groups owned Internet access and at least one computer at home. These students also had many years of experience in using the Internet ranging from two to eleven years. Nevertheless, students in the experimental group were briefed on how to navigate in the *i*ELC discussion platform. This briefing includes the technical know-how of posting and replying of forum posts, and the chat, dialogue and resource tools. Instructors from the experimental groups, moreover, were briefed earlier regarding participation in the activities to be conducted in the *i*ELC discussion platform when the researcher approached them for their consent to participate in this study.

Data Collection

The Motivated Strategies for Learning Questionnaire (MSLQ) (Pintrich, Smith, Garcia & McKeachie, 1991) was used to measure students' practice of self-regulation. However, the MSLQ was adapted to better suit the needs of this study. First, the task value and test anxiety subscales were omitted from the original scale. The exclusion of these subscales did not affect the validity of the entire instrument. Hence, the modified version of the MSLQ adopted in this study had 13 subscales, which were, intrinsic goal orientation, extrinsic goal orientation, control of learning beliefs, self-efficacy for learning and performance, rehearsal, elaboration, organization, critical thinking, metacognitive self-regulation, time and study environment, effort regulation, peer learning and help seeking. These subscales were scored on a 7-point Likert scale, ranging from 1 (not at all true of me) to 7 (very true of me). Second, the modified MSLQ instrument was translated into the Malay language. This measure was to ascertain that the students understood the items and were able to accurately answer the items. Reliability analysis for the modified MSLQ instrument was conducted with Form Four students (n=37) in a school with comparable context as the schools in the actual data collection process. Results indicated a Cronbach alpha value of 0.93.

The MSLQ was used to determine to what extent the students were practicing self-regulation in their learning process. Based on the research design illustrated in Table 2, the data collection process consisted of three major phases; pretest, treatment, and posttest. First, the students from both the experimental and control groups responded to the MSLQ to determine their existing level of self-regulation in their learning process. This measure accounted for the pretest phase.

Subsequently, students in the experimental group utilized the *i*ELC discussion platform for duration of eight weeks. The treatment in this experimental study was participation in the activities conducted in the *i*ELC discussion platform, aimed to stimulate students' practice of self-regulation of the learning process. Students in the control group did not participate in the *i*ELC discussion platform.

After duration of eight weeks, students from both the experimental and control groups again responded to the MSLQ in the posttest phase. The purpose of the posttest is to determine the extent to which they have practiced self-regulation in their learning process. For students in the experimental groups, the posttest were intended to measure how much, if at all, the students had improved their practice of self-regulation after participating in the *i*ELC discussion platform in the identified duration of eight weeks. For students in the control groups, the posttest were intended to measure how much, if at all, the students had improved their self-regulation without engagement in the *i*ELC discussion platform.

FINDINGS AND DISCUSSION

Based on the research design employed in this study, the two-way between-groups analysis of variance (ANOVA) (Campbell & Stanley, 1963) was used to analyze the mean values scored on the MSLQ instrument. The ANOVA analysis was conducted on a.05 level of significance.

According to Pallant (2001, p. 201), "*Two-way* means that there are two independent variables, and *between-groups* indicates that different people are in each of the groups." The two independent variables in this study are pretest (students who were or were not subjected to the pretest) and treatment (students utilizing the *i*ELC discussion platform). The two groups are the students in the experimental and control groups.

Pallant (2001) also emphasized that use of this statistical analysis allows the researcher to analyze the individual and interaction effect of the two independent variables on the dependent variable. The dependent variable is the practice of self-regulation in the *i*ELC discussion platform. The interaction effect determines if there exists an interaction effect between the two independent variables on the dependent variable. The main effect determines if each independent variable possess effect on the dependent variable. Table 3 presents the results of the analysis. These findings were underlined according to the research questions stated in this study.

To account for the first research question, findings indicated that there was no significant pretest effect on the practice of self-regulation in the learning process [$F(1, 98) = .15$, $p = .70$]. This implied that the students who were subjected to pretest and students who were not subjected to the pretest did not differ in terms of their practice of self-regulation in the learning process. In

Table 3. Two-way between-groups ANOVA for self-regulated learning

Source	Type III Sum of Squares	df	Mean Square	F	Sig.	Eta Squared
Pretest	288	1	288	.15	.702	.00
Treatment	34985	1	34985	17.92	<.0005	.16
Treatment * Pretest	7726	1	7726	3.96	.059	.04
Error	191336	98	1952			
Total	6438962	102				

other words, administration of the pretest did not influence the students to perform better for the posttest (Campbell & Stanley, 1963).

To account for the second research question, findings indicated that there was significant treatment effect on the practice of self-regulation in the learning process [$F(1, 98) = 17.92$, $p < .0005$]. This implied that students who were subjected to the treatment had enhanced practice of self-regulation in the learning process more than students who were not subjected to the treatment. This provides clear evidence that participation in the activities conducted in the *i*ELC discussion platform is effective in enhancing students to practice self-regulation in their learning process. This main effect (treatment) revealed a large effect size with eta squared value of.16 (Cohen, 1988). According to Cohen (1988), eta squared values can be interpreted as: 0.01 = small effect, 0.06 = moderate effect, and 0.14 = large effect. The eta squared value is expressed in percentage by multiplying the value by 100 (Pallant, 2001). That is, 16% of the variance in the practice of self-regulation in the learning process was accounted for by participation in the activities conducted in the *i*ELC discussion platform.

To account for the third research question, findings indicated that there was no significant interaction effect between the two independent variables on the dependent variable [$F(1, 98) = 3.96$, $p = .05$]. This implied that there was no significant difference in the effect of pretest on practice of self-regulation in the learning process for both the experimental and control groups.

These findings clearly indicated that participation in the activities conducted in the *i*ELC discussion platform was effective in encouraging practice of self-regulation in the learning process. Findings of descriptive analysis indicated that students who were subjected to the *i*ELC discussion platform possessed better practice of self-regulated learning ($M = 5.34$, $SD = 0.92$) compared to students who experienced the traditional classroom environment ($M = 4.37$, $SD = 0.83$). Hence, it was evident that engagement in the *i*ELC discussion platform is an effective alternative to improve practice of self-regulation in the learning process with respect to engagement in traditional classroom environment. Table 4 presents the mean and standard deviation values for each school, grouped in the experimental and control schools.

Accordingly, findings of this study also establish that participation in technology-enhanced learning environments fosters improved practice of self-regulation in the learning process. Ng (2005) argued that researchers should not always assume that only self-factors contribute to the practice of self-regulation. She found that IT-integrated learning settings were among the determinants to effective practice of self-regulation. This argument offers convincing support that engagement in technology-enhanced learning environments do provide merit to enhance practice of self-regulation in the learning process.

Table 4. Experimental and control schools group statistics

Group	SCHOOL	N	School Mean ± SD	Group Mean ± SD
Exp	Experimental School 1	21	12.31±2.25	5.34±0.92
	Experimental School 2	29	9.41±1.55	
Control	Control School 1	23	10.38±2.05	4.37± 0.83
	Control School 2	29	7.51±1.32	

Moreover, Dabbagh and Kitsantas (2004) argued that improved practice of self-regulation in the learning process is usually more achievable in technology-enhanced learning environments than in traditional classroom environment. Kinzie (1990, p. 2) concurs, arguing that improved practice of self-regulation in the learning process is due to, among others, improved practice of learner control:

Exercising control over one's learning can be in itself a valuable educational experience – instructional decisions are made, the results experienced, and the best tactics for different instructional situations can be discovered in the process. In this way, the exercise of learner control can be thought of as a precursor to the development of self-regulation.

Among other advantages of learning in technology-enhanced learning environments are development of declarative and procedural knowledge (Sitzmann, Kraiger, Stewart & Wisher, 2006), learner control and independent learning (Knowles, 1990), communication between peers, develops understanding of subject content and increase revision of the subject content (Osguthorpe & Graham, 2003).

CONCLUSION

While numerous studies (Curtis & Lawson, 2001; Garcia & Pintrich, 1994; Schunk & Zimmerman, 1994), at least in the field of educational psychology, have emphasized the benefits of encouraging effective practice of self-regulation, less emphasis has been placed on the effectiveness of stimulating practice of self-regulation in the learning process through engagement in technology-enhanced learning environments. Hence, this study was undertaken to observe such situation. Accordingly, this chapter provides empirical evidence to the effectiveness of the *i*ELC discussion platform in improving practice of self-regulation among Malaysian secondary school students. Findings provided evidence that engagement in the *i*ELC environment warrants for self-regulation in the learning process.

REFERENCES

Brooks, D. W., Nolan, D. E., & Gallagher, S. M. (2001). *Web teaching: A guide for designing interactive teaching for the World Wide Web*. New York, NY: Kluwer Academic Plenum Publishers.

Campbell, D. T., & Stanley, J. C. (1963). *Experimental and quasi-experimental designs for research*. Boston, MA: Houghton Mifflin.

Chang, M. (2005). Applying self-regulated learning strategies in a web-based instruction - An investigation of motivation perception. *Computer Assisted Language Learning*, *18*(3), 217–230. doi:10.1080/09588220500178939

Chen, C. S. (2002). Self-regulated learning strategies and achievement in an introduction to information systems course. *Information Technology, Learning and Performance Journal*, *20*(1), 11–25.

Cohen, J. W. (1988). *Statistical power analysis for the behavioral sciences*. Hillsdale, NJ: Lawrence Erlbaum Associates.

Curtis, D., & Lawson, M. (2001). Exploring collaborative online learning. *Journal of Asynchronous Learning Networks*, 5(1), pp. 21-34. Retrieved June 25th, 2007, from http://www.aln.org/publications/jaln/v5n1/pdf/v5n1_curtis.pdf

Dabbagh, N., & Kitsantas, A. (2004). Supporting self-regulation in student-centered web-based learning environments. *International Journal on E-Learning*, *3*(1), 40–47.

de Zwart, H. (2003). *Moodle: An exploration of the possible uses in secondary education*. Retrieved June 7, 2007, from http://moodle.org/other/hansdezwart.html

Driscoll, M. P. (2005). *Psychology for learning and instruction*. Boston, MA: Pearson Education.

Garcia, T. D., & Pintrich, P. R. (1994). Regulating motivation and cognition in the classroom: The role of self-schemas and self-regulatory strategies. In Schunk, D. H., & Zimmerman, B. J. (Eds.), *Self-regulation of learning and performance: Issues and educational applications* (pp. 127–153). Hillsdale, NJ: Lawrence Erlbaum Associates.

Kinzie, M. B. (1990). Requirements and benefits of effective interactive instruction: Learner control, self-regulation, and continuing motivation. *Educational Technology Research and Development*, *38*(1), 5–21. doi:10.1007/BF02298244

Knowles, M. (1990). *The adult learner: A neglected species*. Houston, TX: Gulf Publishing.

McMahon, M. (1997). *Social constructivism and the world wide web - A paradigm for learning*. Paper presented at the ASCILITE conference. Perth, Australia. Retrieved December 10, 2009, from http://www.ascilite.org.au/conferences/perth97/papers/Mcmahon/Mcmahon.html

Ng, L. Y. (2005). *Predictors of self-regulated learning in secondary smart schools and the effectiveness of self-management tool in improving self-regulated learning*. Unpublished doctoral thesis. University Putra Malaysia, Malaysia.

Osguthorpe, R. T., & Graham, C. R. (2003). Blended learning environments: Definitions and directions. *The Quarterly Review of Distance Education*, *4*(1), 227–233.

Osman, K., Halim, L., & Meerah, S. M. (2006). Pembinaan instrumen untuk mengenal pasti tanggapan keperluan semasa guru-guru sains di malaysia. [Instrument development to identify the current perception needs of science teachers in Malaysia]. *Malaysian Journal of Educators and Education*, *21*, 101–113.

Pallant, J. (2001). *SPSS survival manual*. Canberra, AU: McPherson.

Perens, B. (1997). *The open source definition*. Retrieved March 24, 2007, from http://www.opensource.org/docs/definition_plain.html

Pintrich, P. R. (2000). The role of goal orientation in self-regulated learning. In Boekaerts, M., Pintrich, P. R., & Zeidner, M. (Eds.), *Handbook of self-regulation* (pp. 451–502). San Diego, CA: Academic. doi:10.1016/B978-012109890-2/50043-3

Pintrich, P. R., & Smith, D. A. Gracia, T., & McKeachie, W. J. (1991). *A manual for the use of the motivational strategies for learning questionnaire (MSLQ)*. Ann Arbor, MI: University of Michigan, National Centre for Research to Improve Postsecondary Teaching and Learning.

Ruohotie, P. (2002). Motivation and self-regulation in learning. In Niemi, H., & Ruohotie, P. (Eds.), *Theoretical understandings for learning in the virtual university. Hämeenlinna, FI: RCVE.*

Santrock, J. W. (2001). *Educational psychology (International edition)*. New York, NY: McGraw-Hill Companies, Inc.

Schunk, D. H., & Zimmerman, B. J. (1994). Self-regulation in education: Retrospect and prospect. In Schunk, D. H., & Zimmerman, B. J. (Eds.), *Self-regulation of learning and performance: Issues and educational applications* (pp. 305–314). Hillsdale, NJ: Lawrence Erlbaum Associates.

Sitzmann, T. M., Kraiger, K., Stewart, D. W., & Wisher, R. A. (2006). The comparative effectiveness of web-based and classroom instruction: A meta-analysis. *Personnel Psychology*, *59*(3), 623–664. doi:10.1111/j.1744-6570.2006.00049.x

Smart School: *The story so far* (2003). Retrieved July 23, 2009, from http://www.mscmalaysia.my/codenavia/portals/msc/images/pdf/magazines/september_2002/sep_2002_smart_school.pdf

Trochim, W. M. K. (2006). Social research methods. Retrieved January 19, 2010 from http://www.socialresearchmethods.net/kb/exphybrd.php

Vighnarajah, Luan, W. S., & Abu Bakar, K. (2009). Qualitative findings of students' perception on practice of self-regulated strategies in online community discussion. *Computers & Education*, *53*(1), 94–103. doi:10.1016/j.compedu.2008.12.021

Zimmerman, B. J. (2002). Becoming a self-regulated learner: An overview. *Theory into Practice*, *41*(2), 64–71. doi:10.1207/s15430421tip4102_2

KEY TERMS AND DEFINITIONS

Constructivism: Constructivism is a theoretical pedagogy that focuses on construction of knowledge through interaction with the learning environment (Dougiamas, 2006). In the context of this study, constructivism is an integrated component of the proposed *i*ELC discussion platform theoretical and is practiced for optimal student-centered learning.

Moodle: Moodle is acronym for Modular Object-Oriented Dynamic Learning Environment (Dougiamas, 1999). Moodle is based on Open Source Software (OSS) under the GNU Public License (GPL). The Moodle software was designed to operate on PHP scripts, MySQL database and Apache web server. In this study, Moodle refers to the software that was used to develop the *i*ELC discussion platform and the activities conducted in the *i*ELC discussion platform.

Self-Regulated Learning: According to Santrock (2001), self-regulated learning refers to strategies used to produce self-generation and self-monitoring of thoughts, feelings and behavior in order to achieve an objective. In this study, practice of self-regulated learning was measured using the Motivated Strategies for Learning Questionnaire (MSLQ).

The *i*ELC Discussion Platform: The *i*ELC Discussion Platform is acronym for Interactive E-Learning Community Discussion Platform. The proposed *i*ELC discussion platform is form of online discussion platform developed using open source Moodle software. Activities conducted in this *i*ELC discussion platform were aimed to enhance practice of self-regulation as underlined by the Motivated Strategies for Learning Questionnaire (MSLQ).

Chapter 17
Mark-UP: Promoting Self-Monitoring of Reading Comprehension through Online Environment

Mark McMahon
Edith Cowan University, Australia

ABSTRACT

While reading skills are an accepted key skill both for life and study, the capacity to read critically and apply reading concepts to solve problems and develop higher order conceptual understandings requires a high level of cognitive self-regulation that students do not always have. This chapter describes the development of and research into an environment, Mark-UP, designed to promote the self-monitoring inherent in regulating reading comprehension. The environment consists of a range of tools to assist learners in monitoring their comprehension through annotation, discussion, problem-solving and so on. The tool was applied to a class of undergraduate students in Interface and Information Design at an Australian university. The research involved questionnaires of the whole cohort as well as case studies of a number of student experiences with the environment, using interview and analysis of the students' portfolios. The study found that, concerning students with weak academic skills, Mark-UP provided some support for their learning, but for stronger students it replicated cognitive strategies that they had already developed. The product was most effective for those students with moderate existing academic skills as it articulated and modeled strategies for reading that they could apply and go beyond to develop their own cognitive regulatory strategies for reading.

DOI: 10.4018/978-1-61692-901-5.ch017

INTRODUCTION

Care needs to be taken when attempting to define environments to support self-regulation, particularly online learning experiences which by their very nature seem already to make demands on students' abilities to regulate their learning. There is a high drop-out rate for students with poor study skills when they venture online (Loomis, 2000). Brooks (1997, p. 135) claims that students 'who are poor at self regulation easily can be slaughtered in WWW-based courses'.

The reality of some online courses is that they place learners in purely independent mode without providing the toolset required to assist students in becoming better learners. If this is to happen courses need to be designed not just to meet specific unit outcomes but also to scaffold the development of learner's self-regulatory skills. This student-centred focus, characteristic of contemporary educational philosophy, seeks to empower the learner rather than to 'teach' the learner through a traditional learning approach based on knowledge transfer (Jonassen & Land, 2000) and is a frequently cited focus of on-line learning, where students are expected to engage in academic texts with typically little or no direct instruction on their comprehension (Reeves & Reeves, 1997).

One important use of the Internet is as a means of accessing course readings, either in the form of Web pages or as electronic documents, such as PDF resources. It provides an efficient and maintainable means of dissemination. The approach of providing several electronic readings rather than a single text also promotes the multiple perspectives inherent in contemporary approaches to learning, such as those espoused in cognitive flexibility theory (Spiro, Feltovich, Jacobson, & Coulson, 1992). However, it is erroneous to assume that students entering tertiary education are able to engage effectively in readings in a self-regulated way. There is a difference between learning to read and reading to learn. Most students have little difficulty with the building blocks of reading such as phonics, but even by their final year of secondary education, studies have shown that only 40% of students can be identified as 'proficient' at the level of reading that involves engagement 'in higher level, problem solving literacy of the kind required in an information generating and information transforming economy' (Greenleaf, Schoenbach, Cziko, & Mueller, 2001, p. 83).

Reading comprehension itself can be classed as a generic skill. While the purpose of this study is not to attempt to promote reading comprehension skills per se, the metacognitive processes inherent in the task makes for a strong relationship to self-regulation:

The ability to read critically is widely regarded as one of the essential generic skills that should be gained through university education. It is often assumed that students will acquire the ability to read critically simply by virtue of studying at University without active intervention from their teachers. We aspire for our students to read with a critical eye in order to develop their own reasoned and ethical position. However, the reality is that students often read as passive consumers of information. (Wilson, Devereux, Macken-Horarik, & Trimingham-Jack, 2004, p. 341)

Research has shown that metacognitive knowledge and self-regulation facilitate reading comprehension (V. L. Collins, Dickson, Simmons, & Kameenui, 2001) but this is an end-product rather than a process. One cannot assume that simply placing students in a mode of study that requires self-regulation will help to promote it. Rather than throw students 'in at the deep end', mechanisms must be in place which bridge the nexus between supported and self-regulated learning.

Proactive measures need to be taken to assist students in developing the necessary skills to learn independently. It would seem possible that an appropriately designed and implemented on-line environment could both minimise student disorientation in new forms of learning, while maximising the opportunities associated with the flexibility afforded by on-line technologies and their potential to have learners engage in tasks in self-directed ways. It should be noted that such an on-line environment would not be a replacement for face-to-face learning:

Accepting the challenge and embracing these forms of delivery and assessment as a replacement of traditional methods is not always appropriate. Instead, the findings suggest that these approaches offer real benefits to some students in particular situations and therefore should be viewed as worthwhile supplements to offer all students more flexibility and the opportunity to enhance their tertiary education experience by encouraging and supporting self-directed and independent learning skills. (Kehoe, Tennent, & Windeknecht, 2004, p. 55)

Such a product would not have a role purely in distance education, but would have potential for all students who may be grappling with engaging in reading concepts and where an on-line component can be integrated to support students' developing skills as self-regulating learners.

This chapter describes such a product, designed to assist in developing learners' self-regulation through engaging them in monitoring their understandings of their reading abilities. The primary goal was to enhance their metacognitive awareness in order to assist in the development of reading strategies. The study involved design-based research in which an online environment, Mark-UP, provided a range of tools for learners to work with readings both on an individual and shared level. The product was developed within a theoretical framework that delineates the processes inherent in metacognitive development of reading comprehension. This model was explored as it was instantiated in Mark-UP to identify the extent to which the product promoted these processes as it was implemented with a group of undergraduate students at an Australian university.

THEORETICAL FRAMEWORK

Self-Regulation has been described as 'a fusion of skill and will' (Garcia, 1995, cited in Brooks, 1997, p. 139). Two particular models were used to inform the development of a framework which focused primarily on self-regulation at the skill level. Boekaerts (1997) provides a six component model based upon the following notions:

- content domain (conceptual and procedural knowledge, misconceptions and inert knowledge);
- cognitive strategies (such as rehearsal, elaboration, generating questions and so on);
- cognitive regulatory strategies (mental representations of learning goals, defining a plan, monitoring and evaluation, goal achievement);
- metacognitive knowledge and motivational beliefs (beliefs, attitudes and values related to tasks within a domain);
- motivational strategy use (such as coping processes, effort avoidance and so on); and
- motivational regulatory strategies (mental representations of behavioural intention, linking this to an action plan, and maintaining that plan in the face of obstacles).

In an alternative model, Garcia and Pintrich (1994) articulate self-regulation in terms of knowledge and beliefs, strategies used, and outcomes. Each of these is moderated by motivational and cognitive components such as personal beliefs and conceptual knowledge, motivational and cognitive strategies, and quantity and quality of effort.

Common to both models is an integration of both affective and cognitive issues and the interdependence of these was a factor that needed to be accommodated in the research. Even at the cognitive level, however, a common assertion is that self-regulation occurs through a series of levels, from metacognition (self-awareness as a learner) through to strategy development, with a monitoring process that underpins the formulation of learning strategies.

Metacognition itself can be defined as 'knowledge and beliefs about thinking and the factors affecting thinking' which regulate 'the articulation of strategy and knowledge' (Pressley, 1998). As such it is a necessary precursor to self-regulation. Flavell (1987, cited by Boekaerts, 1997) identified three types of metacognition: knowledge of self, knowledge about various cognitive tasks and strategy knowledge. At the strategic level, cognitive self-regulation is manifest through the application of learning strategies such as rehearsal, elaboration, and organisational strategies, as well as memorization through clustering, imagery, use of mnemonics and so on (Weinstein & Mayer, 1986). Lin (2001) identifies strategies such as error detecting, effort and attention allocating, elaborating, self-questioning, self-explanation, constructing visual representations, activating prior knowledge, rereading difficult text sections, and going back to revise as examples of cognitive strategies.

Typically, deeper cognitive processes such as transformation, the creation of something new out of existing information, are more successful than ones which engage in knowledge as a static entity, such as rehearsal (Risemberg, 1996). It is important to note, therefore, that knowledge of learning strategies is not enough to ensure that they take place. Regulation strategies must be implemented to co-ordinate effort and task.

At the core of these is a focus on monitoring. Weinstein and Mayer (1986) describe all metacognitive activities as involving the monitoring of comprehension, and it would appear that this ability to monitor oneself is what distinguishes metacognitive activity from domain-specific cognition. Self-monitoring is an initial step towards the development of cognitive strategies but continuous self-monitoring is also a strategy in itself. Depending on one's theoretical orientation, this component can manifest itself as social cognitive self-observation, Vygotskian inner speech, or behaviourist self-recording (Zimmerman, 1989). Regardless of whether one views cognition itself as an important construct, however, self-monitoring is a pervasive key process to self-regulation.

Such tenets find credence in the broader research on metacognition. Nelson and Narens (1994) in their research on metamemory identified the relationship between the meta-level and the object-level of cognition through a reciprocal flow of control and monitoring. Blakey and Spence (1990) cite Dirkes' synthesis of much of the literature on metacognitive processing to describe this flow in the following features:

- connecting new information to former knowledge;
- selecting thinking strategies deliberately; and
- planning, monitoring, and evaluating thinking processes (Dirkes, 1985).

This approach has informed the development of a theoretical framework to promote the metacognitive development of strategies specific to the domain of reading comprehension while still acknowledging the nature of metacognition as crossing domains. Ultimately however, any learning must be grounded in a context, and the model demonstrated in Figure 1 articulates the framework within the context of Reading Comprehension.

The regulatory strategies inherent in reading comprehension may vary. However, in academic texts it is fair to say that some ideas may be more important than others, depending on the student's needs. Therefore, the regulatory strategies defined in this model are those proposed by Dole, Duffy, Roehler, and Pearson (1991), specifically, summarizing, drawing inference, questioning and determining importance.

Figure 1. A Model of Self-Regulation for Reading Comprehension

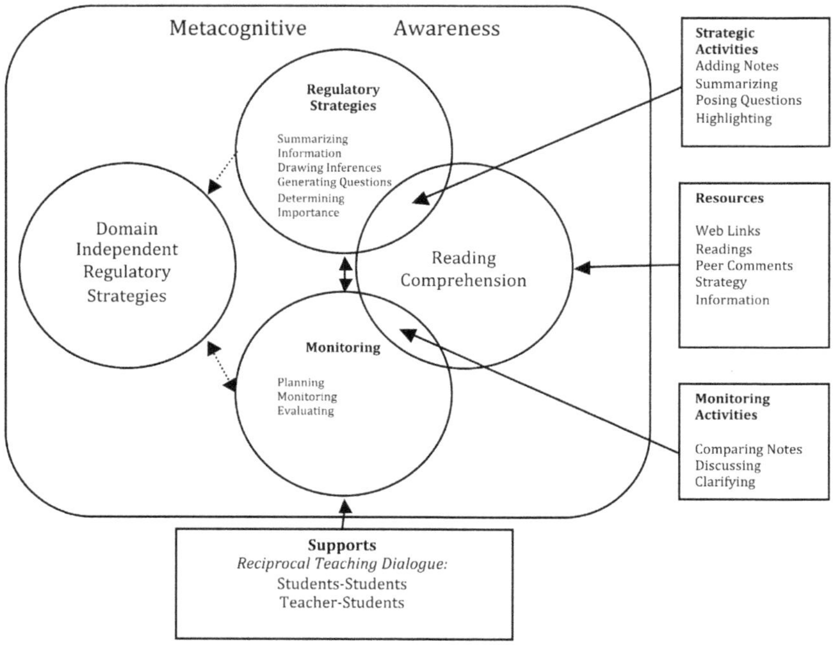

However the strategies themselves are only the outcome of the learning process. The process itself can be seen as the combination of support, activities and resources to support these outcomes (Oliver, 1999). An approach to supporting learners in this model was developed from Palincsar and Brown's (1984) Reciprocal Teaching Approach. In this theory, there are three main components to supporting learning:

- dialogue between students and teacher, each taking a turn in the role of dialogue leader;
- 'reciprocal' interactions where one person acts in response to the others; and
- structured dialogue using four methods: questioning, summarizing, clarifying, predicting.

This model was identified as particularly pertinent to self-regulation due to the monitoring inherent in the methods for structured dialogue. Therefore activities can be identified that are strategic in that they model or require learners to perform the regulatory strategies inherent in reading comprehension but also monitoring activities, which focus on the end users' planning, monitoring and evaluation of their own thinking processes.

MARK-UP

The theoretical framework was manifest in the development and implementation of an online learning environment, Mark-UP. This product used PHP/MySQL for the provision of online readings which could then be summarized, discussed, annotated and so on, with the opportunity for comments to be shared and reviewed through a reciprocal teaching dialogue.

The interface from the end user's perspective can be seen in Figure 2. The bulk of the screen consists of reading pages, with a toolset to the right which allows for page navigation, accessing the various activities defined for that reading

Figure 2. Mark-UP interface from end user perspective

and a legend explaining the meanings of specific annotation icons.

Figure 3 shows part of an example reading with annotations visible. Learner's could add annotations, respond to others and so on. The basic nature of the annotations are identified using different icons, and where there are multiple comments on a section of the text a yellow note identifying the number of comments is displayed. This provided not only discrete comments about particular sections of the reading, but also discussion threads for the social negotiation of the concepts inherent in them.

Annotation, however, is only one strategy embedded in Mark-UP. Course designers with administrative control within the system had the ability to identify a number of tasks for learners that use a range of tools within the product. Some of these tools specifically focus on supporting learners' monitoring of their understandings of

Figure 3. A Section of a Reading in Mark-UP with Annotations

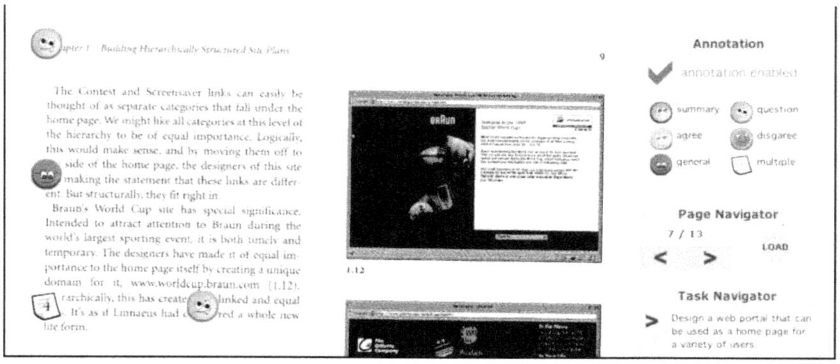

the readings, though attached with each reading when implemented were also design problems, which gave learners opportunities to articulate their understandings of the concepts in the reading, transforming beyond pure comprehension of the text itself. The full toolset within Mark-UP is described in Table 1, with a rationale for how each fits within the regulatory model.

This toolset provided flexibility to the teaching and learning process. Figure 2 shows how a user can access two tasks: a design problem; and marking up the reading. For each of these tasks, any of the tools in Table 1 can be made available as an activity. This is shown from an administrator's perspective in Figure 4.

METHODOLOGY

This particular study was part of a broad iterative piece of research that sought to explore the value of Mark-UP both intrinsically as a product design and its potential for promoting metacognitive regulation. This provided both the narrow and broad research goals inherent in design-based research (Cobb, Confrey, DiSessa, Lehrer, & Schauble, 2003, p. 9). Design-based research was also selected as a methodology because of its capacity to engage in an action-oriented research approach that acknowledges the messiness of authentic settings, while providing a testbed for innovation (Cobb et al., 2003). Ultimately the researcher is a co-participant in design and analysis rather than an experimenter (A. Collins, 1999). It is 'pragmatic as well as theoretical in orientation in that the study of function - both of the design and of the resulting ecology of learning – is at the heart of the methodology' (Cobb et al., 2003, p. 9).

The product was implemented in a cohort of 26 second year undergraduate Interface and Information Design students, studying that subject as part of a degree in Interactive Media at an Australian university. Students used Mark-UP to solve design problems, reflect on their performance throughout the semester and mediate their understandings of the readings through the available toolset within the product. The tools were implemented in a scaffolded way combining initial introductions to each tool with associated strategy instruction leading to a final phase where students used the tools of their choice in an independent manner.

The research included all aspects of planning, monitoring and evaluating evident through the use of Mark-UP, however the specific focus of this chapter is on the use of Mark-UP in supporting self-monitoring to promote metacognitive reading comprehension and then transfer to the domain of design problem solving.

To this end, data was collected both from students perceptions of themselves as learners but also from the nature of self-monitoring evident throughout their use of Mark-UP in the form of

Table 1. Detailed description of Mark-UP tools

Tool	Description	Rationale
Design Problem	This tool enabled the course designer to pose questions and provide a text box for users to complete. Questions could take many forms, for example prompts about a reading, or instructions for the end user to provide concrete examples.	This was a generic tool that allows handling of information types not supported by links and annotations. As well as responding to design problems it provided an opportunity for users to evaluate their progression over a period of time and review previous work to identify their conceptual growth. This was the tool for example that was used by subjects to identify plans and evaluate them later in the semester.
Summary	This tool was designed to allow subjects to summarise a whole reading. As the discussion of the administration of Mark-UP later in this chapter shows, the tool was essentially the same as the Design Problem tool, using a text-box response, but was used in this case for a different purpose. The tool also allowed the course designer to provide a model answer, which subjects could review after having submitted their original summary.	This tool enabled subjects to engage in the strategy of summarising which has been identified as integral to strategic reading. Subjects could be required to summarise a whole reading, or identify key ideas within it. The implementation of a second level of activity by having subjects review and compare their response to a model answer enabled a further level of reflection and monitoring.
Post URL	Subjects could add a link to an external website, including a title and comment. Once completed, they then had an option to review the URLs posted by others and rate them according to a star value (0-5) as well as add comments.	This tool was used to have students reflect on their interpretation of a specific reading and engage in information seeking by finding a website that covers a similar topic, and then discuss the similarities and differences in points of view. The ability to rate other students' links, and compare perspectives, also provided the reciprocal teaching for the self-monitoring processes in which learners engaged.
Annotation	Users clicked on a part of the reading to add an annotation to it, which then appeared as an icon on the screen at the point where they annotated. Annotations took the forms of: • Summary • Questions • Agree • Disagree • General Each type of annotation was represented by a different icon. Learners could view each other's annotations and add to them.	This tool was used to have students engage in the regulatory strategies for reading comprehension as proposed by Dole et al. (1991). Support for self-monitoring was provided by the discussion with peers.
Forum Discussion	This provided a direct link to an on-line discussion board. Subjects could start general discussion threads or respond to existing discussions	The Forum Discussion tool is one common to many on-line learning environments. In this case, it accommodated discussion that was not tied to a specific section of a reading but could be more general in nature. It enabled peer collaboration about a range of issues that were prompted by the reading
Portfolio	Portfolios consisted of a summary of all the subjects' work organised by reading. Students generated their portfolio which they could review and amend before submission.	The value of journaling as a means to enhance self-awareness has been well documented (Brooks, 1997). The Portfolio tool's role was for summative assessment, but most of all it provided an information base for further reflection. Students were required at times to review their understandings as articulated within the portfolio to describe how these had developed.
Review URL	This was an adjunct tool rather than a tool in its own right since it did not require any response from the subject per se, but could be integrated into the above tools, such as having subjects review an URL before finding one which complemented the example provided.	This tool operated as a prompt and enabled the course designer to integrate other sources of information into a Design Problem. Since it was a discrete tool within the project it is mentioned here although it did not actually involve any response from the user, and was not explored as part of this research.

Figure 4. Mark-UP administration environment

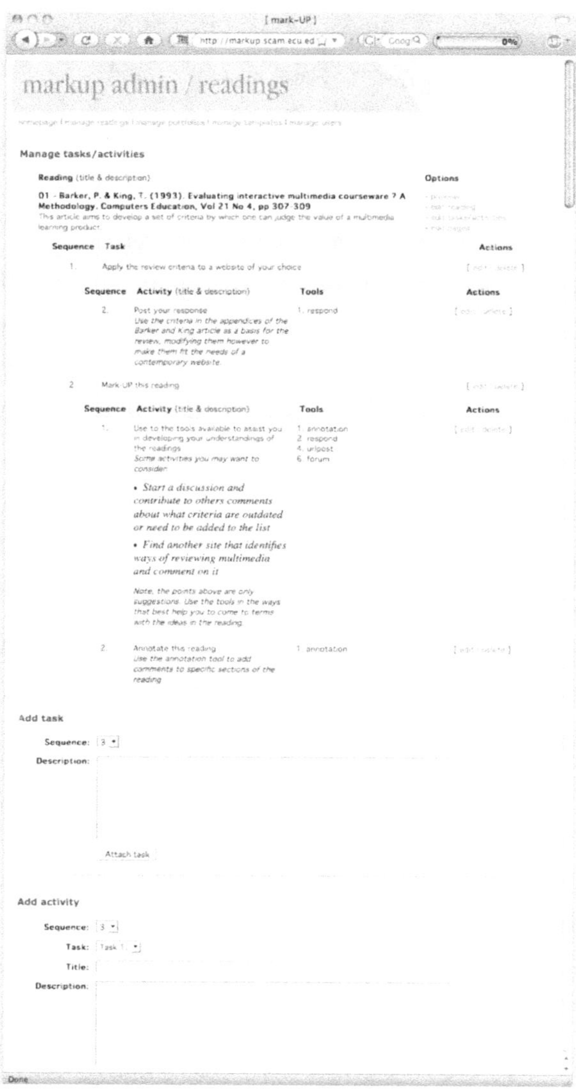

reflective comments, self-assertions and strategies used. This took the form of:

- transcripts of work produced within the product (portfolios);
- questionnaires about students' backgrounds; and
- interviews with 12 students about their experiences with Mark-UP.

These forms were selected for their ability to provide 'thick descriptive datasets' (The Design-based Research Collective, 2003, p. 7). Collins (1999), for example, specifically mentions electronic journals and on-line discussions as innovative approaches to data collection that can do much to explain the nature of activity within real settings. The portfolios provide a detailed journal of all the work that students completed within the environment from the beginning of semester to the end. While the questionnaires were used for the whole cohort of students, this was primarily as a means to ascertain the effectiveness of the product in terms of general student perceptions regarding its implementation and usability. The interviews allowed for deeper exploration of responses, particularly with regard to interrogating causal relationships. Since self-regulation is developed individually and through a lengthy process beyond the scope of this study, a detailed exploration of selected individuals' experiences through the semester was beneficial.

The semi-structured interviews with the students addressed the following broad question types:

1. What factors inherent in the design and implementation of the environment affect its use?
2. What factors inherent in students' backgrounds affect the use of the environment?
3. What are the external environmental factors that affect the use of the learning environment?
4. What are learners' prior experiences of learning metacognitively?
5. What forms of self-monitoring take place when students annotate and transform text-based problems to solve problems?
6. In what way do students apply the processes of metacognitive regulation when scaffolding has been removed?

The first three of these provided insight into the extraneous factors that needed to be qualified in order to properly interrogate the final three questions, which focused specifically on students' metacognition and self-monitoring within the environment and beyond it. The findings focus primarily on the final three questions.

FINDINGS

Zimmerman & Martinez-Pons (1988) identified monitoring in the forms of questioning, comparing ideas, seeking information and organising and transferring their ideas. These formed the basis of a textual analysis of the portfolio content generated by the tools within Mark-UP.

All of the tools offered affordances and limitations in the types of monitoring that they promoted. The Forum Discussion tool certainly appeared weakest both in the variety of monitoring and depth of monitoring evidenced. It was also the least used of the tools. As a forum for sharing ideas in a global sense rather than relating to a specific point in the text, one might have expected responses to be more general in nature, although with more room for opinion and the opportunity to integrate concepts beyond the specific concepts within the reading.

Certainly, many of the comments were quite general. Several failed to integrate specific concepts into their discussion, as was the case with participant 108:

I do agree that it is important to follow guidelines such as the ones shown in this weeks reading but I also feel that if we do not experiment and think outside the square we live in we will not grow. And IMM [Interactive Multimedia] is growing at an amazing speed.

Such platitudinous comments do not demonstrate any clear metacognitive monitoring. While the point made is valid enough, a lack of any rationale, or reference to specific examples, dilutes its value both as an example of monitoring and as a contribution to the reading content.

It appeared that many of the responses were written without a clear sense of audience. While discussion forums typically provide a means for people to share ideas and opinions, and answer each other's questions, many of the responses could be defined as a summary.

Consisting of 37,196 words once collated, made up from 89 individual students' responses, the Summary tool was the most widely used of the other tools. Summary itself is directly tied to reading comprehension as a regulatory strategy. This suggested that it had value for subjects, enabling them to monitor their comprehension of the reading. In terms of the types of monitoring evidenced, one would expect it to demonstrate a somewhat limited range, given its focus on reorganisation of concepts rather than transfer, comparing ideas and so on. Some opportunity for questioning would also be expected, although the lack of feedback would have made such questions predominantly rhetorical in nature.

In fact, these forms of monitoring did dominate the responses. Such organisation generally took one of two forms: summaries that were brief and more personal, being written in the language of the subject; or summaries that were more direct in the manner in which they condensed the ideas within the reading. The latter summaries tended to follow the structure of the reading more closely and were generally longer. A further type was where there were some transformative elements, either through significant prioritization or engaging in a level of critique of the reading. In one case, for example, a student discussed the nature of motivational theory by tying it to teaching strategies while another suggested how the reading could be improved through case studies.

Similarly, the content generated within the Post URL tool demonstrated cognition that went beyond the monitoring process of information seeking and demonstrated some critique in the ratings of the URLs, albeit in a briefer and more limited manner than was evidenced in the Summary tool. While the 'star' rating for websites provided little information, sometimes the sites themselves that were selected and the comments attached to those demonstrated a clear ability to compare the content between the reading and the URL posted.

By far the most complex forms of monitoring appeared to take place within the Annotation and Design Problem tools. The design of the Annotation tool provided several categories that facilitated identification of the types of annotations students could make within the tool. Specifically students identified their annotation as belonging to one of the following types.

- Agree;
- Disagree;
- Summary;
- Question; and
- General Comment

Many students found these categories somewhat arbitrary, and argued they did not effectively characterise the types of comments they were making. Additionally, in identifying a type of comment, the above categories do not actually suggest an underpinning monitoring process. While questioning can be considered indicative of monitoring, since internal feedback is used to generate the question (possibly as a result of cognitive conflict when confronted with an idea or issue that is difficult to reconcile with existing understandings), the categories *agree*, *disagree*, *summary*, and particularly *general comment*, appeared to be too broad to subjects to be aligned to a specific monitoring process.

In fact, the *general comment* category was by far the most widely used within the Annotation tool, with the *agree* and *question* categories making the bulk of the other comment types. In order to explore these annotations in terms of monitoring, therefore, the first step taken was to review a series of annotations to identify the monitoring processes that informed them. These occurred across several readings, as it appeared the nature of the reading itself played a role in the forms of monitoring that was evident. The students perceived some readings as more difficult than others. In one difficult reading, many of the annotations related to the nature of the reading itself. As such it provided a good example of metacognitive regulation of reading comprehension, albeit in an external critical mode rather than an internally reflective one. For example, the main basis for discussion in one series of annotations centred on the first comment, a single line summary of the paragraph. It appears that this initial rephrasing of this paragraph made the concept more accessible to other students. Certainly, the negative comments such as 'complex wording for a simple message' and, 'I think some of the terms could be better explained' indicated a level of frustration with the manner in which the article was written.

Annotation therefore was used to critique readings and make them accessible, as well as to negotiate understandings of the key concepts. Overall, the dominant forms of monitoring evident in the Annotation tool was rephrasing the content of the section (in effect a brief summary) and of clarification of its main ideas. Nevertheless, deeper forms of monitoring where evident, particularly when students discussed the readings in terms of prior knowledge.

In one section of a reading that discussed the nature of epistemology and learning, the first comment demonstrated information seeking as an approach to monitoring. By directly drawing an allusion to a previous reading, the subject was able to both emphasise and expand on the point made. As well as information seeking, monitoring was also manifest in the questioning and answering that followed, allowing students to identify and seek to address inadequacies in their understanding of the material. The questions 'how do you determine what the potential users already know?' and 'How do you build on their existing schemata, when they will differ from individual to individual?' both sought clarity about the issue as well as extended beyond the main concept to the practical implementation of those ideas. This demonstrated monitoring at a high level and created a cumulative body of information that added to the value of the reading itself. It also evidenced an orientation to the transformation of understandings to a new setting.

Some of the more personal annotations also showed a high level of monitoring. On one reading where the attributes of the learner were mentioned as a factor inherent in Instructional Design decision-making, one student posed the question 'how do you create something that caters for everyone and at the same time maintaining the same standards?' As well as a good example of monitoring in the form of questioning itself, several subjects' responses to this annotation were able to use their own experiences to negotiate the issue in question. These related to the use of levels in gaming as a means of catering for individual differences, firstly in a general sense, then in one student's use of a specific example from a game to tie a concrete instantiation of the concept to the more general description. An additional annotation, where a student proposed self-assessment as an initial strategy to facilitate learners being able to select tasks of appropriate difficulty, showed that the comments went far beyond the reading and into the realm of design itself.

This ability to cross domains was one of the key foci of the Design Problem tool. As with annotations, the solutions to design problems demonstrated a range of forms of monitoring, the four main ones being:

- explication of design suggestions that were relevant to the problem, but broad and only loosely tied to the reading concepts;
- re-iteration of reading concepts tied to the context of the problem but without uniquely customising the response to design issues within the website;
- application of reading concepts tied closely to the context of design issues with selected websites; and
- integration of personal comments, alternative examples or other forms of new information into responses that related strongly to both the reading and the problem.

The first two forms of monitoring were the weakest, mostly because they failed to engage both with the organisation of conceptual understandings and transfer to the problem at hand. The first indicated adequate transformation, but weak organisation, while the second demonstrated an ability to organise reading concepts but inadequately apply them to the problem.

As one would expect, the further level of monitoring required to organise and transform reading concepts beyond their initial contexts was challenging for many students. This created a diversity of responses in terms of the types of monitoring evidenced. Most subjects appeared able to engage in monitoring at some level, whether this meant they could reorganise the reading concepts in a way that was personally relevant to them, or they could transform previously formed ideas rather than the reading concepts to tackle the design problem. Most students were able to engage in organisation and transformation at some level. However, it appeared that they were more able to do this when the immediacy of the reading to the design problem was clearer. In some of the more abstract readings, a division between the monitoring activities of organisation and transformation was evident in a tendency for subjects to restate reading concepts in their own words and select appropriate ones, but then only to transform those concepts in a limited way, either by writing in generalities or relying more heavily on previously understood concepts.

This diversity in terms of the level of monitoring required further exploration in order to identify the impact of individuals' existing cognitive and affective attributes in their use of Mark-UP and the value they found in it as an environment for supporting monitoring. This was enabled through the interviews conducted with 12 students who demonstrated a range of preexisting interpreted levels of metacognition. An analysis of the interviews and their portfolios is summarised in Figure 5.

One of the strongest findings in this exploration of the interview subjects' use of Mark-UP to monitor their understandings was the perceived strength of the Annotation tool for this purpose. Being one of the more unique aspects of the product, it was gratifying to find that the majority of students found value in it.

This was not true for everybody however. As Figure 5 shows, Annotation was more popular with subjects who were identified as metacognitively weak or moderate than with those who were interpreted as operating at a high level of metacognition. While Alan's preference for annotation seemed to be in contrast to this, it was interesting that his overall portfolio displayed a lower level of monitoring than some of the others. While the emphasis that annotation placed on external references for monitoring meant that it was very accessible to weak or moderately metacognitive students, it did not always promote the deepest level of monitoring. Nevertheless, nearly all of the interview subjects were able to demonstrate a medium level of monitoring within Mark-UP, and the Annotation tool appeared to be the dominant medium for this.

For the metacognitively strongest students, the Design Problem tool was found to enable a deep level of monitoring and demonstrated their ability to actively monitor their understandings to apply reading concepts to new and unfamiliar tasks. Claire, for example, noted the value of this above all other tools. This was not always true for the metacognitively weak or moderate students, however. While Jake could argue strongly around a specific point in a reading, he was less able to transfer this knowledge across to practical application.

In this sense, it appeared that Mark-UP supported subjects in operating at a level of monitoring at least commensurate with their interpreted level of metacognition. Where there were exceptions to this, subjects nearly always displayed a deeper level of monitoring than their metacognitive level would have suggested. As Figure 5 shows, only Yvette's level of monitoring was interpreted as low. Interestingly, she was also the subject who made least use of the product. Two of the students perceived as metacognitively stronger, Belinda and Claire, also made less use of Mark-UP than the others. In their case, however, they were able to rationalize that within the tendency of the envi-

Figure 5. Summary of interview participants

	Interpreted Level	Monitoring		
Name	of Metacognition	Pref Tool	Portfolio Qual	Interpreted Level
Dean	Low	Annotation	Medium	Medium
Yvette	Low	Annotation	Low	Low
Frances	Low	Annotation	Low	Low
Brian	Low	Annotation	Medium	Medium
Craig	Medium	Summary	Medium	Medium
Sylvia	Medium	Mixed	Medium	Medium
Jake	Medium	Annotation	Medium	Medium
Duncan	Medium	Annotation	Medium	Medium
Claire	High	Design Problems	High	High
Belinda	High	None	High	High
Debbie	High	None	High	High
Alan	High	Annotation	Medium	Medium/High

ronment to replicate existing cognitive regulatory strategies. In interview, Yvette maintained a positive attitude towards annotation. However, there were few actual annotation artifacts to analyse. Rather than be totally attributable to her perceived level of metacognition, it appeared that motivational aspects contributed to her lack of success, particularly since Dean, Frances and Brian were all capable of demonstrating monitoring when tied to an external form of feedback.

The affective components of self-regulation appeared to play a significant role in the benefits derived from the product and the way in which it was used. Duncan acknowledged a tendency to 'laziness' and this undoubtedly impacted on his performance within the product. For example, while much of the content within his portfolio demonstrated implicit monitoring, this did not always appear to be a formalised process. His annotations tended to be less specific than some of the others and draw less directly on experience and other readings, although obviously spurred by the external stimulus of a previous post. Equally, Debbie's choice of summary restricted the range of monitoring she was able to demonstrate compared to her use of annotations. In interview she exhibited a lack of confidence. When combined with the overall high quality of other forms of monitoring demonstrated, this suggested that the cognitive aspects of self-monitoring could be greatly impacted upon by issues such as self-concept and motivation.

In the end, it appeared that the metacognitively strong participants were the ones who were least well serviced by the product. Alan's slightly lower performance compared to his high interpreted level of metacognition provided a reminder that monitoring was only one aspect of metacognition. Other aspects such as his emotional maturity and ability to regulate most aspects of his life were not necessarily captured in his use of the tool. The positive value he placed on the product also gave some confidence in the worth of it. Unfortunately this was not true for Debbie, Belinda and, to a lesser extent, Claire. While some value was found in the Design Problem tool's ability to act as a medium to transfer learning concepts to a practical outcome, the fact that Mark-UP tended to replicate what were already well established cognitive strategies for these subjects made it more of a hindrance than a help. Nevertheless, all three students were able to demonstrate a high level of monitoring in their use of the product.

CONCLUSION

Mark-UP proved to be an effective tool for engaging learners' in monitoring their learning as they undertook reading tasks, particularly annotation and problem solving around the readings. There was some discrepancy between those students perceived as metacognitively weak and those metacognitively strong.

For those learners who are already operating at a high level of self-regulation, environments such as Mark-UP may be somewhat redundant:

> *An important consequence of Self-regulatory behavior is that students who self-regulate find a way to learn. It does not matter if the instructor is a poor lecturer, the textbook is confusing, the test is difficult, the room is noisy, or if multiple exams are scheduled for the same week; self-regulatory learners find a way to excel. (Dembo & Praks Seli, 2004, p. 3)*

However, for metacognitively moderate and weaker learners, the need for environments that promote students' ability to learn independently is real and pressing. For these learners, environments need to be created that scaffold their development of learning strategies and metacognitive application of these. Mark-UP was one such environment that engaged learners effectively in the self-monitoring processes that underpin such strategy development.

To ensure success, the affective dimensions of self-regulation need to be foregrounded in future developments. It has been argued that educational researchers have ignored motivation as an explanation of why students fail to change their learning and study strategies (Simpson & Nist, 2000), and a pervasive finding for this study was the integral aspects of affective dimensions such as self-concept, motivation and volitional control to cognitive self-regulation. The positive message to be drawn however is the role of metacognition in enhancing learners' values of themselves where effective 'strategy use has a direct impact on self-concept, attitudes about learning, and attributional beliefs about personal control' (Borkowski, Carr, Rellinger, and Pressley, 1990, cited by Vandergrift, 2002, p. 571).

Ultimately the best learners are ones who understanding their learning, engage in self-monitoring and develop effective cognitive strategies through a learning experience that is supportive of the needs of the individual as a whole. While it is certainly true that learners do not become metacognitive through brief interventions such as this one, it is contended that continued research, development and utilisation of environments that promote self-monitoring such as Mark-UP that will eventually yield improvements to students' cognitive regulation and strategy use.

REFERENCES

Blakey, E., & Spence, S. (1990). Developing Metacognition. ERIC Digest. Retrieved 19 January, 2002, from http://www.ed.gov/databases/ERIC_Digests/ed327218.html

Boekaerts, M. (1997). Self-Regulated Learning: A new concept embraced by researchers, policy makers, educators, teachers, and students. *Learning and Instruction*, *7*(2), 161–186. doi:10.1016/S0959-4752(96)00015-1

Brooks, D. W. (1997). *Web Teaching: A guide to designing interactive teaching for the World Wide Web*. New York: Plenum Press.

Cobb, P., Confrey, J., DiSessa, A., Lehrer, R., & Schauble, L. (2003). Design experiments in educational research. *Educational Researcher*, *32*(1), 9–13. doi:10.3102/0013189X032001009

Collins, A. (1999). The changing nature of educational research. In Lagemann, E., & Schulman, L. (Eds.), *Issues in Educational Research* (pp. 289–298). San Francisco: Jossey-Bass.

Collins, V. L., Dickson, S. V., Simmons, D. C., & Kameenui, E. J. (2001). Metacognition and Its Relation to Reading Comprehension: A Synthesis of the Research. Retrieved 9 January, 2002, from http://idea.uoregon.edu/~ncite/documents/techrep/tech23.html

Dembo, M., H., & Praks Seli, H. (2004). Students' resistance to change in learning strategies courses. *Journal of Developmental Education*, *27*(3), 2–11.

Dole, J. A., Duffy, G. G., Roehler, L. R., & Pearson, P. D. (1991). Moving from the old to the new: Research on reading comprehension instruction. *Review of Educational Research*, *61*(2), 239–269.

Garcia, T., & Pintrich, P. R. (1994). Regulating motivation and cognition in the classroom: The role of self-schemas and self-regulatory strategies. In Schunk, D. H., & Zimmerman, B. J. (Eds.), *Self Regulation of Learning and Performance: Issues and educational applications* (pp. 127–153). Hillsdale, N. J.: Erlbaum.

Greenleaf, C., L., Schoenbach, R., Cziko, C., & Mueller, F. L. (2001). Apprenticing adolescent readers to academic literacy. *Harvard Educational Review*, *71*(1), 79–127.

Jonassen, D. H., & Land, S. M. (Eds.). (2000). *Theoretical Foundations of Learning Environments*. Mahwah, New Jersey: Lawrence Erlbaum Associates.

Kehoe, J., Tennent, B., & Windeknecht, K. (2004). The challenges of flexible and non-traditional learning and teaching methods: Best practice in every situation? *Studies in Learning, Evaluation. Innovation and Development, 1*(1), 55–62.

Lin, X. (2001). Designing metacognitive activities. *Educational Technology Research and Development, 49*(2), 23–40. doi:10.1007/BF02504926

Loomis, K. D. (2000). Learning styles and asynchronous learning: Comparing the LASSI model to class performance. *Journal of Asynchronous Learning Networks, 4*(1), 23–31.

Nelson, T. O., & Narens, L. (1994). The role of metacognition in problem solving. In Metcalfe, J., & Shiminura, A. (Eds.), *Metacognition* (pp. 207–226). Cambridge: MIT Press.

Oliver, R. (1999). Exploring strategies for on-line teaching and learning. *Distance Education, 20*(2), 240–254. doi:10.1080/0158791990200205

Palinscar, A. S., & Brown, A. L. (1984). Reciprocal teaching of comprehension-fostering and comprehension-monitoring activities. *Cognition and Instruction, 1*(2), 117–175. doi:10.1207/s1532690xci0102_1

Pressley, M., Van Etten, S., Yokoi, L., Freebern, G., & Van Meter, P. (1998). The metacognition of student scholarship: A grounded theory approach. In Hacker, D. J., Dunlosky, J., & Graesser, A. C. (Eds.), *Metacognition in Educational Theory & Practice* (pp. 347–366). New Jersey: Lawrence Earlbaum Associates Inc.

Reeves, T. C., & Reeves, P. M. (1997). Effective dimensions of interactive learning on the WWW. In Khan, B. H. (Ed.), *Web Based Instruction* (pp. 59–66). Englewood Cliffs, New Jersey.

Risemberg, R. (1996). Reading to write: self-regulated learning strategies when writing essays from sources. *Reading Research and Instruction, 35*(Summer 96), 365-383.

Simpson, M. L., & Nist, S. L. (2000). An update on strategic learning: It's more than textbook reading strategies. *Journal of Adolescent & Adult Literacy, 43*(6), 528–541.

Spiro, R. J., Feltovich, P. J., & Jacobson, M. J., & Coulson, R. L. (1992). Cognitive flexibility, constructivism, and hypertext: Random access instruction for advanced knowledge acquisition in ill-structured domains. In T. Duffy & D. H. Jonassen (Eds.), *Constructivism and the Technology of Instruction*. Hillsdale, N.J.: Erlbaum.

The Design-based Research Collective. (2003). Design-based research: An emerging paradigm for educational inquiry. *Educational Researcher, 32*(1), 5–8. doi:10.3102/0013189X032001005

Vandergrift, L. (2002). 'It was nice to see that our predictions were right': Developing metacognition in L2 listening comprehension. *Canadian Modern Language Review, 58*(4), 555–575. doi:10.3138/cmlr.58.4.555

Weinstein, C. E., & Mayer, R. (1986). The teaching of learning strategies. In Wittrock, M. (Ed.), *Handbook of Research on Teaching* (pp. 315–327). New York: MacMillan.

Wilson, K., Devereux, L., Macken-Horarik, M., & Trimingham-Jack, C. (2004). Reading readings: How students learn to (dis)engage with critical reading. In F. Sheehy & B. Stauble (Eds.), *Tranforming Knowledge into Wisdom: Holistic Approaches to Teaching and Learning: Proceedings of the 2004 Annual International Conference of the Higher Education Research and Development Society of Australasia (HERDSA)* (pp. 341-348). Milperra NSW: Higher Education Research and Development Society of Australasia.

Zimmerman, B. J. (1989). Models of self-regulated learning and academic achievement. In Schunk, D. H., & Zimmerman, B. J. (Eds.), *Self-Regulated Learning and Academic Achievement: Theory, Research, and Practice* (pp. 1–25). New York: Springer-Verlag.

Zimmerman, B. J., & Martinez-Pons, M. (1988). Construct validation of a strategy model of student self-regulated learning. *Journal of Educational Psychology*, *80*(3), 284–290. doi:10.1037/0022-0663.80.3.284

KEY TERMS AND DEFINITIONS

Annotation: adding notes to readings to help mediate understandings

Cognitive Self-Regulation: the process by which learners manage their thinking process to become independent learners

Mark-Up: an online environment designed to support the metacognitive development of reading comprehension

Problem-Solving: a key generic skill and learning process that involves students applying understandings to solve a problem.

Reading Comprehension: the ability to extract meaning from written texts

Self-Monitoring: a key reflective process that underpins metacognition and leads to the development of regulatory strategies

Chapter 18
Self-Regulation of Learning Supported by Web 2.0 Tools: An Example of Raising Competence on Creativity and Innovation

Maria Luisa Sanz de Acedo Lizarraga
Public University of Navarre, Spain

Oscar Ardaiz Villanueva
Public University of Navarre, Spain

Maria Teresa Sanz de Acedo Baquedano
Public University of Navarre, Spain

ABSTRACT

Our main purpose in this chapter is to examine the possibility of stimulating self-regulation of learning (SRL) by means of Information and Communication Technologies (ICT), more specifically, Web 2.0 technologies. Web 2.0 is commonly associated with applications that facilitate interactive information sharing and collaboration on the World Wide Web. To that end, the authors first present a theoretical description of the topics that are relevant to this chapter: SRL and ICT. Second, they compare SRL and ICT characterizing features, establishing functional relation between both sets of variables. Third, they define the Web 2.0 and two tools, Wikideas, and Creativity Connector, which were designed by us according to Web 2.0 technology. Fourth, the authors briefly report a pilot intervention they carried out in order to support SRL, using these two applications to perform some tasks that required competence in "creativity and innovation". Lastly, after summarizing these ideas, the authors suggest further study topics that may promote interesting lines of research.

DOI: 10.4018/978-1-61692-901-5.ch018

INTRODUCTION

In the last few decades, we have witnessed the rapid theoretical and empirical development of self-regulated learning (SRL), in both the psychological and educational fields (Boekaerts & Corno, 2005; Mooij, 2009; Sanz de Acedo Lizarraga, Ugarte, Iriarte, & Sanz de Acedo Baquedano, 2003; Waugh, 2003; Zimmerman, 2001). There are many reasons for this development. The most important of these reasons is the following: students who regulate their academic activities have a strong desire to learn and obtain the best learning results (Njiru & Waugh, 2007; Williams & Hellman, 2004).

Research on SRL, which has mostly been carried out in traditional educational contexts has allowed us to (1) better understand this multidimensional process; (2) determine which subjects and environmental variables are involved in the above mentioned process; and (3) discover how and under what circumstances students come to direct and supervise their learning and activate the cognitive, metacognitive, motivational, and behavioural competencies required to maintain and achieve their goals (Zeidner, Boekaerts, & Pintrich, 2000).

We have also observed growing interest in the study of SRL in contexts enriched by ICT. Technological resources are increasingly used at all levels of the educational system, as they are assumed to improve the quality of teaching and learning, and to provide new avenues for thinking, interacting, and working (McLoughlin & Lee, 2009). Their global impact on learning, particularly on SRL, requires further study. Currently, they are thought to have a positive effect because they motivate students to intentionally participate in their learning processes (Banyard, Underwood, & Twiner, 2006). Becker (2000) proposed four key benefits of ICT in education when they are implemented in a responsible way: they increase students' commitment to better performance both in and out of class; they improve writing and research competencies; they foment authentic efforts; and they help students to better learn the academic material.

SRL and ICT are considered essential for balanced development of the key competencies that should be promoted in Primary and Secondary Compulsory Education (Eurydice, 2002). If students do not monitor their learning tasks, such as those that they perform while using new technologies, they will not be able to develop the necessary skills to join the contemporary society of knowledge and computer technology. However, with the help of their classmates, teachers, and educational institution, they may manage to become familiar with and to integrate SRL and ICT into their academic activities. If they are able to meet these goals, they will be successful in their studies, and this success is more likely to be more permanent (Montgomery, 2000; Sosin, Lecha, Agarwal, Bartlett, & Daniel, 2004).

ICT are increasingly more inclusive because most of the educational actions require one-on-one interaction. Consequently new software has been designed to facilitate the teaching and learning processes in specific subjects. Even so, few tools have been designed to directly support SRL, although many may be used for this purpose (Trigano, 2006), as in our case. Based on the Web 2.0 technologies, we have designed two tools, called Wikideas and Creativity Connector, with the aim of helping students to become competent in "creativity and innovation". In other words, we intend to help them exercise their capacity to generate many, varied, original, and detailed ideas, and to allow their implementation (Ardaiz Villanueva, Sanz de Acedo Lizarraga, & Sanz de Acedo Baquedano, 2008). We chose to reinforce this competence because we considered it essential for the students to be able to develop initiative and an entrepreneurial spirit and to learn to express their original ideas using computer resources.

Our goal in this chapter is, therefore, to examine the relationship between SRL and ICT. More specifically, we will investigate two issues:

how SRL can be stimulated by means of Web 2.0 tools and, in turn, how these tools and other ICT can be used reflexively and imaginatively so that their transformation potential could achieve the desired impact on students. In this sense, we offer a brief theoretical review of SRL and the ICT; we establish a conceptual relationship (an interface) between them; we underline the most important characteristics of the Web 2.0; we describe the tools Wikideas and Creativity Connector; we depict an experience of using them to regulate tasks that demand creativity; and, lastly, we formulate some conclusions and lines of research that may clear new pathways to study these fascinating topics.

In this chapter, we attempt to answer the following questions: What are the charactirizing aspects of SRL and the ICT? How can ICT support SRL? How can students learn to regulate the creative process using Web 2.0 tools?

THEORETICAL FRAMEWORKS

Overview of Self-Regulated Learning

The diverse definitions of SRL, particularly those that use a restricted interpretation of this concept, have aspects in common, which means that researchers express some agreement about the nature of this construct (Njiru & Waugh, 2007; Zeidner et al., 2000). SRL has been defined as a cyclic process aimed at achieving the desired learning goals (Locke & Latham, 2002), and also as a series of features possessed by students who control most of the variables that are involved in this process (Mooij, 2009; Zimmerman, 1990, 2000). We believe that both forms of study interact and complement each other.

SRL is a process that requires that individuals focus on their own psychological functioning, activate consciousness and metacognitive knowledge, analyze their strong and weak points with regard to the task, and promote feelings and motivations to enhance achievement of their goals (Flavel, 1979; Hacker, 1998; Zimmerman, 2001). In turn, the self-regulated process is influenced by three kinds of personal beliefs (Pintrich, 1999): (1) self-efficacy or the conviction that one is capable of carrying out certain tasks; (2) task value or one's opinion of the academic tasks, either attractive and useful, or boring and personally meaningless; and (3) goal orientation, either the desire to achieve learning goals related to developing one's capacity and acquiring knowledge, or the desire to achieve performance goals related to explicit or social rewards (Bandura, 1986, 1991).

Moreover, as indicated by diverse social cognitive models, the SRL process has at least three cyclically sequential stages. These stages are: (1) planning, preparation or forethought, which involves analysis of the task and elaboration of plans before carrying out the task (setting goals, selecting strategies, foreseeing difficulties, recovering prior knowledge, task motivation, etc.); (2) monitoring performance, or adjustment while carrying out the plan (controlling cognitive and emotional variables, checking the usefulness of the strategies, self-observing one's work, coordinating time and effort, etc.); and (3) evaluation or self-reflection of learning after concluding that the activities included in the plan (reviewing the results, the process and the affective and motivational reactions to one's self-regulatory efforts, studying the errors made, comparing the effort spent with the results obtained, transferring the achieved learning, etc.). These three stages support students' commitment during the entire learning process, allowing them to practice all of their capacities (Boekaerts & Cascallar, 2006; Paris & Winograd, 2001; Zimmerman, 2000).

Self-regulation can also be investigated as a set of features or attributes that are possessed by self-regulating people. However, which features, skills, or personal dispositions are associated with this process? The investigations that attempt to answer this question state that self-regulating students are characterised by: (1) having analytic and synthetic thinking, being capable of organizing and

Figure 1. Most relevant characteristics of SRL

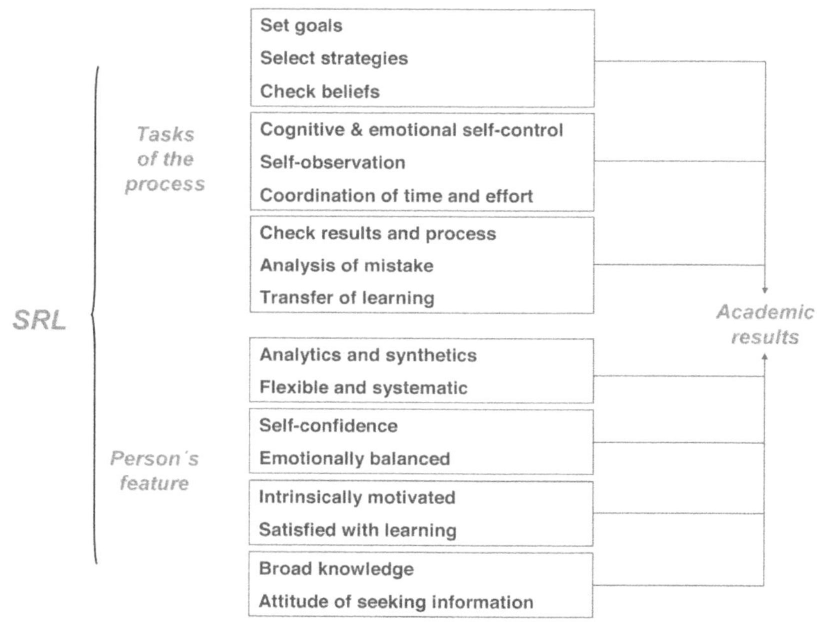

combining information, possessing a flexible and systematic mind to adapt their behaviours, having a critical attitude towards their results, the metacognitive capacity to be aware of their own way of thinking and behaving, and the cognitive skills to approach the tasks with confidence, diligence, and strategic behaviour (Mooij, 2009; Zimmerman, 1990); (2) having a self-assured personality and being confident of their own efficacy, perseverant in the face of difficulties, and emotionally balanced; (3) having strong intrinsic motivation that orients and maintains their behaviours while learning and that generates satisfaction and makes them attribute their academic results to their capacity and effort (internal locus of control) (Harris & Kington, 2002; Winne & Perry, 2000); and (4) having a stable attitude of seeking and managing information, and interest in knowing how digital tools can help them achieve their learning goals (Njiru & Waugh, 2007).

Figure 1 presents a summary of the most relevant characteristics of SRL, underlining the tasks performed in the three stages of the regulating process and the most common traits displayed by people who regulate their actions to achieve their learning goals.

As can be seen in the following sections, the above-mentioned characteristics of SRL can presumably be reinforced with continued use of ICT.

Overview of Information and Communication Technologies

As an umbrella term, ICT cover all products that store, retrieve, manipulate, transmit or receive information electronically in a digital form. For example, personal computers, digital television, e-mail, and robots are included in ICT. These products are changing the work and lifestyle of most youngsters and adults, as well as their ways of interacting and acquiring knowledge (Osimo, 2008). Their influence on learning can be studied from diverse educational perspectives and levels (Punie, Zinnbauer, & Cabrera, 2006). The Organization for Economic Cooperation and Development (OECD, 2005) has queried whether the way that universities teach and students learn is changing, or if simply students are typing up their essays on computers and professors send course reading lists or work assignments by e-mail. The OECD distinguishes four methods by which to incorporate ICT into academic activities at the university level:

- *Web-supplemented courses* focus on classroom-based teaching, but include elements, such as putting a course outline and lecture notes online, use of e-mail and links to online resources.
- *Web-dependent courses* require students to use the Internet for key elements of the program, such as online discussions, assessment, or online project/collaborative work, but without a significant reduction in classroom time.
- In *mixed mode courses*, e-learning elements begin to replace classroom time. Online discussions, assessment, or project/collaborative work replace some face-to-face teaching and learning. Significant campus attendance remains part of the course.
- In *fully online courses*, students can follow courses offered by a university in one city from another town, country, or time zone.

The OECD acknowledges that this variety of course types has not minimised the importance of the "face-to-face" process of teaching and learning, which indicates that, at this moment, online learning in Europe is mostly related to concrete learning modules or segments of a course, and plays only a peripheral role. In our opinion, the problem is not so much whether ICT should play a primary or secondary role in education, because this will depend on many factors, but how to use ICT so that they contribute to the integral development of students. It would be beneficial for the students to achieve a positive exchange between their own experiences and needs and the computer tools available to them. ICT will have a greater impact if they include educational strategies that improve normal and special students' learning, as well as diverse scientific curricular content. Thus, one could say that ICT have significant effects on students' attitudes, motivation, and performance.

According to Drent and Meelissen (2008), ICT can be used as a relevant aspect of a profession (occupational preparation in business), as subject content (separate courses in computer education), and as a means to support teaching and learning (use of word processing software by students to present work to their teachers). These purposes may overlap, as in our investigation. For example, the students were studying a subject related to new technologies, and they used some of these tools to facilitate their learning. By promoting a variety of learning spaces, ICT allow both individual and personalised learning and collaborative and cooperative learning. Thus, ICT support the incorporation of aspects of constructivist cognitive psychology and the formation of a more holistic view of the student.

In our hyper-connected society, where citizens are consumers and producers of information, it is not surprising that new theories of learning are emerging. An example of such theories is "Connectivism", which incorporates technology, social networks, and "peer-to-peer" learning in the explanation of learning (Boud, Cohen, & Sampson, 2001; Siemens, 2004). In contemporary learning, the efforts of teachers and students can be described as acts of connecting specialised information sets in an ongoing fashion to recognise new information and alter the observed landscape of knowledge. With these new technologies, the changes generated in the learning process reside not only in the individual, but also in nonhuman facilities; that is, in a virtual organization that exists as electronically store data (databases). Self-regulation of these new learning styles is an essential skill in the digital era (Voogt, 2003).

SRL SUPPORTED BY ICT

For some years now, scientists have expressed interest in investigating the role of ICT in the process of SRL (Torgerson & Elbourne, 2002). A recent example of this interest is the study carried out by the TELEPEERS Project in nine European universities on how *Technologies of Enhanced Learning Environments* (TELEs) can facilitate SRL. These universities shared experiences, identified intervention opportunities, developed assessment instruments (TELE-SRL and TELESTUDENTS-SRL), and formulated relevant research challenges in this area (Steffens, 2006). The authors who participated in the investigation concluded that, in Europe, many publications investigate the relation of learning with ICT, but very few relate TELEs and SRL. Therefore, it is extremely desirable to investigate whether ICT can support SRL. However, the task will not be easy, and we may need to research in depth this issue.

Figure 2 shows some contributions of ICT to each of the three SRL stages -planning, P; monitoring, M; and evaluation, E- and to the learning process in general. In the first SRL stage -planning, P- ICT can facilitate access, organization, editing, storage, and retrieval of the information needed for the specific desired learning, the definition of goals, and the selection of adequate strategies to achieve such goals. In the second stage -monitoring, M- ICT can facilitate the supervision of task performance, self-observation of one's work, and communication with the group and with the tutor. Lastly, in the third stage -evaluation, E- ICT can facilitate correction of the written work, its grammar and structure, self-assessment and co-assessment of the results and of the process, and application of the learnt material in other situations. Use of ICT that support self-regulatory processes in the three phases (before, during, and after task execution) should enable learners to see positive consequences of their efforts and hence enable them to develop feelings of competence and control (Zimmerman & Tsikalas, 2005). The investigation that we present later facilitates understanding of the integration of the ICT in the processes of SRL. In addition, ICT can support students' learning processes in many other ways, such as motivating them to study, offering flexible options for them to personalize their choice, allowing them to interact at various levels, introducing changes to improve their activities, creating documents when necessary, and developing their cognitive and metacognitive skills.

Just as SRL is influenced by students' beliefs about task difficulty, so are ICT. Students' perceptions about whether or not a certain application is easy to use or whether it is useful to achieve their goals can determine their attitude toward that tool, the frequency of its use, and its incorporation into their learning setting.

It should not be forgotten that SRL supported by ICT can be carried out within and outside of formal education, throughout one's life. This distinction partially explains the instructional change that is being undertaken: teaching is increasingly oriented toward the student and less toward the educator. This change requires self-regulated learning and the incorporation of new technologies.

These aspects corroborate the many parallelisms that can be established between SRL and ICT. If we accept that ICT support the student in the input and performance phases of the learning process, then we can expect that they will also do so in the output phase, thus improving outcomes (Fuchs & Woessman, 2004). ICT, therefore, assist learners in the completion of cognitive tasks and allow them to control, generate and test hypotheses in the context of problem solving (Azevedo, 2005). If students are able to overcome difficulties that they face when interacting with ICT, they will

Figure 2. Contributions of ICT to the SRL stages and to the learning process

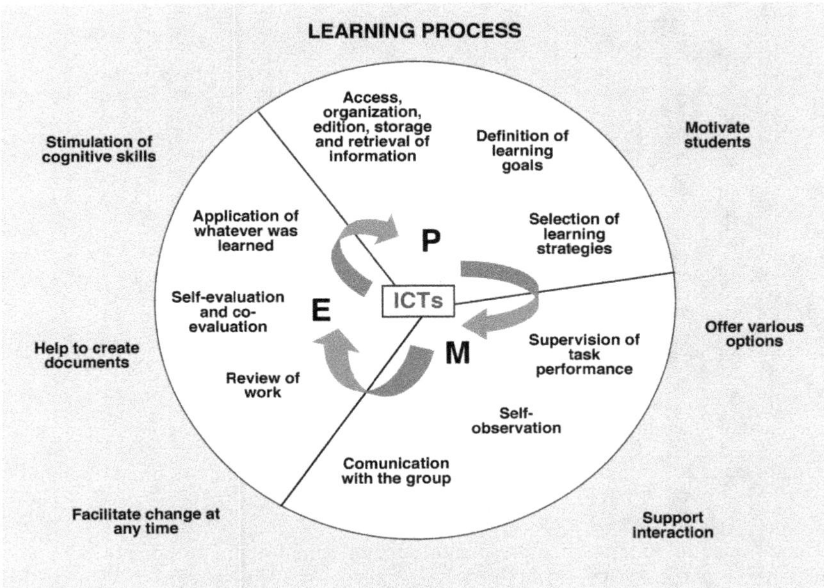

probably improve their performance and progress in self-regulated activities in a computer-based environment. We are convinced that ICT have great potential to support SRL, but they will only be efficient if the students believe that the use of such resources is suitable and they establish a "metacognitive dialogue" with the tools that will help them select and practice the appropriate cognitive processes when working with the tools. This awareness will contribute to enrich the quality of the students' learning experiences mediated by ICT.

SRL USING WEB 2.0 TOOLS

Web 2.0

The Internet has undergone a spectacular change with the appearance of Web 2.0, also known as social software, social computing, participative Web, or people's Web. Web 2.0 is a set of internet services and practices that give a voice to individual users. Such services thereby encourage internet users to participate in various communities of knowledge building and knowledge sharing. This term was coined by O'Reilly in reference to second generation websites (O'Reilly, 2005). According to the opinion of Freire (2007), Web 2.0 could be defined as a system of technologies (Ajax, CSS, Open API, P2P, XML, etc.), applications (Blog, Podcast, RSS feeds, Wiki, etc.), and values (collaboration, conversation, user as a producer of knowledge, user as part of the collective intelligence, simple use, etc.), organised in order to create and disseminate knowledge, share resources, and improve the quality of collaborative work (see the conceptual map of Web 2.0 in Figure 3). Web 2.0 acts more as a meeting point than as a set of traditional websites. Its main difference with the latter lies in the fact that its contents are generated by the users and not only by the owners, as is the case of traditional websites.

Two of the most important applications of Web 2.0 are Wikis and social networks. The Wiki technologies allow users to design, modify, and save websites for subsequent viewing. The social networks allow users to communicate with each

Figure 3. Summarized conceptual map of the Web 2.0

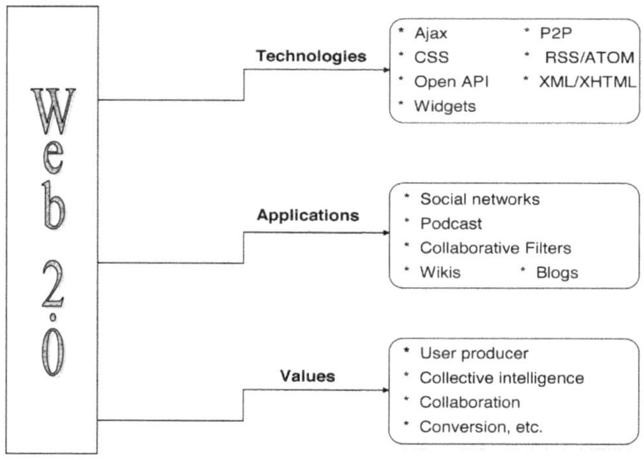

other, thereby forming a community with "collective intellectual capacity" (Leuf & Cunningham, 2001; O'Reilly, 2005).

Web 2.0 technologies are frequently used in education, in part because of the ease with which free content can be placed on a website and shared with an extensive web of users (Crook, 2008). Any student, teacher or educational institution can create Web 2.0 sites and form social networks to support learning (for example, online publication of content and engagement in collaborative learning activities). According to McLoughlin and Lee (2009), in parallel to the Web 2.0 movement, Pedagogy 2.0 is being developed, so as Web 2.0 tools can be positively incorporated into student-focused learning over their entire lifespan.

In view of the advantages of Web 2.0 technologies, as mentioned before, we have designed two tools, Wikideas and Creativity Connector, with the purpose of utilizing SRL in tasks that require creative skills (Ardaiz Villanueva et al., 2008).

Wikideas

The Wikideas tool is mainly based on Wikka, Ajax, and CSS technologies, on PHP and JavaScript programming language, and on a table of relationships organised according to the MySQL database. It is installed on an Apache-type web server. Wikideas, like most Web 2.0 tools, can be easily integrated into other applications -RSS Feeds, Widgets or Rest. It encodes information that can be viewed, depending on the arrangement, by different sizes of user groups. Wikideas extends the Wikka software with new features. First, it incorporates panels which allow users to view a large number of titles corresponding to items stored in the server. Second, Wikideas permits each user to assign a numerical value to each item in the server, and to present items organized according to average values. Third, information presented to users is completely anonymous without any identification or nickname of the content creator, so that there is no bias towards friends or the most productive users. Those new features permit Wikideas to perform various functions, such as generating, communicating, analyzing, assessing, and grouping information or ideas. Figure 4 shows the interface of the initial webpage of this tool from which you can go to panels that facilitate the performance of the four functions mentioned.

It has been stated that ICT tend to restrict students' creativity because they force students to act in a predefined way with limited possibili-

Figure 4. Initial screen that provides access to Wikideas' panels

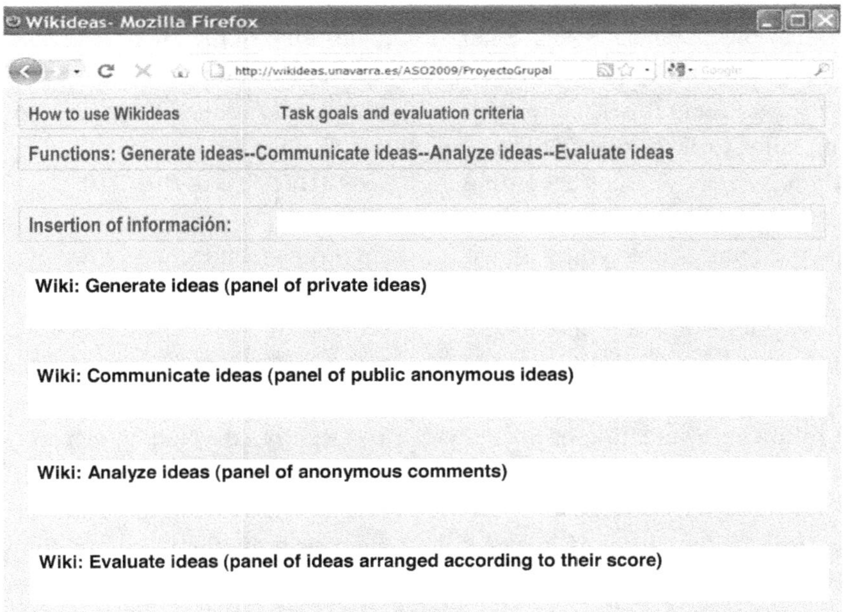

ties of interaction and decision (Selker, 2005). This is not the case for Wikideas, which has been developed having present in the mind a set of "design principles" for tools to support creative thinking (Shneiderman et al, 2006). For example, Wikideas support exploration (first principle), low threshold, high ceiling, and wide walls (second principle), support collaboration (fourth principle), support open interchange (fifth principle), make it as simple as possible (sixth principle), and balance user suggestions with observation and participatory processes (ninth principle). Below we describe each of the functions of this tool.

1. *Generating ideas*. It is a digital, personal, private space on which each user can develop, express, and store their new ideas. This space is similar to a portfolio that gathers the user's thoughts. When viewing and assessing the space, the user can feel satisfied and competent, which notably improves his or her self-esteem.
2. *Communicating ideas*. It is an open and flexible Wiki space. Ideas and information saved on the above-mentioned private space can be published anonymously on Wikideas so that other people can have access to them. This task is carried out as follows: after clicking on the information one wishes to share, one drags and drops it onto the panel of public ideas indicated by an icon that symbolizes shared idea. From that moment on, the information is viewable by other application users.
3. *Analyzing ideas*. Published information can be commented on and examined by other users who can formulate questions about it anonymously. Then, in a low-risk environment, the author can answer these questions by means of the corresponding Wiki site, either redefining the idea or adding a pertinent observation. The ideas that have been commented on by another user are marked on the "idea generator" panel so that their creator can respond to the comments.

4. *Assessing ideas*. Wikideas incorporates a system of idea assessment according to a Likert scale with values of 1, 3, and 5 points. Each user picks a number of ideas, for example three, that they consider the most interesting. They assign a value to each one and the tool subsequently arranges the ideas according to the score, from the highest to the lowest, based on the total points granted by all users. To explain this function, we present an example of our pilot study in Figure 5. The user "X", after providing his identifying code and password on the panel of this function, obtains the following evaluation of his/her ideas given by other users:
    - The total score for each idea, for example, the idea "modifier of images" has received 11 points.
    - The ideas that were generated by use "X" are highlighted with a folder icon located beside their total score. As an example, the idea "seeker of blogs" is highlighted and shown to have 9 points.
    - All of the scores received for an idea from other users (three in this case) are also displayed. For example, the idea "online store of music" has received 3 points from user "X" (the first score in brackets) and 1 point and 3 points from other participants (the second and third scores in brackets, respectively). This scoring scheme indicates that user "X" and a second user have valued this idea highly, whereas another user was not as keen on it.

Together this information allows user "X" to compare his/her ideas with those of other users. He can also compare the evaluations that he gave to an idea with those given by other users.

In general, this tool can be used by large groups (over 100 people) if the computer on which it is installed has adequate capacity. As usual with Web 2.0 applications, all of its functions can be used at any step of the creative process, so that a user can generate ideas after assessing other ideas and can then continue to analyze more ideas.

Creativity Connector

Creativity Connector is a social network tool that connects users who wish to participate in common projects. It is implemented with collaborative filtering systems and graph algorithms, and it uses PHP programming languages. Its main functions are performed when connected to the Wikideas tool using the information saved therein. These functions are calculating the creativity and affinity indexes, grouping ideas, and forming work teams according to common creative interests.

The creativity index is obtained from the number of ideas generated and introduced by each user and their level of elaboration (length of the description of each idea). The affinity index is obtained from the similarity of the values assigned by each user to the diverse ideas. For example, two or three participants who obtain the same score will have the same affinity for that idea; the higher its valuation, the higher will be their affinity index.

Figure 6 shows:

1. A list of ideas (a, j, e, g, b, f, c, i, h, and d) and their assigned global scores (10, 8, 7, 7, 6, 6, 6, 3, 1, and 0, respectively).
2. The individual scores that each user [T (5, 1, 3), S (5, 3, 1,), X (3, 5, 1), Z (5, 3, 1), R (3, 1, 5), and Y (1, 5, 3)] has granted to three ideas according to the Likert scale of 1, 3 and 5 points. This score is introduced for each user in the "assessing ideas" function of the Wikideas tool.
3. The creativity index of each user [T (9), S (8), X (6), Z (4), R (3), and Y (1)].
4. The structure of two groups, one made up of users T, S, and Z, and the other of users

Figure 5. Evaluate ideas screen

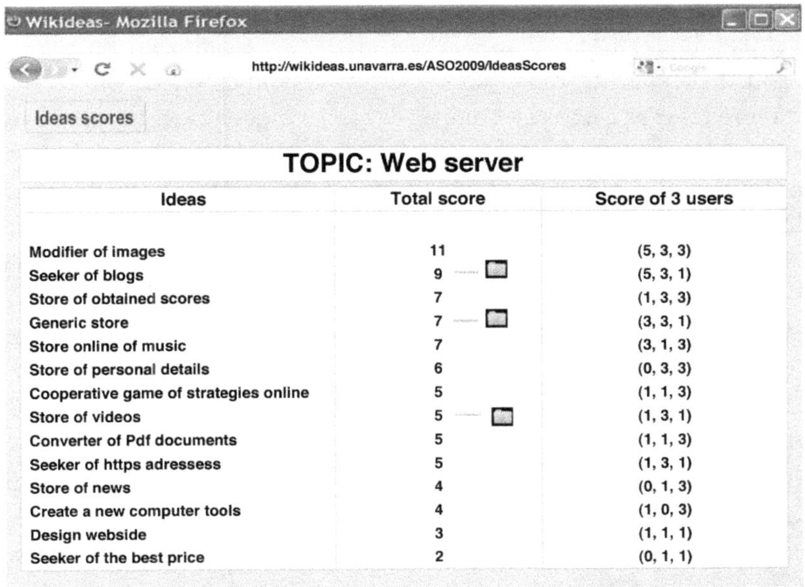

Ideas	Total score	Score of 3 users
Modifier of images	11	(5, 3, 3)
Seeker of blogs	9	(5, 3, 1)
Store of obtained scores	7	(1, 3, 3)
Generic store	7	(3, 3, 1)
Store online of music	7	(3, 1, 3)
Store of personal details	6	(0, 3, 3)
Cooperative game of strategies online	5	(1, 1, 3)
Store of videos	5	(1, 3, 1)
Converter of Pdf documents	5	(1, 1, 3)
Seeker of https adressess	5	(1, 3, 1)
Store of news	4	(0, 1, 3)
Create a new computer tools	4	(1, 0, 3)
Design webside	3	(1, 1, 1)
Seeker of the best price	2	(0, 1, 1)

Figure 6. Grouping of ideas and users according to Creativity Connector

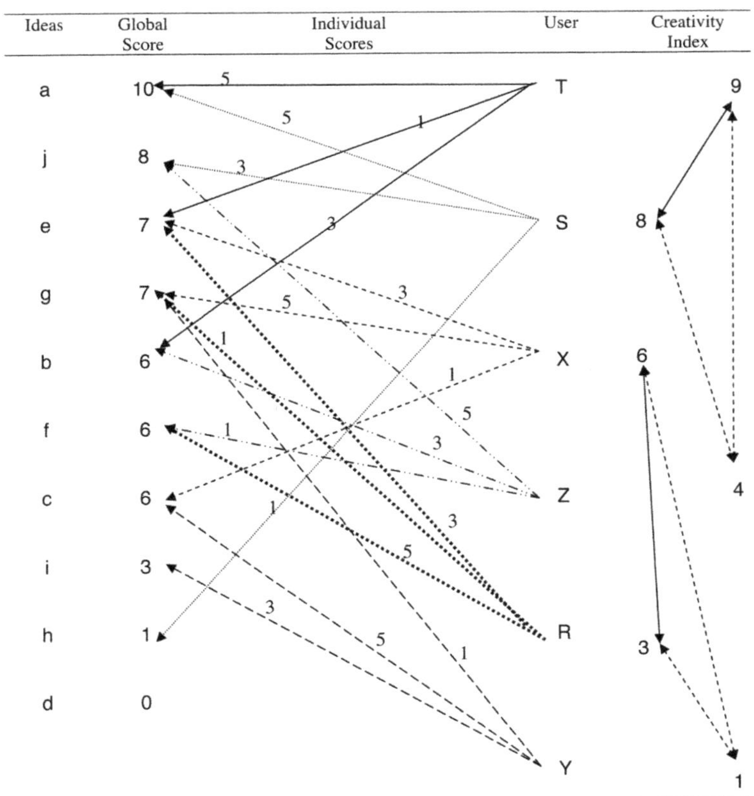

X, R, and Y. Figure 6 also shows that, in the first group users, T and S have more affinity with each other (solid line) than with user Z (dotted line), and, in the second group, users X and R have more affinity with each other than with user Y. Also, note that groups of users with nonadjacent creativity indices can also be formed.

TEACHING SRL WITH WEB 2.0

These proposals assume that SRL can be supported by Web 2.0 technologies. For this purpose, we conducted a pilot study to verify whether the Wikideas and Creativity Connector tools can favour SRL in academic tasks that demand, among other competence, creativity and innovation. The study participants were 37 students of Technical Engineering in Computer Management who were enrolled in the "Extension of operational systems" course. The aim of this course was to familiarise students with web search engine technologies.

In order to optimise the effect of these tools on SRL, we incorporated them into the "Think Actively in Creative Context" method (TACC), implemented by Wallace and Adams (1993) and adapted by Sanz de Acedo Lizarraga, Sanz de Acedo Baquedano, Goicoa Mangado and Cardelle-Elawar (2009). This method is made up of eight phases (gather and organise information, set goals, generate ideas, prioritise contents and actions, develop the project, evaluate, present the work, and learn from the experience) that can be grouped into the three stages of the self-regulating process. The intervention was carried out over five months, including eight weeks for the preparation stage (planning), eight weeks for the performance stage (monitoring), and four weeks for the assessment stage (evaluation). The study subjects participated for four hours per week in the presence of the teacher.

Preparation

1. *Gathering and organizing*. The students, individually and collectively, brought into their working memory the network of knowledge and ideas concerning Web 2.0 tools; that is, they remembered what they knew about this technology. They searched for and managed new information, and they formulated and wrote down research questions pertaining to relevant issues. For example, 'how do you create something like facebook?' 'how do you create databases?' Students had to undertake these selective, reflexive, and creative tasks using mainly Google and Wikideas.
2. *Setting goals*. In the spaces provided by Wikideas (see Figure 4), the students, with the help of the teacher, defined the goal that they hoped to achieve for this subject, which was "to learn to self-regulate academic activities related to the creation of a new computer tool". In addition, they also fixed the evaluation criteria that they would use to assess the obtained results: "to analyze their project within their own group, defend the tool in front of the class, and present it to businesses in the area".
3. *Generating ideas*. The students proposed many ideas (298) about their possible projects. They presented them using the "generate ideas" function, saved them, and exchanged them anonymously with their classmates. This stage stimulated the students' ideational creativity and noticeably motivated them to develop their projects. It was important to encourage all learners to consider several ideas before deciding on the best idea or course of action.
4. *Prioritizing ideas*. After communicating their ideas to their classmates, the students anonymously analyzed and evaluated their classmates' ideas using a score of 1, 3, and 5. The tool then arranged the ideas ac-

cording to their scores, and the Creativity Connector tool subsequently calculated the creativity and affinity indexes, based on which groups of students were formed (a total of 12). These groups then worked to carry out their innovative project. Some of the selected projects were to design a search engine for videos, a search engine for online broadcasting radio stations, an online store that would work according to distributor demand, an information warehouse to store large quantities of data, a specialised web search engine, and a Web application to facilitate social networking of evaluators of a certain market product.

Performance

5. *Developing the Project.* Organised in groups according to their interests, the students monitored the execution of their project, always taking into account that it should display clear signs of creativity and innovation (market utility). During this stage, each group met with the teacher weekly to report on their work, clarify doubts and specify the tasks to perform for the next meeting. To carry out this task, they used other tools, such as Google, Wikis, e-mail, and Blogs.

Evaluation

6. *Evaluating.* The students reflected upon their own work and their participation in a small group using the Wikideas functions to generate, communicate, and evaluate ideas. Then, they sent their conclusions to the members of their group. Later, working together, they drafted a comprehensive assessment of their workgroup.
7. *Presenting the projects.* In various class meetings, each group defended its project, underlining the achievements with regard to self-regulation, creativity, and innovation. They also compared their results with those of other groups and commented on the help provided by the computer tools and the limitations observed in the execution of the project. The groups carried out this activity using additional tools such as Microsoft PowerPoint allowed them to present their work in a highly professional manner in the classroom, and Blogs were used to publish on the Web their project results for potential employers. The creativity of the project was highly valued by all of the groups and by the teacher.
8. *Learning from experience.* The students reflected on how to transfer their learning to other academic, professional, and personal situations. They commented in detail on the aspects of the project and the tools that should be improved in future experiences and on the significant changes achieved in their SRL and creativity and innovation competencies. Lastly, they chose a number of centres and companies to which they wished to present their projects and organised how and when to present the material. All groups explained their projects at the University Laboratory for the Creating of Business.

Results and Discussion of the Study

The qualitative results of this pilot study revealed that it is possible to practice the skills of SRL with the support of Web 2.0 tools. The students expressed this statement in a questionnaire containing 40 "multiple choice" questions with response options of 1 (not at all), 2 (somewhat), 3 (much), and 4 (very much). The questionnaire was administered after the students completed the academic course. The questionnaire collected information about their strategic learning, academic goals (self-regulation and creativity and innovation), contents, computer resources, and teacher. Table 1 presents the results related to the students' evaluation of the support that they received from the Wikideas and Creativity Connector tools. It can be seen that 59 to 73% of the students agreed that these tools helped them in the tasks of accessing information, planning activities, setting goals, generating new ideas, and committing to and feeling motivated about the academic course. In addition, 54 to 73% considered that these ad hoc designed tools, along with other applications, helped them to make decisions on, develop, and supervise the activities required by the design of a new computer tool in order to address the difficulties that emerged, to communicate with classmates and the teacher, to meet the assigned deadlines, and to enjoy work, even though it demanded great effort. Lastly, 54 to 76% stated that the tools supported them in diverse activities during the evaluation stage of the self-regulating process. We also note the percentages of students who found the course to be easier than they had thought (76%), and who commented on the ease with which they analyzed the results with their classmates and the teacher (70%).

We conclude that most of the students had a favourable opinion of the aid provided by the tools and the method. Almost all of them prepared their learning well, discussing what they knew before the course and what they wished to find out about the tools before beginning, keeping the purpose of the course clearly in the mind, generating ideas involving creative thinking, and prioritizing and giving reasons for their choices. They carried out their creative project step-by-step using a wide range of human abilities. In addition, they evaluated the results obtained in depth by self-evaluating their work and their product, sharing their ideas and the results of their efforts, and transferring what they had learned.

Importantly, at the beginning of the study, the students were rather slow in learning, but they soon became familiar with the work setting and the skills required. Wikideas and Creativity Connector had an impact on the students' academic results and they notably reinforced the students' confidence in SRL and in the performance of tasks that demand searching for and managing information as well as creativity and innovation. Therefore, both applications were used in appropriate pedagogical and environmental conditions (Shields & Behrman, 2000). The organizational structure provided by the TACC method, which is a good didactic strategy to make students focus on learning and developing their competencies, may also have contributed to this positive consideration. According to the teacher, these results were better that those obtained in previous courses, which were carried out without an educational environment enriched by virtual tools and a systematic teaching method.

Notably, the students were capable of regulating their activities in a collaborative learning environment and the Web 2.0 tools favoured the practice of SRL cognitive and metacognitive skills. Our perception is similar to that defended by other authors, among them Mooij (2009) and Torgerson and Elbourne (2002). Lastly, it can be stated that the students remained motivated during the entire academic course and enjoyed learning at all times, perhaps more than in previous courses.

CONCLUSION

In this chapter, we have commented on the main aspects of SRL and ICT. We have conceptualised SRL as a series of characteristics that can be summarized as follows: to precisely set the goals one wishes to achieve, to organise the learning environment, to become motivated and excited about what one wants to learn, to permanently control the performance of the activities, to evaluate one's achievements, and to transfer them to other situations. We also considered that, given their nature, ICT constitute very valuable learning resources for the development of one's comprehensive, creative, and critical thinking skills. Learning how to learn and development of digital skills is necessary to face the challenges presented by the new society of knowledge (Lajoie, 2007).

Table 1. Students' perceptions of some of the benefits of the wikideas and creativity connector tools (N = 37)

Wikideas and Creativity Connector helped me:	Number of Students (Percentage)			
	Not at all	Somewhat	A lot	Very much
Preparation				
Access relevant information for my works.	-	-	10 (27%)	27(73%)
Commit to the work project.	-	4 (11%)	11 (30%)	22 (59%)
Plan the learning activities.	-	1 (3%)	12 (32%)	24 (65%)
Set clear goals.	-	5 (14%)	9 (24%)	23 (62%)
Generate more ideas than I had thought at first.	-	-	14 (38%)	23 (62%)
Motivate myself to learn.	-	1 (3%)	9 (24%)	26 (70%)
Performance				
Decide how to develop the project.	-	6 (16%)	11 (30%)	20 (54%)
Supervise the activities to be carried out.	-	2 (6%)	9 (24%)	26 (70%)
Resolve the difficulties that emerged.	-	3 (8%)	9 (24%)	25 (68%)
Communicate with my classmates and the tutor.	-	-	10 (27%)	27 (73%)
Meet the assigned deadlines.	-	5 (14%)	11(30%)	21 (56%)
Feel pleased during the course.	-	-	12 (32%)	25 (68%)
Evaluation				
Reflect on how I worked.	1 (3%)	4 (11%)	10 (27%)	22 (59%)
Correct the errors I committed.	-	-	13 (35%)	24 (65%)
Discuss the results with my classmates and tutor.	-	-	11(30%)	26 (70%)
Transfer what I learned to other activities.	-	6 (16%)	8 (22%)	23 (62%)
Present the project to my classmates.	-	-	17 (46%)	20 (54%)
Consider the course easier than I had thought.	-	-	9 (24%)	28 (76%)

Based on the opinions given by the students and teacher, our strategy has proved successful at integrating the use of technological tools inside the TACC method, which means that the tools were used according to the demands of the tasks of each stage of the method and that the interaction student/computer was efficient and satisfactory. In addition, we have found evidence that the Wikideas and Creativity Connector tools help students in the self-regulation of learning during the execution of a creative and innovative project. We hypothesize these achievements were possible in our study because it allowed for a dynamic interplay between the cognitive, metacognitive, motivational, and behavioural characteristics of the learner; available tools and the learning context.

Interestingly, the students became aware of the advantages of collaborative work in the diverse stages of the method, the aid provided by the tools to carry out almost all the activities, and the possibility to express their creativity by means of new technologies. As one of the students commented: "we have worked more intelligently and originally, but not harder".

Our research had many limitations, mostly due to the design of the study, for example, the study had no control group, neither measures before the intervention nor the sample was selected at random. These limitations reduce our ability to generalise the results obtained to other academic situations. However, these results have allowed us to re-conceptualise and create environments of learning centred on Web 2.0 technologies that can better and more explicitly stimulate the processes of SRL and creativity.

There is much left to be done. In the 21st century, research on SRL will likely be marked by ICT, more specifically by Web 2.0. We must refine and validate the tools that already exist and that have shown a potential to support SRL. Based on our experience, we believe that our future research efforts should focus on designing: (1) tools that support SRL in diverse academic contexts; (2) tools that are adapted to all education levels and learning styles; (3) didactic strategies that lend educational meaning to the diverse digital platforms; (4) theoretical learning models that tap on technological advances; and (5) software that stimulates the creative process.

Several studies have investigated the impact of ICT on SRL. Therefore, it would be desirable to conduct a meta-analysis to compare the existing results, to evaluate them jointly, and to obtain objective criteria of effectiveness. Such a meta-analysis would allow for systematic integration of the information obtained in the different studies carried out about ICT and SRL, and in this way help the field reach a quantitative and synthetic estimation of all of the available studies.

These lines of research would be more fruitful if they were managed by interdisciplinary groups of researchers -psychologists, educators, engineers, etc.- because the field of analysis of SRL and ICT is multidimensional. Consequently, it requires the contribution of varied specialists so that the end product will be reliable, creative, and useful on the theoretical, methodological, and practical level. As stated by Florida (2002, 2005), one should know how to conjugate all three *ts*: technology, talent, and tolerance.

REFERENCES

Ardaiz Villanueva, O., Sanz de Acedo Lizarraga, M. L., & Sanz de Acedo Baquedano, M. T. (2008, September). *Wikideas and Creativity Connector supporting group ideational creativity*. Paper presented at the International Symposium on Wikis, Porto, Portugal.

Azevedo, R. (2005). Computer environments as metacognitive tools for enhancing learning. *Educational Psychologist, 40*(4), 193–197. doi:10.1207/s15326985ep4004_1

Bandura, A. (1986). *Social foundations of thought and action: A social cognitive view*. Englewood Cliffs, NJ: Prentice Hall.

Bandura, A. (1991). Social cognitive theory of self-regulation. *Organizational Behavior and Human Decision Processes, 50*(2), 248–287. doi:10.1016/0749-5978(91)90022-L

Banyard, P., Underwood, J., & Twiner, A. (2006). Do enhanced communication technologies inhibit or facilitate self-regulated learning? *European Journal of Education, 41*(3/4), 473–489. doi:10.1111/j.1465-3435.2006.00277.x

Becker, H. J. (2000). Findings from the teaching, learning, and computing survey: Is Larry Cuban right? *Education Policy Analysis Archives, 8*(51). Retrieved July 26, 2009, from http://epaa.asu.edu/epaa/

Boekaerts, M., & Cascallar, E. (2006). How far have we moved toward the integration of theory and practice in self-regulation? *Educational Psychology Review, 18*(3), 199–210. doi:10.1007/s10648-006-9013-4

Boekaerts, M., & Corno, L. (2005). Self-regulation in the classroom: A perspective on assessment and intervention. *Applied Psychology: An International Review, 54*(2), 199–231. doi:10.1111/j.1464-0597.2005.00205.x

Boud, D., Cohen, R., & Sampson, J. (2001). *Peer learning in higher education: Learning from and with each other*. London, UK: Kogan Page.

Crook, C. (2008). *Web 2.0 technologies for learning: the current landscape –opportunities, challenges and tensions.* Retrieved November 20, 2009, from http://partners.becta.org.uk/upload-dir/downloads/pagedocuments/research/web2technologies_learning.pdf

Drent, M., & Meelissen, M. (2008). Which factors obstruct or stimulate teacher educators to use ICT innovatively? *Computers & Education, 51*(1), 187–199. doi:10.1016/j.compedu.2007.05.001

Eurydice (2002). *Competencias clave* [Key competences]. Madrid, ES: Unidad española de la red Eurydice. Retrieved March 15, 2009 from http://www.eurydice.org.

Flavell, J. H. (1979). Metacognition and cognitive monitoring: A new area of cognitive-developmental inquiry. *The American Psychologist, 34*(10), 906–911. doi:10.1037/0003-066X.34.10.906

Florida, R. (2002). *The rise of the creative class. How it's transforming work, leisure, community and everyday life*. New York, NY: Basic Books.

Florida, R. (2005). *The flight of the creative class*. New York, NY: Harper Collins.

Freire, J. (2007). Los retos y oportunidades de la Web 2.0 para las universidades [Challenges and opportunities of Web 2.0 for universities]. In R. Jiménez & F. Polo (Eds.). *La gran guía de los blogs* [The great blog guide] (pp. 82-90). Barcelona, ES: El Cobre.

Fuchs, T., & Woessmann, L. (2004). Computers and student learning: Bivariate and multivariate evidence on the availability and use of computers at home and at school. *CESifo, Working Paper, 1321*. Munich, DE: CESifo. Retrieved November 20, 2009, from http://ideas. repec.org/p/ces/ceswps/_1321.html.

Hacker, D. J. (1998). Definitions and empirical foundations. In Hacker, D. J., Dunlosky, J., & Graesser, A. C. (Eds.), *Metacognition in educational theory and practice* (pp. 1–23). Mahwah, NJ: Erlbaum.

Harris, S., & Kington, A. (2002). *Innovative classroom practices using ICT in England: The Second Information Technology in Education Study (SITES)*. Slough, UK: National Foundation for Educational Research.

Lajoie, S. P. (2007). Aligning theories with technology innovations in education. *British Journal of Educational* Psychology. *Monograph, Series II, 5*, 27–38.

Leuf, B., & Cunningham, W. (2001). *The Wiki way: Collaboration and sharing on the Internet*. Reading, MA: Addison-Wesley.

Locke, E. A., & Latham, G. P. (2002). Building a practically useful theory of goal setting and task motivation: A 35-year odyssey. *The American Psychologist, 57*(9), 705–717. doi:10.1037/0003-066X.57.9.705

McLoughlin, C., & Lee, M. J. W. (2009). Future learning landscapes: Transforming pedagogy through social software. Retrieved May 20, 2009, from http://www.innovateonline.info/index.php?view=article&id=539

Montgomery, K. (2000). Children's media culture in the new millennium: Mapping the digital landscape. *The Future of Children: Children and Computer Technology, 10*(2), 145–167. doi:10.2307/1602693

Mooij, T. (2009). Education and ICT-based self-regulation in learning: Theory, design and implementation. *Education and Information Technologies, 14*(1), 3–27. doi:10.1007/s10639-008-9066-8

Njiru, J. N., & Waugh, R. F. (2007). Rasch measurement of self-regulated learning in an information and communication technology (ICT)-rich environment. *Journal of Applied Measurement*, *8*(4), 417–437.

O'Reilly, T. (2005). What is Web 2.0: Design patterns and business models for the next generation of software. Retrieved May 20, 2006, from http://www.oreillynet.com/pub/a/oreilly/tim/news/2005/09/30/what-is-web-20.html

OECD. (2005). *E-learning in tertiary education: Where do we stand?* Paris, FR: OECD.

Osimo, D. (2008). Web 2.0 in Government: Why and how? Institute for Prospective Technological Studies (IPTS), JRC, European Commission, EUR 23358 In. Retrieved May 12, 2009, from http://ipts.jrc.ec.europa.eu/publications/pub. cfm?id=1565

Paris, S. G., & Winograd, P. (2001). *The role of self-regulated learning in contextual teaching: Principles and practices for teacher preparation* (CIERA, Archive N° 01-03.) Retrieved July 10, 2009 from http://www.ciera.org/library/archive/2001-04/0104parwin.htm.

Pintrich, P. R. (1999). The role of motivation in promoting and sustaining self-regulated learning. *The Journal of Educational Research*, *31*(6), 459–470. doi:10.1016/S0883-0355(99)00015-4

Punie, Y., Zinnbauer, D., & Cabrera, M. (2006). A review of the impact of ICT on learning. *JRC Technical Notes*. European Commission: IPTS. Working paper prepared for DG EAC. Retrieved July 10, 2009, from http://ftp.jrc.es/EURdoc/JRC47246.TN.pdf

Sanz de Acedo Lizarraga, M. L., Sanz de Acedo Baquedano, M. T., Goicoa Mangado, T., & Cardelle-Elawar, M. (2009). Enhancement of thinking skills: Effects of two intervention methods. *Thinking Skills and Creativity*, *4*(1), 30–43. doi:10.1016/j.tsc.2008.12.001

Sanz de Acedo Lizarraga, M. L., Ugarte, M. D., Iriarte, M. D., & Sanz de Acedo Baquedano, M. T. (2003). Immediate and long-term effects of a cognitive intervention on intelligence, self-regulation, and academic achievement. *European Journal of Psychology of Education*, *18*(1), 59–74. doi:10.1007/BF03173604

Selker, T. (2005). Fostering motivation and creativity for computer users. *International Journal of Human-Computer Studies*, *63*(4-5), 410–421. doi:10.1016/j.ijhcs.2005.04.005

Shields, M. K., & Behrman, R. E. (2000). Children and computer technology: Analysis and recommendations. *The Future of Children*, *10*(2), 4–30. doi:10.2307/1602687

Shneiderman, B., Fischer, G., Czerwinski, M., Resnick, M., Myers, B., & Candy, L. (2006). Creativity support tools: Report from a U.S. National Science Foundation sponsored workshop. *International Journal of Human-Computer Interaction*, *20*(2), 61–77. doi:10.1207/s15327590ijhc2002_1

Siemens, G. (2004). Connectivism: A learning theory for the digital age. E-learnspace, Retrieved March 14, 2009, from http://www.elearnspace.org/Articles/connectivism_self-amused.htm

Sosin, K., Lecha, B. J., Agarwal, R., Bartlett, R. L., & Daniel, J. I. (2004). Efficiency in the use of technology in economic education: Some preliminary results. *The American Economic Review*, *94*(2), 253–258. doi:10.1257/0002828041301623

Steffens, K. (2006). Self-regulated learning in technology enhanced learning environments: Lessons of a European peer review. *European Journal of Education*, *41*(3/4), 353–379. doi:10.1111/j.1465-3435.2006.00271.x

Torgerson, C. J., & Elbourne, D. (2002). A systematic review and meta-analysis of the effectiveness of information and communication technology (ICT) on the teaching of spelling. *Journal of Research in Reading*, *25*(2), 129–143. doi:10.1111/1467-9817.00164

Trigano, P. (2006). Self-regulated learning in a TELE at the Université de Technologie de Compiègne: An analysis from multiple perspectives. *European Journal of Education*, *41*(3/4), 381–395. doi:10.1111/j.1465-3435.2006.00272.x

Voogt, J. M. (2003). Consequences of ICT for aims, contents, processes and environments of learning. In J. van den Akker, W. Kuiper & U. Hameyer (Eds.), *Curriculum landscapes and trends* (pp. 217-236). Dordrecht, NL: Kluwer.

Wallace, B., & Adams, H. B. (1993). *TASC. Thinking actively in a social context. Bicester, UK.* AB: Academic.

Waugh, R. F. (2003). *On the forefront of educational psychology*. New York, NY: Nova Science.

Williams, P. E., & Hellman, C. M. (2004). Differences in self-regulation for online learning between first-and second-generation college students. *Research in Higher Education*, *45*(1), 71–82. doi:10.1023/B:RIHE.0000010047.46814.78

Winne, P. H., & Perry, N. E. (2000). Measuring self-regulated learning. In Boekaerts, M., Pintrich, P. R., & Zeidner, M. (Eds.), *Handbook of self-regulation* (pp. 531–566). Orlando, FL: Academic Press. doi:10.1016/B978-012109890-2/50045-7

Zeidner, M., Boekaerts, M., & Pintrich, P. R. (2000). Self-regulation, directions and challenges for future research. In Boekaerts, M., Pintrich, P. R., & Zeidner, M. (Eds.), *Handbook of self-regulation* (pp. 749–768). San Diego, CA: Academic Press. doi:10.1016/B978-012109890-2/50052-4

Zimmerman, B. J. (1990). Self-regulated learning and academic achievement: An overview. *Educational Psychologist*, *25*(1), 3–17. doi:10.1207/s15326985ep2501_2

Zimmerman, B. J. (2000). Attaining self-regulation: A social cognitive perspective. In Boekaerts, M., Pintrich, P. R., & Zeidner, M. (Eds.), *Handbook of self-regulation* (pp. 13–41). San Diego, CA: Academic Press. doi:10.1016/B978-012109890-2/50031-7

Zimmerman, B. J. (2001). Theories of self-regulated learning and academic achievement: An overview and analysis. In Zimmerman, B. J., & Schunk, D. H. (Eds.), *Self-regulated learning and academic achievement. Theoretical perspectives* (pp. 1–37). Mahwah, NJ: Erlbaum.

Zimmerman, B. J., & Tsikalas, K. E. (2005). Can computer-based learning environments (CBLEs) be used as self-regulatory tools to enhance learning? *Educational Psychologist*, *40*(4), 267–271. doi:10.1207/s15326985ep4004_8

KEY TERMS AND DEFINITIONS

Competence: Combination of capacities, knowledge, attitudes and conduct directed toward efficient problem solving. It is a measure of what a person can do adequately as result of the mobilization of their resources and planning of their actions after completing a learning process.

Creativity: Ability to generate many original, different, detailed, useful and valuable ideas for solving problems and to see connections and possibilities where others do not. It is a way of thinking, a process whose result is a quality product. An example would be conceptualizing something in a way that others have not done before.

Information and Communication Technologies (ICT): A diverse set of technological tools and resources used to communicate, create, disseminate, store, and manage information. These technologies include computers, Internet, broadcasting technologies, and telephony.

Innovation: Transformation and application of new ideas to useful products for society, economy, education, etc. The goal of innovation is change, growth or improvement of some item. Normally, it is the result of research carried out with great effort expended over time by creative groups on an idea, including its development and marketing.

Self-Regulated Learning (SRL): The autonomy and responsibility of students to take charge of their own learning. SRL involves cognitive and metacognitive strategies, intrinsic motivation, and behaviour control. SRL is a vital prerequisite for the successful acquisition of knowledge in school and beyond. It is of particular importance for lifelong learning.

Web 2.0: A term used to describe a second generation of the World Wide Web that is focused on people's ability to collaborate and share information online. Web 2.0 supports mashing, which is the process of building new services from reusable components of other services. Its difference from the traditional web is that its contents are generated by users and not only by owners.

Wiki: A collaborative website that comprises the perpetual collective work of many authors. A wiki allows anyone to edit, delete, or add content that has been placed on the website using a browser interface. One of its more important and popular applications is the creation of collaborative encyclopaedias, such as Wikipedia.

Chapter 19
Exploring the Effects of an Optional Learning Plan Tool in Technology-Enhanced Learning

Antje Proske
TU Dresden, Germany

Susanne Narciss
TU Dresden, Germany

Hermann Körndle
TU Dresden, Germany

ABSTRACT

Self-regulated learners deal with a complex interplay of forethought, performance, and self-reflection processes. This might be a reason why many students struggle with regulating their learning in a technology-enhanced learning environment (TELE). Although TELEs provide various tools supporting self-regulation, research indicates that learners seldom use the tools meaningfully. This contribution investigates whether the provision of an optional metacognitive tool (i.e. a tailored learning plan) affects tool use, learning activities, and posttest performance in the TELE "Studierplatz". To this end, students were instructed to use a learning plan in order to reach a predetermined learning goal. Results show that only 20% of the students used the tool. Furthermore, no significant effects on posttest performance were found. However, learning plan tool use positively affected working on learning goal relevant sections. These results are discussed with respect to current research on tool use in self-regulated learning with TELEs.

DOI: 10.4018/978-1-61692-901-5.ch019

INTRODUCTION

Technology-enhanced learning environments (TELEs) combine (a) multiple sources of information (e.g., instructional material, further readings, internet-links), (b) multiple information presentation formats (e.g., texts, graphics, simulations, learning tasks), (c) multiple modalities (e.g., visual or auditory information), and (d) multiple interaction possibilities (e.g., highlighting and note-taking functions, feedback to learning tasks, communication tools). Due to this, learners may select and examine from a large pool of information only those pieces necessary to meet their personal learning goals. They can process this material in accordance with their individual preferences and strategies at any time and from any place. Thus, TELEs show great potential for fostering students' self-regulated learning (e.g., Azevedo, 2005b; Winters, Greene, & Costich, 2008).

Self-regulated learning is an active, constructive process in which learners set goals and then cyclically adapt their thoughts, feelings, and learning behaviors to attain their personal goals (Pintrich, 2000; Winne & Hadwin, 1998; Zimmerman, 2000). This requires not only the goal-directed use of cognitive, but also of motivational and metacognitive strategies (e.g., Narciss, Proske, & Körndle, 2007). However, the universal access to multiple sources of information, as well as the non-linear structure and interactivity of TELEs, create additional demands on learners in all phases of the self-regulated learning process; the amount and structure of the learning materials can pose learners great difficulties. Learners have to choose permanently not only between various sources of information, but also between distinct information presentation formats, information modalities, and interaction possibilities (e.g., Bannert & Mengelkamp, 2008). During learning, various connections between these multiple materials from multiple sources of information need to be established. This process involves risks due to the non-linearity of TELEs and the learner's own knowledge deficiencies with regard to content and strategy (e.g., Bannert, 2006; Foltz, 1996; Rouet & Levonen, 1996). Thus, learners might be distracted from their learning goals or lose their way in hyperspace; they may work on too much irrelevant information or only consume important information cursorily (e.g., Salomon & Almog, 1998). Some may not even start the learning activities as taking the first step onto this "mountain of information" can be perceived as an insurmountable task (e.g., Narciss et al., 2007).

In order to deal with this problem, many TELEs provide several instructional interventions aimed to support learners in applying efficient cognitive, metacognitive and motivational strategies (e.g., Schraw, 2007). However, empirical evidence on the effectiveness of instructional interventions is limited. The present contribution seeks to explore the effects of including an optional metacognitive tool in a technology-enhanced learning environment called Studierplatz. More specifically, the focus is on (a) how the use of metacognitive tools can be fostered and (b) how metacognitive tool use is related to learning behavior and performance in self-regulated learning with TELEs.

CHALLENGES OF SELF-REGULATED LEARNING WITH TECHNOLOGY-ENHANCED LEARNING ENVIRONMENTS

According to Zimmerman (2000) self-regulated learning involves three cyclic phases including forethought, performance, and self-reflection. These phases require learners to perform several cognitive, metacognitive, and motivational processes:

- Forethought: task analysis, self-motivation beliefs;
- Performance: self-control, self-observation;
- Self-reflection: self-judgment, self-reaction.

The *forethought phase* involves goal specification and planning. It starts with the learners' analysis of the situation, i.e. an orientation on task, instruction, and available resources (Bannert & Mengelkamp, 2008; Pintrich, 2000). Based on this analysis the learners specify learning goals and derive information for learning process planning (Pintrich, 2000; Zimmerman, 2000). Goal specification refers to defining standards for outcomes of learning or performance. It further includes the identification and definition of sub goals which can serve as means of achieving higher-order outcome goals (Latham & Brown, 2006; Seijts & Latham, 2001). Learning process planning involves the selection and ordering of appropriate activities and strategies directed at approaching the learning goal (Narciss et al., 2007; Winne & Hadwin, 1998). The goal setting and planning processes are constrained by a number of motivational beliefs (e.g., self-efficacy, outcome expectations, intrinsic value, goal orientation, cf. Zimmerman, 2000).

During the *performance phase* the learner has to search for information and to judge whether the information found is really relevant in relation to the learning goal. Success at finding relevant information is related to higher posttest performance when self-regulated learning with TELEs (e.g., Gauss & Urbas, 2003; Naumann, Richter, Christmann, & Groeben, 2008). Furthermore, relevant information has to be extracted and elaborated (Bannert & Mengelkamp, 2008; Narciss et al., 2007). These activities have to be constantly monitored and controlled. Here, self-control serves for coordinating and monitoring the activities and strategies selected during the forethought phase (e.g., Kuhl, 2000). Furthermore, the use of learning strategies such as note-taking or using learning tasks for test preparation facilitate learning and guide learning efforts during the performance phase (e.g., Proske, Narciss, & Körndle, 2007; Weinstein, Husman, & Dierking, 2000). Self-observation refers to the learners' tracking of specific aspects of their own performance, the conditions that surround it, and the effects that it produces. Sub goals defined during the forethought phase may facilitate self-observation because they focus on specific processes and events (Zimmerman, 2000).

The *self-reflection phase* demands for evaluating the learning outcome with respect to the learning goal (Bannert & Mengelkamp, 2008; Narciss et al., 2007). This entails processes of self-judgment and self-reaction (Zimmerman, 2000). Self-judgment refers to the learners' judgments and evaluations of their performance on the task and their attributions for performance (Pintrich, 2000; Zimmerman, 2000). It requires comparing the self-monitored performance with a standard or a goal (e.g., to know the main concepts of a topic) in order to assess whether the learning process should continue as is or if some type of change is necessary (Pintrich, 2000; Zimmerman, 2000). Based on this judgment the learners decide if and how to adapt and change their activities and strategies selected during forethought and implemented during the performance phase. These changes are referred to as self-reaction and influence ongoing and following learning activities (Winne & Hadwin, 1998; Zimmerman, 2000).

There is no predetermined order in which all these cognitive, metacognitive, and motivational processes occur, they rather interact recursively with each other (e.g., Pintrich, 2000). Therefore, forethought, performance, and self-reflection can occur simultaneously and dynamically as the learner progresses through the learning process. This implies that the goals and plans of the forethought phase are likely to be changed or updated on the basis of the feedback from the performance and self-reflection phases (e.g., Pintrich, 2000; Winne & Hadwin, 1998). In order to be successful, self-regulated learners have thus to deal with a complex cyclical interplay of forethought, performance, and self-reflection processes which are in need of continuous changes and updates.

This model describes the demands which learners are confronted with when they choose to engage in self-regulated learning. However, success in this complex interplay of self-regulatory activities is guided and constrained by the learner's individual skill and will (i.e. their cognitive and metacognitive competency, as well as their motivation, see for example McCombs & Marzano, 1990). External contextual characteristics (e.g., task characteristics, characteristics of the TELE, see for example Pintrich, 2000; Winne & Hadwin, 1998; Zimmerman, 2000) also influence the self-regulated learning process. This might be a reason why many students struggle with efficiently regulating their learning in a TELE (e.g., Azevedo & Cromley, 2004; Balcytiene, 1999; Schraw, 2007). Furthermore, during learning with TELEs students hardly spontaneously perform key self-regulatory strategies such as planning and monitoring (Winters et al., 2008). Therefore, a focus of research on self-regulated learning with TELEs lies on the development of effective instructional interventions fostering the application of such key self-regulatory strategies.

SUPPORTING SELF-REGULATED LEARNING IN THE TECHNOLOGY-ENHANCED LEARNING ENVIRONMENT STUDIERPLATZ

Instructional interventions can be developed and examined for all phases of the self-regulated learning process, with respect to cognitive, motivational and meta-cognitive strategies (e.g., Azevedo, 2007; Clarebout & Elen, 2009). Friedrich and Mandl (1992) distinguish between direct and indirect instructional interventions. Direct instructional interventions are trainings in which learners are taught skills and strategies for self-regulated learning with TELEs (e.g., Azevedo & Cromley, 2004). By contrast, indirect instructional interventions are inherent characteristics of a TELE. These indirect instructional interventions support learners in coping with the demands of self-regulated learning (e.g., Bannert, 2006). Indirect instructional interventions can be either embedded or non-embedded into a TELE (Clarebout & Elen, 2006).

Embedded instructional interventions are integrated into a TELE in such a way that learners are forced to use them. Orientation and navigation support by means of a structured overview of the available information (e.g., Rouet & Levonen, 1996; Rouet & Potelle, 2005) is one example; another example is prompting metacognitive strategies (e.g., Bannert, 2006; Bannert, Hildebrand, & Mengelkamp, 2009). By contrast, *non-embedded instructional interventions* are provided within a TELE, but their use is optional; learners can decide to use them or not. Non-embedded instructional interventions are also called *tools*. Note-taking tools for information elaboration (Kauffman, 2004; Narciss et al., 2007; van Oostendorp, 1996) or interactive learning tasks for self-observation and self-evaluation (Nadolski, Kirschner, & van Merriënboer, 2006; Proske et al., 2007) are examples of non-embedded indirect interventions.

The provision of indirect instructional interventions into TELEs is considered to promote self-regulated learning as well as motivation. However, research indicates that a successful self-regulation of the learning process is only guaranteed when the learners use these interventions in the way intended by the instructional designer.

In the project *Study 2000* generic authoring tools have been developed and evaluated which facilitate the ergonomically and psychologically sound design of TELEs including indirect instructional interventions (see Narciss & Körndle, 1998; Narciss et al., 2007; Proske et al., 2007, http://studierplatz2000.tu-dresden.de). The authoring tools include:

- The *s2w-compiler* (Study-to-Web Compiler) which supports instructors in presenting multiple learning materials and media within an integrated user interface (http://studierplatz2000.tu-dresden.de/s2w).
- The *EF-Editor* (Exercise Format Editor) which facilitates the construction and im-

plementation of interactive learning tasks (for a detailed description see Proske, Körndle, & Narciss, 2004, http://studierplatz2000.tu-dresden.de/efb). In contrast to test exercises, interactive learning tasks are solved interactively with the additional aid of multiple-try strategies and informative tutoring feedback if required (Narciss, 2008).

The TELEs created with these authoring tools are called *Studierplatz,* i.e. a working space for learning and studying. Studierplatz TELEs are designed to complement instruction in different learning contexts, such as lectures, seminars or project-based courses. Up to now, these tools have been used to design Studierplatz TELEs in various domains such as psychological learning theories, stochastics, or foreign language learning (Körndle, Narciss, & Proske, 2009). They present multiple materials and information on a specific topic. Thus, students are not only able to prepare for lessons, but also to elaborate and repeat information by means of self-regulated learning. In order to support students' key self-regulatory activities, the following indirect instructional interventions are implemented into Studierplatz TELEs:

- embedded interventions: orientation and navigation support;
- non-embedded interventions: learning tools supporting performance; monitoring tools supporting self-reflection; interactive learning tasks and a learning plan tool supporting forethought, performance, and self-reflection.

Orientation and Navigation Support: User Interface of a Studierplatz TELE

In order to initiate and facilitate flexible access to multiple materials and resources a Studierplatz TELE presents multiple materials and information on a specific topic by means of an integrated user interface (see Figure 1).

Figure 1. User Interface of a Studierplatz TELE

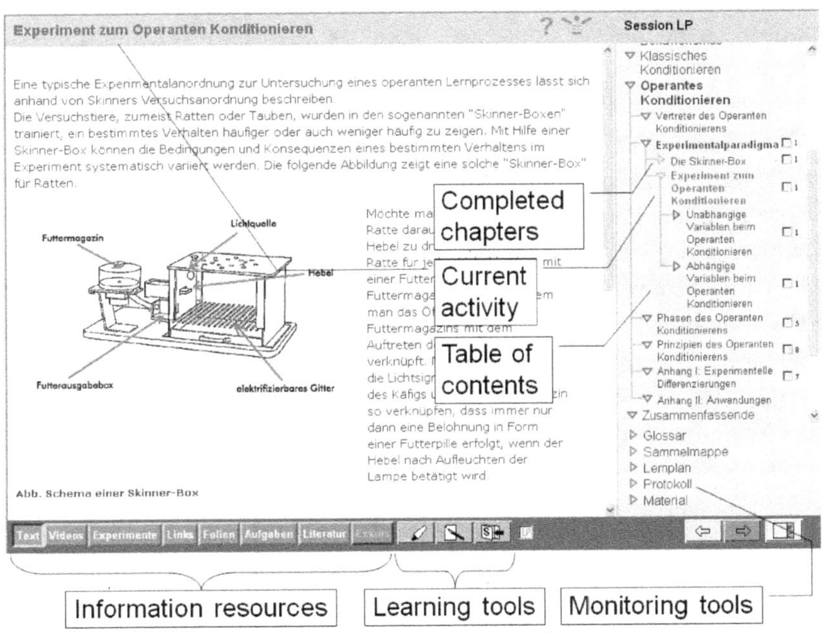

In order to facilitate orientation and navigation students should be informed of the sections they have already worked on and which section they are currently processing (Rouet & Levonen, 1996; Rouet & Potelle, 2005). Furthermore, students should be informed which other information resources are available for learning. Consequently, the Studierplatz interface offers the following information (see Figure 1):

- The working area for the learner can be found on the left side of the user interface.
- Content-related navigational information is provided on the right side of the user interface by a hierarchical table of contents. This table of contents offers a general overview of the Studierplatz content by listing its main sections. In order to get the detailed structure and subsections students have to click on the title of the particular main section.
- A running title above the working area indicates the subsection on which the learner is currently working. This running title is also colored red in the table of contents.
- Information on sections and resources that were already processed is indicated by colored accentuations in the table of contents.
- Resources-related navigational information for a particular section is provided at the bottom of the working area. Whereas a blue button indicates an available resource, a grey button indicates an inaccessible one. The section which is depicted in Figure 1 provides access to learning texts, videos, experimental simulations, internet-links, slides from lecture presentations, original papers, and interactive learning tasks. The excursus is not available for this section.

There is no restriction on how to access all the provided information. Thus, learners by themselves decide if, when, how, and by what learning goal they will process a particular information resource, i.e. they self-regulate their learning activities. All personal data (i.e. learning history, status of interactive learning tasks, etc.) can be stored and retrieved at any further access to the Studierplatz TELE.

Learning Tools: Support for Performance

In order to initiate and facilitate active engagement and constructive processing of the accessible information, a Studierplatz includes learning tools that enable learners to elaborate the information and materials of the TELE. These tools permit the application of widely-used conventional learning strategies with which the students are familiar (Weinstein et al., 2000). However, their use is dependent on the learners' initiative. The learning tools can be activated by clicking on the learning tool buttons at the right bottom of the working area (see Figure 1).

- The highlighting and note-taking tools enable students to individually process the provided text material by highlighting concepts or summarizing parts of the given material.
- The glossary tool enables students to search for relevant terms, their synonyms, and definitions. It includes hyperlinks to all sections of the Studierplatz which contain information on the particular term.
- The integrator tool allows students to rearrange the provided information by individually collecting selected resources into a commented dossier. As the individual dossier can be released in a slide-show, this integrator can also be used as a presentation tool.

Monitoring Tools: Support for Self-Reflection

In order to initiate and facilitate processes of self-evaluation and self-reaction, a Studierplatz TELE includes monitoring tools. The monitoring tools are located downright on the user interface (see Figure 1). If applied, the monitoring tools support the learners in controlling their course of learning (Narciss et al., 2007; Proske et al., 2007).

- The progress report tool lists all learning activities that learners previously have carried out. It includes information on the topic, title, and kind of resource, as well as the date and duration of its processing.
- The learning task report tool visualizes the number of accomplished and unaccomplished interactive learning tasks, assigned to the main sections of the Studierplatz. It includes information on the correctness of the completed learning tasks.

Interactive Learning Tasks: Support for Forethought, Performance, and Self-Reflection

In order to support processes of forethought, performance, and self-reflection, a Studierplatz TELE includes interactive learning tasks. Learners can access the interactive learning tasks via the respective resource button at the bottom of the working area (see Figure 1).

- *Forethought*: Working on interactive learning tasks enables students to assess their prior knowledge. Furthermore, the learning tasks point to concepts and principles which are central to the topic. This information can guide students' goal setting and attention during performance.
- *Performance*: Processing interactive learning tasks initiates active engagement and elaboration of information (Proske et al., 2004). In contrast to test exercises, such learning tasks offer the possibility to solve tasks interactively by providing multiple-try strategies and informative tutoring feedback (Narciss, 2008). Informative tutoring feedback delivers strategically useful information that guides the learner stepwise towards successful task completion by assisting multiple solution attempts.
- *Self-reflection*: The informative tutoring feedback from the learning tasks allows learners to assess their acquired knowledge on the one hand (self-evaluation) and promotes and facilitates self-reaction processes on the other.

Learning Plan Tool: Support for Forethought, Performance, and Self-Reflection

Research shows that success at finding sections containing learning goal relevant information is a predictor of posttest performance when self-regulated learning with TELEs (Gauss & Urbas, 2003; Naumann et al., 2008). To initiate and facilitate planning the learning process, selecting relevant material, as well as evaluating one's own course of learning, a Studierplatz TELE offers a learning plan tool. If students use this tool, it requires them to make instructional decisions; they have to set learning goals, assign sub-goals, search for learning goal relevant materials, select and sequence these materials, as well as to evaluate their learning progress. The learning plan tool can be used in a variety of ways by employing the learning plan site and/or the table of contents (see Figure 2).

- *Forethought*: On the learning plan site, learners can schedule their learning by individually selecting and sequencing learning materials from the resources provided within the Studierplatz TELE. After this, the selected learning materials can be consecutively accessed via this site. Sections included into the individual learning plan are also automatically tagged with a "LP" in the table of contents (see Figure 2). Thus, the materials included into the learn-

Figure 2. Learning plan site of a Studierplatz TELE

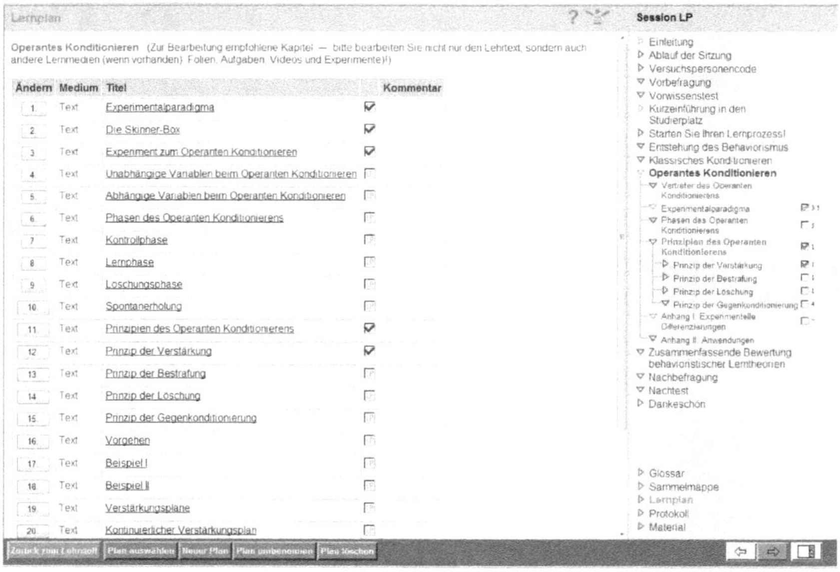

ing plan can also be accessed via the table of contents.

- *Performance*: The indication of the materials included into the learning plan on the learning plan site and within the table of contents facilitates students' self-observation. Learners immediately can track which sections and resources they had already processed and which section they plan to process next. Thus, students do not have to continuously decide on the next steps of their learning process; they simply can follow their previously defined learning plan (cf. Azevedo, 2005a; Bannert & Mengelkamp, 2008).
- *Self-reflection*: The learning plan tool encourages learners to tick off mastered sections. This can be done either on the learning plan site or within the table of contents by clicking on the LP tags.

Mindful using the learning plan tool may contribute to students' application of key self-regulatory processes. Hence, it is considered to be a metacognitive tool (cf. Azevedo, 2005a, 2007).

However, providing tools is a necessary, but not a sufficient condition for efficient self-regulated learning with TELEs (e.g., Clarebout & Elen, 2006, 2008). Preliminary studies with various Studierplatz TELEs show that only ca. 40% of the students use most of the provided tools in a serious manner (Narciss et al., 2007; Proske et al., 2007). Therefore, further research should identify means of fostering the optimal use of non-embedded indirect instructional interventions.

FOSTERING LEARNING PLAN TOOL USE: AN EXPLORATORY STUDY

Compiling an individual learning plan is a challenging task, in particular for students with low prior knowledge on the topic. Thus, providing a learning plan which explicitly specifies sections containing learning goal relevant information should take off cognitive work from the students during learning. The purpose of this study was therefore to explore whether the provision of a tailored learning plan within a TELE will influence self-regulated learning. More specifically, the following research questions were pursued:

1. How do students use a recommended, tailored optional learning plan?
2. How does learning plan use affect navigational behavior and use of learning goal relevant information resources?
3. How does learning plan use affect performance in a posttest?

Method

Participants

Participants of the study were 25 teacher students (76% women, 24% men, mean age = 21 years, Range: 19 - 26 years) attending a Psychology lecture, "Introduction to Psychology", at Dresden University of Technology, Germany. Most participants were in the second year of study (second term; range: first term - fourth term).

Design and Procedure

The study participants were provided with a Studierplatz on the topic of "Classical and Operant Conditioning"[1]. Classical and Operant Conditioning are psychological learning theories which usually are taught in Psychology introductory courses. Both belong to learning theories of Behaviorism. The Studierplatz included four major sections: (a) origins of behaviorism, (b) classical conditioning, (c) operant conditioning, (d) summary on behaviorist learning theories (see Figure 1). Each of the major sections consisted of several subsections.

A precondition for tool use is that learners are in need of the particular tool (Clarebout & Elen, 2006). In order to establish the necessity of using the learning plan tool, the participants of the study were instructed to reach a predetermined learning goal by self-regulated learning. Furthermore, students had only limited time (about 75 minutes) at their disposal. The predetermined learning goal was to learn the section on Operant Conditioning. Thus, the Studierplatz contained not only learning goal relevant, but also learning goal irrelevant sections. It was not possible to learn all the information within the given time limit.

In order to foster learning process planning, the Studierplatz for this study included a tailored learning plan specifying learning goal relevant sections and resources; however, its use was optional. Provided that students used the tailored learning plan they had neither to set sub-goals for their learning nor to find the location of learning goal relevant information. This information was given by the tailored learning plan.

All participants learned self-regulated. That means, all students by themselves decided if, when, and how they used the learning materials and information provided by the Studierplatz. The experimenter instructed the students to focus on the major section on Operant Conditioning and to use the tailored learning plan. Furthermore, students were informed that a test would be administered immediately after studying. Then the students started the TELE.

At the beginning, learning strategy use, TELE experience, motivation, and prior knowledge were assessed as control variables. After this, the students were provided with a section explaining the tools of the Studierplatz including the tailored learning plan. Subsequently, students started learning. Students had the possibility to use the learning plan whenever they wanted. After 60 minutes learning time the students were automatically reminded to answer the posttest. However, they could finish their current learning activities before answering the posttest.

The study applied an *ex post facto* design. Depending on their learning plan tool use, participants were grouped ex post into three different groups:

a. 13 students not using the learning plan (NLP group);

b. 7 students accessing the learning plan site, but not overtly acting on it (LP-site group);
c. 5 students ticking off sections included into the learning plan (LP-tick group).

Material

Depending on the contents of the posttest, 26 subsections of the Operant Conditioning section were identified that contained learning goal relevant information (altogether 26 text sites, 80 interactive learning tasks, 30 slides from lecture presentations, 4 videos, and 2 experiment simulations).

The learning goal relevant (LGR) sections were specified within the tailored learning plan. The learning plan was available via the learning plan site and apparent by LP tags within the table of contents (see Figure 2). Students could access these LGR sections either from the learning plan site or by using the table of contents. Likewise, they could tick off a completed section by checking the respective LP tag either at the learning plan site or within the table of contents. On the top of the learning plan site students were prompted to work not only on the relevant learning texts, but also on slides from lecture presentations, learning tasks, videos, and experiment simulations, when available.

Measures

All students' activities were recorded in log-files. Frequency and time spent on each learning activity were automatically summarized for each log-file.

Learning plan use was assessed by recording the time and duration of accessing the learning plan site. Furthermore, the frequency of ticking off learning goal relevant sections was summarized from the log-files. The learning goal relevant sections could be ticked off by both the learning plan site and the table of contents.

Learning activities refers to number of and time spent on accessed sections and resources within learning goal relevant and learning goal irrelevant sections, respectively. Learning goal relevant sections (LGR) contain all resources which were included into the tailored learning plan, whereas learning goal irrelevant sections (LGI) contain all other resources of the Studierplatz. More specifically, it was recorded whether the students worked on learning goal relevant or learning goal irrelevant

- learning texts,
- interactive learning tasks,
- information resources (videos, lecture presentations, experiment simulations),
- learning tools (highlighting, note-taking, integrator tools, glossary),
- monitoring tools (progress report tool, learning task report tool).

Performance during learning was assessed by the percentage of correctly solved learning tasks in the first attempt. It was used to estimate the quality of students' learning activities while learning self-regulated with the Studierplatz.

Prior knowledge was assessed by a test containing 8 tasks on the topic of operant conditioning. It is expressed in the percentage of correctly solved tasks. *Posttest performance* refers to the difference between correctly solved tasks in the posttest and correctly solved tasks in the pretest. Pre- and posttest consisted both of 8 tasks on basic principles of Operant Conditioning. As there were several tasks in each of the tests with more than one answer, the test results are expressed as percentages of correctly answered sub tasks.

Statistical Analyses

The ex post assignment to the three students groups (NLP group, LP-site group, and LP-tick group) resulted in uneven group sizes. Thus, only nonparametric methods (e.g., Mann-Whitney test; Kruskal-Wallis test) were used to analyze the data. Learning strategy use, experience with TELEs, and motivation were assessed as control variables at the beginning of the session. There were no differences between the three groups in terms of these variables. For all statistical tests an alpha level of .05 was used.

Results

Learning Plan Use

Table 1 presents the means, standard deviations, as well as minima and maxima of learning plan use. A total of 48% of the students attended to the learning plan (LP), i.e. they either accessed the learning plan site and/or ticked off learning goal relevant sections (see Table 1). Thirty six percent of the study participants took a look on the learning plan site, but 78% out of these did not use the learning plan further (LP-site group). Students of the LP-site group for the first time accessed the learning plan site ca. after 20 minutes learning time (total learning time was ca. 75 minutes).

The time spent on the learning plan site was rather short as represented by a mean of a half a minute (see Table 1). It should be noted that this time only represents the time that was spent on the learning plan site. Here the students got an overview about the learning goal relevant sections. Students also had the possibility to get the same information when paying attention to the LP tags within the table of contents. Unfortunately, in this study it was not possible to measure the time that students expended for looking at the table of contents.

Only 20% of the study participants ticked off sections that were included into the learning plan

Table 1. Descriptive statistics of learning plan site use (LP-site group, n = 7)

	M	*SD*	*Min*	*Max*
Time on LP site [a]	0.49	0.71	0.07	2.07
Moment of first LP site access [a]	19.43	16.37	1.71	42.88

Note.[a] Measures represent time in minutes.

(LP-tick group). Moreover, only two students of the LP-tick group also accessed the LP site (one student after half a minute, the other after nearly 20 minutes working time). The other students of the LP-tick group used the "LP" checkboxes in the table of contents to evaluate their learning process (see also Figure 2, on the right side of the user interface). On average they ticked off 15 ($M = 15.40$, $SD = 9.89$) sections of the learning plan (ca. 62% of the learning goal relevant sections).

Figure 3 shows the sections in which the 9 students (7 students of the LP-site group and two students of the LP-tick group) worked prior and after accessing the learning plan site (click 1 to 10). It can be seen that 89% of the students studied learning goal irrelevant (LGI) sections before accessing the learning plan site. Although the learning plan site showed which sections of the Studierplatz contain learning goal relevant information (a) only half of the students shifted to these sections and (b) only few students remained at these sections until the 10th navigational click after accessing the learning plan site.

Learning Activities

In a first step, total working time on relevant (LGR) and irrelevant (LGI) sections was analyzed for all participants of the study. Total working time was about 75 minutes from which only 22 minutes (31%) were employed for studying LGR sections. Nearly 40% of the students studied less than 10 minutes relevant information. Students on average accessed LGR sections after nearly one half of total working time.

Table 2 presents descriptive statistics for accessing LGR sections for the non-learning plan tool (NLP), LP-site, and LP-tick groups. Kruskal-Wallis tests revealed that the groups statistically differed in terms of total working time ($H(2) = 8.4, p < .05$) and time on LGR sections ($H(2) = 7.1, p < .05$).

Mann-Whitney-U pairwise comparisons showed that the LP-tick group significantly ex-

Figure 3. Sections of the Studierplatz on which students learnt prior and after their first access to the learning plan site

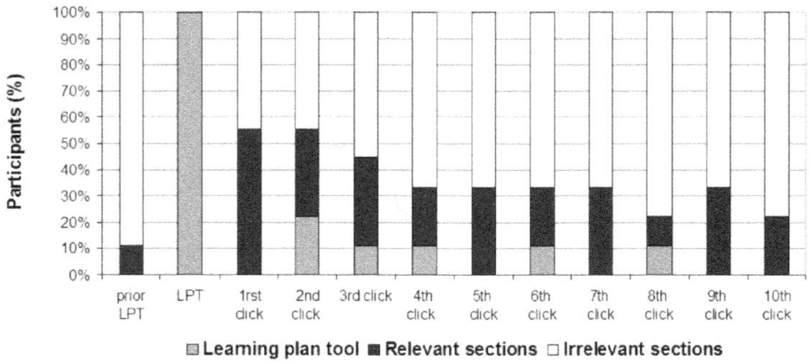

ceeded the NLP group ($U = 10.0, p < .05$) and the LP-site group ($U = 0.0, p < .01$) in terms of total working time. The same was true for time on LGR sections (LP-tick to NLP: $U = 9.0, p < .05$; LP-tick to LP-site: ($U = 4.0$, $p < .05$). Furthermore, the LP-tick group accessed LGR sections earlier during their course of learning. However, this difference failed to reach statistical significance.

The working time, frequency and duration on information resources and tools of goal relevant sections (LGR) can be taken from Table 3. The LP-tick group achieved higher values in all variables. This indicates that the LP-tick group accessed more LGR information and worked longer on LGR sections. These differences were statistically significant for the variables (a) time on learning tools ($H(2) = 10.5, p < .01$), (b) number of learning tools ($H(2) = 10.8, p < .01$), as well as (c) total number of learning goal relevant resources and tools ($H(2) = 5.3, p < .05$).

Note. Information resources = videos, slides from lecture presentations, experiment simulations; Learning tools = highlighting, note-taking, integrator tools, glossary; Monitoring tools = progress report tool, learning task report tool

Mann-Whitney tests were used to follow up these findings. It appeared that the LP-tick group used the learning tools within the LGR sections longer than the NLP group ($U = 4.0, p < .01$), as well as the LP-site group ($U = 4.0, p < .05$). In addition, the LP-tick group accessed the learning tools within the LGR sections significantly more frequently than the other groups (NLP: $U = 3.5, p < .01$, LP-site: $U = 4.0, p < .05$). The LP-tick group also accessed significantly more LGR resources and tools than the other groups (NLP: $U = 6.5$, $p < .01$, LP-site: $U = 3.5, p < .05$). In contrast to

Table 2. Descriptive statistics of time on learning goal relevant sections for the NLP, LP-site, and LP-tick groups

	NLP (n = 13)		LP-site (n = 7)		LP-tick (n = 5)	
Time in minutes	*M*	*SD*	*M*	*SD*	*M*	*SD*
Total working time	69.26*	9.29	64.43*	6.26	80.84*	2.33
Time on LGR sections	29.93*	21.80	20.07*	20.05	65.35*	27.66
Moment of first LGR section access	32.09	21.44	30.33	20.40	14.55	25.28

* p < .05

Table 3. Descriptive statistics of resources and tool use within learning goal relevant sections for the NLP, LP-site, and LP-tick groups

	NLP (n = 13)		LP-site (n = 7)		LP-tick (n = 5)	
Resources/Tools	*M*	*SD*	*M*	*SD*	*M*	*SD*
	Time (min)					
Text	10.13	8.69	9.41	8.28	17.88	13.10
Learning tasks	5.53	8.45	3.82	9.00	13.72	12.87
Information resources	1.85	2.99	0.77	1.63	3.06	3.91
Learning tools	0.24**	0.70	0.59**	1.52	1.93**	1.54
Monitoring tools	0.00	0.00	0.02	0.05	0.08	0.13
Total	18.48	16.19	15.00	18.95	39.46	22.48
	Frequency of access					
Text	11.00	8.25	13.29	7.20	20.40	5.98
Learning tasks	8.23	12.50	3.00	7.10	12.00	11.77
Information resources	0.92	1.61	1.57	3.05	4.00	5.43
Learning tools	0.92**	2.75	1.71**	4.11	5.80**	5.31
Monitoring tools	0.00	0.00	0.14	0.38	0.40	0.55
Total	21.08*	18.09	19.71*	15.91	42.60*	8.76
	Duration per site (min)					
Text	0.66	0.48	0.64	0.36	0.84	0.45
Learning tasks	0.32	0.39	0.37	0.63	0.72	0.71
Information resources	1.09	2.12	0.37	0.72	1.11	1.30

** $p < .01$, * $p < .05$

these results, no statistically significant differences between the three groups were found for resources and tool use of learning goal irrelevant sections.

With respect to performance during learning no statistically significant differences were found between the groups. However, as can be seen in Figure 4, the LP-tick group solved little more interactive learning tasks correctly and abandoned fewer tasks of the LGR sections than the other two groups (see Figure 4).

Posttest Performance

Posttest performance refers to the difference between correctly solved tasks in the posttest and correctly solved tasks in the pretest. As presented in Table 4, all groups achieved better at the posttest than at the pretest. This indicates that students of all groups succeeded in gaining knowledge on operant conditioning when learning with the Studierplatz. It can also be seen that the LP-tick group improved their performance more than the other two groups. However, this difference marginally failed to reach statistical significance ($H(2) = 4.5, p = .10$).

DISCUSSION

The aim of this study was to explore whether the provision of a given metacognitive tool in terms of a tailored learning plan will affect tool use, learning activities, as well as posttest performance. To this end, students were instructed to reach a specific learning goal by learning with a Studierplatz TELE. Furthermore, they were advised to use a

Figure 4. Learning task solution behavior of the NLP, LP-site, and LP-tick groups with respect to learning goal relevant and irrelevant sections of the Studierplatz

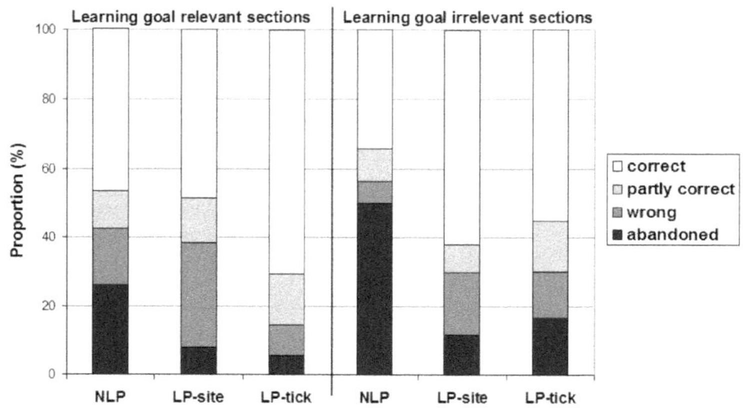

tailored learning plan which was included as a tool within the Studierplatz. This tailored learning plan was expected to facilitate students' success at finding learning goal relevant information within the TELE (Bannert & Mengelkamp, 2008; Gauss & Urbas, 2003; Naumann et al., 2008) which in turn should positively affect learning behavior and posttest performance.

The results of the study confirm predictions from current research on tool use in self-regulated learning with TELEs (Narciss et al., 2007; Proske et al., 2007). In line with these findings, learning plan tool use was low, even though students received an explicit recommendation to use it and worked with time constraints; only 48% of the students attended to the learning plan tool. Moreover, the tool on average was accessed the first time after one third of total working time. The mean time spent on the learning plan site was with about one minute really short compared to ca. 75 minutes total working time. These results indicate that most of the students did not use the learning plan as intended by the instructional designer (Clarebout & Elen, 2006, 2009).

However, this study also shows that 20% of the study participants (the LP-tick group) extensively and intensely used the learning plan tool; they evaluated their self-regulated learning process by ticking off sections they have mastered. Contrary to our expectations most of them did not draw on the learning plan site, but rather used the LP checkboxes in the table of contents for their self-reflection processes. It might be that the design of the learning plan site was not as good for the students' individual purposes as expected. Another explanation is that the eye-catching LP checkboxes within the table of contents distracted learners' attention from the learning plan site. These issues

Table 4. Descriptive statistics of posttest performance for the NLP, LP-site, and LP-tick groups

	NLP (n = 13)		LP-site (n = 7)		LP-tick (n = 5)	
	M	*SD*	*M*	*SD*	*M*	*SD*
Pretest: Correctly solved tasks (%)	30.07	13.58	29.87	15.49	23.64	20.93
Posttest: Correctly solved tasks (%)	51.05	20.17	58.44	4.86	65.45	25.23
Posttest performance (Difference post-pre)	20.98	19.77	28.57	13.31	41.82	13.79

deserve further attention, because a metacognitive tool has to be designed in such a way that it appears useful for the students and that they can use it optimally (Clarebout & Elen, 2006, 2009).

In addition, the results reveal that students of the LP-tick group (a) spent more time on learning goal relevant sections and (b) used the learning tools of the Studierplatz (e.g., note-taking tool, highlighting tool, and glossary) longer and more frequently within learning goal relevant sections than students of the other two groups. These results show that using the tailored learning plan during the performance phase positively affects learning activities. The learning plan guided the course of learning of the LP-tick group and encouraged students of this group to an active engagement and constructive processing of learning goal relevant information.

The results of this study also show that the learning activities of students who only accessed the learning plan site did not differ from the learning activities of the NLP-group which did not attend the learning plan. Therefore, an important prerequisite for a possible benefit of a tailored learning plan within TELEs is that the students not only superficially inspect the metacognitive tool but rather use its potential and functionalities goal-directed.

Although the descriptive data of posttest performance suggest that the LP-tick group performed better on the posttest, no significant differences in posttest performance were found among the three groups. This is not in line with prior research of self-regulated learning with TELEs in which finding learning goal relevant information and using tools was associated to higher posttest performance (e.g., Gauss & Urbas, 2003; Proske et al., 2007). By contrast, this result is in accordance with research on metacognitive prompts in TELEs in which students using metacognitive tools also did not perform better than controls on recall and knowledge tests. However, they performed better on transfer tests (Bannert, 2006; Bannert et al., 2009). As the posttest in this study consisted only of recall and knowledge tasks, this result indicates that the investigation of metacognitive tools in TELEs should also incorporate transfer tasks as criteria for tool effectiveness (Bannert et al., 2009). Another explanation for the results is the small sample of this study which made it difficult to statistically analyze the data of the learning plan users. Taking into account that only some of the study participants will use the provided tools (Clarebout & Elen, 2009; Narciss et al., 2007; Proske et al., 2007), the investigation of tool use in self-regulated learning with TELEs requires designing studies with larger sample sizes which guarantee that a comprehensive number of tool users can be examined in greater detail.

LIMITATIONS OF THE STUDY

Two limitations concerning the generalizability of the present study should be pointed out. First, the sample was small. Furthermore, the intervention period was relatively short with only about 75 minutes working time. Yet, this study indicates that providing an optional learning plan could be an effective means to support self-regulation during learning. This result therefore justifies conducting further studies with larger sample sizes and longer intervention periods.

Second, the present study included primarily time and frequency measures in order to characterize the learning behavior. These measures are quantitative and do not qualitatively characterize the different self-regulatory learning activities performed by the students. Some relevant characteristics of learning plan use such as paying attention to sections with LP tags within the table of contents even could not be measured. Thus, future studies should also consider qualitative measures of learning activities, such as thinking-aloud protocols or eye-tracking (Bannert & Mengelkamp, 2008).

CONCLUSION

The findings of the study do raise some issues. First, a tailored learning plan tool should be implemented into a TELE so that it (a) is adequate to a given learning task and (b) can be used optimally by the learner. It is further to ensure that students are knowledgeable of and willing to properly use the metacognitive tool (Clarebout & Elen, 2009). Future research should thus investigate whether (a) the automatic display of the tailored learning plan at the beginning of the course of learning, (b) TELE integrated instruction on how to use the learning plan (e.g., Clarebout & Elen, 2009), (c) automatically prompting students to use the learning plan tool (Bannert, 2006) and/or (d) direct training in using metacognitive tools (e.g., Azevedo & Cromley, 2004; Bannert et al., 2009) will lead to a more frequent and goal-directed learning plan use.

Second, the results of the study show that the design of the learning plan has to be investigated in more detail. For example, most students of the LP-tick group preferred to tick off completed sections within the table of contents. Furthermore, the learning plan could not be modified by the students. Thus, future research should investigate the conditions under which students make use of a tailored learning plan, students' needs on a suggested course of learning, as well as students' acceptance of a tailored learning plan.

Third, learning plan use and its relationships to learner characteristics has to be investigated. For example, learners with high skills in self-regulated learning are expected to benefit more from indirect non-embedded instructional interventions than learners with low skills in self-regulation (e.g., Schraw, 2007).

However, if the tool is applied, the results of this study show that a tailored learning plan can facilitate students' self-regulated learning within TELEs. It (a) guides students' course of learning, (b) encourages them to active engagement and constructive processing of learning goal relevant information, and (c) initiates self-evaluation processes. Nonetheless, it is up to the learner to use the metacognitive tool. Therefore, even though these results are promising, many questions on how to design a useful learning plan tool and on how to initiate and facilitate learning plan tool use within TELEs are not answered yet and need to be investigated in further research.

REFERENCES

Azevedo, R. (2005a). Computer environments as metacognitive tools for enhancing learning. *Educational Psychologist*, *40*(4), 193–197. doi:10.1207/s15326985ep4004_1

Azevedo, R. (2005b). Using hypermedia as a metacognitive tool for enhancing student learning? The role of self-regulated learning. *Educational Psychologist*, *40*(4), 199–209. doi:10.1207/s15326985ep4004_2

Azevedo, R. (2007). Understanding the complex nature of self-regulatory processes in learning with computer-based learning environments: An introduction. *Metacognition and Learning*, *2*(2), 57–65. doi:10.1007/s11409-007-9018-5

Azevedo, R., & Cromley, J. G. (2004). Does training on self-regulated learning facilitate students' learning with hypermedia? *Journal of Educational Psychology*, *96*(3), 523–535. doi:10.1037/0022-0663.96.3.523

Balcytiene, A. (1999). Exploring individual processes of knowledge construction with hypertext. *Instructional Science*, *27*(3), 303–328. doi:10.1007/BF00897324

Bannert, M. (2006). Effects of reflection prompts when learning with hypermedia. *Journal of Educational Computing Research*, *35*(4), 359–375. doi:10.2190/94V6-R58H-3367-G388

Bannert, M., Hildebrand, M., & Mengelkamp, C. (2009). Effects of a metacognitive support device in learning environments. *Computers in Human Behavior*, *25*(4), 829–835. doi:10.1016/j.chb.2008.07.002

Bannert, M., & Mengelkamp, C. (2008). Assessment of metacognitive skills by means of instruction to think aloud and reflect when prompted. Does the verbalisation method affect learning? *Metacognition and Learning*, *3*(1), 39–58. doi:10.1007/s11409-007-9009-6

Clarebout, G., & Elen, J. (2006). Tool use in computer-based learning environments: Towards a research framework. *Computers in Human Behavior*, *22*(3), 389–411. doi:10.1016/j.chb.2004.09.007

Clarebout, G., & Elen, J. (2008). Tool use in open learning environments: In search of learner-related determinants. *Learning Environments Research*, *11*(2), 163–178. doi:10.1007/s10984-008-9039-2

Clarebout, G., & Elen, J. (2009). Benefits of inserting support devices in electronic learning environments. *Computers in Human Behavior*, *25*(4), 804–810. doi:10.1016/j.chb.2008.07.006

Foltz, P. W. (1996). Comprehension, coherence, and strategies in hypertext and linear text. In Rouet, J.-F., Levonen, J. J., Dillon, A., & Spiro, R. J. (Eds.), *Hypertext and cognition* (pp. 109–136). Hillsdale, NJ: Lawrence Erlbaum.

Friedrich, H. F., & Mandl, H. (1992). Lern- und Denkstrategien - ein Problemaufriss. [Learning and thinking strategies - a problem outline] In Mandl, H., & Friedrich, H. F. (Eds.), *Lern- und Denkstrategien. Analyse und Intervention* (pp. 3–54). Göttingen: Hogrefe.

Gauss, B., & Urbas, L. (2003). Individual differences in navigation between sharable content objects - an evaluation study of a learning module prototype. *British Journal of Educational Technology*, *34*(4), 499–509. doi:10.1111/1467-8535.00346

Kauffman, D. F. (2004). Self-regulated learning in web-based environments: Instructional tools designed to facilitate cognitive strategy use, metacognitive processing, and motivational beliefs. *Journal of Educational Computing Research*, *30*(1), 139–161. doi:10.2190/AX2D-Y9VM-V7PX-0TAD

Körndle, H., Narciss, S., & Proske, A. (2009). Developing and evaluating tools for web-based learning and instruction. In M. W. Greenlee (Ed.), *New issues in experimental and applied psychology. A Festschrift for Alf Zimmer* (pp. 127-164). Lengerich: Pabst Science.

Kuhl, J. (2000). A functional-design approach to motivation and self-regulation: The dynamics of personality systems and interactions. In Boekaerts, M., Pintrich, P. R., & Zeidner, M. (Eds.), *Handbook of self-regulation* (pp. 111–169). San Diego, CA: Academic Press. doi:10.1016/B978-012109890-2/50034-2

Latham, G. P., & Brown, T. C. (2006). The effect of learning vs. outcome goals on self-efficacy, satisfaction and performance in an MBA Program. *Applied Psychology*, *55*(4), 606–623. doi:10.1111/j.1464-0597.2006.00246.x

McCombs, B. L., & Marzano, R. J. (1990). Putting the self in self-regulated learning: The self as agent in integrating will and skill. *Educational Psychologist*, *25*(1), 51. doi:10.1207/s15326985ep2501_5

Nadolski, R. J., Kirschner, P. A., & van Merriënboer, J. J. G. (2006). Process support in learning tasks for acquiring complex cognitive skills in the domain of law. *Learning and Instruction*, *16*(3), 266–278. doi:10.1016/j.learninstruc.2006.03.004

Narciss, S. (2008). Feedback strategies for interactive learning tasks. In J. M. Spector, M. D. Merrill, J. J. G. van Merriënboer & D. M. P. (Eds.), *Handbook of Research on Educational Communications and Technology* (3rd ed., pp. 125-144). Mahwah, NJ: Lawrence Erlbaum.

Narciss, S., & Körndle, H. (1998). Study 2000 - Problems and perspectives for the development of multimedia tools for teaching and learning in the Internet. *European Psychologist, 3*(3), 219–226. doi:10.1027/1016-9040.3.3.219

Narciss, S., Proske, A., & Körndle, H. (2007). Promoting self-regulated learning in web-based learning environments. *Computers in Human Behavior, 23*(3), 1126–1144. doi:10.1016/j.chb.2006.10.006

Naumann, J., Richter, T., Christmann, U., & Groeben, N. (2008). Working memory capacity and reading skill moderate the effectiveness of strategy training in learning from hypertext. *Learning and Individual Differences, 18*(2), 197–213. doi:10.1016/j.lindif.2007.08.007

Pintrich, P. R. (2000). The role of goal orientation in self-regulated learning. In Boekaerts, M., Pintrich, P. R., & Zeidner, M. (Eds.), *Handbook of self-regulation* (pp. 451–502). San Diego, CA: Academic Press. doi:10.1016/B978-012109890-2/50043-3

Proske, A., Körndle, H., & Narciss, S. (2004). The Exercise Format Editor: A multimedia tool for the design of multiple learning tasks. In H. M. Niegemann, D. Leutner & R. Brünken (Eds.), *Instructional design for multimedia learning* (pp. 149-164). Münster, Germany: Waxmann.

Proske, A., Narciss, S., & Körndle, H. (2007). Interactivity and learners' achievement in web-based learning. *Journal of Interactive Learning Research, 18*(4), 511–531.

Rouet, J.-F., & Levonen, J. J. (1996). Studying and learning with hypertext: Empirical studies and their implications. In Rouet, J.-F., Levonen, J. J., Dillon, A., & Spiro, R. J. (Eds.), *Hypertext and cognition* (pp. 9–23). Hillsdale, NJ: Lawrence Erlbaum.

Rouet, J.-F., & Potelle, H. (2005). Navigational principles in multimedia learning. In Mayer, R. E. (Ed.), *The Cambridge handbook of multimedia learning* (pp. 297–312). New York: Cambridge University Press.

Salomon, G., & Almog, T. (1998). Educational psychology and technology: A matter of reciprocal relations. *Teachers College Record, 100*(2), 222–241.

Schraw, G. (2007). The use of computer-based environments for understanding and improving self-regulation. *Metacognition and Learning, 2*(2), 169–176. doi:10.1007/s11409-007-9015-8

Seijts, G. H., & Latham, G. P. (2001). The effect of distal learning, outcome, and proximal goals on a moderately complex task. *Journal of Organizational Behavior, 22*(3), 291–307. doi:10.1002/job.70

van Oostendorp, H. (1996). Studying and annotating electronic text. In Rouet, J.-F., Levonen, J. J., Dillon, A., & Spiro, R. J. (Eds.), *Hypertext and cognition* (pp. 137–147). Hillsdale, NJ: Lawrence Erlbaum.

Weinstein, C. E., Husman, J., & Dierking, D. R. (2000). Self-regulation interventions with a focus on learning strategies. In Boekaerts, M., Pintrich, P. R., & Zeidner, M. (Eds.), *Handbook of self-regulation* (pp. 727–747). San Diego, CA: Academic Press. doi:10.1016/B978-012109890-2/50051-2

Winne, P. H., & Hadwin, A. F. (1998). Studying as self-regulated learning. In Hacker, D. J., Dunlosky, J., & Graesser, A. C. (Eds.), *Metacognition in educational theory and practice* (pp. 277–304). Mahwah, NJ: Lawrence Erlbaum.

Winters, F., Greene, J., & Costich, C. (2008). Self-regulation of learning within computer-based learning environments: A critical analysis. *Educational Psychology Review, 20*(4), 429–444. doi:10.1007/s10648-008-9080-9

Zimmerman, B. J. (2000). Attaining self-regulation: A social cognitive perspective. In Boekaerts, M., Pintrich, P. R., & Zeidner, M. (Eds.), *Handbook of self-regulation* (pp. 13–39). San Diego, CA: Academic Press. doi:10.1016/B978-012109890-2/50031-7

KEY TERMS AND DEFINITIONS

Indirect Instructional Intervention: Indirect instructional interventions are inherent characteristics of a TELE which support learners in coping with the demands of self-regulated learning. Indirect instructional interventions can be either embedded or non-embedded into a TELE.

Learning Plan Tool: The learning plan tool of a Studierplatz TELE requires students to set learning goals, assign sub-goals, search for learning goal relevant materials, select and sequence these materials, as well as to evaluate their learning progress.

Metacognitive Tool: A metacognitive tool requires students to make instructional decisions regarding their learning. It supports the application of key self-regulatory processes of forethought, performance, and self-reflection.

Non-Embedded Instructional Intervention: Non-embedded instructional interventions are tools included within a TELE whose use is optional for the learner.

Self-Regulated Learning: Self-regulated learning is an active, constructive process in which students are required to set goals for their learning and then attempt to monitor, regulate, and control their cognition, motivation, and behavior.

Studierplatz: A Studierplatz is a technology-enhanced learning environment on a particular topic. It can be created by using the authoring tools s2w-compiler and EF-Editor (http://studierplatz2000.tu-dresden.de). A Studierplatz TELE automatically includes indirect instructional interventions supporting orientation and navigation, as well as key self-regulatory processes of forethought (e.g., learning plan tool), performance (e.g., learning tools), and self-reflection (e.g., interactive learning tasks).

ENDNOTE

1 We would like to thank Stevka Peters, Rüdiger Krauße, Anja Eichelmann, and Sandra Friedel for preparing the Studierplatz, as well as collecting and analyzing the data.

Chapter 20

Reference Course Model:
Supporting Self-Regulated Learning by Cultivating a University-Wide Media Culture

Per Bergamin
Swiss Distance University of Applied Sciences, Switzerland

Marco Bettoni
Swiss Distance University of Applied Sciences, Switzerland

Simone Ziska
Swiss Distance University of Applied Sciences, Switzerland

Cindy Eggs
Swiss Distance University of Applied Sciences, Switzerland

ABSTRACT

Since our mission is the collaborative cultivation of a university-wide media culture, in this chapter the authors propose to look at the relation between Self-regulated learning (SRL) and Technology-Enhanced Learning Environments (TELE) from the point of view of a learning organization. The goal is to clarify how to embed TELE-technologies in educational institutions in a collaborative way that sustains and continuously improves the quality of teaching and learning at a university. Our solution is focused around the concept of "university-wide media culture", a corporate culture for new media that we hope to develop by means of a collaborative instrument called the "Reference Course Model". The authors begin by screening and summarizing what they consider to be relevant aspects of components of the SRL theory (models, learning strategy, prompting) and continue by introducing the concepts of media culture, media literacy and their relation to TELE and SRL; based on this they then present their idea of what they call a "Reference Course Model", explaining its theoretical foundation and developing its conceptual features. Finally, they conclude by showing how they have implemented this model in their university and reflect on the experiences collected to-date.

DOI: 10.4018/978-1-61692-901-5.ch020

INTRODUCTION

Learning and teaching at a distance makes high demands to the learners, the teacher and the organisation in charge. From our experience at the Swiss Distance University of Applied Sciences, we agree with Garrison (2000) that one major challenge that theory and practice of distance education have to deal with, today, is a collaborative approach to learning (as opposed to individual learning) before, during and after the teaching and learning process.

On the other hand, since our students learn mostly part-time (because of family obligations and a nearly full-time employment), maintain a close connection with their profession and adopt a very practice-oriented attitude, they need to be more autonomous, more flexible in planning, more motivated, more able to organize their learning resources, more involved in setting their learning goals and more active in their learning. In short, they need (and wish) a higher ability and more opportunities to control their own learning. Furthermore in a study (Bergamin, Ziska & Groner, 2009) we have found three relevant factors of flexibility for university students: flexibility of time, flexibility of teacher contact and flexibility of content.

It was because of this challenge of supporting self-regulated learning (SRL) on one side, and at the same time promoting a practice of collaboration among teachers and among students on the other side, that projects aiming at cultivating a university-wide media culture were launched at our university.

SELF-REGULATED LEARNING

In the German and English speaking area we find different terms for self-regulated learning such as self-steered, independent, self-determined, autonomous, self-organized, self-directed learning etc. (Götz, 2006; Schreiber, 1998). This multiplicity hinders reaching a clear and consistent definition (Artelt, 2000). In our view one of the still most concise definitions originates from Knowles (1975, p.18): "...a process in which individuals take the initiative, with or without the help from others, in diagnosing their learning needs, formulating goals, identifying human and material resources, choosing and implementing appropriate learning strategies, and evaluating learning outcomes". One of the most remarkable points in this understanding of learning lies in the emphasis on the active role of the learners (Fischer & Mandl, 2002). But from the perspective of the teaching organisation, respectively of the teacher, we ought also to mention, that in practice a fully self-regulated as well as a fully externally regulated learning is impossible to realize. The student's learning actions are more appropriately conceived as moving over a continuum between the two poles self-regulation and external regulation (Schreiber, 1998; Artelt, 2000).

Therefore, what counts from an organisational perspective is the degree of expression of different characteristics, such as the orientation of the learning experience (learner orientation vs. teacher orientation), activity level of the learners (active learners vs. passive learners), time flexibility of the learners (flexible learning times vs. fixed learning times), freedom of decision concerning learning goals (learning goal autonomy vs. predetermined learning goals), design of the learning experience (decisional options vs. strict planning), assessment of the learning success (self-assessment vs. external assessment) during a learning process.

Models

Scientifically established models of SRL, mostly coming from a cognitive approach, try to describe the process of self-regulation, to explain the learning processes taking place and to relate the characteristics involved with the learning achievements. Between these concepts there are similarities but also differences (Boekaerts & Corno, 2005). One of the most fundamental differences lies in the allocation of traits respectively of situation specific variables to self-regulated learning actions. The first mentioned group of models can be called component models (see Boekaerts, 1997, 1999; Pintrich, 1999; 2000). In this approaches learning actions or more precisely "learning strategies" are regarded as traits of learners which can be observed and registered independently of situations. They are mainly developed by individuals in the adolescence and consolidated later on (Winne & Perry, 2000). The second group can be subsumed under the term of process models. In these approaches learning strategies are supposed to be context specific, they are applied differently in different tasks and are acquired in social learning (Zimmerman, 1998; 2000; Winne & Hadwin, 1998). As already brought up, the component models describe which elements are needed for a successful learning process. However they do not take into account in which phase which component is essential. According to Boekaerts (1999), self-regulated learning can be characterised basically through the following components:

a. Cognitive components which embrace conceptual and procedural knowledge as well as knowledge about task specific strategies and their application conditions (conditional knowledge).
b. Motivational components which serve for the initiation and the maintenance of learning activities and also incorporate assessments of achievements and beliefs concerning the effectiveness of personal learning
c. Metacognitive components which consist in part of knowledge about own abilities and the individual learning process, in part of planning, monitoring and regulation of personal actions according to the aimed learning goals.

Process models can be regarded as complementary. They often describe an „ideal" process of self-regulated learning. Self-regulative competence is in this context a circular process which is associated with motives and beliefs of self-efficacy. Feedback of previous learning experiences is used to prepare, to design and to monitor an upcoming learning process. For instance Zimmermann (1998) distinguishes here three phases.

In the "forethought" phase the individual prepares him- or herself for the proper action through goal setting and strategic planning. Self-related, motivational reflections about self-efficacy and outcome expectation as well as intrinsic interest and learning goal orientation moderate this process (Pintrich & De Groot, 1990, Zimmerman & Kintsantas, 1997). The "performance" phase is characterized especially by attention and action. The two volitional control processes of self-control and self-observation (Zimmermann, 2000) serve for the optimisation of perceptual and learning behaviour. The third phase of "self-reflection" consists of two main processes. The process of self-evaluation refers to the evaluation of the personal achievements and their causal attribution. Self-related reactions appraise the learning performances and the following actions on an affective level. Such process models exist in different variants. For instance Pintrich (2000) proposes four phases: planning, monitoring, control and evaluation.

As these two types of models show, the concept of self-regulated learning describes a very challenging learning strategy (Wirth & Leutner, 2008) in which students have to plan their learning process on their own, set their goals for themselves, activate their previous knowledge, search for learning resources and work on learning contents on their own according to their own pace. But another important function of the learning process is an active monitoring of the learning progress respectively of the learning outcomes. This means that the learners have to constantly adapt their learning activities to new requirements. They have to be able to cope with learning challenges, to motivate themselves and to manage and maintain their attention (Fischer & Mandl, 2002; Simons, 1992).

Research in self-regulation during the last 30 years has shown that especially metacognition respectively metacognitive learning strategies can be considered as a central component of SRL (Artelt, 2000; Borkowski, Chan & Muthukrishna, 2000; Schraw, Crippen & Hartley, 2006).

Strategic Processes in Form of Learning Strategies

As far as learning strategies and their impact on learning performances are concerned, a consistent view is still missing (Krapp, 1992; Artelt, 2000). The following classifications have been widely adopted in the literature:

- Primary and secondary strategies (Danserau, 1985)
- General and specific strategies (Friedrich & Mandl, 1992, Klauer, 1988)
- Higher and subordinate strategies (Leutner & Leopold, 2003)
- Deep and surface strategies (Schmeck, 1988).

In research about the relation between learning strategies and learning success, the most used classification originates from the "approaches to learning" conceptions and divides the learning strategies in five groups (Weinstein & Mayer, 1986): rehearsal, organization, elaboration, metacognition and resource management. Based on this classification Wild and Schiefele (1994) formulated three types of learning strategies:

- cognitive strategies (elaboration, organisation and rehearsal)
- support strategy (time management, configuration of the learning environment, effort)
- metacognitive strategies (planning, monitoring / control, evaluation / adaptation).

Without going deeper into the similarities and differences of the concepts mentioned, we can assume that higher functions of learning like planning, controlling and adapting are ascribed to metacognition (Veenman, van Hout-Wolters & Afflerbach, 2006). While metacognitive strategies, on one side, empower the learners to monitor and improve their progress, cognitive strategies, on the other side, serve to attain learning progress, as for example to create knowledge.

In this context it is also important to mention that metacognitive knowledge, the monitoring of learning actions and of learning outcomes and the related self-regulation play an important role in relation to the learning performance. Various investigations at an empirical level also showed that metacognitive strategies are tightly related to learning performances. Correlation studies confirm the postulated positive correlations between cognitive, metacognitive, motivational variables and learning success, but they don't shed enough light on the direction and the mechanisms of their causal and functional interaction (Boekaerts, 1999; Pintrich, 1999, Leutner & Leopold, 2002). However, the fact can be emphasized that metacognitive competences are partially independent from intelligence, therefore they constitute an entity which can be fostered rather well (Veenman, 1993; Veenman & Beishuizen, 2004).

In this context the following question arises: is there also evidence on an empirical level that corresponding metacognitive learning strategies can be trained and/or be stimulated to come to a better learning performance? A first quick answer can already be given. On the one hand there are intervention studies in which both strategy training and corresponding learning success has been observed. On the other hand some studies showed also that interventions more on a methodological level than training e.g. the introduction of journals or prompts could also lead to a better learning performance.

Fostering Strategic Knowledge

Within the framework of programs for improving student's ability to learn from texts by means of strategic measures Palincsar and Brown (1984) found positive effects on both text comprehension as well as on the transfer of acquired strategies. Bielaczyc, Pirolli and Brown (1995) proved the efficacy of strategy training for problem solving as regards the use of self-explanation and self-regulation strategies and also as regards the learning success in computer programming. Leutner, Barthel and Schreiber (2001) found positive effects in a training which consisted in fostering motivation strategies and their regulation. They reported also positive effects on the utilisation of learnt strategies as well as on motivation and text comprehension. Further Perels, Gürtler and Schmitz (2005) observed positive effects of a combined self-regulation and problem solving training on self-reported self-regulation competences.

Based on the empirical research findings in the transfer of metacognitive strategies Veenman, van Hout-Wolters and Afflerbach (2006) summarize three main points as indicators of successful strategy trainings:

- Strategy instruction should be embedded in a content-based learning context
- The learners need to be informed about the usefulness of the learning strategies which were part of the training program in order to increase their willingness to engage in using the strategy
- The trained strategies should be practised extensively.

Hence we can state at least in principle that strategy trainings have proven to be effective instruction measures. At the same time it should also be noted that training programs which combine learning strategies of higher order, which are not linked directly to the learning content but more to planning and to organizing learning processes (goal setting, self-observation, self-estimation) combined with learning factors that help to maintain learning processes (motivation information, information processing, saving and recall and transfer) are more effective than programs regarding the comprehension of the learnt content. Beside the knowledge about learning strategies, another important issue is the strategy use itself. This means that training should also include sequences which allow to learn both to distinguish in which situations special learning strategies are effective and to be able to transfer them spontaneously to other situations. For the learner herself the correct and autonomous application of strategies and the transfer to new contexts is essential. This shows that a neat and extended practice is essential to avoid overlap effects of newly learnt strategies.

Prompting

Under the concept of "prompting" we understand an instruction measure which encourages learners during the learning process to activate cognitive, metacognitive and motivational learning processes. In contrast to the above described approach of fostering strategic knowledge it is assumed here that learners already know learning strategies which support the learning process effectively but often do not use them spontaneously. Prompting provides in this sense are an opportunity to foster self-regulation strategies "indirectly".

The encouragement of self-explanation constitutes the main part of prompting studies. In a seminal work Chi, De Leeuw, Chiu and LaVancher (1994) show that, when reading a text, prompting for self-explanation through tutors positively influences deeper text comprehension. In a newer study Schworm and Renkl (2006) showed that self-explanations which have been evoked through prompts are more effective compared with self-generated but externally presented explanations. Nowadays the effects of prompts are often also studied in connection with computer based learning environments. For instance positive impacts of prompts had already been proved by Renkl, Atkinson and Maier (2000) in learning transfer tasks, by Aleven and Koedinger (2002) in learning solution examples in the context of problem solving, by Azevedo, Cromley and Seibert (2004) for learning complex science topics (human circulatory system) and by Gerjets, Scheiter and Schuh (2005) for learning with exemplars. Other studies point especially to the effects of metacognitive prompts. Lin and Lehman (1999) compared the impact of cognitive, metacognitive and emotion focused prompts during learning in the context of a computer based learning environment in biology. What came out was a superiority of metacognitive prompts as regards learning success as well as regarding transfer performances. Davis and Linn (2000) compared effects of metacognitive prompts to activity oriented prompts during learning in a computer based learning environment with natural science content. The superiority of metacognitive prompts was also confirmed. In this context not only effects on self-explanation were shown but also on planning. Moreover Bannert (2003) demonstrated in an experiment that during learning in a hypermedia learning environment the utilisation of metacognitive strategies is more frequent and also the application of learnt content is better when learners get metacognitive prompts by an information sheet which was allocated beside the computer compared to learner which didn't get this information.

As we can see from the studies presented here, metacognitive prompts are an efficient and an economic method to stimulate metacognitive processes during learning. A positive correlation appeared especially with self-explanation and with the planning of learning processes. Interesting is the fact that externally guided presentations appeared to be more effective than self-regulated ones. It is indeed to ask if this is a contradiction. We assume at the moment that the pre-knowledge regarding the content could play an important and facilitating role. We presume also that prompting can not only foster knowledge about learning strategies but also knowledge about the usefulness of strategies and through this also promote the transfer to new situations.

Basically, on the background of the findings presented above it can be asserted that the method of prompting is an efficient way for stimulating particularly metacognitive processes esp. in promoting their utilisation when this does not happen spontaneously. But in our view the most effective way to foster self-regulated learning is a combination of direct promotion (training) and indirect promotion (prompts). Therefore we have also to look at factors of the learning environment and the related organizational forms.

Environmental Factors

Based on Bandura's social cognitive perspective (1986) Zimmermann (1989; 2000) postulates that SRL is a triadic interaction of variables of the person (self), the behaviour (action) and the environment. In his concept he assumes a cyclic process between these variables. While the regulation of behaviour consists of self-monitoring processes and the alignment of strategic options for acting, the regulation of processes internally allocated in the person is in relation with cognitive and affective states. The regulation of the environment consists in its observation and accommodation by the individuals (Figure 1).

There are other authors who give also environmental factors an important role in SRL. Friedrich and Mandl (1997) point out, that learning takes always place in more or less structured learning environments. They point to persons, institutions, media as well as instructional arrangements like methods, exercises, learning sequences etc. as relevant environmental factors. In further approaches it is similarly postulated that the environment influences also situational factors of learning processes and motivation (Nenninger, 1999; Perels, Gürtler & Schmitz, 2005, Wosnitza, 2000). Based on our own investigations (Bergamin, Ziska & Groner, 2009) we

Figure 1. Interaction between personal and environmental factors (Adapted from Zimmerman, 1989)

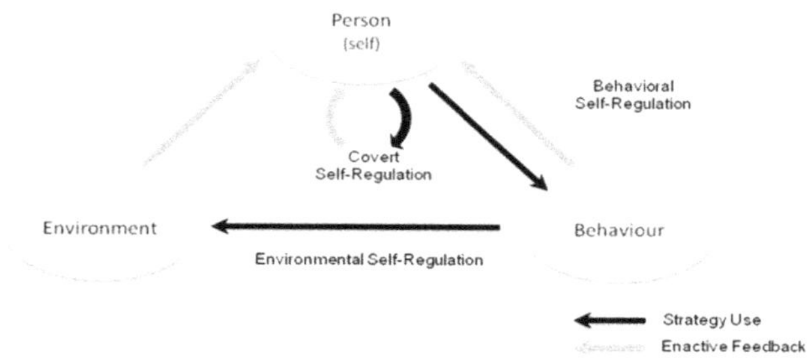

assume that additionally factors of flexibility (time, content, teacher contact) play a major role.

If we assume, as already described in a previous section, that instructional interventions with media have high potential to facilitate SRL, then the question arises how the findings presented above could be integrated in the cultivation of entire organisational processes.

MEDIA LITERACY AND MEDIA CULTURE IN AN ORGANSATIONAL PERSPECTIVE

After the initial enthusiasm for teaching and learning with new media in the nineties and a subsequent phase of disillusionment (Schnotz, Seufert & Bannert, 2001), the application of ICT in education has now become more precise and purposeful. Educational institutions understand now that ICT solutions - for example Technology-Enhanced Learning Environments (TELEs) - should not be used in all situations merely in order to justify their purchase or because others are working with them, instead they have to be applied only where they facilitate the student's learning process. This however, presupposes a proficient use of TELEs by both lecturers and learners or, more generally, an appropriate level of media literacy.

In our view - in line with Sutter and Charlton (2002) – we distinguish and concentrate on five fundamental dimensions of media literacy: understanding media, controlling media, using media, designing and evaluating media. Trying to clarify what skills are behind this multidimensional description, basically four types can be distinguished (Kübler, 1999):

- **Cognitive skills:** Knowledge about structures, organizational forms and functioning, as well as about processing and content of media
- **Analytical and evaluative skills:** Abilities to assess and evaluate media – especially their content - based on a variety of criteria
- **Social reflexive skills:** Abilities that each person, families and even groups should develop with the media or in terms of their content. They can be trained and learned by experiencing, observing and becoming aware of individual uses, habits, needs, temptations and compensations (see also emotional intelligence).

The question is now, how these skills can be developed in the context of the organization of a university and by means of organizational development. This involves educational issues and arguments within the framework of sustainable development (Euler, 2004).

There are various proposals for the development of media literacy for university lecturers (e.g., Schulmeister, 2005). We present three of them, which represent different perspectives: a) the individual perspective (individual competence), b) the teaching perspective (teaching methods) and c) the sustainability perspective (sustainable change).

The Individual Perspective

In Albrecht (2003) the main focus is put on the qualification of teachers. It is assumed that during the development of teachers' media literacy (in various publications the term e-competence is used instead of media literacy - this is because of the restriction in the use of media to the new ICTs), four constitutive aspects of the entire action field of eLearning are involved.

For every aspect, skills to make educationally appropriate decisions are needed. For the aspect of technology, the issues at stake are the evaluation and use of basic ICT technologies and of eLearning technologies, e.g. learning platforms, groupware, authoring tools, etc. For the aspect of course organisation, it is a matter of decisions concerning the implementation of teaching methods like for example the enrichment of face to face lectures with media, on-site teaching alternated with virtual forms of teaching, or delivering a "pure virtual teaching". But it is also a matter of defining fundamental didactical strategies, like for example putting the focus either on guided, on self-controlled, or on collaborative learning. Just at his point it is very important to set a link to the principles of SRL for example by recognizing that it is important to build up the knowledge of teachers about SRL and also to cultivate a positive attitude towards the effectiveness of relevant learning and teaching actions. And last but not least, it is also important to assess both appropriate learning materials and a learning environment in which relevant exercises are supported.

This approach, which stresses the promotion of skills by offering qualification opportunities for the individuals involved in eLearning and SRL, has the advantage of simple practicability. However, for a sustainable skills development program, additional measures are also needed as we will show in the next sections.

The Teaching Perspective

Bremer (2006) formulates to this end a virtualization strategy. She assumes that with the introduction of eLearning, strategic decisions have to be taken. Accordingly and in view of the growing virtualization of teaching methods especially at universities, three typical concepts which differ in details can be formulated. They reach from the enrichment of classroom teaching up to completely internet-based, virtual classes (Figure 2).

When it comes to the issue of the real implementation of strategic concepts, questions arise not only concerning what university lecturers have to be capable of, but also how the desired goals can be achieved. The background of this question can be found in the context of Weinert's definition of competence, according to which competence is not restricted to some skills on a cognitive level, but entails also an acting orientation connected with motivational, volitional and social dispositions and abilities. Accordingly we propose here (see Figure 2) that, in addition to the qualification in the course design phase, a good support system should be introduced in the realisation phase consisting of media-didactic advice and training, technical advice and support of the course production process in relation with implementing methods for fostering SRL. In the implementation phase we furthermore suggest to assess and improve its use in teaching by means of institutionalised evaluation processes.

Finally, we propose that user requirements of the prospective learner must be taken into consideration in every step of the development by progressively optimizing the usability (Groner, Raess & Sury, 2008).

The quality of the entire development process of eLearning and SRL is not just a matter of increasing the proficiency of the teachers, but one of fostering the competence of the organization

Figure 2. Concepts of virtualization of university teaching (Adapted from Bremer, 2004)

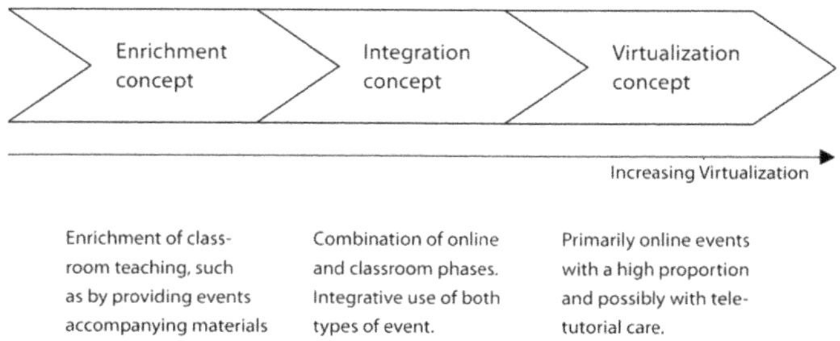

as a whole. This can be achieved by means of appropriate supportive offerings (e.g. for ensuring the quality of processes, contents, learning forms, etc.) made at the organisational level, in addition to strategic decisions.

The Sustainability Perspective

If we assume that the qualifications and parallel supportive offerings outlined in the previous sections have been introduced, then we have reached the point where the issue of sustainability comes into play, i.e. the aim of consolidating respectively changing the mentioned processes into permanent activities. This however is not a matter of stability of the activity, but rather one of durability of structures, which leads to a sustained change in teaching. The goal is to use outcomes and insights not only in the participating units but also in other units both during and after a course project.

Euler (2004) proposes the establishment of a culture for the development of e-skills at universities. In our view this means integrating new media and SRL in the daily teaching and learning process and the consequent adjustment of habits and attitudes of teachers and learners. This dimension also includes the sustainable conservation and further development of the achieved outcomes. In this sense, one can also speak of a learning organization.

Media Culture Involving SRL as a Basic Topic

Our experience shows that, for the development and sustainable implementation of media literacy, all three perspectives should be taken into account in reference to the concept of the virtualisation of teaching and learning (Bremer, 2006): taken together these elements and their relations constitute what we mean by the term “media culture”. At the same time it is also important to link these perspectives with the development of methodological approaches. That’s why we propose, while developing media literacy to strongly link the related activities with the methodological approach of SRL. Therefore we basically consider three kinds of actions:

1. training actions for developing knowledge and competence in media literacy and SRL
2. providing support services
3. organisational learning via research in the use of media to support SRL and the consequent transfer of findings to support the daily teaching and learning.

REFERENCE COURSE MODEL

Taking into account the above mentioned concepts, perspectives and actions, we now present a model called the “Reference Course Model” (RCM) and

Figure 3. The didactic tetrahedron and its relationship factors

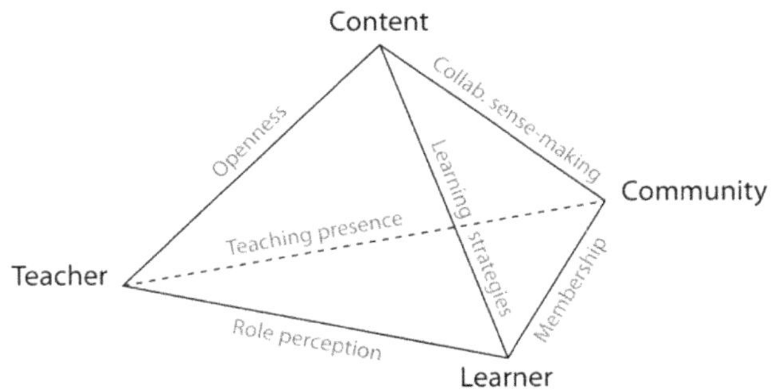

some first experiences, showing how at the Swiss Distance University of Applied Sciences media literacy and media culture are being developed and linked to SRL.

The term "reference" means in our understanding a reference system for the implementation of online courses in a blended learning scenario by means of so-called reference courses. A "reference course" is a generic TELE course, technically implemented as a course template on the learning platform (a Moodle course); this template is given to individual lecturers as a starting point for implementing their own individual courses. What we call a "Reference Course Model" then, is the approach or system that specifies principles, structures and procedures needed for producing such a template or "reference course".

The conceptual basis of teaching at the Swiss Distance University of Applied Sciences is constituted by the "Didactic tetrahedron" (Bergamin & Brunner-Amacker, 2007). In that approach (Figure 3), we assume that, in addition to the interaction of the three elements "Teacher", "Learner" and "Content" - as in the classical teaching triangle displayed and revised for the digital future by Haugan & Hopmann (2004) - an additional, fourth element plays a constitutive role in teaching and learning: the element of "Community".

Against this background we were able to define a first important component of the Reference Course Model: four principles used as a foundation for designing the structures and procedures needed for producing a reference course. These principles are:

a. prequalification of lecturers by means of workshops
b. gradual implementation of e-learning in the classroom, by using a variety of different tools and a mix of approaches
c. teaching experience and application of the model are considered as parallel, collaborative processes; they are supported by an online exchange in a community of teachers dealing with the topic of course development and revision
d. a suitable "media culture as corporate culture" must necessarily be cultivated also by sharing experience, knowledge and best practice.

The main features of a reference course that is produced within the context of our Reference Course Model are:

a. *Didactic standards for teaching:* Beside content transfer, the standards include also principles of self-regulated learning. In particular the main issue to consider is the strategic knowledge of the students. This is achieved by doing appropriate exercises twice to three times each semester in the first

Figure 4. Process for the sustainable use of online teaching materials and activities

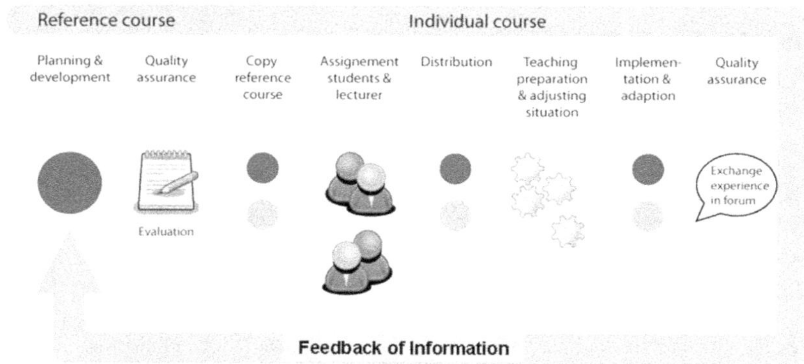

two academic years. This is supplemented through a facultative offer for fostering learning strategies.

b. *Guidelines and methodological standards for materials (texts, exercises, examples, etc.):* All the respective texts and exercises contain prompts either as a self-request for reflection or as cues to promote metacognitive competence.

c. *Recommendations for interaction and communication in the courses:* By disseminating through the course platform in the "course view" for teachers hints and tricks as well as opportunities for support.

Through this design, the qualification of lecturers and developers of the reference course as well as the organizational processes of planning, realization, implementation and evaluation (quality assurance) of the courses are integrated in a feedback loop (see Figure 4).

In particular, our model is constituted by a set of 10 measures:

1. Training of the reference course developers by the Institute for Research in Open, Distance, and eLearning (IFeL) as well as by experts
2. Qualification of lecturers through the continuous education program of the Swiss University of Applied Sciences
3. Drafting and revision of the reference courses by experts in the domain
4. Monitoring the planning of the reference course by staff from the Quality Assurance Division
5. Support of the reference course developer by an external technical and educational service
6. Evaluation of the reference course by the IFeL Research Institute
7. Release of individual courses (including student assignment) by a central service
8. Providing information about the individual courses by the faculties
9. Discussion, analysis and systematization of experiences in the course implementation by the faculty
10. Feedback of the experiences and knowledge acquired in the development of the individual course implementation for planning and development of the next reference courses.

Through the continuous reuse, revision and adaptation of reference courses, a "university-wide media culture" is emerging. Within this "media culture", emphasis is placed on the efficiency by self-regulated learning principles in the development of online teaching material, on the openness in the didactical and methodical usage as well as on stability of both the technical and organizational processes.

CONCLUSION

The reference courses that are produced by the process defined in our Reference Course Model can be considered as a kind of "boundary object" in the sense that Leigh Star conceived when she coined this term (Star & Griesemer, 1989): they support communication and serve to coordinate the perspectives of all stakeholders involved in the process of designing, realising and implementing their own individual technology enhanced online courses within a blended learning scenario.

As a consequence of this approach we introduced at the same time also a strategic opportunity on a methodological level: SRL. Therefore our courses are conceptualised and assessed not merely as units of teaching but also as opportunities for collaborative processes that can foster the development of a university-wide media culture and by that satisfy the SRL needs of our students: being more autonomous by supporting the setting of learning goals, promoting the planning of learning phases, helping organizing learning resources and maintaining motivation.

Delfino and Persico (2007) have shown that online collaborative learning can be designed in such a way that it encourages both the individual learners and the virtual community to gradually take control of their own learning. This kind of design – we claim here – can emerge and be more easily cultivated thanks to the collaborative framework and nurturing provided by our Reference Course Model.

REFERENCES

Albrecht, R. (2003). *E-Learning an Präsenzhochschulen. Die Implementierung von E-Learning aus hochschuldidaktischer Perspektive (E-Learning at Presence Universities. The Implementation of E-Learning from the perspective of University Didactics)*. Berlin, DE: Verlag im Internet GmbH.

Aleven, V., & Koedinger, K. R. (2002). An effective metacognitive strategy: Learning by doing and explaining with a computer-based Cognitive Tutor. *Cognitive Science, 26*(2), 147–179. doi:10.1207/s15516709cog2602_1

Artelt, C. (2000). *Strategisches Lernen (Strategic Learning)*. Münster, DE: Waxmann.

Azevedo, R., Cromley, J. G., & Seibert, D. (2004). Does adaptive scaffolding facilitate students' ability to regulate their learning with hypermedia? *Contemporary Educational Psychology, 29*(3), 344–370. doi:10.1016/j.cedpsych.2003.09.002

Bandura, A. (1986). *Social foundations of thought and action: A social cognitive theory*. Englewood Cliffs, NJ: Prentice-Hall.

Bannert, M. (2003). Effekte metakognitiver Lernhilfen auf den Wissenserwerb in vernetzten Lernumgebungen (Effects of metacognitive learning aids on the acquisition of knowledge in interconnected learning environments). *Zeitschrift fur Padagogische Psychologie, 17*(1), 13–25. doi:10.1024//1010-0652.17.1.13

Bergamin, P., & Brunner-Amacker, B. (2007). Medienkompetenz: Der Schlüsselfaktor zur Umsetzung von Blended Learningszenarien an Hochschulen (Media Literacy: The Key Factor for Implementing Bleanded Learning Scenarios at Universities). In P. Bergamin & G. Pfander (Eds.), *Medien im Bildungswesen* (pp. 15-31). Bern, CH: hep.

Bergamin, P., Ziska, S., & Groner, R. (2009). Structural equation modelling of factors affecting success in Students performance in ODL Programs: Extending quality Management Concepts. Paper presented at ICDE 23rd World Conference on Open Learning and Distance Education, Maastricht, June 2009.

Bielaczyc, K., Pirolli, P. L., & Brown, A. L. (1995). Training in self-explanation and self-regulation strategies: Investigating the effects of knowledge acquisition activities on problem solving. *Cognition and Instruction, 13*(2), 221–252. doi:10.1207/s1532690xci1302_3

Boekaerts, M. (1997). Self-regulated learning: A new concept embraced by researchers, policy makers, educators, teachers, and students. *Learning and Instruction, 7*(2), 161–186. doi:10.1016/S0959-4752(96)00015-1

Boekaerts, M. (1999). Self-regulated learning: Where are we today. *International Journal of Educational Research, 31*(6), 445–457. doi:10.1016/S0883-0355(99)00014-2

Boekaerts, M., & Corno, L. (2005). Self-regulation in the classroom: A perspective on assessment and intervention. *Applied Psychology: An International Review, 54*(2), 199–231. doi:10.1111/j.1464-0597.2005.00205.x

Borkowski, J. G., Chan, L. K. S., & Muthukrishna, N. (2000). A process-oriented model of metacognition: Links between motivation and executive functioning. In Schraw, G., & Impara, J. C. (Eds.), *Issues in the measurement of metacognition* (pp. 1–43). Lincoln, NE: Buros Institute of Mental Measurements.

Bremer, C. (2004). E-Learning Strategien als Spannungsfeld für Hochschulentwicklung, Kompetenzansätze und Anreizsysteme (E-Learning Strategies Area of Tension for University Development, Literacy Approaches and Incentive Systems). In Bremer, C., & Kohl, K. (Eds.), *E-Learning Strategien - E-Learning Kompetenzen an Hochschulen (E-Learning Strategies – E-Learning Literacy at Universities)* (pp. 9–30). Bielefeld, W.: Bertelsmann Verlag.

Bremer, C. (2006). Qualitätsicherung im eLearning: Implementierungsansätze für die Hochschule (Quality Assurance in eLearning: Implementation Approaches for the University). In Sindler, A. (Eds.), *Qualitätssicherung im eLearning (Quality Assurance in eLearning)* (pp. 185–2002). Münster, DE: Waxmann.

Chi, M. T. H., De Leeuw, N., Chiu, M.-H., & LaVancher, C. (1994). Eliciting self-explanations improves understanding. *Cognitive Science, 18*(3), 439–477.

Dansereau, D. F. (1985). Learning strategy research. In Segal, J. W., Chipman, S. F., & Glaser, R. (Eds.), *Thinking and learning skills, 1: Relating instruction to research* (pp. 209–239). Hillsdale, NJ: Erlbaum.

Davis, E. A., & Linn, M. C. (2000). Scaffolding students' knowledge integration: Prompts for reflection in KIE. *International Journal of Science Education, 22*(8), 819–837. doi:10.1080/095006900412293

Delfino, M., & Persico, D. (2007). Designing and Running Online Collaborative Courses That Support SRL Development. In Beishuizen, J., Carneiro, R., & Steffens, K. (Eds.), *Self-regulated Learning in Technology Enhanced Learning Environments: Individual learning and Community of Learners* (pp. 26–39). Aachen: Shaker.

Euler, D. (2004). Einfach, aber nicht leicht – Kompetenzentwicklung im Rahmen der Implementierung von E-Learning an Hochschulen (Simple but not easy – Competence Development in the Context of the Implementation of E-learning in Universities). In Bett, K., Wedekind, J., & Zentel, P. (Eds.), *Medienkompetenz für die Hochschullehre (Media Literacy for E-Learning in Universities)* (pp. 55–71). Münster, DE: Waxmann.

Fischer, F., & Mandl, H. (2002). Lehren und Lernen mit neuen Medien (Teaching and Learning with new media). In Tippelt, R. (Ed.), *Handbuch Bildungsforschung (Handbook of educational research)* (pp. 623–637). Opladen, DE: Leske & Budrich.

Friedrich, H. F., & Mandl, H. (1992). Lern- und Denkstrategien – ein Problemaufriss (Learning and Teaching Strategies – A Problem Breakdown. In Mandl, H., & Friedrich, H. F. (Eds.), *Lern- und Denkstrategien: Analyse und Intervention (Learning and Thinking Strategies: Analysis and Intervention)* (pp. 3–54). Göttingen, DE: Hogrefe.

Friedrich, H. F., & Mandl, H. (1997). Analyse und Förderung selbstgesteuerten Lernens (Analysis and Promotion of Self-regulated Learning). In Weinert, F. E., & Mandl, H. (Eds.), *Psychologie der Erwachsenenbildung, Enzyklopädie der Psychologie (Psychology of Adult Education)* (*Vol. 4*, pp. 237–293). Göttingen, DE: Hogrefe.

Garrison, R. (2000). Theoretical Challenges for Distance Education in the 21st Century: A Shift from Structural to Transactional Issues. *International Review of Research in Open and Distance Learning*, *1*(1), 1–17.

Gerjets, P. H., Scheiter, K., & Schuh, J. (2005). Instruktionale Unterstützung beim Fertigkeitserwerb aus Beispielen in hypertext-basierten Lernumgebungen (Instructional Support for Skills Acquisition from Examples in Hypertext-based Learning Environments). *Zeitschrift fur Padagogische Psychologie*, *19*(1-2), 23–38. doi:10.1024/1010-0652.19.12.23

Götz, T. (2006). *Selbstreguliertes Lernen. Förderung metakognitiver Kompetenzen im Unterricht der Sekundarstufe (Self-regulated Learning. Promoting metacognitives Skills in Teaching at Secundary Level)*. Donauwörth, DE: Auer.

Groner, R., Raess, R., & Sury, P. (2008). Usability: Systematische Gestaltung und Optimierung von Benutzerschnittstellen (Usability: Systematic Design and Optimization of User Interfaces). In Batinic, B., & Appel, M. (Eds.), *Lehrbuch Medienpsychologie (Textbook Media Psychology)* (pp. 426–448). Berlin, DE: Springer.

Haugan, J., & Hopmann, S. (2004). Die allgemeine Didaktik im Zeitalter von Online-Lernen: Digitale Zukunft.... analoge Vergangenheit (General Didactics in Online-Learning Times: Digital Future... Analogical Past. In Rinn, U., & Meister, D. M. (Eds.), *Didaktik und Neue Medien (Didactics and New Media)* (pp. 72–83). Münster, DE: Waxmann.

Klauer, K. J. (1988). Teaching for learning-to-learn: A critical appraisal with some proposals. *Instructional Science*, *17*(4), 351–367. doi:10.1007/BF00056221

Knowles, M. S. (1975). *Self-directed learning. A guide for learners and teachers*. New York, NY: Association Press.

Krapp, A. (1992). Konzepte und Forschungsansätze zur Analyse des Zusammenhangs von Interesse, Lernen und Leistung (Concepts and Research Approaches for the Analysis of the Relationship between Interest, Learning and Performance). In Krapp, A., & Prenzel, M. (Eds.), *Interesse, Lernen und Leistung (Interest, Learning and Performance)* (pp. 9–52). Münster, DE: Aschendorff.

Kübler, H.-D. (1999). Medienkompetenz – Dimensionen eines Schlagwortes (Media Literracy – Dimensions of a Catchword. In Schell, F., Stolzenbrug, E., & Theunert, H. (Eds.), *Medienkompetenz. Grundlagen und pädagogisches Handeln (Media Literacy. Basics and Pedagogic Acts* (pp. 25–47). München, DE: KoPäd.

Leutner, D., Barthel, A., & Schreiber, B. (2001). Studierende können lernen, sich selbst zum Lernen zu motivieren: Ein Trainingsexperiment (Students can learn to motivate themselves for learning: a Training Experiment). *Zeitschrift fur Padagogische Psychologie*, *15*, 155–167. doi:10.1024//1010-0652.15.34.155

Leutner, D., & Leopold, C. (2002). Selbstreguliertes Lernen: Lehr-/Lerntheoretische Grundlagen (Self-regulated Learning: Teaching/Learning Basics). In Witthaus, U., Wittwer, W., & Espe, C. (Eds.), *Selbstgesteuertes Lernen: Theoretische und praktische Zugänge (Self-regulated learning: Theoretical and practical approaches)* (pp. 43–67). Bielefeld, DE: Bertelsmann.

Leutner, D., & Leopold, C. (2003). Selbstreguliertes Lernen als Selbstregulation von Lernstrategien - ein Trainingsexperiment mit Berufstätigen zum Lernen aus Sachtexten (Self-regulated Learning as a Self-regulation of learning Strategies – a Training Experiment with Employed People About Learning from Subject Texts). *Unterrichtswissenschaft*, *37*(1), 38–56.

Lin, X., & Lehman, J. D. (1999). Supporting Learning of Variable Control in a Computer-Based Biology Environment: Effects of Prompting College Students to Reflect on Their Own Thinking. *Journal of Research in Science Teaching*, *36*(7), 837–858. doi:10.1002/(SICI)1098-2736(199909)36:7<837::AID-TEA6>3.0.CO;2-U

Nenniger, P. (1999). On the role of motivation in self-directed learning. The 'Two-Shells-Model of motivated self-directed learning' as a structural explanatory Concept. *European Journal of Psychology of Education*, *14*(1), 71–86. doi:10.1007/BF03173112

Palincsar, A. S., & Brown, A. L. (1984). Reciprocal teaching of comprehension-fostering and comprehension-monitoring activities. *Cognition and Instruction*, *1*(2), 117–175. doi:10.1207/s1532690xci0102_1

Perels, F., Gürtler, T., & Schmitz, B. (2005). Training of self-regulatory and problem solving competence. *Learning and Instruction*, *15*(2), 123–139. doi:10.1016/j.learninstruc.2005.04.010

Pintrich, P. R. (1999). The role of motivation in promoting and sustaining self-regulated learning. *International Journal of Educational Research*, *31*(6), 459–470. doi:10.1016/S0883-0355(99)00015-4

Pintrich, P. R. (2000). The role of goal orientation in self-regulated learning. In Boekaerts, M., Pintrich, P. R., & Zeidner, M. (Eds.), *Handbook of Self-Regulation* (pp. 451–502). San Diego, CA: Academic Press. doi:10.1016/B978-012109890-2/50043-3

Pintrich, P. R., & De Groot, E. V. (1990). Motivational and self-regulated learning components of classroom academic performance. *Journal of Educational Psychology*, *82*(1), 33–40. doi:10.1037/0022-0663.82.1.33

Renkl, A., Atkinson, R. K., & Maier, U. H. (2000). From studying examples to solving problems: Fading worked-out solution steps helps learning. In L. Gleitman & A.K Joshi (Eds.), *Proceeding of the 22nd Annual Conference of the Cognitive Science Society* (pp. 393-398). Mahwah, NJ: Erlbaum.

Schmeck, R. R. (1988). An introduction to strategies and styles of learning. In Schmeck, R. R. (Ed.), *Learning strategies and learning styles* (pp. 3–19). New York, NY: Plenum Press.

Schnotz, W., Seufert, T., & Bannert, M. (2001). Lernen mit Multimedia: Pädagogische Verheissungen aus kognitionspsychologischer Sicht (Learning with Multimedia: Pedagogical Promises from a cognitive psychology viewpoint). In Silbereisen, R. K., & Reitzle, M. (Eds.), *Psychology 2000. Bericht über den 42. Kongress der Deutschen Gesellschaft für Psychologie (Report on the 42 Congrss of the german Society for Psychology)* (pp. 457–467). Lengrich, DE: Papst.

Schraw, G., Crippen, K. J., & Hartley, K. (2006). Promoting self-regulation in science education: Metacognition as part of a broader perspective on learning. *Research in Science Education*, *36*(1-2), 111–139. doi:10.1007/s11165-005-3917-8

Schreiber, B. (1998). *Selbstreguliertes Lernen (Self-regulated Learning)*. Münster, DE: Waxmann.

Schulmeister, R. (2005). Welche Qualifikationen brauchen Lehrende für die „Neue Lehre"? Versuch einer Eingrenzung von eCompetence und Lehrqualifikation (Which Qualifications Need Teachers for the „New Doctrine"? Attempt at Scoping eCompetence and Teachers Qualifications. In Kerrres, M., & Keil-Slawik, R. (Eds.), *Hochschulen im digitalen Zeitalter: Innovationspotentiale und Strukturwandel (Universities in the Digital Age: Innovation Potentials and Structural Change* (pp. 215–234). Münster, DE: Waxmann.

Schworm, S., & Renkl, A. (2006). Computer-supported example-based learning: When instructional explanations reduce self-explanations. *Computers & Education*, *46*(4), 426–445. doi:10.1016/j.compedu.2004.08.011

Simons, P. R. J. (1992). Lernen selbständig zu lernen - ein Rahmenmodell (Learning to learn autonomously- A Frame Model. In Mandl, H., & Friedrich, H. F. (Eds.), *Learning and Thinking Strategies: Analysis and Intervention* (pp. 251–264). Göttingen, DE: Hogrefe.

Star, S. L., & Griesemer, J. (1989). Institutional ecology, "translations" and boundary objects: amateurs and professionals in Berkele's museum of vertebrate zoology 1907-1939. *Social Studies of Science*, *19*(3), 387–420. doi:10.1177/030631289019003001

Sutter, T., & Charlton, M. (2002). Medienkompetenz – und einige Anmerkungen zum Kompetenzbegriff (Media Literacy – and some Notes on the Concept of Literacy). In Groeben, N., & Hurrelmann, B. (Eds.), *Medienkompetenz. Voraussetzungen, Dimensionen, Funktionen (Media Literacy: Premises, Dimensions, Functions* (pp. 129–147). Weinheim, DE: Juventa.

Veenman, M. V. J. (1993). *Intellectual ability and metacognitive skill: Determinants of discovery learning in computerized learning environments*. Amsterdam, NL: University of Amsterdam.

Veenman, M. V. J., & Beishuizen, J. J. (2004). Intellectual and metacognitive skills of novices while studying texts under conditions of text difficulty and time constraint. *Learning and Instruction*, *14*(6), 619–638. doi:10.1016/j.learninstruc.2004.09.004

Veenman, M. V. J., van Hout-Wolters, B., & Afflerbach, P. (2006). Metacognition and learning: Conceptual and methodological considerations. *Metacognition and Learning*, *7*, 3–14. doi:10.1007/s11409-006-6893-0

Weinstein, C. E., & Mayer, R. E. (1986). The teaching of learning strategies. In Wittrock, M. C. (Ed.), *Handbook of research in teaching* (pp. 315–327). New York, NY: Macmillan.

Wild, K.-P., & Schiefele, U. (1994). Lernstrategien im Studium: Ergebnisse zur Faktorenstruktur und Reliabilität eines neuen Fragebogens (Learning Strategies in academic Studies). *Zeitschrift für Differentielle und Diagnostische Psychologie*, *75*(4), 185–200.

Winne, P. H., & Hadwin, A. F. (1998). Studying as self-regulated learning. In Hacker, D. J., Dunlosky, J., & Graesser, A. C. (Eds.), *Metacognition in education theory and practice* (pp. 277–304). Mahwah, NJ: Erlbaum.

Winne, P. H., & Perry, N. E. (2000). Measuring self-regulated learning. In Boekaerts, M., Pintrich, P., & Zeidner, M. (Eds.), *Handbook of self-regulation* (pp. 531–566). San Diego, CA: Academic Press. doi:10.1016/B978-012109890-2/50045-7

Wirth, J., & Leutner, D. (2008). Self-Regulated Learning as a Competence. Implications of Theoretical Models for Assessment Methods. *The Journal of Psychology, 216*(2), 102–110.

Wosnitza, M. (2000). *Motiviertes selbstgesteuertes Lernen im Studium. Theoretischer Rahmen, diagnostisches Instrumentarium und Bedingungsanalysem (Erziehungswissenschaft 5) (Motivated, self-regulated Learning in university Studies. Theoretical Framework, Diagnostic Instruments and Conditional Analyses)*. Landau, DE: Verlag Empirische Pädagogik.

Zimmerman, B. J. (1989). A social cognitive view of self-regulated academic learning. *Journal of Educational Psychology, 81*(3), 329–339. doi:10.1037/0022-0663.81.3.329

Zimmerman, B. J. (1998). Developing self-fulfilling cycles of academic regulation: Ananalysis of exemplary instructional models. In Schunk, D. H., & Zimmerman, B. J. (Eds.), *Self-regulated learning: From teaching to self-reflective practice* (pp. 1–19). New York, NY: Guilford Publications Inc.

Zimmerman, B. J. (2000). Attaining self-regulation: A social cognitive perspective. In Boekaerts, M., Pintrich, P. R., & Zeidner, M. (Eds.), *Handbook of self-regulation* (pp. 13–39). San Diego, CA: Academic Press. doi:10.1016/B978-012109890-2/50031-7

Zimmerman, B. J., & Kitsantas, A. (1997). Developmental phases in self-regulation: Shifting from process to outcome goals. *Journal of Educational Psychology, 89*(1), 29–36. doi:10.1037/0022-0663.89.1.29

KEY TERMS AND DEFINITIONS

Didactic Tetrahedron: A constructivist model of teaching and learning based on the assumption that learning is in its essence a fundamentally social phenomenon. For this reason in addition to the three conventional elements "subject", "learner" and "teacher", the didactic tetrahedron model considers "community" as the forth main constitutive element of teaching and learning scenarios.

IFeL: The acronym stays for "Institut für Fernstudien- und eLearningforschung" (Institute for Research in Open-, Distance- and eLearning), a research institute with a double affiliation: the Swiss Distance University of Applied Sciences (FFHS) and the Distance Learning University Foundation Switzerland (Stiftung Universitäre Fernstudien Schweiz, FS-CH). It works at the interface between educational, medial and information technology issues to implement distance education concepts and blended learning scenarios (www.ifel.ch).

Prompting: An instruction measure which encourages learners during the learning process to activate cognitive, metacognitive and motivational learning processes.

Reference: Terms used in the compound terms "reference course" and "reference course model" (see next) where it indicates a reference system for the implementation of online courses in a blended learning scenario by means of course templates.

Reference Course (RC): A generic TELE course, technically implemented as a course template on the learning platform (a Moodle course); this template is given to individual lecturers as a starting point for implementing their own individual courses.

Reference Course Model (RCM): The approach or system that specifies principles, structures and procedures needed for producing a "reference course".

Self-Regulated Learning (SRL): A process in which individuals take the initiative, with or without the help from others, in diagnosing their learning needs, formulating goals, identifying human and material resources, choosing and implementing appropriate learning strategies, and evaluating learning outcomes (Knowles, 1975).

Swiss Distance University of Applied Sciences (FFHS): Switzerland's public 'open' university, based in Brig (Valais). The academic divisions of FFHS are a Business School, the faculty of Computer Science and the Engineering faculty (original German name: Fernfachhochschule Schweiz, FFHS, www.ffhs.ch).

University-Wide Media Culture: A corporate culture in the academic context of a university, in which new media are an essential part of the mission, vision and strategy and are implemented in its organisational structures and behaviour by integrating three perspectives for the development of media literacy: individual (competence), teaching (methods) and sustainability (change).

Chapter 21
Fostering Self-Regulated Learning in e-Health

Sisira Edirippulige
University of Queensland, Australia

Rohana B. Marasinghe
Sri Jayewardenepura University, Sri Lanka

ABSTRACT

Compared to other fields (such as media, banking and communication), the integration of information and communication technologies (ICT) in health has been slow. Among other factors, the lack of systematic education has been identified as a significant barrier. The use of ICT in healthcare delivery is widely known as e-Health. Evidence shows that if used in right context, e-Health can be efficient and cost effective. While designing e-Health curriculum, there are a number of factors to be considered. Due to the specific nature of the subject matter and the learners, the traditional teaching methods and pedagogical constructs may not be suitable. The choice of education methods must be based on the capacity of achieving the learning outcomes. E-Learning has proven to be an effective way of delivering education, particularly for rural and remote learners. Based on blended learning model, E-Health teaching at the Centre for Online Health University of Queensland, Australia has shown its capacity to provide a unique learning experience to students. While designing e-Health curriculum, a particular attention has been paid to aspects such as flexibility of learning processes, students' control in learning, self observation and self evaluation. These are, in fact, core principles of self-regulated learning (SRL) that have been incorporated in the teaching and learning process of e-Health. This chapter sets out to examine in details the elements of SRL embedded in e-Health teaching and the role of SRL in maximizing the learning outcomes.

DOI: 10.4018/978-1-61692-901-5.ch021

INTRODUCTION

The challenges faced by the healthcare systems around the world are critical. Increasingly aged population, changes in the disease patterns and the cost inflation have put enormous pressure on the already overburdened health systems. Solutions to these problems need long term financial investments and policy implementations. Evidence is emerging that the use of information and communication technologies (ICT) in health practices can offer a range of benefits. A new discipline known as e-Health aims to explore alternative ways of providing health services by using ICT. Studies have shown that e-Health can effectively be used in clinical, administrative, educational and research purposes in healthcare settings. Despite the growing evidence, the integration of e-Health in mainstream healthcare practices has been slow, compared to other fields such as banking, mass media and commerce. Among other reasons, a lack of systematic education to provide knowledge and to develop relevant skills in e-Health has been identified as a main barrier.

Dissemination of e-Health education may require specific considerations due to the nature of the discipline, cohort of recipients/students and the particularity of the knowledge and skills to develop. The conventional classroom-based teaching may not be the best method to provide e-Health education. Therefore new educational models and pedagogical constructs must be explored to meet the requirements of e-Health education. Providing e-Health education using e-learning modalities may have particular advantages, as students may have opportunity to familiarize themselves with online technologies that can be used in their future health practices.

The developments in the information and communication technologies have brought about significant changes in education. Health and medical education is exploring the potential of ICT in delivering and disseminating education in more flexible and effective ways. Online learning (or e-learning) has been effectively used in various aspects of medical and health education. E-learning can potentially be used for providing e-Health education. In essence, online e-Health education means teaching how to use ICT for healthcare delivery by using ICT.

The University of Queensland Centre for Online Health (COH) has been involved in teaching e-Health for the last 10 years. While designing the E-Health teaching programs, a significant attention has been paid to the specific nature of the discipline, learning cohort and skills to be developed. Delivery mode of the e-health education has been chosen to suit the busy health professionals. Thus, E-health education is provided online using flexible delivery mode. Learning process is designed in such a way that the learner will have control over their own learning. Self-observation and self-evaluation are key elements in the learning process.

It is fair to say that the designers of e-Health curriculum, consciously or unconsciously, have incorporated elements that are well versed within the theory of self-regulated learning. As Pintrich and Zusho (2002) defined, 'self-regulated learning is an active process whereby learners set goals for their learning and monitor, regulate and control their cognition, motivation and behavior, guided and constrained by their goals and the contextual features of the environment'. The following discussion will focus on the features of SRL in e-Health curriculum and the role of SRL to facilitate effective learning outcomes.

WHAT IS e-HEALTH?

A number of different terms have been associated with this new discipline. For example, terms such as telehealth, telemedicine and health informatics have been used interchangeably. It is apparent that the definitions are continuing to evolve. In a broader sense, e-Health connotes the use of ICT for the delivery of healthcare, health administration, health education and research (Edirippulige & Wootton, 2006).

Research has shown that e-Health may help make health systems more efficient while providing cost savings (Bendixen, Levy, Olive, Kobb & Mann, 2009; Smith & Gray, 2009). Evidence is also emerging on the safety, reliability and quality of care provided by using e-Health (Hailey, Roine & Ohinmaa, 2002; Marcin, Nesbitt & Cole, 2005). Studies have shown the provider and consumers satisfaction of e-Health applications (Saffle, Edelman, Theurer, Morris, & Cochran, 2009; Giordano, Scalvini, Zanelli, Corrà, Longobardi, Ricci, Baiardi, & Glisenti, 2009).

Research and Practice

Regardless of the growing body of research evidence for the effectiveness of e-Health, its practical use remains limited. It is evident that the health sector has been somewhat reluctant to embrace new ICTs. This is particularly true for the use of e-Health in clinical practice.

Researchers have suggested a number of reasons for this discrepancy. Lack of funding for infrastructure, networks and equipment has been one of such factors. Some authors have pointed out the unresolved privacy and confidentiality issues as a main barrier for the wider use of e-Health (Sharma, Xu, Wikremasinghe & Ahmed, 2006). Research has also found that the lack of systematic education to provide knowledge and relevant skills in e-Health has been a main constraint (Edirippulige, Smith, Beattie, Davies & Wootton, 2007; Edirippulige, Smith, Young & Wootton, 2006; Lamb & Shea, 2006). It is vital that the health professionals are appropriately educated to understand the main conceptual theories of e-Health. Education must ensure that the health practitioners have appropriate skills to use e-Health in their routine practices. Education gives health professionals confidence and appreciation. Therefore, the wider adoption of e-Health in health practice will require a significant change-management process in the health systems. Systematic e-Health education is an integral part of this process.

e-Health Teaching at the University of Queensland

The University of Queensland, Centre for Online Health, is a research institution exploring the potential of e-Health in delivering healthcare. The Centre has been involved in a large number of research studies to test and evaluate the effectiveness of e-Health modalities in the delivery of care, health administration and education. For example, the Centre has been instrumental in designing, developing and running a series of multidisciplinary sub-specialist services using e-Health/telemedicine tools in the areas of diabetes, endocrinology, burns, cardiology, dermatology, oncology, orthopaedics, gastroenterology, neurology and paediatric surgery. The studies have shown the effectiveness, reliability and potential for cost-saving of e-Health applications (Smith, Youngberry, Mill, Kimble & Wootton, 2004).

E-health teaching is another important activity of the Centre for Online Health, which has an active e-Health education program. Over the last 10 years a large number of students have completed graduate and undergraduate level studies in e-Health (Table 1). There is a well established postgraduate research program (Masters and PhD). In addition, the Centre conducts Continuing Professional Development (CPD) courses (short courses) related to e-Health.

Developing the e-Health Teaching Programs

The 'change of practice' has been at the heart of e-Health teaching at the UQ Centre for Online Health. Designers of e-Health education believe that it is not the bare knowledge which would change the traditional practices of medical and health practitioners, but it is the transformation of their attitudes and behaviors coupled with

Table 1: Number of Students in e-health programs

Year	Disciplinary area	Number of Post-graduate students	Number of Undergraduate students
2004	Allied health/Nurse/psychology/IT	13	9
2005	Occupational therapy /Physio/Speech/Nursing/IT/education	10	26
2006	Occupational therapy/Physio/Speech/Nursing/IT	15	38
2007	Occupational therapy /Physio/Speech/Nursing/IT/education	12	52
2008	Occupational therapy /Physio/Speech/Nursing/IT/Sports science	14	98
2009	Occupational therapy /Physio/Speech/Nursing/IT	10	129
Total		74	352

new knowledge and skills that would lead to a new way of practicing. In order to achieve these objectives, learners need to see the potential of e-Health and appreciate the ability of this new tool in their practices. Such change of attitude would lead the healthcare practitioners to adopt e-Health.

Another important consideration has been the diversity of student cohort. The cohort of e-Health students are adult students and busy practitioners; the teaching and learning methods must suit the needs of this population. The majority of the postgraduate students are full time workers; therefore, they have to fit their studies into their busy schedules and daily commitments. Moreover, a large number of students live and practice in rural and remote locations. For them, the flexibility of learning is an important feature. Importantly, a large number of students work in the health and medical sectors. In terms of professional background, they are general practitioners (GPs), nurses, physiotherapists, mental health workers or social workers (Table 1). Therefore, the link between the learning experience and the applicability of it in practice is a vital consideration.

Emanating from those considerations, flexible delivery of the e-Health teaching was conceived to be an important requirement. Thus the distance education has been the optimal delivery mode to offer students flexibility in order to combine their studies with work and other commitments. E-Health courses at the Centre for Online Health are offered mainly online. Students have the flexibility of accessing the course material from anywhere at their own pace.

Providing e-Health education by means of e-learning methods has another advantage. E-Health education via e-learning methods allow students to have an opportunity to appreciate the potential of online techniques for distance communication. The exposure to online technologies while learning e-Health can give them an opportunity to develop skills to use these technologies and appreciate the use of such tools in their practices.

As Figure 1 depicts, e-Health courses are delivered using an online learning platform, namely Blackboard. This e-Learning platform enables not only the delivery of course material and online assessments, but is also used for student/lecturer communication and interactions. Weekly online lectures, discussion forums, real time chat sessions and online reflective journals are some of the learning activities used in e-Health teaching. The use of such tools in e-Health education indirectly offers the learners an opportunity to develop skills in using these technologies and appreciate the potential of them for distance communication and interaction. These competencies may encourage them to use such technologies in their regular practices.

e-Health teaching is inspired by various curriculum models such as problem-based learning (Bligh, Prideaux & Parsell, 2001). In the course

Figure 1. Self-regulated learning model for e-Health learning

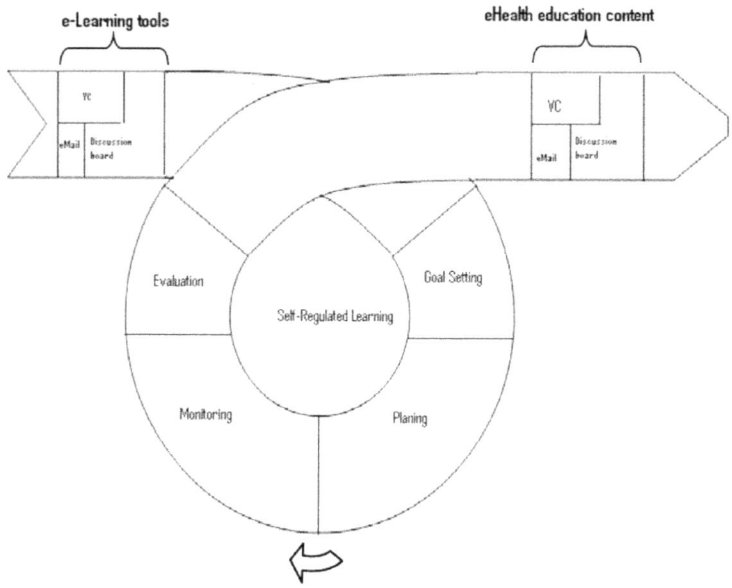

of studies, we use various scenarios and case studies which are relevant to the topics covered. We make sure that the scenarios and case studies are relevant to students' fields of practice so that they can relate the study to their experience.

The design of E-Health education has also ensured to incorporate a strong inter-professional element (Oandasan & Reeves, 2005). We believe that the aspect of inter-professionalism, defined as 'a partnership where members from different domains work collaboratively towards a common purpose' (MacIntosh & McCormack, 2001), is an important element in health practice. Particularly, the practice of e-Health always involves a team of professionals from different fields. Our experience shows that the e-Learning environment is a desirable place for interaction, communication and collaboration. We exploit the fact that the student cohort of e-Health programs at the Centre for Online Health consists of diverse professional backgrounds. The experience that students gain while learning e-Health paves path to work inter-professionally, which is an important element of e-Health practice.

As mentioned, online techniques provide a unique capability for enhanced learning in e-Health. However, we also believe that the development of practical skills in e-Health is a vital element for practitioners to use e-Health in their routine practice. Having hands-on skills to use e-Health techniques is key to wider use. E-Health education at the Centre for Online Health includes an in-house practical component where students attend a week-long practical sessions. During this practical, students learn how to use relevant equipments, test and trial applications and observe real-life e-Health consultations.

Main Components of e-Health Teaching

The term 'Blended learning' would best explain the e-Health teaching at the COH. Both online and face to face teaching methods are used in the learning process. The components of e-Health learning at the COH can mainly be divided into two: online components and the e-Health practical.

Online Components

A substantial proportion of teaching is done online using the UQ e-learning platform –Blackboard. Blackboard is used for the delivery of course material, student administration, communication, assessment and evaluation. Various online tools are used, such as text based material, online lectures, voice over powerpoints, video clips, email, discussion boards, chat rooms and weekly reflective journals. Formative and summative assessments are used for assessing the level of knowledge and competence.

Practical Component

E-health practical is designed to provide students an opportunity to gain hands-on skills and exposure to e-Health practices. The practical activities are based on the current projects at the COH. During the practical, students familiarise themselves with a range of technologies, equipments and systems. They also test, set up and use e-Health systems using scenarios provided to them. The practical is also an opportunity for students to observe real-life e-Health practices, where clinicians provide consultations using e-Health systems. During the practical, students have an opportunity to meet with clinicians and researchers who are engaged in e-Health practices.

SELF-REGULATED LEARNING IN e-HEALTH TEACHING

Three main pillars of self-regulated learning are embedded in e-Health teaching and learning. Those are:

1. Promotion of meta-cognition
2. Promotion of strategic action
3. Promotion of motivation to learn

Promotion of Meta-Cognition

As theorists of self-regulated learning suggest, learning guided by meta-cognition- (thinking about one's thinking) leads to better learning outcomes (Butler & Winne, 1995). To achieve this goal, it is important to give learners an opportunity to play a critical role in the learning process. Being cognizant of the fact that the learners are adults and they are busy healthcare professionals, it is important to give them the opportunity to play a main role in the learning process. They take charge of the learning process and the role of the coordinator is to facilitate the learning.

We present below some examples taken from the students' online discussion, to highlight:

1. How student establish the relevance of e-Health learning to their practices;
2. How student evaluate their capabilities and expect to use various e-learning tools;
3. How students use blended learning strategy to enhance their learning;
4. How peer-learning has taken place;
5. How students manage their own learning.

How Student Establish the Relevance of E-Health Learning to Their Practices

I decided to take on e-Health course because I want to broaden my knowledge within this area as well as learning how ICT works (both its benefits and limitations). I was hesitant to take on an external subject at first but I am happy to see the enthusiasm of everybody on the discussion boards.

Having grown up in remote Queensland on a sheep and cattle property, I realize the importance of e-Health and the promotion of health in rural communities. This connection is one reason why I chose to study e-Health this semester.

How Student Evaluate Their Capabilities and Expect to Use Various e-Learning Tools

I am in my second year of Occupational Therapy. I am interested to learn about ways in which I can assist my future clientele in accessing therapy via technology. More importantly though, I am keen to learn ways in which I can use technology to assist me in my occupation, especially seeing as I am rather computer illiterate.

My experience with e-health thus far is similar to many others as I have used this tool primarily for personal use and university assignments. I too have used the internet to find information on symptoms and treatment methods on numerous occasions with varying success. I have always found online databases to be a more credible and reliable source of information than the internet and so have relied on this resource for university research.

How Students Use Blended Learning Strategy to Enhance Their Learning

From the lecture, readings and other student' posts it seems that the definition that would define e-Health most fully would be, e-Health is a broad umbrella term which covers health informatics, telemedicine, telehealth and telecare.

How Peer-Learning Has Taken Place

It is great to read variety of opinions from the class. I would not be able to read through all these material myself. Thanks everyone for posting and sharing thoughts and ideas of the definition of telemedicine, telehealth and e-Health. I agree that the definitions are still in a transition state.

How Students Manage Their Own Learning

I also agree with the concept 'if in doubt - document' As long as something is on paper we are showing good and responsible practice.

I think many companies are probably running programs online such as Webex, That they assure to be completely secure... but as you said Amanda, this program can be used worldwide? So it is therefore using the internet, as opposed to an intranet???.

Louise and Jillian have already highlighted a number of important factors including the maintenance of patient confidentiality, security of electronic health records and data networks used by health professionals and the difficulty of ensuring only certain health staff can access these records.

The learning process is strucutred on a weekly basis. Each week, students are provided with an online lecture and relevant reading to do. The discussion board provides the platform for students to communicate with each other and the lecturers. Each week, a discussion topic is uploaded to invite students to engage in the discussion. The objective of the discussion is to give students opportunity to brainstorm (Figure 2 and 3).

A learning activity known as weekly reflection is also designed to encourage students' metacognition. Three short questions are uploaded for students to reflect on the week's learning material and issues emanating from it. Often these questions stem from problem-based scenarios and students can relate the key issues learnt in the topic to the content of the scenarios. Answers are invited in a brief and focused form (max 200 words). The emphasis is on the reflection.

E-health education puts a significant emphasis on reflection. We believe that the key to better learning outcomes is reflection. Students

Figure 2. The discussion board encourages students' interaction

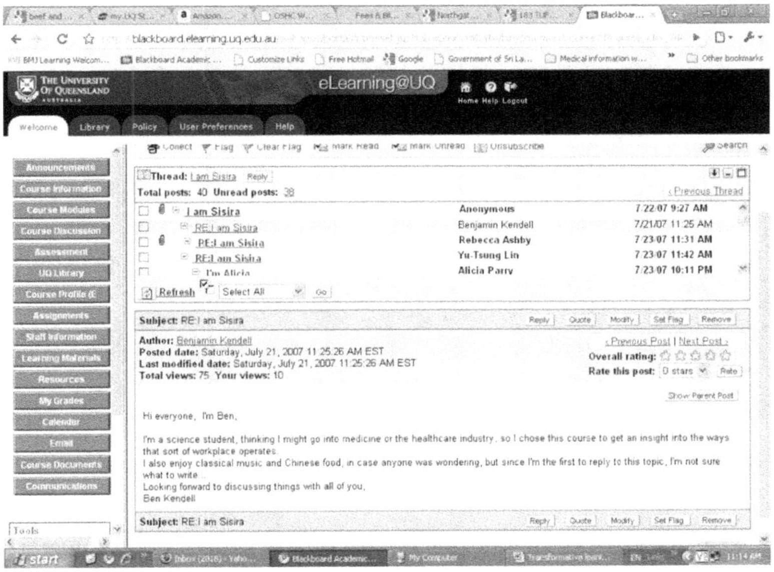

are encouraged to reflect on their own practice and relevance of the study in their practice. The scenarios and weekly learning material are based on the cases that students can relate to. The prime objective is that students make reflection a habit in their future practice.

Promoting Strategic Action (Planning, Monitoring and Evaluating)

Strategic action is another important element for students to achieve their goals. This includes planning, monitoring, and evaluating personal progress against a standard (Winne & Perry, 2000). E-health teaching encourages students to develop a strategic action in their learning.

This element has a particular relevance to the e-Health project. As a part of the e-Health program, students have to design and undertake a research to evaluate an e-Health application. Students are encouraged to choose a study relevant to their routine practices. Since the majority of students are working in the health sector, in most cases they select a research question relating to their work. This year-long project gives students an ample opportunity for working independently. Based on student's work-place, the study examines how a particular e-Health application can improve their practices. With the coordinator's guidance, students plan the project. However, students are expected to take charge of the work and their input is assessed.

Students are requested to keep records of their actions related to the development of the project. This gives them an opportunity to reflect on their planning and progress. Project design journals are required to be submitted with the final version of the project.

Motivation to Learn

Motivation to learn is an important component for any student; but this is more relevant to those who study remotely and while working full time (Perry, Phillips & Hutchinson, 2006; Zimmerman, 1990). Sustaining motivation is not an easy task especially when learners are inundated with their day to day work. This is why a particular atten-

Figure 3. Weekly discussion topics guides students learning needs

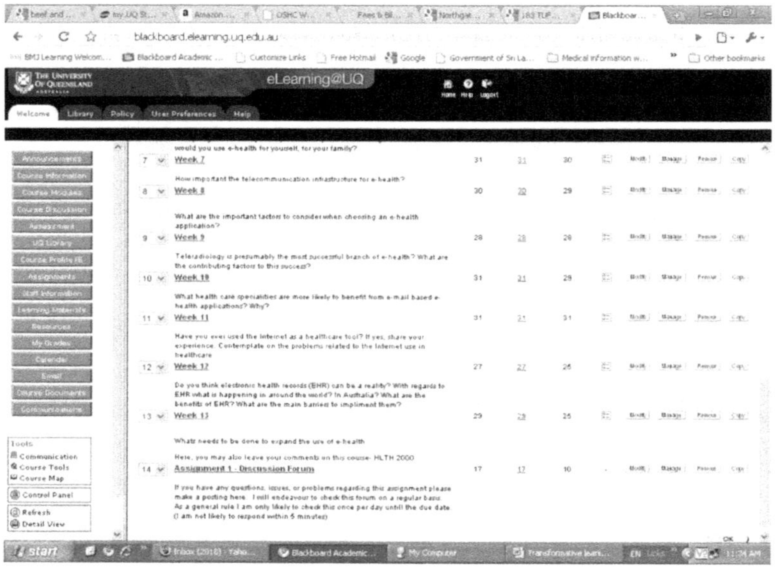

tion must be paid to maintaining the motivation, particularly when adult learners are concerned.

Self motivation is an important part of the sustaining drive to achieve desired goals. Health professionals who enroll in e-Health programs are motivated as they commence their studies with enthusiasm and curiosity. The challenge for educators is to help sustain the level of motivation.

In e-Health context, the learning activities are one of the most important elements for sustaining student motivation. Activities are designed to attract attention of students, and challenge them. Adult learners are usually motivated learners. However, challenging and stimulating learning material is a key aspect for sustaining the enthusiasm and curiosity. As mentioned, learning activities in e-Health courses stem from problem-based learning. Students have ample opportunity to relate the learning process to their practices. During the study, students bring their own issues and problems related to their practices and attempt to find solutions. Students share their views on these issues on weekly online forum. Postings of students on the forum show that they find the process of learning interesting and stimulating. Undoubtedly, achieving desired goals is a factor that maintains motivation to learn. Systematic progress in the learning process provides confidence and competence that keep student moving towards the desired goals.

Further, the applicability of knowledge and skills gained from learning in practical life is an important incentive for motivating the learner. Especially adult learners who commit to undertake e-Health education expect that they can use the knowledge and skills gained through the education in their routine practices. E-Health courses are designed to provide health practitioners the necessary knowledge and practical skills for applying it in their practices. While online components provide knowledge and understanding, in-house e-Health practical is particularly designed to provide the skills which can be useful in routine practice. During the practical, students have an opportunity to familiarize themselves with commonly used e-Health systems and equipment. Students are also required to set up and test e-Health applications for clinical purposes.

E-Health practical has been an attractive element in the program. A study evaluating the quality of e-Health practical showed that students highly rated the practical. The survey showed that the majority of students (88%) agreed that they obtained the necessary hands-on skills during the practical. Over 80% of participants agreed that the practical provided them an opportunity to explore ways of using different technologies in delivering healthcare (Edirippulige, Smith, Armfield, Bensink & Wootton, 2009). e-Health practical is an important element in sustaining motivation for learners to achieve their desired goals.

In addition, traditionally accepted educational principles such as alignment of learning objectives, activities and assessments, timely feedback, updated content etc. are also very important elements for keeping students motivated. E-Health teaching considers those factors vital for motivating students and obtaining a stimulating learning experience.

Finally, we can summarize the main elements of self-regulated learning attributed to the e-Health teaching/learning as follows

1. Encourage goal setting and self monitoring of progress towards the goals
2. Incorporate opportunities for directed and self-directed reflection
3. Develop and foster habit of self-reflection
4. Move away from highly structured assessment tasks
5. Providing learner to make important decisions about their learning process
6. Provide necessary feedback to motivate learner

CONCLUSION

Despite growing evidence for the effectiveness of e-Health, its use in mainstream practice is still limited. The lack of systematic education has been often referred to be one serious barrier. Whether it is conventional or novel, the choice of educational method must be based on the learning outcomes. Self-regulated learning has been established as an effective education construct; thus this chapter examined how e-Health teaching at the Centre for Online Health, University of Queensland, Australia, has included self-regulated learning in the design of the e-Health program.

e-Health teaching at the Centre for Online Health is based on a blended approach. The choice of blended learning is attributed to a variety of important considerations, all aiming to provide a unique and stimulating learning experience for students. Considering the uniqueness of student cohort, their geographic locations and the skills to be developed, both e-Learning methods and conventional in-house practical are used in the teaching and learning process. Confirming the findings of Winne and Stockley (1998), online technologies have proven to be effective tools for e-Health teaching. Elements of self-regulated learning are highly relevant to e-Health learning since self-regulated learning develop motivation to sustain learning, ability to monitor one's own thinking as well as use of learning strategies.

REFERENCES

Bendixen, R. M., Levy, C. E., Olive, E. S., Kobb, R. F., & Mann, W. C. (2009). Cost effectiveness of a telerehabilitation program to support chronically ill and disabled elders in their homes. *Telemedicine Journal and e-Health, 15*(1), 31–38. doi:10.1089/tmj.2008.0046

Bligh, J., Prideaux, D., & Parsell, G. (2001). PRISMS: new educational strategies for medical education. *Medical Education, 35*(6), 520–521. doi:10.1046/j.1365-2923.2001.00984.x

Butler, D. L., & Winne, P. H. (1995). Feedback and self-regulated learning: A theoretical synthesis. *Review of Educational Research, 65*(3), 245–281.

Edirippulige, S., Smith, A. C., Armfield, N., Bensink, M., & Wootton, R. (2009). Teaching e-Health practical skills: Results of a cross-sectional study. Paper presented at the *International Conference Successes and Failures in Telehealth*, Brisbane.

Edirippulige, S., Smith, A. C., Beattie, H., Davies, E., & Wootton, R. (2007). Pre-registration nurses: an investigation of knowledge, experience and comprehension of e-health. *The Australian Journal of Advanced Nursing*, *25*(2), 78–83.

Edirippulige, S., Smith, A. C., Young, J., & Wootton, R. (2006). Knowledge, perceptions and expectations of nurses in e-health: results of a survey in a children's hospital. *Journal of Telemedicine and Telecare*, *12*(Suppl. 3), 35–38. doi:10.1258/135763306779380255

Edirippulige, S., & Wootton, R. (2006). Telehealth and Communication. In M. Conrick (Ed.), *Health Informatics, Transforming Health care with Technology*, Melbourne, AU: Thomson.

Giordano, A., Scalvini, S., Zanelli, E., Corrà, U., Longobardi, G. L., & Ricci, V. A. (2009). Multicenter randomised trial on home-based telemanagement to prevent hospital readmission of patients with chronic heart failure. *International Journal of Cardiology*, *131*(2), 192–199. doi:10.1016/j.ijcard.2007.10.027

Hailey, D., Roine, R., & Ohinmaa, A. (2002). Systematic review of evidence for the benefits of telemedicine. *Journal of Telemedicine and Telecare*, *8*(Suppl 1), 1–7. doi:10.1258/1357633021937604

Lamb, G. S., & Shea, K. (2006). Nursing education in telehealth. *Journal of Telemedicine and Telecare*, *12*(2), 55–56. doi:10.1258/135763306776084437

MacIntosh, J., & McCormack, D. (2001). Partnerships identified within primary health care literature. *International Journal of Nursing Studies*, *38*(5), 547–555. doi:10.1016/S0020-7489(00)00095-X

Marcin, P. M., Nesbitt, T. S., Cole, S. L. (2005). Changes in Diagnosis, treatment and clinical improvement among patients receiving telemedicine consultations. *Telemedicine and e-Health, 11*(1), 36-43.

Oandasan, I., & Reeves, R. (2005). Key elements for interprofessional education. Part 1: The learner, the educator and the learning context. *Journal of Interprofessional Care*, *19*(Supplement 1), 21–38. doi:10.1080/13561820500083550

Perry, N. E., Phillips, L., & Hutchinson, L. R. (2006). Preparing student teachers to support for self-regulated learning. *The Elementary School Journal*, *106*(3), 237–254. doi:10.1086/501485

Pintrich, P. R., & Zusho, A. (2002). Student motivation and self-regulated learning in the college classroom. In Smart, J. C., & Tierney, W. G. (Eds.), *Higher Education: handbook of theory and research* (*Vol. XVII*, pp. 55–128). New York, NY: Agathon Press.

Saffle, J., Edelman, L., Theurer, L., Morris, S. E., & Cochran, A. (2009). Telemedicine evaluation of acute burns. Is accurate and cost-effective? *The Journal of Trauma Injury Infection and Critical Care*, *67*(2), 358–365. doi:10.1097/TA.0b013e3181ae9b02

Sharma, S. K., Xu, H., Wikremasinghe, N., & Ahmed, N. (2006). Electronic healthcare: issues and challenges. *International Journal of Electronic Healthcare*, *2*(1), 50–65. doi:10.1504/IJEH.2006.008693

Smith, A. C., & Gray, L. C. (2009). Telemedicine across the ages. *The Medical Journal of Australia*, *190*(1), 15–19.

Smith, A. C., Youngberry, K., Mill, J., Kimble, R., & Wootton, R. (2004). A review of three years experience using email and videoconferencing for delivery of post-acute burns care to children in Queensland. *Burns*, *30*(3), 248–252. doi:10.1016/j.burns.2003.11.003

Winne, P. H., & Perry, N. E. (2000). Measuring self-regulated learning. In Pintrich, P., Boekaerts, M., & Seidner, M. (Eds.), *Handbook of self-regulation* (pp. 531–566). Orlando, FL: Academic Press. doi:10.1016/B978-012109890-2/50045-7

Winne, P. H., & Stockley, D. B. (1998). Computing technologies as sites for developing sel-regulated learning. In Schunk, D. H., & Zimmerman, B. J. (Eds.), *Self-regulated learning: from teaching to self-reflective practice* (pp. 106–136). London, UK: The Guilford Press.

Zimmerman, B. J. (1990). Self-regulated learning and academic achievement: An overview. *Educational Psychologist*, *25*(1), 3–17. doi:10.1207/s15326985ep2501_2

KEY TERMS AND DEFINITIONS

E-Health: Delivery of healthcare using information and communication technologies (ICT), particularly the Internet.

E-Learning: The use of ICT for delivery of education and teaching and learning management.

Inter-Professional Learning: Different professional groups learning complementary knowledge, skills and attitudes around a common focus or problem. This practice allows each group to reflect on what they bring to the learning, and to identify complementarity in the application of such learning.

Telemedicine: Any medical activity occurred at distance. The term telemedicine is currently used to describe the delivery of medical services at a distance using ICT.

Chapter 22
Informal Self-Regulated Learning in Corporate Organizations

Wim Veen
Delft University of Technology, The Netherlands

Jan-Paul van Staalduinen
Delft University of Technology, The Netherlands

Thieme Hennis
Delft University of Technology, The Netherlands

ABSTRACT

Sharing knowledge is one of the most challenging tasks modern companies have to deal with. A vast amount of knowledge exists within organizations; however it is often difficult to find and to judge its value. As a consequence, learning and knowledge building seem to be a lonely activity, separated from everyday work. Transfer of knowledge acquired in formal courses has little impact and effect on day-to-day work. That is why training often has a low return on investment. Knowledge management systems have also proven to be ineffective as they fail to present the knowledge employees are looking for. So how can we improve learning in organizations using ICT? To find an answer to this question we might learn from the generation that has grown up with modern communication technologies. This Homo Zappiens has shaped new ways of communication and information sharing including attitudes and views leading to collective knowledge building strategies. Prominent characteristics of Homo Zappiens include their preference for images and symbols as an enrichment of plain text, their seemingly effortless adoption of technology and their cooperation and sharing in networks. This generation seems to take exploration and learning, discovering the world, into their own hands. Homo Zappiens shows us we can increasingly rely on technology to connect us and allow us to organize as a group. In a networked society, the individual has more room for contributing its unique value, and innovation and knowledge reside in a network, rather than in each separate individual. Realizing that we need a flexible structure for organizing ourselves and the world around us, we can look at Homo Zappiens for a clue. This chapter describes self-regulated learning within a network (Networked Learning) and presents a model for it. It also presents experiences with the model at the multi-national corporation IBM, where a Technology-Enhanced Learning Environment (TELE) was built and introduced.

DOI: 10.4018/978-1-61692-901-5.ch022

Figure 1. Learning activities and their relevance

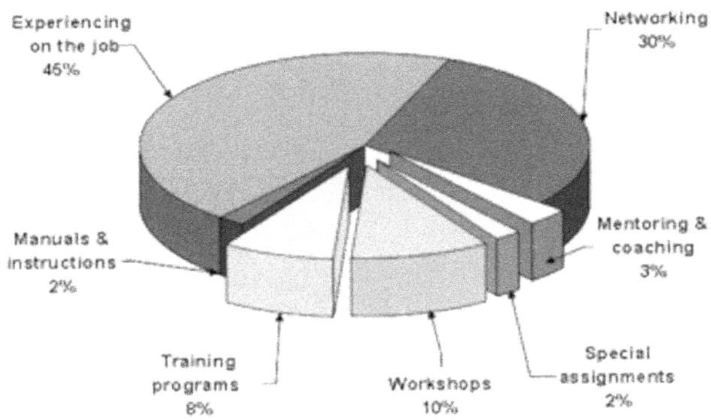

INTRODUCTION

The sharing of knowledge between employees is one of the most challenging tasks modern companies nowadays face (Siemens, 2005; Siemens 2006). Organizations have access to vast amounts of tacit and tangible knowledge; but for an employee this knowledge is often difficult to find and its actual value cannot always be judged. As a consequence, active learning and knowledge acquisition seem to be lonely activities, both distinctly separated from regular, everyday work. Knowledge transfer that occurs in formal courses has little impact on the day-to-day work of employees (Weistra, 2005), as shown in Figure 1. That is why training often has a low return on investment. Advances in ICT, through the years, have provided us with new possibilities and opportunities for improving learning. But technology enhanced learning in companies currently often supports a rather traditional single actor learning, such as first generation e-learning where printed matter has been digitalized into hypertexts, or available data-bases that are poorly used, or blended learning scenarios using learning platforms (Siemens, 2006). The learning in these situations remains an individual act with no interpersonal communication or connection to the daily working practice. New concepts of learning are needed to improve training and learning of the employees, through the use of technology.

For inspiration on how to organize learning alternatively, we can look to the new generation of students that have had their views on the world around them shaped by modern communication technologies (Tapscott, 1998; Collis & Moonen, 2001; Oblinger & Oblinger, 2005; Veen & Vrakking 2006). Due to their preference for television and Internet technology, this generation has been called Homo Zappiens (Veen & Vrakking 2006), or the Net Generation (Oblinger & Oblinger, 2005). Prominent characteristics of Homo Zappiens include their preference for images and symbols, their seemingly effortless adoption of technology, and their cooperation and sharing in networks. Through the use of technology, Homo Zappiens learns to develop new skills and exhibits new behavior that may show us a way of how higher education will be organized in the future. Homo Zappiens learns to participate in society through networks, switching between streams of information, and learning to cooperate and share in getting relevant information. Their strategies seem to fit companies' needs as they have to deal with challenges of knowledge (co-)creation and knowledge sharing. This is not only an issue that matters the organization as a whole, but is also highly relevant for each employee within the organization. Knowledge sharing in virtual social networks, solving problems that are closely related to daily work, demand the attitude youngsters have developed from early childhood on. They have been using social networks in Facebook, MySpace, YouTube, or any other profiling platforms. In those networks they have learned that each individual matters, and is valued and treated respectfully. Online social networks are flat, that is they do not discriminate on who you are, and foster the attitude that sharing is winning (Veen & Vrakking, 2006) It is this sociability that makes people help each other, solving problems collaboratively. These attitudes and views on how to share and build knowledge can be of great help designing new learning infrastructures for organizations and companies alike.

This chapter describes a Technology-Enhanced Learning Environment (TELE), called Yuno that has been based on the Homo Zappiens values and norms. These values are expressed in the Networked Learning Model that is described in a following section. Yuno allows for self-regulated learning, and was developed for the multi-national corporation IBM. A case is presented about learning in the corporate sector, where most of the learning takes place with regards to the work that people do. This work-related learning always has production as a purpose: production of a new service, production of a more efficient procedure, a new product, et cetera. The context of the learning will also be often related to corporate goals and vision, such as sustainability, profitability, and innovation. When work-related learning originates from the employee's initiative, it is by definition self-regulated.

In this chapter we present a case where learning is informal, time- and place-independent, and is not assessed by exams, tests, or the acquisition of a certificate. Yuno was created with the use of the Networked Learning Model, a model based on experiences with Networked Learning and inspired by the behavior of Homo Zappiens. These latter two concepts will be further explained in the following sections, after which the Networked Learning Model will be described. At the end of the chapter, Yuno is presented.

A NEW GENERATION OF SELF-REGULATED LEARNERS

Today we are confronted with an educational challenge, which is related to motivating and retaining talented employees. The characteristics of today's and future's worker is changing (Cross, 2007; Siemens, 2005; 2006; Veen & Vrakking, 2006). Self-learning has bloomed; discovering online possibilities is a skill now developed from early childhood through advanced adult life. Even online reading has become collaborative, interactive, nonlinear and relational, engaging multiple voices (Davidson & Goldberg, 2009). For example, in the past, process industries needed skilled workers to build products. Nowadays these industries seek employees who can co-create solutions by designing products collaboratively with customers using 'building blocks' of a multitude of companies from over the world. Skills of making a product do not longer suffice such companies, many other skills are needed, including designing skills, communication skills, and managerial skills (Veen, 2009). This also applies to other sectors such as banking and insurance. These companies can only thrive through employees who are able to judge situations independently and can come up with new solutions as they communicate with clients. Younger employees, to speak in general terms, use technology in a functional manner, not touching what they can't use, and increasingly, they seem to take exploration and learning, discovering the world, into their own hands (Veen & Vrakking, 2006). These attitudes coincide with companies' needs of today (Allen & Velden, 2007).

Given the rise of interactive network technology, the logical direction for education and learning to develop, as society ever more embraces the uniqueness of each individual, is for the process of learning to become more natural. As people can learn from any new experience, and what they learn is often unpredictable, there isn't a real way in which the optimum level of development for each individual can be reached through structured education. Accepting that the outcome of learning cannot be controlled up front, would lead to a more natural form of learning. Technological advances have made it possible for people to start learning this self-regulated, non-linear way; our most natural way of learning (Siemens, 2006), by pursuing personal goals in changing learning environments.

The outcome of this personalized learning often leads to storage of knowledge in companies' databases. This understanding of storing knowledge does not lead to lively shared knowledge which is always linked to people and context. In his book Knowing Knowledge, Siemens (Siemens, 2006) discusses how knowledge is no longer a product, but rather a process of interaction and negotiation. Veen and Vrakking (2006) agrees on this line of thought and defines learning as the process of searching for meaning and knowledge as the process of communication about giving meaning to information. As a consequence, learning is a social process of negotiating references, and individuals learn by externalizing their knowledge, sharing their input into the social network of peer learners, experts, and others that might be involved in the process. Collective learning explicitly aims at collaborative performance and innovative knowledge co-creation within a team. In practice, this means that learners should be facilitated to communicate in organizations. Informal communication and activities within a group of people from diverse backgrounds, helps them to share and produce knowledge relevant for business goals. Well-known examples of this kind of informal learning have been implemented at Google and Nokia. Younger employees are used to participate in society through networks, anticipating that different situations may require different roles and developing the competence to quickly switch between roles just as they switch between streams of information (Veen & Vrakking, 2006).

Technology has taken dominance over society as a means of providing organization to our lives. Immersion, interactivity, and communication are critical characteristics of the technologies Homo Zappiens has been using from early childhood on. This has important implications for learning and educational organizations. The impact of this can be seen in the behavior of Homo Zappiens.

In a world where technology is available, the most important skills are those that enable us to use that technology to enrich our lives (Laurillard 1993; Harper, Rodden, Rogers & Sellen, 2008). That is why Homo Zappiens adopts the use of new technology at an amazing pace. At the same time they seem to be paying less attention to learning mathematical skills, grammar and memorization; rather, they rely on calculators and search engines to provide them with the same information. The cause of this is not so much a disinterest for 'old concepts', but much more often a form of prioritizing skills.

On the Internet, communities of people gather information that is relevant to them and recommend it to others within the community (Wenger, 1998). Through a form of internal recommendation, information is filtered based on perceived value and importance. Cooperation thus provides a mechanism for distributing the increasingly larger task of determining which information is valuable.

People used to keep their most prized knowledge and competences private, and thus scarce, but nowadays an increasing number of young people let others know about their knowledge and skills (Oblinger & Oblinger, 2005). In an organizational system that promoted competition, keeping knowledge private made sense. Yet in a network, where negotiation and communication are the most important elements (Steeples & Jones 2002), the need for privacy is an outdated concept, and the need for attention becomes key.

Homo Zappiens is currently entering the labor market, and one of the issues to consider is how we can take advantage of this generation's networking attitudes and skills to enhance business outcomes while avoiding a disintegration of the workforce along the demographic structure of the organization. Networked Learning is a mechanism for companies to develop a more effective learning environment for their employees. It is based on the idea to improve trainings' outcomes by bringing learners together in a learning and knowledge sharing network. The innovative and creative nature of team learning can be used strategically within organizations. This is the aim of the concept of Networked Learning. Networked Learning is to be understood as a different way of learning which is different to ICT-supported ways of traditional learning. The focus of Networked Learning is on connectedness of people and artifacts in a network with the use of technology to integrate delivery of knowledge with interaction, communication and application. The monitoring of one's own activities while acquiring knowledge and the self-evaluation of one's performance (Zimmerman, 1990), makes Networked Learning by default a self-regulated learning process. It is a way to find other experts and to share knowledge with them. In this understanding we suggest the following definition for Networked Learning:

Networked Learning is learning through establishing and maintaining connections: between people, resources and technical systems, using tools which promote knowledge sharing and knowledge construction.

Recent approaches (Goodyear 2005; Siemens, 2006; Rosenberg, 2006) define Networked Learning as a new way of learning. The term network applies in the first place on a human network in which individuals are connected, having access to personal information as well as (shared) artifacts. The network exists through the help of technology, which supports the integration of the delivery of knowledge on the one hand, and the interaction, communication and application on the other. Networked Learning is today more evident because it finds its existence in explicit network structures, such as global networks like the Internet, the web (Siemens, 2008), or collaborative group support systems which exist within certain organizations and companies. The other way around, the use of ICT, in particular the Internet is a basic prerequisite for Networked Learning. Unfortunately, it seems the use of ICT for learning in numerous companies is still in the first phase of substitution. E-learning as it appears in the corporate sector is providing printed matter in a digital format. E-learning materials stem from the printed stuff; the only difference is that the learner now has to switch his computer on in order to read it. It is focused on individual learning by reading. In fact, it is hardly more than a computer-assisted form of traditional learning, mainly used for distance learning, as an overview shows (Darby, 2002).

Socially networked collaborative learning extends some of the most established practices, virtues, and dispositional habits of individualized learning. These include taking turns in speaking, posing questions, listening to and hearing others out. Networked Learning, however, goes beyond these conversational rules to include correcting others, being open to being corrected oneself, and working together to fashion workarounds when straightforward solutions to problems or learning challenges are not forthcoming. It is not that individualized learning cannot end up encouraging such habits and practices. But they are not natural to individual learning, which leans on a social framework that stresses competition and hierarchy rather than cooperation, partnering, and mediation. [...] The power of ten working interactively will almost invariably outstrip the power of one looking to beat out the other nine (Davidson & Goldberg, 2009, p. 30).

Networked Learning is happening now—not in the future. Since the current generation of college student has no memory of the historical moment before the advent of the Internet, Davidson and Goldberg (2009) suggest that Networked Learning as a practice is no longer exotic or new but a commonplace way of socializing and learning. For many, it seems entirely unremarkable. Global business more and more relies on collaborative practices where content is accretive, distributed, and participatory. In other areas too—from the arts to the natural and computational sciences and engineering—more and more research is being enacted collaboratively.

A New York Times article from 2008 even suggested that a future Nobel Prize winner might not be an oncology researcher at a distinguished university but a blogging community where multiple authors, some with no official form of expertise, actually discover a cure for a form of cancer through their collaborative process of combining, probing, and developing insights online together (Davidson and Goldberg, 2009, p. 13).

THE NETWORKED LEARNING MODEL

Self-regulated learning emphasizes the role of self as an agent in establishing learning goals and tactics. It acknowledges that perceptions of the self and the task influence the learning process (Paris & Winograd, 2003). Self-regulated learning refers to the number and kinds of decisions made by learners, sometimes in collaboration with other persons. These include planning, making strategies, choosing learning activities, evaluating of progress and reflecting on errors and successes (Hout-Wolters, Simons & Volet, 2000; Zimmerman 2000). Self-regulation in learning is cyclical, because after each process, the outcomes of efforts provide the basis for further development (Zimmerman, 1998). Similarly, in networked learning models, history of interaction and processes, both social and learning, form the basis for making connections. Autonomy and control over learning is central to both self-regulated and networked learning. An online network with peers and social software offer learners the possibility to interact, and effectively collaborate during the preparatory, executive, and concluding phases of self-regulated learning.

Networked Learning is a way of learning that is self-regulated, work-related and mostly informal. Employees learn when needed, just-in-time, they are actively seeking for solutions, hence producing new knowledge (Veen, Lukosch & De Vries, 2009). These aspects of Networked Learning apply to many forms of learning, from novice training to expert learning. To aid us in constructing tools for Networked Learning, a Networked Learning Model was designed by combining the best of the existing definitions for it. The model itself is rooted on various existing educational theories and concepts. The theory of social constructivism is useful for a model of Networked Learning because it is based on the premise that, by reflecting on own experiences, and sharing these with others, learners construct their own understanding of the world they live in. Learning within this theory is considered as an active process in which learners construct new ideas based upon current and past knowledge (Bruner, 1991). The underlying idea of learning communities can also be found in the Activity theory. In its third generation it includes a network of interacting activity systems (Hout-Wolters, Simons & Volet, 2000). The actor-network theory drives our attention to the point that social networks are almost always mediated technically by some form of material medium, for example the cyberspace as one of those. The learner is seen as an integral part of a network.

Those networks can also be seen as communities of practice. A community of practice is thus defined as a learning group in which new insights can be transformed into knowledge through mutual engagement around a joint enterprise (Wenger, 1998). It gives some useful insights in the encouragement of change and learning and with this shows some implications for facilitating

Networked Learning. Communities of practice imply the characteristics of situatedness, knowledge sharing and collective learning as well as the focus on practice, which are basic for Networked Learning. Engagement theory states that learners have to be engaged by several steps in order for effective learning to occur. These specific steps lead to creative, meaningful and authentic learning, which is the aim of Networked Learning, too. Technology in this way of understanding is seen as a facilitator for all aspects of engagement.

Networked Learning wants to address the challenges of organizational knowledge and transference. Thus the theory of Connectivism (Siemens, 2005) is very helpful to understand how decisions are made and how knowledge is acquired and shared. Knowledge that resides in databases needs to be connected to the right people in the right context in order to be classified as learning. In other words, it shows that information flow within an organization is an important element in organizational effectiveness. For the flow of information, trust is the needed utility, as is pointed out in Stephenson's Theory of Trust. Different people have different roles in different networks. These roles have to be identified to become a trustful source or target for the flow of knowledge.

Homo Zappiens, being an increasing part of the upcoming labor force, will rely heavily on technical and distributed social networks. The Networked Learning model is derived from the behavior of Homo Zappiens and the way they learn and participate in fluid, online networks and communities. Networked Learning refers to a context in which internet-based information and communication technologies are used to promote connections: between participants; between participants and experts; between a learning community and its learning resources, so that participants can extend and develop their understanding and capabilities in ways that are important to them, and over which they have significant control. These connections vary from face-to-face to distribute, across a variety of media, and with various degrees of time shifting. Effective self-regulated learning depends on many things, such as the organizational context (for example, support by management), the learners and teachers themselves (motivation), the learning environment, instructional goals and methods, the materials offered, and assessment. Most of these are implicitly part of the Networked Learning model.

The Networked Learning Model has four complementary areas that play an important role in knowledge development, as shown in Figure 2. Each of the elements that are connected to these areas is selected on the basis of their relevance for this development process in which the technology is a major facilitator for processes of communication, information retrieval and information sharing. These areas are: Profiling, Connectedness, Knowledge and Business Development. 'Profiling' is the area describing a collection of social and organizational aspects of how individual employees function in an organization. 'Profiling' is the area of the Digital Me, where the learner takes control over his/her professional development by using technology in a functional and interactive manner. 'Connectedness' stands for the connection among people and between people and resources. It relates to social networks and the way interaction and human relations are relevant for people to perform in communities. These communities are fluent; you can take part for some time depending on the purpose of the community. Communities are based on peer references and are not limited to office hours. Reciprocity and peer reflection provides the support and motivation needed for self-regulated learning. 'Connectedness' is the area of the Digital We. 'Knowledge' is the area that defines content and information in the Networked Learning Model. This content is distributed and discontinuous, stored in databases. Learners have to aggregate bits and pieces (modules) into a meaningful whole. They do this collaboratively, sharing their expertise with others. 'Business Development' is the area that describes

Figure 2. A Model for Networked Learning: enabling people to connect, share and co-create

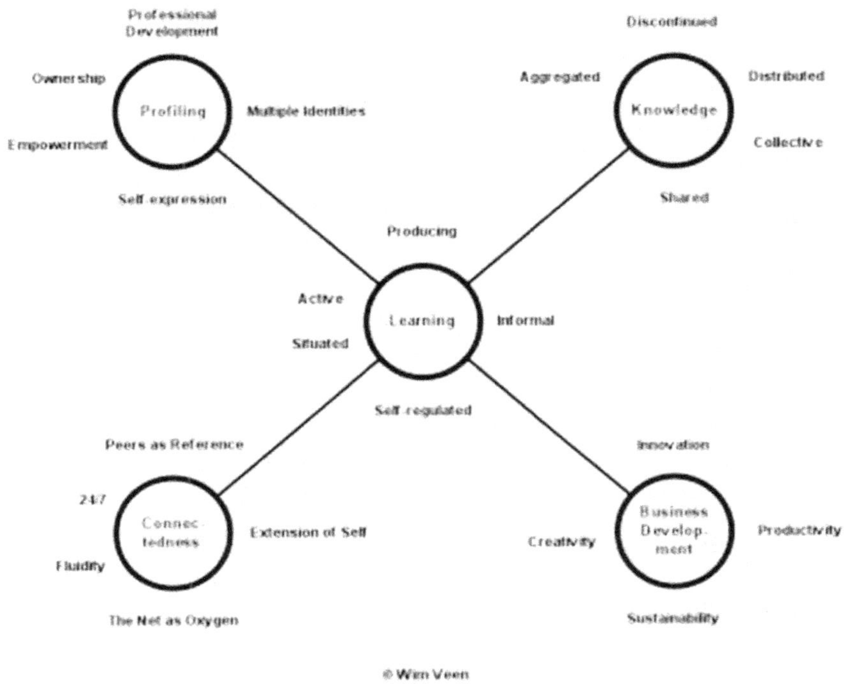

the major companies' business goals. These goals are the reference framework in which learning takes place; it provides the organizational context.

The elements that constitute the Networked Learning Model are not new or innovative. It is the combination of the areas and the elements represented in this model that makes it innovative and different from the existing Computer Supported Collaborative Learning approaches.

Different principles of self-regulated learning are present in the networked learning model. Paris and Winograd (2003) describe 3 principles of self-regulated learning: awareness of thinking, the use of strategies, and sustained and situated motivation. We will discuss them below with respect to the networked learning model.

Awareness of thinking is supported in the networked learning model in several ways. First of all, awareness emerges from the production of, and reflecting on knowledge. In ICT networks, both the history of activity and the impact or value of the activities can be logged and evaluated. So by sharing your ideas, questions, and thoughts in the network, a process of reflection and assessment is initiated, through interaction between agents and through automation (i.e. analysis of activities and contributions). Reflection is especially relevant in networked learning, because it assumes that learning is a distributed activity and that knowledge resides in the network. It is therefore necessary for learners to be aware of their own knowledge, as well as to be able to find and filter relevant knowledge in their network. Usually, this is mediated with interactive discussion, content management, and collaboration tools. A second important aspect in the process of creating awareness is self-expression and profiling. Networked learners identify themselves in multiple ways, depending on the context. Expressing oneself in multiple ways is a process of identification, and the learner becomes aware of his or her identity, both through expression and interaction with others.

The use of strategies involves a person's growing repertoire of strategies. One of the central aspects of the control and regulation of cognition is the actual selection and use of various cognitive strategies for memory, learning, reasoning, problem solving, and thinking (Pintrich, 2004). The emphasis should be on "being strategic", rather than "having a strategy" (Paris & Winograd, 2003). The use of strategies is fundamental in Massive Online Role Playing Games, such as World of Warcraft. Before deciding upon tactics, players first collaborate and discuss about strategy by considering different options. All the meta-cognitive aspects of creating strategies - declarative knowledge (what the strategy is), procedural knowledge (how the strategy operates), and conditional knowledge (when and why a strategy should be applied) (Paris, Lipson & Wixson, 1983) - are considered in these types of gameplay. The Networked Learning Model is based on the online behavior of gamers, which is characterized by reciprocity and playful exchange of knowledge in a distributed, self-organizing system (Hertz, 2001).

The third principle of SRL concerns motivation. Learning requires effort and choice. Learners base their efforts on the goal of the activity, the perceived "return on investment", the difficulty of the task, and their own ability to be able to complete it. It is therefore intertwined with awareness and reflection, because that influences the choice of the learner in any given situation. Reasons for failure or ineffective learning include setting the wrong goals, avoiding risk or failure, and the attribution of performance to external forces. In successful communities and networks, inhibitors of effective self-regulated learning are managed and countered by the community as a whole, as well as by the technology available. Additionally, both networked and self-regulated learning imply "personalized cognition and motivation" as an important premise for learning (Hickey, 1997). In the Networked Learning Model, motivation is always "personalized", because it is situated in the actual professional environment of the learner and learning goals emerge from problems faced on the work floor.

AN INNOVATIVE TOOL FOR NETWORKED LEARNING: YUNO AT IBM

At IBM employees looked for answers to their questions by consulting colleagues, not databases. As a consequence, finding solutions for challenges of individuals is more a matter of finding the right people and sharing and building knowledge collaboratively, than it is typing in keywords in search engines providing hundreds of thousands of hits. Facilitating the way people tend to learn on a daily basis has become our focus in the project of networked learning. Based on the Networked Learning Model, a networked learning tool --called Yuno-- was built and implemented at IBM and connected to the IBM infrastructure for experimental use. In a period of 6 months, Yuno has been used and tested amongst a group of 50 professionals. These professionals were technical consultants, business and sales managers. During the design phase, IBM and TU Delft researchers focused on four aspects:

- Interconnectedness: supporting the connection between persons, and between persons and resources.
- Peers as a reference: supporting the construction of knowledge, experience, and expertise inter-subjectively through interaction.
- Profiling of virtual identity: automatic and manual expression of passion and talent as well as the possibility to adopt different roles in different contexts, providing the support for having different identities.
- Self-regulated learning system: based on the online behavior of gamers, Yuno supports distributed management and self-

Figure 3. Screen shot of the Networked Learning tool

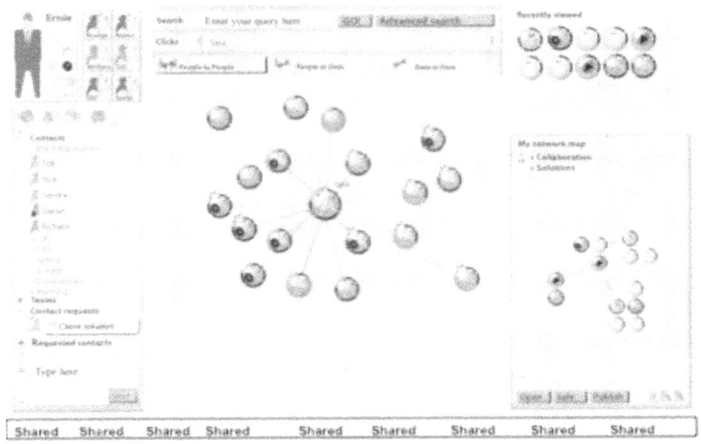

organization through simple but effective tools as tagging and rating. Reciprocity, status and reputation form crucial elements of the design of the portal.

Van Hout-Wolters, Simons & Volet, (2000) provide an extensive list with 45 learning functions needed for effective learning. These functions are divided in preparatory, executive, and closing learning functions, similar to the categories of self-regulated learning processes described by Schunk and Zimmerman (2003). Yuno supports effective self-regulated learning by offering an online place to support and improve interaction between colleagues and resources and aligning this with their existing processes. It offers support for many of the described learning functions.

Preparatory functions that are supported by Yuno include finding missing prior knowledge, getting an overview, trickling curiosity, and finding and interpreting connections (for example between prior knowledge expressed in the profile and recommended content). Executive learning functions are present in Yuno by offering support in selecting information (see Figure 3), getting an overview, sharing practice and applying knowledge, as well as monitoring and evaluating contributions/learning outcomes. It does not explicitly support for the monitoring of learning processes and the reflection on them. This happens in a more implicit, distributed manner, where reflection is made easier by showing the history of one's activities and through peer evaluation. Closing learning functions, such as judgments of results, rewarding, and evaluation processes are embedded in the rating and peer-evaluation functionalities of the tool.

Yuno helps employees finding peers who are experts in their field of interest, helping employees to organize these peers into a community and solving the problem with the help of their peers in a highly self-organizing manner. When a user types in a search entry, he/she is shown a network of related items, people, and resources (Figure 3). Rather than textual representation, a choice is made for iconic and graphical representation, in line with Homo Zappiens' preferences for images and symbols as an enrichment of text. The way knowledge is represented, embedded in a network rather than in a linear way, allows learners to explore more effectively and discover new knowledge in their network.

Contextualized learning is supported in two ways. First of all, the environment itself is rooted in the organizational processes. Employees use the internal social network to find people and resources; Yuno offers the extra graphical and functional layer to it that improves collaboration and creation of knowledge. Secondly, Yuno supports the adding of metadata to resources *and* its users, making contextualization of learning and assembly of resources better possible for individual users. Unfortunately, not all functionalities that have been primarily designed could be implemented due to limitations of time and resources.

The outcomes of the test of the Yuno tool showed that it functioned as a vehicle for the transfer of the Networked Learning concept onto the different organizational levels of IBM. Findings have been:

- Yuno, qualified as a meta search and information rating engine, was considered a timely and refreshing application. This can be considered a good qualification from people who are IT-savvy and work with technological innovations on a daily basis.
- Not all participants used the IBM databases or engines outside their intranet environment. This certainly strengthened the performance of Yuno as a tool, because these IBM databases contain huge amounts of information that now could be accessed and used in a usable format, thus extending personal networks.
- The former search action patterns of most participants appeared to be rather traditional. They relied on their contact base of friends, current colleagues, and former colleagues that all were within reach through email and telephone. These traditional strategies appeared hard to change and a tool such as Yuno provided quick wins and early successes to convince them to do things differently.

Yuno supports self-regulated learning in a number of ways. Through Yuno, the employee in the Networked Learning Model gains ownership over his or her own learning and professional development. Employees are no longer dependent on acquiring certificates, but gain control of their own development with regards to their work, which ultimately benefits the company. The network provides functions for filtering information, building reputation and sustaining interaction between professionals. Users become aware of their own and other people's expertise and knowledge by contributing and reviewing content. They can consult experts and online coaches, and there is the ability to assume different roles, based on organizational role, contributions and expertise. Peers are able to reflect on contributions and actions by a user, and all relevant activities are stored in a log file that can be analyzed. Even though this has not been done, it offers opportunities for learners to get better insight in their learning patterns, and to receive recommendations and support for making an effective learning strategy.

Also, when searching for answers, the employee participates in temporary communities in which a sharing culture is created, which will lead to increasing social cohesion within the company. This cohesion can lead to participation and increased sense of community, which is a crucial element in any successful online community. It should be noted that, even though advantages of sharing can be clearly communicated, it is still difficult to incentivize sharing behavior and to elicit true participation. It requires well-developed technology, which is embedded in existing processes, well marketed and supported by management.

Addressed to Homo Zappiens, Yuno is a fluid environment that can be consulted 24/7 and offers tools and content allowing learning to happen in small chunks. This lowers the barriers for learning, and learning can happen more spontaneously and without much effort. Self-regulated learning occurs when learners are, within a certain context, sufficiently motivated to do a certain task as well as have a strong enough perception of their ability to conclude the task in a positive way. Both motivation and perception are elements in networked learning in general, and the Yuno tool in specific. Motivation in networked learning is always contextualized, because learning emerges from problems or challenges encountered on the workplace. Information about the size and difficulty of a task (such as a learning object) can be retrieved in the network through interaction and discussion, or by looking at statistics and metadata about the task (or resource). In Yuno, metadata and additional information about learning objects are continuously added through interaction and logging activities, enabling more effective learning.

Self-generated learning in Yuno can be improved by offering guidance in finding relevant materials, aligning the Yuno tool with incumbent technologies and processes, improving feedback and content creation tools, improving the filtering, recommendation and searching algorithms, showing insightful statistics that promote self- and community-awareness, and developing better tools that specifically help users in making strategies in a collaborative setting, such as a "personalized learning plan".

Ultimately, the pilot was not successful from the perspective of developing measurable innovation, which was obviously a too ambitious goal for the project setting. But it did contribute to the understanding of the Networked Learning concept and the acceptance of this concept by the IBM employees as a good basis for collaboration and innovation in the field of life long learning in daily practice. The Networked Learning Model alone could never have achieved that result.

FUTURE RESEARCH DIRECTIONS

Through Networked Learning, the employee gains ownership over his own learning and professional development. The employee no longer is dependant on certificates, but is emancipated by himself taking the reins of his own development, which benefits the company. Next to that, the employee functions in temporary communities, in which a sharing culture is stimulated. This leads to (stronger) social cohesion within the company.

Most fundamental is the understanding that Networked Learning is about a process and not always a final product (Davidson & Goldberg, 2009). Education in organizations will evolve due to new generations of employees demanding different pedagogical approaches and technological environments for corporate training. In the future, learning will be about externalizing the knowledge of participants. The major goal of learning will be to co-create new knowledge. Learning will become an on-going work-based activity for many employees, being eager to participate in online communities focusing on their work-related needs. Technology will continue to evolve and mixed realities and virtual realities will further enhance our opportunities to reduce scarcity of presence. Employees will be able to meet, work together and discuss using spectacles and facilities such as caves where they can shape ideas and design products or models.

Life long learning will become an integral part of the working life of individuals as networks will continue to exist between industry and higher education. This view might sound like a scenario unlikely to come true to its full extension. But looking at the way how working teams are nowadays more often organized in an ad-hoc manner or how coalitions shift allegiance with the shifting of political tides, we can already see how society has been increasingly incorporating the concept of flexible structures to the organization of dynamic reality. There are strong developments such as globalization, virtual communities sharing knowledge, and specialization among scientific institutions at a global level, indicating towards structures that will make any university that will not participate in this world-wide progress an isolated regional or local school. These developments of globalization and sharing knowledge at a global level will enhance the chances that institutions will adopt technology to enable their global orientation. Homo Zappiens is a major driving force behind these changes. We have to improve learning theory and practice, taking into account new learning skills, and develop new technologies (or apply existing ones) such that these skills are fostered in self-regulating learning networks.

Future research into Networked Learning will need to focus on three key areas in order to make the most of this new way of learning. First of all, the analysis of social networks will provide us with more insights into the behavior and interests of its members. This will give us clues about setting up our own corporate networks for self-regulated learning. Next, a further understanding of collective knowledge creation is needed to provide our employees with the tools they need to effectively (co-)create and acquire knowledge they require. Finally, organizational changes are needed to cope with the needs for self-regulated Networked Learning that employees have. Future research can determine which changes are required.

CONCLUSION

The Networked Learning Model can be considered as a useful strategy for dealing with the large amounts of information available in knowledge intensive companies. It is an answer to the challenges of informal, goal oriented learning on the workplace. Further work is now to be done on the evaluation of the learning model and the tool. The tool can be applied in contexts and organizations where human intervention, tools and resources must be combined in order to create successful products and services or to unlock 'wisdom of the crowd' (or database) and to support knowledge creation and sharing.

The Yuno tool, developed based on the Networked Learning Model, provides a structured way to collective learning. It is a challenge to find knowledge linked both to persons and to situations and allows capturing part of reality in a framework in order to reduce the complexity and allow closer examination. The concept maps made with help of the tool can be used companywide to be applied to new projects or upcoming problems. The Yuno tool supports active, self-regulated learning by giving learners network of people and peer-reviewed content resources to tap into, in the context of their work. In their network, learners are supported in making decisions through consultation with experts and coaches, plan their learning, and assess their activities through interaction with peers. The tool does not specifically offer functions that support decision-making or self-reflection, but assumes this to happen in the network.

The Networked Learning Model and the Yuno tool have been built for a worldwide knowledge intensive company, but can also be useful for any other business where many experts work together in a network. Associations of companies to get to know about the knowledge and the resources of each other can also use it. They just have to take on a networked perspective.

REFERENCES

Allen, J., & van der Velden, R. (Eds.). (2007). *The Flexible Professional in the Knowledge Society: General Results of the REFLEX Project*. Maastricht, NL: Research Centre for Education and the Labour Market, Maastricht University.

Bruner, J. (1991). *Acts of meaning*. Cambridge, MA: Harvard University Press.

Collis, B., & Moonen, J. (2001). *Flexible learning in a digital world*. London, UK: Kogan Page Limited.

Cross, J. (2007). *Informal Learning*. San Fransisco, CA: Pfeifer.

Darby, J. (2002). Networked Learning in Higher Education: The Mule in the Barn. In Steeples, C., & Jones, C. (Eds.), *Networked Learning: Perspectives and Issues* (pp. 17–26). Berlin, DE: Springer.

Davidson, C. N., & Goldberg, D. T. (2009). *The Future of Learning Institutions in a Digital Age*. Cambridge, MA: MIT Press.

Fullan, M. G. (2001b). *Leading in a Culture of Change*. Toronto: Wiley, John & Sons.

Goodyear, P. (2005). Educational design and networked learning: Patterns, pattern languages and design practice. *Australasian Journal of Educational Technology*, *21*(1), 82–101.

Harper, R. Rodden, T. Rogers, Y., Sellen, A. (2008). *Being Human: Human-Computer Interaction in the Year 2020*. Microsoft, Cambridge, UK. http://research.microsoft.com/hci2020

Herz, J. C. (2002). Gaming the system. What Higher Education Can Learn from Multiplayer Online Worlds. In: *Devlin, M., The Internet and the University: 2001 Forum,* 169-191. Washington, EDUCAUSE Publications.

Hickey, D. (1997). Motivation and contemporary socio-constructivist instructional perspectives. *Educational Psychologist*, *32*, 175–193. doi:10.1207/s15326985ep3203_3

Laurillard, D. (1993). *Rethinking University Teaching: A Framework for the Effective Use of Educational Technology*. London, UK: Routledge.

Oblinger, D., & Oblinger, J. (Eds.). (2005). *Educating the Net Generation*. Washington, DC: Educause.

Paris, S. G., Lipson, M. Y., & Wixson, K. K. (1983, Jul). Becoming a strategic reader. *Contemporary Educational Psychology*, *8*(3), 293–316. doi:10.1016/0361-476X(83)90018-8

Paris, S. G., & Winograd, P. (2003). *The role of self-regulated learning in contextual teaching: Principles and Practices for Teacher Preparation.* A commissioned paper for the Department of Education Project, Office of Educational Research and Improvement, Washington DC. Available at: www.ciera.org/library/archive/2001-04/0104prwn.pdf.

Pintrich, P. R. (2004). A Conceptual Framework for Assessing Motivation and Self-Regulated Learning in College Students. *Educational Psychology Review*, *16*(4), 385–407.

Rosenberg, M. J. (2006). *Beyond E-learning. Approaches and Technologies to Enhance Organizational Knowledge, Learning and Performance*. San Fransisco, CA: Pfeifer.

Schunk, D. H., & Zimmerman, B. J. (2003). Self-regulation and learning. In: W. M. Reynolds, G. E. Miller & I. B. Weiner. *Handbook of Psychology, Volume 7*, 59-78. 7 ed. Wiley.

Siemens, G. (2005). *Connectivism: A Learning Theory for the Digital Age.* Retrieved August 4, 2008, from http://www.elearnspace.org/Articles/connectivism.htm.

Siemens, G. (2006). *Knowing Knowledge*. Winnipeg: Complexive Inc.

Tapscott, D. (1998). *Growing up Digital: The Rise of the Net-Generation*. New York, NY: McGraw-Hill.

Van Hout-Wolters, B., Simons, P. R. J., & Volet, S. (2000). Active learning: self-directed learning and independent work. In Simons, P. R. J., van der Linden, J., & Duffy, T. (Eds.), *New learning*. Dordrecht: Kluwer Academic Publishers.

Veen, W. (2009). *Homo Zappiens, leven, werken en leren in een digiaal tijdperk.* [*Homo Zappiens, growing up, living, and working in a digital era*]. Amsterdam, NL: Pearson Education.

Veen, W., Lukosch, H., & De Vries, P. (2009): Improving organizational learning through Networked Learning. *In: Proceedings of the 5th Conference of Professional Knowledge Management, Solothurn (CH). March 25-27, 22-31.*

Veen, W., & Vrakking, B. (2006). *Homo Zappiens: Growing Up in a Digital Age*. London, UK: Network Continuum Education.

Weistra, H. (2005). Leerenergie en de voorwaarden voor (in)formeel leren (Energy for learning and the prerequisites for (In)formal Learning). [Education and Development]. *Opleiding en Ontwikkeling, 10*(18), 17–21.

Wenger, E. (1998). *Communities of Practice: Learning, Meaning and Identity*. Cambridge, UK: Cambridge University Press.

Zimmerman, B. J. (1990). Self-regulated learning and academic achievement: An overview. *Educational Psychologist, 25*, 3–17. doi:10.1207/s15326985ep2501_2

Zimmerman, B. J. (1998). Academic studying and the development of personal skill: A self-regulatory perspective. *Educational Psychologist, 33*, 73–86. doi:10.1207/s15326985ep3302&3_3

Zimmerman, B. J. (2000). Attaining self-regulation: A social cognitive perspective. In Boekaerts, M., Pintrich, P. R., & Zeidner, M. (Eds.), *Handbook of self-regulation* (pp. 13–40). San Diego: Academic Press. doi:10.1016/B978-012109890-2/50031-7

ADDITIONAL READING

Csikszentmihalyi, M. (1990). *Flow: The psychology of optimal experience*. New York, NY: Harper Perennial.

De Vries, P., & Brall, S. (2008). *Microtraining as a Support Mechanism for Informal Learning. In: Elearningpapers of Elearningeuropa*. Retrieved November, 2008, from http://www.elearningpapers.eu/

Dewey, J. (1998). *How we think (revised edition)*. Boston, MA: Houghton Mifflin Company.

Diepstraten, I. (2006). *De Nieuwe Leerder: Trendsettende leerbiografieën in een kennissamenleving (The New Learner: Trendsetting Learning Biographies in a Knowledge Society)*. Tilburg, NL: F&N Boekservice.

Fullan, M. G. (2001a). *The New Meaning of Educational Change*. New York, NY: Teachers College Press, Columbia University.

Gee, J. P. (2003). *What video games have to teach us about learning and literacy*. New York, NY: Palgrave Macmillan.

Jonassen, D. H. (1997). Instructional Design Models for Well-Structured and Ill-Structured Problem-Solving Learning Outcomes. *Educational Technology Research and Development, 45*(1), 65–94. doi:10.1007/BF02299613

Livingstone, S., & Haddon, L. (2009). *EU Kids Online: Final report*. London School of Economics and Political Science, London. (EC Safer Internet Plus Programme Deliverable D6.5)

Prensky, M. (2001). *Digital Natives, Digital Immigrants*. Omaha: NCB University Press.

Prensky, M. (2001). *Digital game-based learning*. New York, NY: McGraw-Hill.

Rogers, E. M. (2003). *Diffusion of Innovations* (4th ed.). NewYork, NY: Free Press.

Seely Brown, J., & Duguid, P. (2001). *The Social Life of Information: Learning in the Digital Age*. Boston, MA: Harvard Business School Press.

Senge, P. M. (1990). *The Fifth Discipline: The Art & Practice of The Learning Organization. Chatham*. Kent: Century Business.

Siemens, G. (2008). *New structures and spaces of learning: The systemic impact of connective knowledge, connectivism, and networked learning*. Presentation held at the Universidade do Minho, Encontro sobre Web 2.0. Braga, Portugal.

Smith, P. J. (2003). Workplace Learning and Flexible Delivery. *Review of Educational Research, 73*(1), 53–88. doi:10.3102/00346543073001053

Steeples, C., & Jones, C. (Eds.). (2002). *Networked Learning: Perspectives & Issues*. Berlin, DE: Springer.

Surowiecki, J. (2004). *The Wisdom of Crowds: Why the Many are Smarter than the Few and How Collective Wisdom Shapes Business, Economies, Societies, and Nations*. New York: Doubleday.

KEY TERMS AND DEFINITIONS

Connectivism: Upcoming learning theory put forward by George Siemens. His theory takes the influence of technology on learning processes into account.

Corporate Learning: Learning taking place in organizations focusing on professional development of employees.

Homo Zappiens: Generations who have grown up in a world where the scarcity of information and communication has been reduced to a large extent leading to an extension of the physical existence with a virtual one.

Informal Learning: Learning that is not managed or organized in structured institutions, courses or classes, initiated by learners and taking place in communities of individuals. Informal learning in companies can be facilitated by appropriate technologies and organizational support.

Networked Learning: Learning taking place within communities of individuals who are connected through ICT infrastructures in which knowledge co-creation takes place through negotiation and externalization or sharing.

Profiling: Developing a personal profile of skills, expertise and interests, often shared in an online environment.

Work-Related Learning: Learning focusing on the development of individual expertise for work-related tasks at short and middle term.

Chapter 23
Face-to-Face and Web-Forum Interventions Promoting SRL Skills at University

Barbara De Marco
University of Milan Bicocca, Italy

Nicoletta Businaro
University of Milan Bicocca, Italy

Eleonora Farina
University of Milan Bicocca, Italy

Ottavia Albanese
University of Milan Bicocca, Italy

ABSTRACT

Based on recent findings about Self-Regulated Learning (SRL), we outline three educational interventions aimed at fostering students' learning competence. Our particular focus is on the interaction between collaborative learning in Technology Enhanced Learning contexts and the development of SRL competencies. Two interventions carried out by our research team involved collaborative activities conducted both face-to-face and in web-based learning environments, aimed at promoting the SRL skills of first year university students. Based on the outcomes of these two projects, a further project for different departments was undertaken. This last intervention was designed to facilitate collaborative reflection on the components and processes of SRL through e-tivities and discussion forums. Our research suggests that collaboration in analyzing and working on the different competencies involved in self-regulated learning is an optimal means of enhancing the self-regulation competencies of university students.

DOI: 10.4018/978-1-61692-901-5.ch023

INTRODUCTION

This chapter starts from two assumptions. First, that collaborative learning promotes self-regulated learning (Nevgi, Virtanen & Niemi, 2006; Kramarski & Gutman, 2006; Fischer & Baird, 2005). Second, the specific features of web-enhanced learning environments can contribute to the development of self-regulated learning processes at different levels: cognitive, metacognitive, motivational and behavioural (Salovaara, 2005; Dabbagh & Kitsantas, 2005; Barnard, Paton & Lan, 2008).

In the first part of the chapter we outline the Self-Regulated Learning (SRL) model which underpinned our interventions aimed at fostering students' learning competencies. This section also provides a brief review of the current literature on the interaction between collaborative learning in web-enhanced learning environments and the development of self-regulation skills.

The second part of the chapter provides an overview of two different interventions carried out by our research team to develop the SRL skills of first year university students. The first was conducted face-to-face and involved theoretical and practical sessions on a range of SRL processes. The second analyzed the impact on the development of SRL skills of collaboration on specific academic topics via an ICT platform. Based on the outcomes of these two projects, a third intervention for students of different departments was designed with the aim of facilitating collaborative reflection on the specific components and processes of SRL through e-tivities and discussion forums within an ICT platform. This last project and its outcome are described in detail in the final part of the chapter.

Our goal is to suggest good practices that may be implemented with students at university in order to enhance their SRL skills. The outcome of our interventions leads us to recommend the third form of intervention in particular as an effective means of promoting SRL among students at university.

SELF-REGULATED LEARNING: THE PROCESSES INVOLVED AND HOW THEY MAY BE ENHANCED

According to the leading theoretical models, Self-Regulated Learning (SRL) is made up of several components (Pintrich, 2000; Zimmerman, 2008). Although a single empirically validated model of Self-Regulated Learning that describes all the components and the interactions between them has not yet been identified, theorists agree that cognitive, metacognitive, motivational, affective and behavioural/environmental components all play a role in the development of a self-regulated learner (Boekaerts & Corno, 2005). At a cognitive level, a student needs to be familiar with, and implement, a range of strategies: both general study strategies, such as rehearsal, elaboration, integration and organization (Salovaara, 2005; Pintrich, 1999; Cornoldi, De Beni & Gruppo MT, 2001) and domain-specific or problem solving strategies, i.e. individual tactics that students use to perform well in a task (Schraw, Crippen & Hartley, 2006). Implementation of a given strategy is not necessarily automatic, often requiring the learners to metacognitively reflect on what they are doing. Learners can do this by planning, monitoring and evaluating their activity, so as to choose the most suitable strategies in accordance with the task features (Veenman, Van Hout-Wolters & Afflerbach, 2006; Albanese, 2003; Albanese, Doudin & Martin, 2003; De Beni, Moè & Cornoldi, 2003).

Intrinsic motivation, self-efficacy and appropriate achievement goals help the student to persevere in the implementation of proper study methods, even when they are perceived as demanding (Bandura, 1989; Elliott & McGregor, 2001; Dweck, 1999). Students' degree of perseverance is also influenced by the ability to self-regulate negative emotions and to optimize the advantages provided by positive emotions (Pekrun, Goetz, Titz & Perry, 2002). Finally, SRL involves adequate time and effort management, the choice of a suitable physical context for study and ability to seek help as appropriate (Pintrich, 1999).

In summary, self-regulated learners are familiar with and use a variety of strategies and methods. They think about their own cognitive functioning. They are able to reflect on the motivation and goals driving their learning activity and to manage the social and physical context they operate in.

Learning to learn and being adequately motivated appear to be critical in reducing university drop-out (Albanese, De Marco & Fiorilli, 2008; Vallerand, Fortier & Guay, 1997). In addition, the development of self-regulation competencies has been demonstrated to be a predictor of academic achievement (Zimmerman, 2008). These findings encourage researchers to develop instruments and activities aimed at fostering self-regulation skills.

Self-regulated learning can be promoted by a range of activities, such as exercises in problem solving and metacognitive reflection, completion of SRL questionnaires followed by careful analysis of the individual profile emerging, or participation in structured SRL development programs (Albanese, Farina, Fiorilli & Minosso, 2007; Albanese, De Marco & Fiorilli, 2008; Cornoldi, De Beni & Gruppo MT, 2001). Furthermore, SRL can be developed by promoting a collaborative approach to study in which students cooperatively build their knowledge through reciprocal teaching, peer cooperation and teacher-student collaboration; therefore the individual learning emerges as the result of a collective process (Kaye, 1995). Students receive feedback on their performance and concrete examples of alternative ways to perform a task, they can discuss topics and ideas, develop metacognitive competencies such as self-evaluation, share appropriate goals and jointly plan activities. Other students provide support in managing time and effort and contribute to maintaining individual motivation, since the individual works not only for him/herself but for the whole group, aiding collective and personal knowledge growth (Cacciamani & Giannandrea, 2004; Schraw, Crippen & Hartley, 2006; Varisco, 2008).

Although collaborative learning contexts are effective in fostering self-regulated learning, face-to-face collaboration requires organizational constraints (time and space setting limit). In fact students and tutors need to decide when and where they meet and this could became a significant boundary, increasing students' drop-out. Moreover, much of the content of a discussion may be lost if not adequately recorded. Web-based learning environments provide an excellent solution to some of these difficulties. In particular, recent studies have shown that the integration of face-to-face and distance collaborative learning, i.e. web-enhanced and blended learning, seems to promote more efficient learning, not only in terms of content but also in terms of competencies (Ligorio, Cacciamani & Cesareni, 2006; Narciss, Proske & Koerndle, 2007). Face-to-face meetings reduce the risk of depersonalization and the perception of loneliness and distance between participants. On the other hand, collaboration via ICT platforms can reduce the time and space setting limit associated with face-to-face collaborative activities and guarantee continued availability of content and discussions. The features (hyperlinks, regular text, ppt, audio and video presentations) and tools (such as email, forums, advice boards) provided by ICT platforms support communication and cooperation. Moreover the features and the tools foster different SRL skills: the development of different strategies of study, careful reflections on learning goals, the self-evaluation, the help-seeking and the efficient time management in line with commitments, especially when the ICT platforms and study program facilitate the joint expertise and knowledge building (Dabbagh & Kitsantas, 2005). In addition, a self-regulated approach to study in online courses is associated with higher achievement (Barnard, Paton & Lan, 2008).

The links just outlined between development of self-regulation skills, blended learning, coopera-

Table 1. Features of three interventions

Intervention	Aim and Means	Time	Participants	Drop-out
Face-to-face meetings on personal skills in study method (optional)	Promoting SRL skills through theoretical and practical session	10 meetings (in four months)	30	50%
Blended course on academic contents. ICT platform: phpBB (optional with benefit= shortened exam program)	Promoting SRL skills through forums on specific academic topics, dealing with traditional lessons	Three months	106	7%
Blended course on personal skills in study method. ICT platform: Docebo (optional)	Promoting SRL skills through learning objects, e-tivities and forums	Three months	40	25%

tive learning and achievement prompt us to think the practices implemented by universities in order to enhance student performance. Following the theoretical framework presented here, we revised two different interventions planned by our team. Based on the reflections about these two projects, we designed a new web-based intervention aimed at facilitating the development of SRL skills.

FOSTERING SELF-REGULATED LEARNING: EMPIRICAL RECOMMENDATIONS

SRL is a key aspect of achievement at university, facilitating student performance. However, the importance of promoting the various components of SRL has been underestimated at higher levels of education. In fact the primary focus has been on the specific academic disciplines, and consequently too little attention has been paid to training in cross-domain competencies relating to study methods. Unfortunately, individual capability to study in the most appropriate way is not a given. We thus present a series of cross-domain optional interventions designed primarily to enhance the self-regulation skills of first-year university students. The first was a face-to-face intervention aimed at fostering SRL development through collaborative reflection on study method. Our second intervention took the form of a blended course, whereby students both attend traditional lessons and discuss their knowledge on dedicated web-forums. In this case, we wanted to examine the relationship between individual perceptions of improvement in self-regulated learning and participation in web-forum activity. Our third intervention was designed on the basis of the results of the prior two interventions and involved *ad hoc* e-tivities and online discussion about different aspects of SRL on web-forums.

Fostering SRL via a Face-to-Face Course: Reflecting on Personal Skills

The first project entitled "Study methods at university" was a face-to-face intervention aimed at developing the SRL skills of first-year university students from different faculties. Our specific objective was to facilitate reflection by the students on the weaknesses of, and the factors influencing, their study methods. Moreover we wanted to motivate them to try more effective methods while availing of the support of the other participants and the tutors. The course offered "a time and place" where students could contribute to both collective and personal knowledge by sharing practical information about academic courses, as well as their ideas, concerns and advices in relation to study methods.

The purpose and structure of the project were presented during official academic lecture periods. Initially, thirty students enrolled in the course, but only half of these actually attended the face-to-face sessions with tutors.

The students were divided into two groups and allocated one tutor each. They were first asked to fill in four self-evaluation questionnaires. The questionnaires allowed us to assess SRL competencies such as time and environment management, knowledge and use of cognitive and metacognitive strategies, goals achievement, emotions, self-motivation and implicit theories of intelligence.

The Approach to Study Questionnaire (De Beni, Moè & Cornoldi, 2003) was used to assess 5 areas of SRL skills: organization (the ability to fix times, places, conditions, aims and means for proficient learning activities), elaboration (the ability to thoroughly assimilate and analyse concepts), self-evaluation (the ability to assess study processes and the degree to which concepts have been acquired), strategies (the ability to consciously adopt effective procedures) and metacognitive sensitivity (the ability to reflect on the processes and characteristics of one's own approach to study and on the variables that may influence it). The Achievement Goals Questionnaire (Elliot & McGregor, 2001; translated into Italian by Albanese, De Marco & Fiorilli, 2008) was administered to evaluate four related goals: mastery approach, mastery avoidance, performance approach and performance avoidance. The Emotion in Studying Questionnaire (Mega, Moè, Pazzaglia, Rizzato & De Beni, 2007) allowed us to assess positive and negative emotions relating to self, current academic performance and study processes. Finally, the Implicit Theories of Intelligence Scale (Faria & Fontaine, 1997; translated into Italian by Pepi, Faria & Alesi, 2007) provided a measure of conceptions of intelligence.

The project lasted four months, the students met twice a month with a tutor for about two hours. During these sessions, they discussed both their strengths and weaknesses with regard to study methods and were assisted by the tutor to identify implicit factors affecting their learning activities and they were motivated to adopt new strategies. They had the opportunity to discuss specific cognitive, metacognitive and emotional/motivational topics on the basis of stimuli presented by the tutor, and to take part in individual, pair and group activities. Each tutor modified and adapted the activities to meet student requirements, focusing on unsatisfactory and inadequate competencies identified by the participants themselves. Topics covered included how to organize academic materials in preparation for an examination, time and space conditions for study, the factors contributing to loss of concentration, information selection strategies, the importance of self-confidence both in terms of emotions and with regard to evaluation of acquired knowledge and self-evaluation. The need to improve metacognitive sensitivity was also emphasized, and individual strengths and weaknesses with regard to learning processes were outlined.

We analysed questionnaires' results comparing individual scores in each SRL area (repeated measures ANOVA) and we identified categories about SRL in students discussions (strategies, motivation, concentration, marks, content organization, retention, time management, information selection). We outlined students weakness about SRL (see Albanese, Farina, Brambilla & Businaro, 2008). First-year university students reported that their greatest difficulty was organizing course materials in such a way as to have enough time to prepare for an exam, although they were expected to be completely autonomous in this regard. They also reported inability to keep up with their course of study, even if they had prepared a study plan. In addition, they suffered from loss of concentration. Students were in the habit of rehearsing materials mentally rather than verbally, and they reported difficulty in formulating clear arguments during oral exams. In response, the tutors prompted participants to organize their studies on a day-by-day basis, they helped students to find solutions for lack of concentration, identifying the sources of distraction and they suggested exercises and role playing to improve the capacity to rehearse verbally the materials.

Analyzing students' discussion according the SRL categories and questionnaires' results it emerged that there was a discrepancy between students' declarative knowledge and the real application of strategies, such as repetition, and that even when participants adopted new strategies, they were not able to transfer them to different contexts. With regard to the emotions characterizing the study process, the participants reported experiencing positive emotions such as curiosity, interest and hope, and negative emotions such as worry and anxiety (Albanese, Farina, Brambilla & Businaro, 2008). With regard to the goals of their learning activity, the participants predominantly chose mastery goals, preferring to achieve adequate mastery of the course material than to merely perform well in exams. This attitude usually corresponded to a dynamic view of intelligence (Dweck, 1999; Albanese, Fiorilli & Farina, 2004), whereby the participants believed they could be empowered through engagement in study processes, and considered efforts as opportunities to show their capabilities.

We also took note of the students' beliefs about the features of an efficient learning method. They thought that it was essential to master the materials, to remember the most important facts/concepts, to be able to formulate discourse relating newly acquired information to prior knowledge. Most of the participants reported inability to study effectively with respect to their ideal standards.

Finally, the tutors encouraged the students to review their motivation for attending university, and their choice of Faculty, providing them with an opportunity to discuss their doubts in this regard.

Solutions and Recommendations (a)

Participants who completed the study method course reported benefits in terms of improved study methods and SRL competencies (Albanese, Farina, Brambilla & Businaro, 2008; Farina, Businaro, De Marco & Albanese, 2009), but the drop-out rate was very high (almost 50%), as previously stated. We hypothesized that the causes of drop-out were practical difficulties in attending the face-to-face study methods course during scheduled lecture-periods, with many students having the additional commitment of part-time jobs. Moreover, it is likely that the students considered the study methods sessions as a further burden. In fact, reflecting on and actively changing learning methods require considerable effort, as opposed to maintaining established study habits which have become more and more reinforced through repeated use. Finally, doubts about the efficacy of new study methods, and the possibility that new methods might not immediately yield positive results, may also have discouraged students from persevering with the study methods course.

Fostering SRL through Blended Learning: Reflecting on the Contents of an Academic Subject

The second project was called "Study developmental psychology through an ICT platform". It was an intervention developed for first-year university students, offering them the possibility to cooperate and build their knowledge together through web forum on a phpBB platform. In parallel, we wanted to verify the impact of collaboration through web forum on the development of SRL skills. The activity proposed to students on the Developmental Psychology course was participation in online activities designed to go into the contents of course lectures in greater depth. The students who decided to participate had a shortened program for the exam.

The 106 students who participated were divided into four different virtual rooms to ensure that active participation would be possible for everybody: each student could post one or more contributions to each thread, creating discussions rich in content but not too long to be read comfortably. Ninety-nine students completed the program (drop out rate 7%). The whole program lasted nearly 3 months and took the form of a blended learning course. The students participated daily in the blended learning activities. During face-to-face lessons the professor explained the principal topics of the discipline; outside of lecture periods the students had access to online learning objects (i.e. lecture slides and additional reading material) and could participate in web-forum discussions on the topics covered during lectures.

A tutor was available to provide technical assistance and to brief the students on how to cooperate on the web-forums, but the students were free to chose the conditions and the extent of their participation. At the beginning and at the end of the on-line activity, we asked the students to fill in four self-report questionnaires. Three of them were the same administered in the previous intervention: the Approach to Study Questionnaire, the Achievement Goals Questionnaire and the Implicit Theories of Intelligence Scale. An additional instrument was used to analyze ICT skills and individual conceptions of learning and teaching with ICT: an Italian questionnaire, the E-learning: New Technologies for Learning and Teaching Questionnaire, developed by Antonietti and colleagues (Antonietti, Catellani, Ciceri & Gilli, 2004). This instrument assesses knowledge and use of new technologies as well as the perceived influence of e-learning on achievement, on social relationships and on the individualization of learning processes.

In the first step of the program (1 month), participants introduced themselves to the group, debated the choice of name for their group and were free to open new discussion topics. The aim of this first phase was to create positive relations between the members of each group and to allow them to adequately practice writing on the ICT platform and to learn netiquette. It also provided a good setting to foster metacognitive reflection, both on use of the ICT platform and on individual approaches to study. The proliferation of perspectives and topics led students to note individual differences about what was relevant to know. Similarly, students perceived the need to plan and organize the web-forum discussions so as to facilitate selection, reading and participation.

Thus, after nearly four weeks, the students in each of the four virtual classes flagged the need to re-organize the topics in the virtual space. They agreed with the tutor's proposal to close the previously opened threads and cooperated to select a more limited number of relevant topics. In line with the participants' proposals and comments, the tutor opened discussions on twelve developmental psychology topics, as well as a forum to facilitate socialization between group members and a forum where students could post technical queries or flag problems. In addition, two of the four virtual classes had the opportunity to exchange opinions about study methods on a dedicated forum. This second step lasted nearly eight weeks and the number of messages grew in parallel with the development of the various topics throughout the face-to-face lectures.

At the end of the course, all the participants were asked to write a report outlining the advantages and the disadvantages of their online experience.

Although analysis of the pre- and post-results of the self-report questionnaires (MANOVA, repeated measures) did not show significant relevant differences in SRL competencies (see De Marco, 2009; Cesareni, Albanese, Cacciamani, Castelli, De Marco, Fiorilli, 2008; Albanese, De Marco & Fiorilli, in press), content analysis of the students' reports showed that an improvement was perceived in all the SRL competencies in correlation with participation in the blended learning activity (see De Marco & Albanese, in press1; De Marco & Albanese, in press2; De Marco, 2009).

Solutions and Recommendations (b)

Students attending the second project reported an advance in all the SRL competencies analyzed. The perceived improvement, however, was restricted to the study of the contents of the courses they were attending. This finding led us to suppose that ICT environments like web forum could facilitate self-regulated study of the contents in blended learning courses instead of general SRL competencies. In other words, students showed difficulties in the transfer of learned competencies.

Fostering SRL through Blended Learning: Reflecting on Personal Skills

The first project described above enabled students to acquire SRL competencies that could be transferred to different contexts. Unfortunately during face-to-face meetings the drop-out rate was high. The project concerned cross domain competencies, not explicitly linked to content learning, and maybe the training was perceived as too demanding for the advantages it could offer.

The second project was successful in engaging students' interest and commitment, and led to perceived gains in self-regulated learning strategies, but students did not seem to be able to transfer the self-regulation competencies acquired to traditional courses of study.

The third project was developed with the aim of minimizing the limits previously encountered by integrating the best outcomes of the two earlier projects. In detail we focused our intervention on SRL skills while avoiding the perception of demanding activities in terms of contents and organization.

As in the first project mentioned above, the purpose and structure of the SRL course were presented during a regular academic lecture, and 40 students decided to take part in the project. A preliminary face-to-face meeting was held to provide participants with the necessary technical information to access the ICT platform. Most of the participants were already in the habit of taking part in e-learning activities.

We chose an ICT platform, Docebo, which provided access both to web forum and to notices, learning materials and a course calendar.

The notices advised the students about the ongoing e-tivities, where to find the learning materials and how to use them. There were theoretical learning objects presenting the principal concepts for each topic as well as e-tivities providing suggestions, problem-solving and reflection activities, readings and case studies, which students could read and try to apply to their regular courses of academic study. Students had the opportunity to exchange opinions and share their knowledge of the learning objects on dedicated forums. Finally, the calendar allotted a fixed time to each topic, so participants were aware in advance of how the e-tivities were sequenced. The ICT platform and all the learning materials were designed to allow students explore different aspects of SRL via a friendly yet stimulating interface. A further aim of the project was to facilitate collaboration between participants, so a "social" forum was provided where students could meet and talk about themselves and their hobbies.

The project lasted three months. In the first week, students were given the opportunity to familiarize themselves with the on-line environment and get to know the other group members. Participants were first asked to complete two on-line self-report questionnaires: the Motivated Strategies for Learning Questionnaire (Pintrich, Smith, Garcia & McKeachie, 1993) which assess students' motivational orientations (goal orientation, task value, control beliefs, self-efficacy, anxiety) and their use of different learning strategies (rehearsal, elaboration, organization, critical thinking, metacognitive self-regulation, time and study environment, effort regulation, peer learning, help-seeking) and the Emotion in Studying Questionnaire (Mega, Moè, Pazzaglia, Rizzato & De Beni, 2007) previously described. With just two instruments we filled (and sent by e-mail) a complete personal profile for each student without asking them to fill in several questionnaires. In the profiles we outlined strengths and weaknesses of each student with regard to SRL.

The students were invited to take part in co-operative online learning activities. Participants worked on each topic for about two weeks, depending on the time available to them.

We identified the main areas of SRL for the students to work on, based on the findings of the earlier projects. Specifically, we focused on the personal features involved in learning activities such as personal conceptions of intelligence; cognitive styles in terms of tendency to prefer some modes of learning over others; causal attribution style for success and failure; cognitive and metacognitive strategies (planning, monitoring, self-evaluation); positive and negative emotions associated with learning activities and outcomes, with particular emphasis on anxiety; personal motivation and goals; environment management; cooperation and help-seeking strategies to build knowledge. Participants could use the forums to share their opinions, difficulties and personal experience in relation to each of the proposed topics.

We provided summaries of participant contributions, outlining the main concepts which could be helpful to other students, we responded to queries and doubts expressed, we proposed alternative materials relevant to the topics in hand. We also added further topics to meet specific expectations and requirements on the part of the students.

The content analysis of web forum discussion showed that the project promoted a metacognitive approach to study, informing participants of the value of reflecting on their personal beliefs about the role of student and of redefining personal learning processes, thereby moving from a hetero-regulated to a self-regulated study method. For example, students reported that they had become aware of the need to modify cognitive styles and strategies to fit different contexts and learning goals, and to evaluate efficacy by means of control procedures. They also pointed out the importance of ability to objectively evaluate task difficulty, without underestimating or overestimating it, and of ability to tolerate frustration due to failure in meeting goals (see Farina, Businaro, De Marco & Albanese, 2009). The forum discussions highlighted the participants' belief that success in study was determined above all by commitment and effort. Students underlined the salience of intrinsic motivation (interest, curiosity, competence,...) and extrinsic motivation (marks, rewards,...) in driving their learning activity. Another significant challenge for the students was ability to manage anxiety and its influence on performance. Practical exercises were provided to stimulate reflection on factors leading to anxiety (degree of competence, fear of the judgement of other, imperfect knowledge,...) as well as reappraisal of previous life experiences in order to cope better with future events.

Solutions and Recommendations (c)

This project was specifically designed to promote students collaboration. The learner was helped to acquire self-regulation competencies through interaction with other people. It was therefore relevant for participants to reflect on the benefits of help-seeking and to identify sources of social support (peer and adults) helpful for learning.

Although a part of students restricted their involvement to reading the materials and participating in individual e-tivities without participating in forum discussions (lurkers), the overall level of student involvement in the project and the positive feedback received about the opportunity to have flexible access to an ICT platform where opinions and topic knowledge could be shared with other people, lead us to be confident that this approach is effective in developing SRL competencies (Farina, Businaro, De Marco & Albanese, 2009). Students confirmed that they were applying the newly acquired competencies to their regular academic studies; they found the whole project to be useful and motivating and asked for it to be extended for the rest of the academic year.

FUTURE RESEARCH DIRECTIONS

Individual competencies in SRL are too often taken for granted at higher levels of instruction, especially at university level. However, a growing number of researchers are bringing this assumption into question (Mega, Cazzaniga & De Beni, 2007). Nonetheless, the teaching of SRL at university is frequently left to the initiative of a small number of professors who believe in its importance and work to foster it. The need for more structured interventions involving students from different courses and disciplines is now beginning to be perceived (Albanese, De Marco & Fiorilli, 2008; Nevgi, Virtanen & Niemi, 2006; Mega, Cazzaniga & De Beni, 2007). To this end, in-depth study of the processes involved in SRL is required, along with research into the best practices to develop SRL.

The small number of participants and the optional nature of our projects only permit us to suggest some interventions that seem to be useful for the development of SRL as well as motivating for students. Our aim is to provide some examples of good practice in SRL development, but more specific studies must be done to understand how the specific features of the ICT used actually influence each of the SRL competencies.

According to Winne (2005), new assessment techniques must be found for SRL in general and for each competence involved in SRL in particular. An in-depth analysis of each competence would allow the causal relationships between competencies and interventions to be explored. It would also facilitate development of an empirically-based model, outlining how the individual components mutually influence one another. Such a model would allow more structured activities to be planned and tailored to meet individual student needs.

CONCLUSION

This chapter aimed to illustrate the work carried out in the field of SRL development and the results obtained. In particular, we wished to highlight the advantages offered by ICT in fostering SRL skills in higher-level students.

The outcome of many years' experience leads us to be confident in recommending the use of new technologies in future SRL development interventions at tertiary level.

We strongly believe that cooperation and collective reflection on study methods are key elements of effective intervention on SRL. Unfortunately cooperative learning and knowledge building experiences at university are limited, due to the large number of students attending courses and the significant amount of time that has to be dedicated to the academic courses themselves. The specificity of the university courses rarely gives the opportunity to reflect systematically on cross-domain aspects such as the processes involved in studying.

ICT allows us to overcome these practical limitations, at least in part, and provides the optimal setting for the development of SRL skills at university. Students can freely participate in discussions of their interest when they have time, they can decide how much time to dedicate to each topic and they always have the opportunity to exchange view-points with their peers. More experienced students are able to avail of this opportunity, because they have the base knowledge required to participate in the activity without continuous supervision by teachers.

The final project presented above is, in our opinion, an example of an effective intervention to develop SRL. The presence of e-tivities allows students to train themselves in the specific skills involved in the different SRL processes. The multimedia features of e-tivities stimulate student involvement. The e-tivities must be demanding, but not require too much time to complete, in order to minimize loss of attention and drop out.

The web-forums serve to create a supportive and collaborative climate and to develop help-seeking competencies. They also offer students the opportunity to metacognitively rethink their study methods, by comparing what they do with other best practices. Each post stays on the web-forum and may be read flexibly in line with individual student constraints and needs. The calendar enables students to organize and plan topics and activities together, thereby fostering time management skills.

In sum, availability of activities and information about SRL in blended or technology enhanced collaborative courses constitutes an effective means of developing SRL skills. We therefore hope that this research field will be more and more enriched by new proposals targeted at improving such interventions with university students.

ACKNOWLEDGMENT

The Authors together outlined and discussed the topics that the present chapter is devoted to. Ottavia Albanese and Eleonora Farina planned the structure of the chapter and they revised it. The introduction, the paragraphs concerning the processes and the empirical implications of SRL, the blended learning intervention on the contents of academic subjects were written by Barbara De Marco. Nicoletta Businaro reported the face-to-face, the blended learning intervention on personal SRL skills and the conclusion.

Correspondence concerning this chapter should be addressed to Barbara De Marco, Department of Human Science, University of Milan Bicocca, Piazza dell'Ateneo Nuovo 1, 20126 Milan. Electronic mail may be sent to barbara.demarco1@unimib.it.

REFERENCES

Albanese, O. (2003). *Percorsi metacognitivi (Metacognitive Paths)*. Milan, IT: Franco Angeli.

Albanese, O., De Marco, B., & Fiorilli, C. (2008). Processi di autoregolazione e obiettivi di apprendimento nell'attività di studio. Una ricerca per orientare all'università (Self-regulation processes and learning goals in study activity. A research for the vocational guidance at university). In F. Petruccelli (Ed.), *Dalla scuola all'università: una scelta di vita (From school to university: a life choice)* (pp. 123-140). Milan, IT: Franco Angeli.

Albanese, O., De Marco, B., & Fiorilli, C. (in press). Autoregolazione e riflessione metacognitiva: dati empirici sul corso di Psicologia dello Sviluppo erogato in modalità di Blended Learning (Self-regulation and metacognitive reflection: empirical data about the academic blended course of Developmental Psychology). In Cacciamani, S. (Ed.), *Knowledge Building Communities: ripensare la scuola e l'Università come comunità di ricerca (Knowledge Building Communities: rethinking the school and the university in terms of research community). Naple, IT: Scriptaweb.*

Albanese, O., Duodin, P. A., & Martin, D. (2003). *Metacognizione ed educazione. Processi, apprendimenti, strumenti (Metacognition and education. Processes, learning, tools)*. Milan, IT: Franco Angeli.

Albanese, O., Farina, E., Brambilla, P., & Businaro, N. (2008). *Attività di studio all'Università: un progetto di Ateneo per l'orientamento delle matricole (Study methods at University: a vocational guidance project for first year students).* Paper presented at the 10° National Conference Orientamento alla scelta: ricerca, formazione, applicazioni (Vocational Guidance: research, education, training). Florence, IT.

Albanese, O., Farina, E., Fiorilli, C., & Minosso, F. (2007). Lo sviluppo delle abilità di autoregolazione a partire dalla scuola superiore (The self-regulated skills development during high school). In S. Soresi (Ed.), *Orientamento alle scelte. Rassegne, ricerche strumenti ed applicazioni (Vocational guidance: review, research, tools and training)* (pp.22-36). Florence, IT: Giunti.

Albanese, O., Fiorilli, C., & Farina, E. (2004). Autoregolazione nell'attività di studio e concezione dell'intelligenza (Self-regulation in studying and the implicit theories of intelligence). *Giornale Italiano di Psicologia dell'Orientamento (Italian Journal of Psychology Vocational Guidance), 5*(2), 14-29.

Antonietti, A., Catellani, P., Ciceri, M. R., & Gilli, G. M. (2004). I corsi online all'università: aspetti cognitivi e psicosociali (Online courses at university: cognitive and psychosocial aspects). In C. Scurati (Ed.), *E-learning/università. esperienze, analisi, proposte (E-learning/university: experiences, analysis, proposals)* (pp. 243-283). Milan, IT: Vita e Pensiero.

Bandura, A. (1989). Human Agency in Social Cognitive Theory. *The American Psychologist, 44*(9), 1175–1184. doi:10.1037/0003-066X.44.9.1175

Barnard, L., Paton, V. O., & Lan, W. Y. (2008). Online self-regulatory learning behaviors as a mediator in the relationship between online course perceptions with achievement. *International Review of Research in Open and Distance Learning, 9*(2), 1–11.

Boekaerts, M., & Corno, L. (2005). Self-regulation in the classroom: A perspective on assessment and intervention. *Applied psychology: an international review, 54*(2), 199-231.

Cacciamani, S., & Giannandrea, L. (2004). *La classe come comunità di apprendimento (The class as a learning community).* Rome, IT: Carocci.

Cesareni, D., Albanese, O., Cacciamani, S., Castelli, S., De Marco, B., & Fiorilli, C. (2008). Tutorship styles and Knowledge Building in an Online Community: Cognitive and Metacognitive Aspects. In B.M. Varisco (Ed.), *Psychological, pedagogical and sociological models for Learning and assessement in virtual communities* (pp. 13-56). Milano, IT: Polimetrica.

Cornoldi, C., De Beni, R., & Gruppo, M. T. (2001). *Imparare a studiare 2 (Learning to study 2)*. Trento, IT: Erickson.

Dabbagh, N., & Kitsantas, A. (2005). Using web-based pedagogical tools and scaffolds for self-regulated learning. *Instructional Science, 33*(5-6), 513–540. doi:10.1007/s11251-005-1278-3

De Beni, R., Moè, A., & Cornoldi, C. (2003). *Questionario di Approccio allo Studio (The Approach to Study Questionnaire)*. Trent, IT: Erickson.

De Marco, B. (2009). *Favorire un approccio autoregolato allo studio: una ricerca con studenti universitari in situazione di blended learning (Promoting a self-regulated approach to study: a research about students attending a blended course at university).* Unpublished doctoral dissertation, University of Milan Bicocca, Milan, IT.

De Marco, B., & Albanese, O. (in press1). Riflettere all'interno di comunità virtuali per sviluppare competenze autoregolative nell'attività di studio: un'indagine con approccio quali - quantitativo (Reflecting in virtual communities to promote self-regulated competencies in studying: a quali-quantitative research approach). *Qwerty.*

De Marco, B., & Albanese, O. (in press2). Collaborare online per migliorare il metodo di studio: una ricerca con matricole universitarie (Online collaboration improving study method: a research about first year university students). In M. Pieri, & D. Diamantini (Eds.), *Ubiquitous learning.* Milan, IT: Guerini e Associati editore.

Dweck, C. S. (1999). *Self-theories: Their role in motivation, personality, and development.* Philadelphia, PA: Taylor & Francis.

Elliot, A. J., & McGregor, H. A. (2001). A 2×2 Achievement Goal Framework. *Journal of Personality and Social Psychology, 80*(3), 501–519. doi:10.1037/0022-3514.80.3.501

Faria, L., & Fontaine, A. M. (1997). Adolescent's personal conceptions of intelligence: The development of a new scale and some exploratory evidence. *European Journal of Psychology of Education, 12*(1), 51–62. doi:10.1007/BF03172869

Farina, E., Businaro, N., De Marco, B., & Albanese, O. (2009). *Attività di studio all'Università: autoregolazione ed emozioni (Study method at University: self-regulation and emotions)*. Paper presented at the 11° National Conference Orientamento alla scelta: ricerca, formazione, applicazioni (Vocational guidance: research, education, training). Padua, IT.

Fisher, M., & Baird, D. E. (2005). Online learning design that fosters student support, self-regulation, and retention. *Campus-Wide Information Systems, 22*(5), 88–107. doi:10.1108/10650740510587100

Kaye, A. R. (1995). Computer supported collaborative learning. In Heap, N., Thomas, R., Mason, R., & Mackay, H. (Eds.), *Information technology and society*. London, UK: Sage.

Kramarski, B., & Gutman, M. (2006). How can self-regulated learning be supported in mathematical e-learning in environments? *Journal of Computer Assisted Learning, 22*(1), 24–33. doi:10.1111/j.1365-2729.2006.00157.x

Ligorio, M. B., Cacciamani, S., & Cesareni, D. (2006). *Blended learning: dalla scuola dell'obbligo alla formazione adulta (Blended learning: since high school to adult training)*. Rome, IT: Carocci.

Mega, C., Cazzaniga, S., & De Beni, R. (2007). *Differenti profili degli studenti universitari: quale relazione con il rendimento accademico? (Different university students profiles: are they related to achievement?)*. Paper presented at the XXVI National Conference AIRIPA. I disturbi dell'apprendimento (Learning disabilities). Belluno, IT. Available at http://www.airipa.it/convegno_airipa/atti_prec.html

Mega, C., Moè, A., Pazzaglia, F., Rizzato, R., & De Beni, R. (2007). Emozioni nello studio e successo accademico. Presentazione di uno strumento (Emotions and academic achievement. Presentation of a test). [Psychology Italian Journal]. *Giornale Italiano di Psicologia, 34*(2), 451–464.

Narciss, S., Proske, A., & Koerndle, H. (2007). Promoting self-regulated learning in web-based learning environments. *Computers in Human Behavior, 23*(3), 1126–1144. doi:10.1016/j.chb.2006.10.006

Nevgi, A., Virtanen, P., & Niemi, H. (2006). Supporting Students to Develop Collaborative Learning Skills in Technology-Based Environments. *British Journal of Educational Technology, 37*(6), 937–947. doi:10.1111/j.1467-8535.2006.00671.x

Pekrun, R., Goetz, T., Titz, W., & Perry, R. P. (2002). Academic emotions in students' self-regulated learning and achievement: a program of quantitative and qualitative research. *Educational Psychologist, 37*(2), 91–106. doi:10.1207/S15326985EP3702_4

Pepi, A., Faria, L., & Alesi, M. (2007). La scala delle concezioni personali dell'intelligenza: un confronto tra le caratteristiche psicometriche della versione portoghese e di quella italiana (The implicit theories of intelligence scale: comparing psychometric properties of Portuguese and Italian version). [Bulletin of Applied Psychology]. *Bollettino di Psicologia Applicata, 251*(1), 33–44.

Pintrich, P. R. (1999). The role of motivation in promoting and sustaining self-regulated learning. *International Journal of Educational Research, 31*(6), 459–470. doi:10.1016/S0883-0355(99)00015-4

Pintrich, P. R. (2000). The role of goal orientation in self-regulated learning. In Boekaerts, M., Pintrich, P. R., & Zeidner, M. (Eds.), *Handbook of self regulation* (pp. 452–502). San Diego, CA: Academic Press. doi:10.1016/B978-012109890-2/50043-3

Pintrich, P. R., Smith, D. A. F., Garcia, T., & McKeachie, W. J. (1993). Reliability and predictive validity of the Motivated Strategies for Learning Questionnaire (MSLQ). *Educational and Psychological Measurement*, *53*(3), 801–813. doi:10.1177/0013164493053003024

Salovaara, H. (2005). An exploration of students' strategy use in inquiry-based computer-supported collaborative learning. *Journal of Computer Assisted Learning*, *21*(1), 39–52. doi:10.1111/j.1365-2729.2005.00112.x

Schraw, G., Crippen, K. J., & Hartley, K. (2006). Promoting Self-Regulation in Science Education: Metacognition as Part of a Broader Perspective on Learning. *Research in Science Education*, *36*(1-2), 111–139. doi:10.1007/s11165-005-3917-8

Vallerand, R. J., Fortier, M. S., & Guay, F. (1997). Self-Determination and Persistence in a Real-Life Setting: Toward a Motivational Model of High School Dropout. *Journal of Personality and Social Psychology*, *72*(5), 1161–1176. doi:10.1037/0022-3514.72.5.1161

Varisco, B. M. (2008). *Psychological, pedagogical and sociological models for learning and assessment in virtual communities of practice*. Milan, IT: Polimetrica.

Veenman, M. V. J., Van Hout-Wolters, B. H. A. M., & Afflerbach, P. (2006). Metacognition and learning: conceptual and methodological considerations. *Metacognition and Learning*, *1*(1), 3–14. doi:10.1007/s11409-006-6893-0

Winne, P.H. (2005). Key issues in modelling and applying research on self-regulated learning. *Applied psychology: an international review*, *54*(2), 232-238.

Zimmerman, B. J. (2008). Investigating self-regulation and motivation: Historical background, methodological developments, and future prospects. *American Educational Research Journal*, *45*(1), 166–183. doi:10.3102/0002831207312909

ADDITIONAL READING

Alexander, P. A., & Winne, P. H. (2006). *Handbook of educational psychology* (2nd ed.). Mahwah, NJ: Lawrence Erlbaum Associates.

Antonietti, A., Colombo, B., & Lozotsev, Y. (2008). Undergraduates' metacognitive knowledge about the psychological effects of different kinds of computer-supported instructional tools. *Computers in Human Behavior*, *24*(5), 2172–2198. doi:10.1016/j.chb.2007.10.004

Azevedo, R., & Cromley, J. G. (2004). Does training of self-regulated learning facilitate student's learning with hypermedia? *Journal of Educational Psychology*, *96*(3), 523–535. doi:10.1037/0022-0663.96.3.523

Boekaerts, M., Maes, S., & Karoly, P. (2005). Self-regulation across domains of applied psychology: is there an emerging consensus? *Applied psychology: an international review*, *54*(2), 149-154.

Boekaerts, M., Pintrich, P., & Zeidner, M. (2000). *Handbook of self-regulation*. San Diego, CA: Academic Press.

Cacciamani, S. (2001). Costruire conoscenza attraverso le nuove tecnologie: Knowledge Forum e i possibili scenari formativi (Building knowledge through new technologies: Knowledge Forum and the possible educational scenarios). [Pedagogic Guideline]. *Orientamenti Pedagogici*, *1*(1), 84–97.

Fiorilli, C. (2003). Le concezioni degli insegnanti sull'intelligenza e l'importanza del trasformare le credenze in conoscenze (The implicit theories of intelligence and the importance to change the theories in knowledge). In O. Albanese (Ed.), *Percorsi metacognitivi (Metacognitive Paths)* (pp. 90-114). Milan, IT: Franco Angeli.

Garrison, D. R., & Kanuka, H. (2004). Blended learning: Uncovering its transformative potential in higher education. *The Internet and Higher Education*, *7*(2), 95–105. doi:10.1016/j.iheduc.2004.02.001

Gilman, R., & Anderman, E. M. (2006). The relationship between relative levels of motivation and intrapersonal, interpersonal, and academic functioning among older adolescents. *Journal of School Psychology*, *44*(5), 375–391. doi:10.1016/j.jsp.2006.03.004

Kitsantas, A., & Zimmerman, B. J. (2006). Enhancing self-regulation of practice: the influence of graphing and self-evaluative standards. *Metacognition and Learning*, *3*(1), 201–212.

Ligorio, M.B., & Veermans, M. (2005). Perspectives and patterns in developing and implementing international web-based Collaborative Learning. *Environments, computers & education, 45*(3), 271-275.

Meece, J. L., Anderman, E. L., & Anderman, L. H. (2006). Classrooms goal structure, student motivation, and academic achievement. *American review of psychology, 57*(1), 487-503.

Niemi, H., Nevgi, A., & Virtanen, P. (2003). Towards self-regulation in web-based learning. *Journal of Educational Media, 28*(1), 49–71.

Pintrich, P. R., & Schunk, D. (1996). *Motivation in education: theory, research, and applications*. Englewood Cliffs, NJ: Prentice Hall Merrill.

Sawyer, K. (2006). *The Cambridge Handbook of the Learning Science*. New York, NY: Cambridge University Press.

Scardamalia, M., Bereiter, C., McLean, R. S., Swallow, J., & Woodruff, E. (1989). Computer supported intentional learning environments. *Journal of Educational Computing Research*, *5*(1), 51–68.

Veenman, M. V. J. (2007). The assessment and instruction of self-regulation in computer-based environments: a discussion. *Metacognition and Learning*, *2*(2-3), 177–183. doi:10.1007/s11409-007-9017-6

Veenman, M. V. J., Van Hout-Wolters, B. H. A. M., & Afflerbach, P. (2006). Metacognition and learning: conceptual and methodological considerations. *Metacognition and Learning*, *1*(1), 3–14. doi:10.1007/s11409-006-6893-0

Vohs, K. D., Baumeister, R. F., & Ciarocco, N. J. (2005). Self-regulation and self-presentation: regulatory resource depletion impairs impression management and effortful self-presentation depletes regulatory resources. *Journal of Personality and Social Psychology*, *88*(4), 632–657. doi:10.1037/0022-3514.88.4.632

Winne, P. H. (1995). Self-regulation is ubiquitous but its forms vary with knowledge. *Educational Psychologist*, *30*(4), 223–228. doi:10.1207/s15326985ep3004_9

Winne, P. H. (2005). A perspective on state-of-the-art research on self-regulated learning. *Instructional Science*, *33*(5-6), 559–565. doi:10.1007/s11251-005-1280-9

Zimmerman, B. J. (1986). Special issue on self-regulated learning. *Contemporary Educational Psychology*, *11*(1), 305–427.

Zimmerman, B. J. (1989). A social cognitive view of self-regulated academic learning. *Journal of Educational Psychology*, *81*(3), 329–339. doi:10.1037/0022-0663.81.3.329

Zimmerman, B. J. (2002). Becoming a self-regulated learner: an overview. *Theory into Practice*, *41*(2), 64–70. doi:10.1207/s15430421tip4102_2

Zimmerman, B. J., & Martinez-Pons, M. (1988). Construct validation of a strategy model of student self-regulated learning. *Journal of Educational Psychology*, *80*(3), 284–290. doi:10.1037/0022-0663.80.3.284

Zimmerman, B. J., & Schunk, D. H. (Eds.). (2001). *Self-regulated learning and academic achievement: theoretical perspectives* (2nd ed.). Mahwah, NJ: Lawrence Erlbaum Associates.

KEY TERMS AND DEFINITIONS

Collaborative Learning: Method of instruction in which learners with different competencies work together in groups to gain a higher level of knowledge.

Emotion: Mental and physiological subjective state associated with a stimulus that can influence personal thoughts, feelings and behavior.

E-tivities: Activities created with the tools offered by the new technologies and made available to students through the world wide web.

ICT Platform: Learning environment that comprehends virtual rooms, web forums, chatting, calendars, repository, e-tivities and other technological supports for learning.

Information and Communication Technologies (ICT): Ensemble of technological instruments and capabilities that support the transmission of information and communication between people.

Metacognition: Personal knowledge and reflection on individual cognitive processes and all related aspects.

Self-Regulated Learning: Autonomous and active learning guided by cognitive and metacognitive strategic action, by motivation, by emotional control and by management of the study environment.

Study Method: Set of strategies and methodologies usually applied by a person while he/she studies.

Web Forum: Online discussion site, to which access may be regulated by password and restricted to selected users, on which contributors can post messages within threads dedicated to specific topics.

Chapter 24
SRL/SDL and Technology-Enhanced Learning:
Linking Learner Control with Technology

Jane Pilling-Cormick
Hamilton-Wentworth District School Board, Canada

ABSTRACT

When exploring the central role control plays in implementing technology-enhanced learning initiatives, it is essential to take into consideration self-regulated learning (SRL) and self-directed learning (SDL). Pilling-Cormick & Garrison's (2007) work provides a research framework which includes a comprehensive overview of how SRL and SDL are integrally related. In this chapter, the connection is taken one step further by using the framework to explore SRL/SDL Technology-Enhanced learning. Implications for practice are derived from three exploratory studies using technology-enhanced learning (handheld, web-based, and online) with a focus on learner control. Solutions and recommendations arise, including considerations for designing instruction with a focus on learner control as it relates to technology.

INTRODUCTION

This chapter focuses on crucial issues in using a learner-centered approach to technology from both a theoretical and practical viewpoint. Knowing that there is a link between self-directed learning (SDL) and self-regulated learning (SRL) (Pilling-Cormick & Garrison, 2007), this chapter builds on that work with a specific emphasis on that connection. The intent of the chapter is not to give a comprehensive overview of exploratory studies, but is rather a review of lessons learned

DOI: 10.4018/978-1-61692-901-5.ch024

from three studies conducted in a Canadian urban school board where learner control in technology-enhanced learning environments was the focus. The first study, the *handheld study*, involved 11 and 12 year old learners in a classroom setting where they were using handheld computers to learn to read and write. The second, the *web-based study*, focused on a group of 9 to 13 year old struggling readers who were using a web-based program to improve their reading skills. The third, the *online study,* involved a group of 16 to 18 year olds from across the school board, who were enrolled in an online course. In all three exploratory studies, important lessons arose about how educators can increase learner control and how essential this is to successful technology integration.

BACKGROUND

Technology is changing the way people live and learn. Jarvis (2009) describes a "new educational ecology where learners may take courses from anywhere and instructors may select any learners, where courses are collaborative and public, where creativity is nurtured as Google nurtures it, where making mistakes well is valued over sameness and safety, where education continues long past age 21, where tests and degrees matter less than one's own portfolio of work, where the gift economy may turn anyone with knowledge into teachers, where the skills of research and reasoning and scepticism are valued over the skills of memorization and calculation, and where universities teach an abundance of knowledge to those who want it rather than manage a scarcity of seats in a class" (p. 210). As Jarvis admits, his view of a new educational ecology may be utopian. The reality is that there could eventually be a movement toward such a view of education. Educators need to be aware of that possibility and the role learner control plays in this scenario. As Candy (2004, p. 39) claims, there is a lot of literature about the use of digital technologies in education, but little on "the powerful role of the teacher, trainer or facilitator and the relatively powerless and dependent role of the student, trainee, or learner". Issues rising from the research done on learner control and implications for living in a knowledge society help provide opportunities for successful technology-enhanced initiatives. As Coiro, Knoebel, Lankshear and Leu (2008) recommend, researchers need to "pay greater attention to what new technologies mean to users and less attention to specific new technologies per se" (p. 413). This, in turn, means going beyond the bells and whistles of what some new and wonderful technical application can do. In SRL/SDL Technology-Enhanced learning, the focus goes beyond the features of technology to discovering ways learners and educators successfully adjust to technology restrictions and the role control plays. Exploring the SRL/SDL connection, as it relates to learner control, and looking at the ways education is changing become essential.

SRL/SDL Connection

When addressing control in technology-enhanced learning, both SRL and SDL play a crucial role. As Pilling-Cormick & Garrison (2007) indicate, there is a link between SRL and SDL with both addressing issues of responsibility and control. They explored this link by focusing on the covert (person) and overt (behaviour and environment) aspects linking SRL and SDL. By reviewing selected SDL and SRL models and related concepts, they came up with a comprehensive and coherent framework which would then have useful and lasting value to practitioners. Both SRL and SDL are processes involving setting goals and priorities for learning. Learners, in effect, are determining what to learn. Both involve making sense of learning. For this chapter, SDL is defined as a process in which individuals determine their priorities and choose from various resources available. They play an active role in developing a system of meanings to interpret events, ideas or circumstances

(Pilling-Cormick, 1996). SRL is defined as an active, constructive process whereby learners set goals for their learning and then attempt to monitor, regulate, and control their cognition, motivation, and behaviour, guided and constrained by their goals and the contextual features in the environment" (Pintrich 2000, p. 453). SRL and SDL essentially deal with the same concepts – external management practices and internal monitoring processes. For the purpose of the chapter, the connection between the two concepts is referred to as SRL/SDL Technology-Enhanced learning. In order to have a more complete picture of how to provide kinds of technology that build support for taking on responsibility, there needs to be a focus on learner control-the link between SRL and SDL.

Role of Learner Control in Technology-Enhanced Learning

Various issues arise when considering the knowledge society of today. Education is changing rapidly. Youth today learn differently, with a major focus being learner control. As Richardson (2006) indicates, learners will need to know how to manage the information that they consume. Learners will be required to collect, store and retrieve relevant information throughout their lives and they need the skills to do so effectively and efficiently. Warlick (2004) and November (2001) point to the increasingly central role the learner plays in directing or being responsible for learning. When discussing the learner of today, Oblinger & Oblinger (2005) refer to first person learning. Today's learners do better when they can actively construct their own learning. Using technology allows for some of the control these learners seek. As Warlick (2004) indicates, the best thing educators can do for children is to teach them how to teach themselves. Livingston (2006) also claims "researchers agree that 1-to-1 is where the magic happens when it comes to learning with computers" (p. 165). Jarvis (2009) similarly promotes giving people control with technology by stating that "now the internet allows us to speak to the world, to organize ourselves, to find and spread information, to challenge old ways, to retake control" (p. 11). "With a rise in online learning, we are seeing a definite shift away from an "other-directed" to a "self-directed" approach, with students taking on more responsibility for their learning" (Pilling-Cormick, 2005, p. 9). Garrison & Anderson (2003) stress that "e-learning draws attention to fundamental responsibility and control issues" (p. 14). As they further emphasize, "SDL is emerging as an important conceptual model towards understanding issues raised by technology that has the potential to transfer enormous control to the learner" (p. 15). There are restrictions within which technology-enhanced initiatives operate such as curriculum constraints, school operating procedures, and computer software limitations. Technology-enhanced education often puts the learner in control, whether the learner is moving through a computer program, entering text in a handheld, or taking an online course. The connection between technology and control is evident and requires further investigation.

Control can take on various meanings. For the purposes of this chapter and the investigation of SRL/SDL Technology-Enhanced learning, the definition adopted specifically takes into consideration technology use. Pilling-Cormick and Garrison's (2007) research framework provides an excellent starting point for investigating technology related control. Part of the framework includes Pintrich's (2000) definition of SRL, stressing the crucial role of control by claiming "all models of self-regulated learning assume that attempts to monitor and control one's own learning through various adaptive cognitive, motivational, or behavioural regulatory strategies are basically positive for learning and achievement" (p. 492). Pilling-Cormick and Garrison (2007) also focus on the role of control within SDL. In this chapter, control is the extent to which learners can direct their learning; it influences all other aspects of teaching and learning (Pilling-Cormick, 1999).

Control is situational. Candy (1991) states that the degree of learner control is a continuum; various instructional situations appear along that continuum. By using a situational definition, the more control the learner has in a learning situation, the more the learner directs the learning. In Pilling-Cormick's (1999) Self-Directed Learning Process (SDLP) model, the process of SDL is the interaction between learner and educator taking place within the varying context of control. In the model, four factors influence the amount of control learners have: social constraints, environmental features, learner characteristics, and educator behaviour. With technology use, these four factors will influence how much control learners have when using the particular type of technology. For instance, previous experience with computers is a learner characteristic which will often determine how comfortable learners are about taking control when using technology. The more experience the learner has, the more likely one would assume he or she is to take on increasing amounts of control.

INVESTIGATING LEARNER CONTROL IN TECHNOLOGY-ENHANCED LEARNING

Using the SRL/SDL framework as a basis, the researchers completed an exploratory study in three settings where learners used technology-enhanced learning with an element of learner control: a grade 6 handheld application, an elementary web-based program, and an online learning secondary school level setting. The researchers used a case study approach (Gall, Gall & Borg, 1999). From the SRL/SDL theoretical framework, a number of elements were identified as being supportive of increased learner control. Each of the three studies tracked these elements. To enhance the credibility of the exploratory studies, the authors collected data throughout the studies looking for consistent patterns and used several data gathering methods to ensure a triangulation methodology (McMillan, 2000). Data collection included learner focus groups, educator interviews, initial and final educator and learner surveys. Guided questions in both learner focus groups and educator interviews specifically included control issues arising from the SRL/SDL Technology-Enhanced learning research framework. Data analysis used was interpretational analysis (Gall, Gall & Borg, 1999) where themes and patterns emerged.

Grade 6 Handheld Study

The purpose of the handheld study was to explore ways the handhelds enabled learners to take on increased amounts of control over their learning as they worked on improving their literacy skills. The researchers did not expect learners to immediately take responsibility for all their learning. Instead, exploring ways they do take on elements of control became the focus. The learners worked on various activities within the class where they had control of their pacing, trouble shooting for problems and content that they entered into the handhelds. The study took place over one academic year. Forty-one 11 and 12 year old learners in two Grade 6 classes and one educator at one elementary school participated. Learners received a handheld computer (palm pilot), literacy support software and a wireless keyboard. Programs placed on the handhelds by the educator included a word processor (Free-Write), a graphic organizer used in the pre-reading process (Pico-Map), and a tool for animating (Sketchy).

Web-Based Study

The overall intent was to use technology to help build literacy skills with struggling elementary learners. Learners used the IBM program Learning Upgrade for one academic year. The program used a game format where learners progressed in a linear fashion from one level to the next. The study focused on discovering ways learners controlled their learning when using the program, within the constraints of the software. Educators gave learners opportunities to modify the speed at which they completed the program, but there was no flexibility within the program itself for learners to choose different topics to learn. The lack of flexibility of the program format prevented learners from taking full control of how they progressed through the program. Still, when learners progressed through the levels, they needed to make decisions about controlling their learning. They could not deviate from the linear progression within the program. Yet, the program structure allowed learners to pace their own learning and troubleshoot on their own. Participants included fifty nine 9 to 13 year old struggling readers from 2 elementary schools and three educators.

Online Learning Study

The focus of the online learning study was to track ways learners controlled their learning when taking English and Philosophy courses completely online. There were no scheduled face-to-face meetings with the educator at any time and all interaction took place over the internet. Learners were responsible for various aspects of their learning including pacing, troubleshooting, and successful submission of assignments. In the online learning study, sixty-three 16 to 18 year old learners and four educators from across an urban school board participated in completely online courses. The courses were one semester long (5 months). Learners used their own computers or those available at their home schools to complete the course work. These courses followed the same curriculum as courses offered within a school building. Geographically, the learners were scattered throughout the school board.

LESSONS LEARNED FOR SRL/SDL TECHNOLOGY-ENHANCED LEARNING

When designing instruction, the focus is on the approach to learning and how to design instruction to meet the characteristics of that approach. When promoting SDL, it is critical to consider whether instruction is designed to provide opportunities for learners to be SD (Costa & Kallick, 2004). The way educators teach becomes different when using technology, especially when learner control is the aim. As Jarvis (2009) claims "memorization is not as vital a discipline as fulfilling curiosity with research and reasoning when learners recognize what they don't know, form questions, seek answers, and learn how to judge them and their sources" (p. 215). When Tapscott (2009) states educators cannot "throw technology into the classroom and hope for good things" (p. 148), he is stressing the importance of designing instruction specifically for technology use. He further addresses the role of learner control by stating educators need to "use technology for a student-focused, customized, collaborative learning environment" (p. 148). This section focuses on lessons learned from the mentioned exploratory studies where learners take on increasing amounts of control. Emerging trends stress the importance of interacting with others, encouraging SRL/SDL Technology-Enhanced learning, and considering learner characteristics.

Interaction in SRL/SDL Technology-Enhanced Learning

It is a myth that SDL involves learning in isolation and never interacting with the educator (Chene, 1983; Ericksen, 1984; Hiemstra, 1988; Brookfield, 1990; Hiemstra & Sisco, 1990; Candy, 1991; Collins, 1991; Galbraith, 1991). Similarly, SRL/SDL Technology-Enhanced learning does not imply that learners learn only by themselves. Because of the prominent role of this interaction, it becomes necessary to look at what types of interaction encourage learner control when using technology. In the research studies undertaken, a number of valuable practical implications appear when researchers looked at ways the initiatives foster the development of relationships. In SRL/SDL Technology-Enhanced learning, the importance of face-to-face contact must be recognized.

Incorporating Face-to-Face Contact

In SRL/SDL Technology-Enhanced learning, face-to-face contact is often necessary for enabling learner control. As Stafford, Miller and Ollivierre (2006) claim, technology will not replace educators as "teachers still play a significant role in ensuring effective learning by structuring and framing the activity of the learner when technology is utilized in the classroom" (p. 163). As Maeroff (2003) states, "learning that makes extensive use of the Internet must face up to the dark side of solitary education" (p. 41). He goes on to say that able instructors of online courses keep in mind the potential role of interaction in breaking down isolation that might undermine the morale of learners. The intent is to "keep students who are out of view from feeling out of touch" (p. 41). Some learners may require more face-to-face contact just to be in a better position to take on more control. As Siemens & Yurkiw (2003) indicate, establishing a connection with learners allows the educator to ensure learners that he or she is "there" and that everything is working. Educators must realize the humanity of learners when using technology.

In the handheld study, learners completed various written activities where elements of control were evident when learners were in charge of what they were doing on their individual handhelds. Learners controlled what they were typing. They also had to use control when they experienced hardware difficulties or problems with assignment completion. The educators found that a blended approach to technology use worked well, incorporating face-to-face contact with technology. One technique the educator used was to beam instructions for a written task to each learner. If a learner could still not begin the task, the educator relied on face-to-face contact which encouraged the learner to take on more control.

In the web-based study, even though the progress through the levels in the program was linear, being able to advance required control on the learner's part. The speed of progression or pace was determined by the learner. Yet if they attempted to progress exclusively using the program for assistance, a number of learners faced difficulties controlling their learning. According to the educator, the human contact provided the necessary stimulus for learner control.

In the online study, the learners were spread widely across a geographical area, which meant that all of the interaction between educator and learners was online. To encourage and facilitate learner control, results indicated that the connection between educator and learners was vital. Interestingly, both learners and educators expressed the need for face-to-face contact. Other learners similarly indicated that even though e-mail was available for interacting with the educator, they found it hard to get questions clarified enough for their satisfaction. When exclusively relying on online communication, there was a perceived lack of educator/learner interaction appearing. To encourage more control, educators developed opportunities to make contact with learners. Some educators began providing online chats. Others set specific times to be available online. A time was set up where the learners actually did meet the educator in a face-to-face meeting. If learners are to be directing their learning or having more control when using technology, it would become more difficult for them to do so if they required more educator assistance and they perceived it to be unavailable.

Promoting Learning Together

In SRL/SDL Technology-Enhanced learning, encouraging learners to work together is another essential element. As Chene (1983) indicates, it is important to use others to test progress and receive constructive feedback in a SD approach. When making decisions about learning, it is often extremely helpful to acquire the opinions of others. As Oblinger & Oblinger (2005) suggest, the importance of interaction is not new. Learning science consistently shows that learners learn more when they interact with material, each other and faculty. In a SD environment, co-operation rather than competition is also essential. "The teacher and the taught must be prepared to suspend critical judgment and to explore together" (Candy, 1991, p. 337). At the same time, group work often requires co-operation and making compromises which can be a challenge when directing learning. In an online setting, learning through collaboration can require participants to take intellectual risks and it then becomes the responsibility of the course instructors to proactively design and nurture a community culture where participants are honest and supportive (Elbaum, Mcintyre & Smith, 2002). "The interactive nature of the best e-learning captivates students and makes them almost forget that the setting is a classroom of one" (Maeroff, 2003, p. 42). Exploring ways learners and educators use technology to facilitate working together and the subsequent effect on learner control becomes helpful.

In the handheld study, the learners worked either as a group or entire class. When writing stories, the learners worked together to gather ideas. They made decisions about their own learning because they controlled which ideas to incorporate in their individual stories. When they needed help, they would go around the room to sit with their friends. The educator took note that the learners did not waste time with this interaction, but in fact stayed quite focused on the task. Other activities required more formal group work. One task was to use literature circles for reading. Learners form groups and each person in the group takes on a predetermined role. One will be the reader, one the director, another the word finder and another the illustrator. They used the Free-Write program on the handhelds to record what they did in each role. Each person had to control his or her learning in this task because each had a specific role that was individual to each and these all had to be combined to come up with a final product. Another group task was having the learners complete a survey using a program on the handhelds. The educator did note that the learners problem solved a lot more when in a group setting.

When using the web-based programs, learners had control primarily of the pacing. It is important to recognize that this can be a frustrating constraint for learners who want more flexibility. Working together towards the goal of successfully completing the program is one way of helping learners be successful and consequently recognize the value of directing their way through a computer software program. They did work together to some extent. Educators reported that learners helped each other with problems and as a result, the educators did not see many learners with their hands up asking for help. Encouraging this type of interaction when using a mostly linear piece of computer software is important because working together to troubleshoot would make it easier for many learners to progress. At the same time, having a linear piece of software limits complete control because there is only one way of moving through the program.

In the online study, part of the online English course required learners to work in groups and to connect online to complete assignments. The group work exercise required learners to control their own learning in a number of ways. Learners needed to make decisions about when to virtually meet and how to fit their learning into the learning others were completing so that they could combine the work into one finished product. It is necessary to recognize that online group work can become difficult, just because of the dependence on other people. Even though interaction with others is encouraged when using SRL/SDL Technology-Enhanced learning, group work appeared to be somewhat of a limitation for some when controlling their learning.

Encouraging SRL/SDL Technology-Enhanced Learning

Part of instructional design in SRL/SDL Technology-Enhanced learning is to consciously encourage more learner control. Without control, it is extremely difficult for SRL or SDL to take place. Emerging trends across the three studies led to strategies for specifically encouraging learner control development, including helping learners draw connections, preparing learners ahead of time for taking control when using technology, encouraging learners to follow individual learning plans, managing technology-enhanced learning activities in a way that fosters control, and designing assessment that allows for learner input.

Drawing Connections

For effective SRL/SDL Technology-Enhanced learning, it becomes vital for learners to discover how their learning fits together so that they can make wise decisions when using the technology. As Clayton-Pederson & O'Neill (2005) claim, there is a need for using technology to "capture what students know and are able to integrate in their learning" (p. 9.12). "Educators fail to move students toward the self-sufficiency of knowing how to learn when they don't stress the importance of examining evidence, of determining how it is that we know what we know, and of recognizing where to turn to enlarge one's knowledge" (Maeroff, 2003, p. 107). Without drawing connections, it becomes difficult for learners to determine their learning needs and make choices about individual learning. As Larisey (1994) indicates, some learners need help in making connections. In SRL/SDL Technology-Enhanced learning, it is important to explore how and if learners make connections between topics they are learning and how educators can structure technology-enhanced learning activities to help learners make these connections.

In the handheld study, the learners typically used the handhelds in one classroom setting. Drawing connections between the activities done in that class and what the learners previously knew helped learners control their learning by putting them in a better position to make decisions about what they still needed to learn. Learners drew on their previous experiences by looking at certain topics presented. Based on their past experience, they decided what they could write and read about. They then used the software to communicate all of this through the handheld. They also used Pico-Map to put what they were learning together, helping them to draw connections. Looking at websites allowed them to collect information to elaborate on supporting details about their previous experiences and understanding of concepts. According to the educator, pulling all these pieces together put them in a position of feeling like they had more control and were better able to see the overall purpose of their learning.

Learners worked in a resource room, in the web-based study, and not in their regular classes. To help direct learning, it was important for learners to consider the usefulness of what they were learning in the resource room, as it applies to other classes they attend. Educators noted that when learners returned to their other classes where no technology was in use, they stated they had already learned about a topic on the computer program in the resource room. In the online study, one learner remarked that many of the regular classes he had previously taken did not focus directly on what could be learned in the future. In the online course, there were many assignments with potential for later use. By using assignments where learners draw connections, educators encouraged learners to see how learning fits together and how using technology helps to do this. They used the internet to do the research and there was an element of control because the learner chose what sites to visit. Including assignments that encourage these types of connections is important in SDR/SDL Technology-Enhanced learning.

Preparing Learners Ahead of Time for Taking Control

Part of successful SRL/SDL Technology-Enhanced learning is preparing learners ahead of time for an approach where they will have a degree of control. If learners understand why they are learning a topic or are being encouraged to assess their own learning, they are often in a better position to monitor progress. Larisey (1994) stresses that some learners have difficulty trying to understand what a self-assessment activity has to do with the subject area they are studying. Introducing learners ahead of time to the concept of controlling their learning makes it more likely for them to recognize the benefits of using technology to help them take on responsibility for learning.

When using the handhelds, the educator prepared the learners by bringing in a technology expert. The IT consultant from the school board visited the classroom and reviewed how to use the handhelds with the learners. She explained what handhelds were, how to charge them, where the learners would be storing them, and theft concerns. She also reviewed how the learners would be working independently and discussed how the technology would allow them to do this by controlling their learning. According to the educator, without this preparation ahead of time, it would have been more difficult for learners to proceed. They had not used handhelds before and did not normally direct their learning. The preparation was structured and not independent, yet these learners needed this introduction to successfully begin learning on their own with this new type of technology.

In the web-based study, the educators prepared the learners by explaining the rationale behind using the web-based literacy program, instead of just taking the learners directly to the computers. By emphasizing the individual aspect of literacy development, the learners begin to see that the approach focuses on the individual learner. The educators also distributed a survey designed to gather initial thoughts. Using the survey allowed learners to have input right from the beginning, which was another important part of encouraging them to take control. In the online study, time management was a control issue. Some learners had difficulties completing assignments. The school board set dates for course completion and were inflexible. There was no introduction to the interface and learners did not know how to submit assignments. This affected their progress and ability to make decisions about their learning. Educators then began to hold pre-program interface orientations where they met face-to-face with learners to discover how to effectively use the technology and take more control. To be able to take control, a degree of structured learning was then necessary for some learners.

Promoting Independent Learning

In SRL/SDL Technology-Enhanced learning, it is important to promote independent learning and the importance of learner control. Helping learners develop an attitude about learning that fosters independence is a vital part of SDL (Hiemstra, 1988; Brockett & Hiemstra, 1991) and is subsequently vital when successfully using technology. By talking about the value of an independent, SD approach, instructors develop a positive attitude toward directing learning. Motivating learners to learn individually can become a challenge. Yet the use of technology can be the stimulus many learners need.

In the handheld study, the educator promoted independent learning when completing various writing tasks such as using Free-Write to respond to questions, Sketchy to write journal entries with illustrations about what they wanted to be when they left school, and organizing the writing of a poem with Pico-Map. Because learners were using the handhelds to complete individual tasks, they were required to use an element of control. The learners explored each task and decided which step to take next. The educator encouraged the learners to follow this individual approach and reported that learners liked figuring things out for themselves. The learners were self-motivated and the educator did not need to stimulate or motivate them. Learner excitement about using technology is often an essential part of SRL/SDL Technology-Enhanced learning and can be enough of a stimulus for learners to take some control.

In the web-based study, the educators stressed the value of moving individually to improve reading skills and ways the technology would help with specific areas of reading where they may be having difficulties. By introducing this individual aspect and explaining the rationale behind the approach, the educators encouraged the learners to work independently where they would consequently be taking on responsibility for progressing through the program. Similarly, in the online study, the nature of the course allowed independent learning by requiring learners to control the speed of completion. There was not a lot of flexibility in the content of the course, yet learners could control elements within the constraint of the mandatory content. The use of technology allowed for elements of control, if not complete control, over content learned.

Managing the Learning Activity

The way educators manage learning activities often supports SRL/SDL Technology-Enhanced learning, although curriculum and content restrictions can prevent full learner control. As Garrison & Anderson (2003) point out, one of the challenges of learners taking control is the contradictory situation which often exists where the learner is expected to assume responsibility for activities and an outcome over which they have little input. The more the content is fixed in linear format, the greater the tendency is for the educator to teach by presenting information. Yet when digitized content is made available, discovery learning and construction become more dominant instructional modes (Schnitz & Azbell, 2004). The digitized content would then provide more opportunities for learners to control learning. Part of managing learning in SDL is providing courses of study that support an independent approach (Hiemstra & Sisco, 1990; Hammond & Collins, 1991). When using technology, having class activities that reflect this flexibility is a good starting point. There are often constraints within which educators must work which can restrict the amount of control that both educators and learners have, yet educators often have some control over lesson planning. For effective SRL/SDL Technology-Enhanced learning, it is important to look at ways to promote learner control within these constraints.

In the handheld study, the educator had some flexibility and could choose software that allowed for more learner control. Sketchy enabled learners to create graphics to add to their writing, allowing the pieces of writing to differ in terms of ways the messages the learners were writing appeared. Learners had some control because they chose what to create using Sketchy. The educator also gave the learners free time on the handheld where they were able to play and explore, giving them the opportunity to determine the way they should use time.

In the web-based study, the educators concentrated on the pacing of individual learners, which is an aspect where some learner control is possible. Learners reported they couldn't go back and reread difficult portions. There was also no way of saving work part way through, making it difficult to direct learning. The format of the program made it difficult for educators to place learners at different levels according to their skills. When taking on increased control, learners may need some assistance with this initial program placement. Accommodations would have to be in place so that learners could perhaps branch out on their own if they did complete work early.

Designing Control within Assessment

In SRL/SDL Technology-Enhanced learning, where the learner takes on increasingly more responsibility, developing opportunities for learners to do meaningful self-assessment becomes crucial. When incorporating more learner control into the design of the assessment, it is important to recognize that determining success might be a challenge for some learners. In order to have control, learners need to be able to tell when they have successfully learned a topic or skill. Often a frustrating part for learners is educators encouraging a SD approach throughout the course, and then evaluating using a totally other-directed approach where learners have no say in what is happening. Sometimes there is penalization for not following a preset idea of what the instructor feels is the right answer. Garrison & Anderson (2003) promote giving learners opportunities for discussing outcome expectations, learning activities and means of assessment. By including learners in this process and providing choice where appropriate, learners have a sense of control and therefore, take responsibility for the quality of the educational outcome. As Costa & Kallick (2004) claim, the intent of assessment when using a SD approach should be to support learners in becoming self-directing with a focus on becoming able to self-evaluate more. By ensuring learners have some say in developing the evaluation scheme, there is more chance that the evaluation supports the type of learning.

In the handheld study, the educator began assessing when she asked the learners to synch and she marked the work they transferred. They all got an A because they did quite well with minimal assistance from the educator or from each other. Encouraging learners to take control often results from having a non-threatening component to the initial evaluation. During the course, the educator evaluated both informally and formally at the end of each handheld activity. By tracking success, the educator enabled learners to see their progress, encouraging them to control their learning. She also told them if a project was going to be evaluated as a report card mark. By telling learners ahead of time, the educator was being up front, allowing learners to make a choice about how much effort they wanted to put into getting a good mark on that particular assignment.

Some learners were unable to understand the justification for marks in the online study because they felt they could not carry on a conversation with the educator. Essentially learners were in control of determining if they understood instructions correctly. Having checks in place to ensure that this comprehension was accurate was helpful. Another learner indicated a conflict between expectations of the educator and learner when he indicated that one assignment did not receive a grade when indications were that it was a significant assignment. One suggestion was that there be an opportunity to resubmit assignments to encourage revisiting concepts. This suggestion would put the learner in control because he or she would then have the opportunity to decide to resubmit.

Learner Characteristics in SRL/SDL Technology-Enhanced Learning

Being aware of learner characteristics is a critical part of effective SRL/SDL Technology-Enhanced learning. As Hughes (2005) indicates, it is important to ask questions about the learner's readiness for using technology, access to and familiarity with the technology, and individual learning styles when supporting online learners. Educators should take into account learner characteristics to be in a better position to encourage learners to take control. Some learners are more comfortable taking control than others. Being aware of learner capability for taking control and discovering how comfortable learners are being co-learners when using technology become important factors when exploring the role of control.

Knowing the Learner Capability for Taking Control

Having some knowledge of how much control learners are capable of taking is often a good starting point for SRL/SDL Technology-Enhanced learning. As Pilling-Cormick (1996) identifies, learners may learn their preference for traditional patterns of instruction. Some prefer direct instruction and, for whatever reason, they do not want to take control. As Candy (1991) states, educators should respect this right of learners wanting to be taught in a certain way in a particular situation. While recognizing a preference for direct instruction as influencing the amount of control a learner wants to undertake, educators should still make efforts to show these learners the benefits of directing one's learning. Candy (1991, p. 136) argues that "years of passivity in educational settings deprive people of confidence to take charge." There is an assumption that those who want more direction are "victims of an educational system that has systematically deprived them of the opportunity to be self-directed" (Candy, 1991, p. 375). As Candy states, this is not surprising when one considers the predominant methods of instruction found in much formal instruction. In online learning, "those who can teach themselves, in effect, take control over their own learning." (Maeroff, 2003, p. 106).

Part of the handheld study was using technology for reading activities. The educator found that some of the learners had strong opinions about reading on the handhelds. When introducing elements of control, the educator had to be aware of learner preference for reading with books. About half the class said they preferred reading a book. To effectively use technology for reading, the educator had to recognize this preference and used a modified approach. Some reading was done on the computer and other reading with books. A plan was put in place to encourage learners to move away from using books so much. Exploring with learners the benefits of being able to control reading more when using technology is another option.

In the web-based study, learners controlled the pacing of their learning. Some learners were able to more effectively pace than others. By knowing the characteristics of the learners, the educators reported that they recognized when to step in to help with this pacing. They knew from past experience with the learners who needed a quiet learning environment or who had the stigma of working in the resource room. They helped the learners, but their assistance was geared to the learner and the difficulty the learner had. Just in time technical support is a crucial component of classroom management when using technology in the classroom (Livingston, 2006). If educators generally know who their learners are, they will be able to put supports in place to help them to be successful with an independent approach. Educators are then in a much better position to identify learners who might have problems with control because they do not exhibit the typical characteristics of someone who is capable of taking control.

When learning online, the educators found that the learners had differing abilities for being good time managers. Educators reported the concern about time management quite quickly and made adjustments so that there was some structure within which learners could proceed. Being able to effectively manage time is definitely part of controlling learning. One learner commented that what helped him was having a calendar provided by the educator to make sure he got assignments completed by certain dates. Even though this learner labelled the course as using an individual approach to learning, he suggested that the learners were still in secondary school and needed some guidance around time management. An educator claimed procrastination was the learners' largest problem and that the learners had shared that thought with the educator. According to the educator, despite numerous emails, announcements, and assurances, the lack of face-to-face communication reduced their adherence to work requests. Being aware of the learners' difficulties with time management, and consequently with control, made it easier for the educators to develop ways to help learners develop these skills.

Comfort in Being a Co-Learner

In SRL/SDL Technology-Enhanced learning, it is important for educators and learners to be co-learners. Pilling-Cormick (1996) emphasizes this by claiming learners in every situation are different and bring various experiences from which educators can learn. Schuttenberg and Tracy (1987) stress modelling in their description of the colleague role. Galbraith (1991) includes the facilitator's role as that of role model in a learner-centered approach. Learners usually come to a learning situation with respect for the educator. Even though the educator may attempt to promote an informal atmosphere in the classroom, there is still that image of the educator. As Garrison and Anderson (2003) suggest, "the purpose of establishing a secure environment is to facilitate critical thinking and inquiry" (p. 54). They suggest that when using technology, the educator can be an excellent model by encouraging questioning of his or her own comments. The educator then sets the tone by showing it is acceptable and not a personal attack to critically reflect. By becoming a co-learner, the educator is sending a strong, promotional message to the learner. However, not all learners and educators are ready to take on that role. In some cases, it is ingrained that the educator is the educator and the learner is the learner who receives teaching from the educator. It then becomes valuable to track whether this co-learning is actually happening in technology-enhanced learning.

In the handheld study, the educator reported that she was definitely a co-learner. When learners discovered something new, she encouraged them to show her what they had done. When she first used a projector, she was learning as she went along and modeled that it was fine to make mistakes. By modeling that she was a learner, it was more likely for learners to view themselves as such and in turn, feel more comfortable taking on control. In the web-based study, one educator claimed she learned new information about different topics, discovered how to move around the screen, and was then in a better position to assist learners as they made decisions. Another reported that the web-based program was innovative and he continually liked to explore new ways of using computers. Encouraging educators to perceive themselves as co-learners becomes an essential part of effective SRL/SDL Technology-Enhanced learning. Learners perceiving educators as learners gives more of a feeling of sharing which could in turn, encourage learners to take on more control.

The educators in the online study modelled co-learning through online chats. One educator broadened his learning by discovering new ways to teach and modelled learning because he was dealing with new content and a different delivery method. One learner stated that his educator was very available and open to questions by responding to questions in the 'Virtual Office,' and to emails. The educator also took part in the discussions. By participating in these interactions, he was participating as a co-learner and modelling this to the learners. The learners would then observe him critically thinking about his own learning and making decisions about what he was learning. This would encourage learners to more likely respond in a similar manner to what they are learning.

CONCLUSION

With an increased emphasis on technology and learning, it is only natural to expect to see more learner control. SDL is specifically identified as a skill necessary for developing 21st Century learners (Partnership for 21st Skills, 2006; State Educational Technology Directors Association, 2007; Partnership for 21st Century Skills, 2007). As Garrison and Anderson (2003) claim, "SDL is emerging as an important conceptual model towards understanding issues raised by technology that has the potential to transfer enormous control to the learner" (p. 15). To ensure educators are meeting the needs of the 21st Century learner, it is necessary to carefully explore the role of control in technology use. Pilling-Cormick & Garrison's (2007) framework provides a starting point for investigating SRL/SDL Technology-Enhanced learning. Implications for practice learned from the three exploratory studies (handheld, web-based and online) provide emerging trends for instructional design including a focus on interaction, building opportunities for learner control, and considering the importance of learner characteristics.

Using the SRL/SDL Technology-Enhanced learning framework is essential for investigating the connection between control and technology. Without some form of learner control, it is extremely difficult for both learners and educators to be truly successful with technology. Recognizing that technology does not always allow for full learner control, it becomes vital to discover ways to operate within the constraints to give learners some degree of control. More research is needed to explore the internal monitoring and external management issues of technology-enhanced learning. To do this, exploration of the SRL/SDL link needs to continue. There is a need for participatory research studies emphasizing practical applications in the classroom, which, as Unsworth (2008) indicates, would be an important complement to the more traditional, university-based, funded studies which typically require an extended preparatory period prior to implementation. For a more comprehensive picture of how technology enhances learning, educators need to explore ways to encourage degrees of control in technology-enhanced learning environments.

REFERENCES

Brockett, R. G., & Hiemstra, R. (1991). *Self-direction in adult learning: Perspectives on theory, research, and practice*. New York, NY: Routledge.

Brookfield, S. (1990). *The skilful teacher*. San Francisco, CA: Jossey-Bass.

Candy, P. (1991). *Self-direction for lifelong learning*. San Francisco, CA: Jossey-Bass.

Candy, P. (2004). *Linking thinking: Self-directed learning in the digital age*. Canberra, AU: Department of Education, Science and Training. Retrieved December 14, 2009, from http://www.dest.gov.au/sectors/training_skills/publications_resources/profiles/linking_thinking.htm

Chene, A. (1983). The concept of autonomy in adult education: a philosophical discussion. *Adult Education Quarterly*, *31*(1), 38–47. doi:10.1177/0001848183034001004

Clayton-Pederson, A., & O'Neill, N. (2005). Curricula designed to meet 21st century expectations. In D. G. Oblinger & J. L. Oblinger (Eds.), *Educating the net generation* (pp. 9.1-9.16). Retrieved September 3, 2009, from http://www.educause.edu/ir/library/pdf/pub7101.pdf

Coiro, J., Knoebel, M., Lankshear, C., & Leu, D. (2008). Central issues in new literacies and new literacies research. In Coiro, J., Knoebel, M., Lankshear, C., & Leu, D. (Eds.), *Handbook of research on new literacies* (pp. 1–21). Mahwah, NJ: Lawrence Erlbaum Associates.

Collins, M. (1991). *Adult education as vocation*. New York, NY: Routledge.

Costa, A. L., & Kallick, B. (2004). *Assessment strategies for self-directed learning*. Thousand Oaks, CA: Corwin Press.

Elbaum, B., McIntyre, C., & Smith, A. (2002). *Essential elements: Prepare, design and teach your online course*. Madison, WI: Atwood Publishing.

Ericksen, S. C. (1984). *The essence of good teaching: Helping students learn and remember what they learn*. San Francisco, CA: Jossey-Bass.

Galbraith, M. W. (1991). The adult learning transactional process. In Galbraith, M. W. (Ed.), *Facilitating adult learning. A transactional process* (pp. 1–32). Malabar, FL: Robert E. Krieger.

Gall, J. P., Gall, M. D., & Borg, W. R. (1999). *Applying educational research: A practical guide*. (4th ed.) Don Mills, ON, CDN: Longman.

Garrison, D. R., & Anderson, T. (2003). [[st] *century: A framework for research and practice*. New York, NY: RoutledgeFalmer.]. *E-learning*, 21.

Hammond, M., & Collins, R. (1991). *Self-directed learning: Critical practice*. New York, NY: Nichols/GP Publishing.

Hiemstra, R. (1988). Self-directed learning: Individualizing instruction. In Long, H. B. (Eds.), *Self-directed learning: Application and theory* (pp. 99–124). Athens, GA: University of Georgia, Adult Education Department.

Hiemstra, R., & Sisco, B. (1990). *Individualizing instruction: Making learning personal, empowering, and successful*. San Francisco, CA: Jossey-Bass.

Hughes, J. A. (2005). Supporting the online learner. In T. Anderson & F. Elloumi (Eds.), *Theory and practice of online learning* (pp. 367-384). Retrieved December 14, 2009, from http://cde.athabascau.ca/online_book/pdf/TPOL_book.pdf

Jarvis, J. (2009). *What would google do?*New York, NY: Harpers Collins.

Larisey, M. M. (1994). Student self-assessment: A tool for learning. *Adult Learning*, *5*(6), 9–10.

Livingston, P. (2006). *1-to-1 learning: laptop programs that work*. Washington, DC: International Society for Technology in Education.

Maeroff, G. I. (2003). *A classroom of one*. New York, NY: Palgrave Macmillan.

McMillan, J. (2000). *Educational research: fundamentals for the consumer*. (3rd ed.) Don Mills, ON, CDN: Addison Wesley Longman.

November, A. (2001). *Empowering students with technology*. Glenview, IL: Pearson Professional Development.

Oblinger, D. B., & Oblinger, J. L. (2005). Is it age or IT: First steps toward understanding the net generation. In D. B. Oblinger & J. L. Oblinger, (Eds.) *Educating the net generation* (pp. 2.1-2.20). Retrieved September 3, 2009, from http://www.educause.edu/ir/library/pdf/pub7101.pdf

Partnership for 21st Century Skills (2006). *Results that matter: 21st century skills and high school reform*. Retrieved September 3, 2009, from http://www.21stcenturyskills.org/documents/RTM2006.pdf

Partnership for 21st Century Skills (2007). *Beyond the three Rs: Voter attitudes toward 21st century skills*. Retrieved September 3, 2009, from http://www.21stcenturyskills.org/documents/P21_poll-report_singlepg.pdf

Pilling-Cormick, J. (1996). *Development of the self-directed learning perception scale*. Unpublished doctoral dissertation, University of Toronto, Toronto, Ontario, Canada.

Pilling-Cormick, J. (1999). The self-directed learning process model. In Long, H. B. (Eds.), *Contemporary ideas and practices in self-directed learning* (pp. 89–102). Norman, OK: Public Managers Center, University of Oklahoma.

Pilling-Cormick, J. (2005). Learn how to teach online: 5 tips that work! *Teach*, Nov/Dec 2005, 8-10. Retrieved December 14, 2009, from http://www.teachmag.com/teach_archives/05/teachmag_nd_05.pdf

Pilling-Cormick, J., & Garrison, D. R. (2007). Self-directed learning and self-regulated learning: Conceptual links. *Canadian Journal of University Continuing Education, 33*(2), 13–32.

Pintrich, P. R. (2000). The role of goal orientation in self-regulated learning. In Boekaerts, M., Pintrich, P. R., & Zeidner, M. (Eds.), *Handbook of self-regulation* (pp. 451–502). San Diego, CA: Academic Press. doi:10.1016/B978-012109890-2/50043-3

Richardson, W. (2006). *Blogs, wikis, podcasts and other powerful web tools for classrooms*. Thousand Oaks, CA: Corwin Press.

Schnitz, J. E., & Azbell, J. W. (2004). Instructional design factors and requirements for online courses and modules. In Cavanaugh, C. (Ed.), *Development and management of virtual schools: Issues and trends* (pp. 158–177). Hershey, PA: Information Science Publishing.

Schuttenberg, E. M., & Tracy, S. J. (1987). The role of the adult educator in fostering self-directed learning. *Lifelong Learning: An omnibus of practice and research, 10*(5), 4-9.

Siemens, G., & Yurkiw, S. (2003). The roles of the learner and the instructor in e-learning. In Piskurich, G. M. (Ed.), *Preparing learners for e-learning* (pp. 123–138). San Francisco, CA: John Wiley & Sons Inc.

Stafford, B., Miller, L., & Ollivierre, M. (2006). A science-based development and implementation model for online and CD-ROM curriculum programs. In McKenna, M. C., Labbo, L. D., Kieffer, R. D., & Reinking, D. (Eds.), *International handbook of literacy and technology* (*Vol. II*, pp. 163–178). Mahwah, NJ: Lawrence Erlbaum Associates.

State Educational Technology Directors Association. (2007). *Maximizing the impact: The pivotal role of technology in a 21st century education system*. Retrieved September 3, 2009, from http://www.setda.org/web/guest/maximizingimpactreport

Tapscott, D. (2009). *Grown up digital: How the net generation is changing your world*. Toronto, ON, CDN: McGraw-Hill.

Unsworth, L. (2008). Multiliteracies and metalanguage: Describing image/text relations as a resource for negotiating multimodal texts. In Coiro, J., Knoebel, M., Lankshear, C., & Leu, D. (Eds.), *Handbook of research on new literacies* (pp. 67–102). Mahwah, NJ: Lawrence Erlbaum Associates.

Warlick, D. F. (2004). *Redefining literacy for the 21st century*. Worthington, OH: Linworth Publishing Inc.

KEY TERMS AND DEFINITIONS

Control: Extent to which learners can direct their learning and influences all other aspects of teaching and learning. Control is situational. The more control the learner has in a learning situation, the more the learner directs the learning.

Handheld: A device used by students (ie. Palm) which can be held in a hand like a "small computer" upon which various pieces of software can be loaded.

Online: A learning activity or course that occurs entirely via the internet with no scheduled face-to-face meetings with the educator.

Self-Directed Learning Process (SDLP) Model: Depicts the process of SDL as the interaction between learner and educator taking place within the varying context of control. In the model, four factors influence the amount of control learners have: social constraints, environmental characteristics, learner characteristics and educator characteristics. With technology use, these four factors will influence how much control learners have when using the particular type of technology.

Self-Directed Learning: A process in which individuals determine their priorities and choose from various resources available. They play an active role in developing a system of meanings to interpret events, ideas or circumstances.

Self-Regulated Learning (SRL): An active, constructive process whereby learners set goals for their learning and then attempt to monitor, regulate, and control their cognition, motivation, and behaviour, guided and constrained by their goals and the contextual features in the environment.

SRL/SDL Technology-Enhanced Learning: Incorporating elements of both SRL and SDL when using technology in learning with a focus on developing increased learner control.

Web-Based: A computer program which operates online over the web.

Compilation of References

Abraham, R. (2007). Mobile Phones and Economic Development: Evidence from the Fishing Industry in India. *Information Technologies and International Development, 4*(1), 5–17. doi:10.1162/itid.2007.4.1.5

Accascina, G. (2000). Information technology and poverty alleviation, in SD Dimensions, Food and Agricultural Organisation of the United Nations (FAO), Rome. Retrieved September 2009 from www.fao.org/sd/CDdirect/CDre0055h.htm.

Ackoff, R. (1969). Systems, organizations, and interdisciplinary research. In Emery, F. (Ed.), *Systems thinking* (pp. 330–347). Middlesex, England: Penguin Books.

Adital (Agência de Informação Frei Tito sobre a América Latina). (2008). *Notícias da América Latina e Caribe*. Retrieved January 1st, 2009, from http://www.adital.com.br

Adria, M., Bakardjieva, M., Poitras Pratt, Y., & Mitchell, D. (2006). *The Constructive Role of Researchers in the Social Shaping of Technology in Communities*. Paper presented at 3rd Prato International Community Informatics Conference. Retrieved December 2006 from http://www.ccnr.net/?q=node/123

Advomatic (2004). The Internet as Third Place. Retrieved July 3, 2009, from http://www.advomatic.com/thirdplace

Ajzen, I. (1991). The theory of planned behavior. *Organizational Behavior and Human Decision Processes, 50*(2), 179–211. doi:10.1016/0749-5978(91)90020-T

Ajzen, I., & Fishbein, M. (1980). *Understanding attitudes and predicting social behavior*. Englewood Cliffs, NJ: Prentice-Hall.

Alampay, E.A. (2006). Beyond access to ICTs: Measuring capabilities in the information society. *International Journal of Education and Development using Information and Communication Technology, 2*(3), 4-22.

Alampay, E.A. (2006). Beyond access to ICTs: Measuring capabilities in the information society. *International Journal of Education and Development using Information and Communication Technology (IJEDICT) 2*(2), 4-22.

Albirini, A. *(2006). Cultural perceptions: The missing element in the implementation of ICT in developing countries*. International Journal of Education and Development using ICT, 2*(1). RetrievedDecember21, 2008, from*http://ijedict.dec.uwi.edu/index.php

Alfaro Moreno, R. (2006). *Otra brújula: Innovaciones en comunicación y desarrollo*. Lima, Peru: Calandria.

Alverez, M. I., Roman, F., Dobles, M. C., Umana, J., Zuniga, M., Garcia, J., et al. (1998). Computers in schools: A qualitative study of Chile and Costa Rica. *Education and Technology Series: Special Issue*. Washington: The World Bank. Retrieved September 30, 3008, from http://www-wds.worldbank.org/external/default/WDSContentServer/WDSP/IB/2000/02/24/000094946_99031910575711/Rendered/PDF/multi_page.pdf

Amin, S. (1976). *Unequal development: an essay on the Social Formation of Peripheral Capitalism*. Sussex: Harvester Press.

Amin, S. (2001). *Il capitalismo del nuovo millennio*. Milano, Italy: Punto Rosso.

Anand, S., & Sen, A. (1994). *Sustainable human development: concepts and priorities*. New York: United Nations Development Programme.

Anderson, J., Boyle, C., & Reiser, B. J. (1985). Intelligent Tutoring Systems. *Science*, *228*(4698), 456–462. doi:10.1126/science.228.4698.456

Anderson, J. W. (1997) Is the Internet Islam's 'Third Wave' or the 'End Of Civilization'? Globalizing politics and religion In the Muslim world. *Middle East Studies Association Bulletin*. Retrieved July 16, 2008 from http://www.press.umich.edu/jep/archive/anderson.html.

Andrew, T., & Petkov, D. (2003). The need for a systems thinking approach to the planning of rural telecommunications infrastructure. *Telecommunications Policy*, *27*(1-2), 75–93. doi:10.1016/S0308-5961(02)00095-2

Angrist, J., & Lavy, V. (2002). New evidence on classroom computers and pupil learning. *The Economic Journal*, *112*, 735–765.doi:10.1111/1468-0297.00068

Appadurai, A. (1996). *Modernity at large: Cultural dimensions of modernity*. London, Minneapolis: University Of Minnesota Press.

Ariely, D. (2008). *Predictably irrational. The hidden forces that shape our decisions*. London: Harper.

Arora, A., & Athreye, S. (2001). The Software Industry and India's Economic Development, *WIDER Discussion Paper No. 20*. Helsinki: WIDER/UNU.

Arvanitis, S. (2005). Computerization, Workplace Organization, Skilled Labour and Firm Productivitiy: Evidence for the Swiss Business Sector. *Economics of Innovation and New Technology*, *14*(4), 225–249. doi:10.1080/1043859042000226257

Asian Development Bank website (n.d.). Retrieved January 2009 from http://www.adb.org

Assumpção, R. O. (2001). *Além da inclusão digital: O projeto sampa.org*. Unpublished masters dissertation. University of São Paulo, São Paulo, SP. Retrieved December 22, 2008, from http://referencias.onid.org.br/media/arquivos/dissertacao_Rodrigo.pdf

Avgerou, C. (2008). Information Systems in Developing Countries: a critical research review. *Journal of Information Technology*, *23*, 133–146. doi:10.1057/palgrave.jit.2000136

Avgerou, C. (1998). How can IT enable economic growth in developing countries? *Information Technology for Development*, *8*(1), 15. doi:10.1080/02681102.1998.9525288

Avgerou, C. 2000: Recognising Alternative Rationalities in the Deployment of Information Systems, *Electronic Journal on Information Systems in Developing Countries* 23(3) 7, 1–17.

Axelrod, R. (1984). *The Evolution of Cooperation*. New York: Basic Books.

Baily, M. N., & Lawrence, R. Z. (2001). Do We Have a New E-conomy? *American Economic Review. Papers and Proceedings*, *91*, 308–312.

Bakia, M. (2001). *The costs of computers in classrooms: Data from developing countries*. Washington, DC: The Consortium for School Networking. Retrieved June 18, 2008, from http://classroomtco.cosn.org/cic.pdf

Bala, V., & Goyal, S. (2000). *A Strategic Analysis of Network Reliability*. Oxford, UK: Review of Oxford Economic Design.

Bandura, A. (1986). *Social foundations of thought and action. A social cognitive theory*. Englewood Cliffs, NJ: Prentice-Hall.

Bandura, A. (1997). *Self-efficacy: The exercise of control*. New York: WH Freeman & Company.

Bandura, A., Caprara, G. V., Barbaranelli, C., Pastorelli, C., & Regalia, C. (2001). Sociocognitive self-regulatory mechanisms governing transgressive behaviour. *Journal of Personality and Social Psychology*, *80*(1), 125–135. doi:10.1037/0022-3514.80.1.125

Bar, F., Cohen, S., Cowhey, P., DeLong, B., Kleeman, M., & Zysman, J. (2000). Access and innovation policy for the third-generation internet. *Telecommunications Policy*, *24*, 489–518. doi:10.1016/S0308-5961(00)00047-1

Barros, C. (2008). Games e redes sociais em lan houses populares: um olhar antropológico sobre usos coletivos e sociabilidade no "clube local". *III Simpósio Internacional de Administração e Marketing V Congresso de Administração da ESPM.* São Paulo.

Barroso, M. (2007, 25-27 October). *Waves in the forest - Radio communication and livelihoods in Brazilian Amazonia.* Paper presented during the World Congress on Communication for Development. Rome.

Bartel, A., Ichniowski, C., & Kathryn, S. (2007). How Does Information Technology Affect Productivity? Plant-level Comparisons of Product Innovation, Process Improvement and Worker Skills. *The Quarterly Journal of Economics, 122*(4), 1721–1758. doi:10.1162/qjec.2007.122.4.1721

Barzilai-Nahon, K. (2006, November-December). Gaps and bits: Conceptualizing measurements for digital divides. *The Information Society, 22*(5), 269–278. doi:10.1080/01972240600903953

Bayes, A., von Braun, J., & Akhter, R. (1999). *Village Pay Phones and Poverty Reduction: Insights from a Grameen Bank initiative in Bangladesh. Discussion papers on Development Policy No. 8. Center for Development Research.* Bonn: ZEF.

BBC. (2009). Deals drive mobile money services. Technology news of Monday, 16 March 2009. Retrieved on April 25, 2009 from http://news.bbc.co.uk/2/hi/technology/7945878.stm

Beardsley, S., von Morgenstern, B., Enriquez, L., & Kipping, C. (2002). Telecommunications sector reform - A prerequisite for network readiness. In Kirkman, G., Sachs, J., Schwab, K., & Cornellius, P. (Eds.), *The Global Information Technology Report 2001-2002: Readiness for the networked world* (pp. 118–136). Oxford, UK: Oxford University Press.

Bechara, M. (2008). Banda larga: Os espaços públicos de acesso à internet. In CGI.br (Comitê Gestor da Internet no Brasil). *Pesquisa sobre o uso das tecnologias da informação e da comunicação 2007* (pp. 47-50). São Paulo.

Bell, P., Ready, P., & Rainie, L. (2004). *Rural Areas and the Internet.* Washington, D.C.: Pew Internet and American Life Project.

Bengtsson, J. (1999). Skarpnäck – Visionen som kraschade. Dagens Nyheter 991103.

Beninger, J. (1986). *The Control Revolution: Technological and Economic Origins of the Information Society.* Cambridge, MA: Harvard University Press.

Benkler, Y. (2006). *The Wealth of Networks: how social production transforms markets andfreedom.* New Heaven, UK: Tale University Press.

Bennetts, P., Wood-Harper, A., & Mills, S. (2000, April). A holistic approach to the management of information systems development: A review using soft systems approach and multiple viewpoints. *Systemic Practice and Action Research, 13*(2), 189–205. doi:10.1023/A:1009594604515

Berger, G. (2005). *Modernization and Africa's emerging engagement with the Information Society.* Paper presented at *AfroGEEKS: Global Blackness and the Digital Public Sphere.* University of California, Santa Barbara. May 19-25.

Bernard, A. (2004). Philippe Dreyfus, in TechnoScience.net. Retrieved September 2009 from http://www.techno-science.net/?onglet=glossaire&definition=7447.

Berry, J., Harris, K., & Jones, S. (2002). The future of community-based UK online centres: Discussion paper to Dfes. Retrieved July 3, 2009 from http://www.local-level.org.uk/uploads/Future%20of%20community%20based%20centres.pdf

Besson, É. (2008). *France Numeric 2012: Plan de développement de l'économie numérique.* Paris, Secrétariat D'État Chrgé de la Prospective, de L'Évaluation des Politiques Publiques et du Développement de L'Économie Numérique, Républic Française. Retrieved November 29, 2008, from http://francenumerique2012.fr/html/france_2012.html

Bieber, M., McFall, B., Rice, R. E., & Gurstein, M. (2007). Towards systems design for supporting enabling communities. *Journal of Community Informatics, 3*(1). Retrieved December 22, 2008, from http://ci-journal.net/index.php/ciej/article/view/281/31

Biggs, P., Kelly, T., Lee, S-H., Lozanova, Y., Nemoto, T., Sund, C., Hamdi, M., & Minges, M. (2007). ITU/UNCTAD 2007 World Information Society Report: Beyond WSIS.

Bingham, N., Valentine, G., & Holloway, S. L. (1999) Where do you want to go tomorrow? Connecting children and the Internet. *Environment and Planning D: Society and Space,* 17, pp. 655-672.

Black, M. (2002). *La cooperazione allo sviluppo internazionale.* Roma, Italy: Carocci.

Blackmore, J., Hardcastle, L., Bamblett, E., & Owens, J. (2003). *Effective use of information and communication technology (ICT) to enhance learning for disadvantaged school students.* Centre for Education and Change; Institute of Disability Studies, Deakin University and Institute of Koorie Education. Retrieved August 3, 2008, from http://www.dest.gov.au/sectors/school_education/publications_resources/profiles/effective_use_technology_enhance_learning.htm

Blaikie, P. (2000). Development, post-, anti-, and populist: a critical review. *Environment and Planning, 32,* 1033–1050. doi:10.1068/a3251

Blair, T. 'Forward' In Leer, A (2000). *Masters of the wired world: Cyberspace speaks out* (pp. viii-ix) *London*: Financial Times.

Blanchard, A. (2004). The Effects of Dispersed Virtual Communities on Face-to-Face Social Capital. In Huysman, M., & Wulf, V. (Eds.), *Social Capital and Information Technology*. Cambridge: The MIT Press.

Boafo, K. S. T. (1991). Communication Technology and Dependent Development in Sub-Saharan Africa. In Sussman, G., & Lent, J. A. (Eds.), *Transnational Communications: Wiring the Third World* (pp. 103–124). Thousand Oaks, CA: Sage.

Boafo-Arthur, K. (2003). Tackling Africa's Developmental Dilemmas; Is globalization the answer? *Journal of Third World Studies*. Retrieved July 29, 2008 from http://findarticles.com/p/articles/mi_qa3821/is_200304/ai_n9173427

Boerzel, T. (1998). Le reti di attori pubblici e privati nella regolazione europea. In *Stato e Mercato, n. 3*. Bologna, Italy: Il Mulino.

Bohman, J. (2005). *2009 from http//:plato.stanford.edu./entires/critical-theory*. Critical Theory. In Stanford Enyclopedia of Philosophy. Retrieved October.

Bond, P. (2006). *Looting Africa. The economics of exploitation*. London: Zed Books.

Bourdieu, P. (1985). The forms of capital. In Richardson, J. G. (Ed.), *Handbook of Theory and Research for the Sociology of Education*. New York: Greenwood Press.

Boyle, M., Kay, J., & Pond, B. (2001). Monitoring in support of policy: An adaptive ecosystem approach. In Munn, T. (Ed.), *Encyclopedia of Global Environmental Change* (*Vol. 4*, pp. 116–137). London: John Wiley and Sons.

Brachman, J. M. (2006). High-tech terror: Al-Qaeda's use of new technology. *The Fletcher Journal of International Affairs*, *30*(2), 149–164.

Bredarioli, C. M. M. (2008). Comunicação em rede e novos agentes socializadores: recepção e práticas culturais no consumo de Internet em lan-houses. In *Intercom – Sociedade Brasileira de Estudos Interdisciplinares da Comunicação XXXI Congresso Brasileiro de Ciências da Comunicação*, Natal, RN. Retrieved December 23, 2008, from www.intercom.org.br

Bridges.org. (2001). Spanning the Digital Divide: Understanding and Tackling the Issues. Retrieved October 6, 2008 from http://www.bridges.org/publications/65

Broad, R., & Cavanagh, J., (2006). The hijacking of the development debate. How Friedman and Sachs got it wrong. *World Policy Institute*, 21-30.

Brooks, T. (Ed.). (2008). *The global justice reader*. London: Blackwell Publishing.

Brown, W. (in press). *Leveraging Information and Communications Technologies for Development of the Bottom of the Pyramid ICT Market in South Africa.*

Brown, S. A., & Venkatesh, V. (2005). Model of adoption of technology in households: a baseline model test and extension incorporating household life cycle. *Management Information Systems Quarterly*, *29*(3), 399–426.

Brubaker, P. K. (2007). *Globalization at what price? Economic change and daily life*. Cleveland, OH: Pilgrim Press.

Bryden, J. (1994). Towards sustainable rural communities: From theory to action. In Bryden, J. (Ed.), *Towards sustainable rural communities: The Guelph Seminar Series* (pp. 211–233). Guelph, Canada: University of Guelph School of Rural Planning and Development.

Brynjolfsson, E., & Hitt, L. M. (2002). Computing Productivity: Firm-Level Evidence. *The Review of Economics and Statistics*, *85*(4), 793–808. doi:10.1162/003465303772815736

Budde, P. (2007). *Telecoms market review and statistics*. Gambia: Paul Budde Communication Pty Ltd.

Bunge, M. (1998). *Social science under debate*. Toronto: University of Toronto Press.

Burke, J. (1985). *The day the universe changed*. Boston: Little, Brown and Company.

Burns, M. (2006a). *Using ICT for teacher professional development in Namibia*. Infodev Working Paper No. 1. Retrieved October 7, 2008, from http://www.infodev.org/en/Publication.499.html

Burns, M. (2006b). *Improving teaching quality in Guinea with interactive radio instruction*. Infodev Working Paper No. 2. Retrieved November 12, 2008, from http://www.infodev.org/en/Publication.500.html

Burrell, J., & Toyama, K. (2009). What Constitutes Good ICTD Research? *Information Technologies and International Development*, *5*(3), 82–94.

Butcher, G., & Yaron, G. (2006). *Scoping study: Monitoring and evaluation of research communication*. Prepared for the Research Communication M&E Steering Group, London. Retrieved December 5, 2008 from http://www.healthlink.org.uk/PDFs/scoping.pdf

Byrne, A. (2007, November). Working towards evidence-based process: Evaluation that matters. *Mazi, 13*. Retrieved December 5, 2008 from http://www.communicationforsocialchange.org/mazi.php?id=13

Cachia, R., Kluzer, S., Cabrera, M., Centeno, C., & Punie, Y. (2007). ICT, Social Capital and Cultural Diversity: Report on a Joint IPTS-DG INFOSO Workshop held in Istanbul (Turkey), 25 April 2007. Retrieved July 3, 2009 from http://www.kennisland.nl/binaries/documenten/rapporten/e-inclusion-eur23047en.pdf

Caldeira, F. (2005). *Exclusão social: uma aventura histórica pela busca de um conceito*. Paper presented in 4th Geography Week, FCT/Unesp, Presidente Prudente, SP.

Camacho, K., Villalobos, V., & Shrader, E. (2000). Building an ICT Evaluation Framework: IDRC Workshop Proceedings. El Tirol, Costa Rica. Retrieved February 9, 2005 from http://www.bellanet.org/leap/docs/Building_an_ICT_Evaluation_Framework.doc?OutsideInServer=no

Canadian International Development Agency website (n.d.). Retrieved January 2009 from www.acdi-cida.gc.ca

Caprara, G. V., Regalia, C., & Bandura, A. (2002). Longitudinal impact of perceived self-regulatory efficacy on violent conduct. *European Psychologist*, *7*(1), 63–69. doi:10.1027//1016-9040.7.1.63

Capurro, R. (1986). *Hermeneutik der fachinformation*. Freiburg: Alber.

Capurro, R. (1991). Informatics and hermeneutics. In Budde, R., Floyd, C., Keil-Slawik, R., & Züllighoven, H. (Eds.), *Software Development and Reality Construction*. Secaucus, NJ: Springer-Verlag.

Capurro, R. (2000). Hermeneutics and the phenomenon of information. In Mitcham, C. (Ed.), *Metaphysics, epistemology, and technology*. New York: JAI.

Cardoso, F. H. (1972). Dependency and Development in Latin America. *New Left Review*, *74*, 83–95.

Carney, J. A. (1993). Converting the wetlands, engendering the environment: The intersection of gender with agrarian change in The Gambia. *Economic Geography*, *69*(4), 329–348. doi:10.2307/143593

Carney, D., Drinkwater, T., Rusinow, K., Neefjes, S., Wanmali, S., & Singh, N. (1999, February). *Livelihoods approaches compared. A brief comparison of the livelihoods approaches of the UK Department for International Development (DFID), CARE, Oxfam and the United Nations Development Program (UNDP).* FAO E-Conference and forum on Operationalizing Participatory Ways of Applying a Sustainable Livelihoods Approach. Rome: FAO.

Carpentier, N., & Servaes, J. (Eds.). (2004). *Deconstructing WSIS: Towards a Sustainable Agenda for the Future Information Society*. Bristol, UK: Intellect Books.

Carroll, T. (2009). 'Social Development' as Neoliberal Trojan Horse: The World Bank and the Kecamatan Development Program in Indonesia. *Development and Change*, *40*(3), 447–466. doi:10.1111/j.1467-7660.2009.01561.x

Castel, R. (1998). *As metamorfoses da questão social: uma crônica do salário*. (I.D. Poleti, Trans.). Petrópolis: Vozes. (Original work published 1995).

Castells, M. (2001). *The Internet Galaxy: reflections on the Internet*. Oxford, UK: Blackwell.

Castells, M. (Ed.). (2004). *The network society*. Northampton, MA: Edward Elgar.

Castells, M. (1999). *A Sociedade em Rede - a era da informação: economia, sociedade e cultura*. São Paulo: Paz e Terra.

Castells, M. (1996). *The rise of the network society*. Malden, MA: Blackwell.

Castells, M. (2000). Materials for an exploratory theory of the network society. *The British Journal of Sociology*, *51*(1), 5–24. doi:10.1080/000713100358408

Castells, M. (2004). *La città delle reti*. Venezia, Italy: Marsilio Editori.

Castells, M. (2001). *The Rise of the Network Society*. Oxford: Blackwell.

Castells, M. (2004a). Informationalism, networks, and the network society: a theoretical blueprint. In Castells, M. (Ed.), *The network society: A cross-cultural perspective* (pp. 3–45).

Castells, M. & Tubella, I. (Dir.). (2007). *La transición de la sociedad red.* Generalit de Catalunya: Editorial UOC/Ariel.

Cazeloto, E. (2008). *Inclusão digital. Uma visão crítica.* São Paulo: Senac.

CGI.br. (Comitê Gestor da Internet no Brasil). *Pesquisa sobre o uso das tecnologias da informação e da comunicação 2007*. (2008). São Paulo.

Chacko, J. G. (2005). Paradise lost? Reinstating the human *development* agenda in ICT policies and strategies. *Information Technology for Development*, *11*(1), 97–99. doi:10.1002/itdj.20005

Chambers, R. (1997). *Whose reality counts? Putting the first last*. London: IT Publications.

Chambers, R. (2005). *Ideas for development*. London: Earthscan.

Chambers, R. (1980). The small farmer is professional. *Ceres*, (March-April): 19–23.

Chang, H.-J. (2007). *Bad Samaritans: Rich Nations, Poor Policies, and the Threat to the Developing World*. London: Random House.

Chapman, J. (2004). *System failure: Why governments must learn to think differently*. London: Demos.

Chapman, R., Slaymaker, T., & Young, J. (2005). *Livelihood approaches to communication and information in support of rural poverty elimination and food security* [ODI]. London, Rome: DFID and FAO.

Chen, S., & Ravallion, M. (2008). The Developing World Is Poorer Than We Thought, But No Less Successful in the Fight against Poverty. *World Bank Policy Research Working Paper,* 4703. Retrieved December 12, from http://econ.worldbank.org

Chen, W., & Wellman, B. (2003). *Charting and bridging digital divides: Comparing socio-economic, gender, life stage and rural-urban internet access and use in eight countries*. AMD Global Consumer Advisory Board. Retrieved on September 12, 2005 from http://www.amd.com/us-en/assets/content type/

Chigona, W., & Mbhele, F. (2008). Does the internet contribute towards eliminating social exclusion? *South African Computer Journal*, *41*, 75–85.

Chomsky, N., Mitchell, P. R., & Schoeffel, J. (2002). *Understanding Power: The Indispensable Chomsky*. New York: The New Press.

Cisler, S. (2000). Subtract the digital divide. *San Jose Mercury*. Retrieved May 15, 2005 from http://www0.mercurycenter.com/svtech/news/indepth/docs/soap011600.htm

Civille, R., Gurstein, M., & Pigg, K. E. (2001). *Access to What? First Mile Issues for Rural Broadband. Report prepared for the Computer Science and Telecommunications Board. Broadband Last Mile Technology Project*. Washington, D.C.: National Research Council.

Clarke, M. (2003). *E-development? Development and the New Economy. UNU/WIDER Policy Brief No. 7*. Helsinki: WIDER.

Clarke, M. (2006). Are the Development Policy Implications of the New Economy, New? All that is Old is New Again for E-development. *Journal of International Development*, *18*, 639–648. doi:10.1002/jid.1254

Cline-Cole, R., & And Powell, A. (2004). ICTs, 'Virtual Colonisation' and political economy. *Review of African Political Economy*, *31*(99), 5–9. doi:10.1080/0305624042000258388

Coco, A., & Short, P. (2004). History and habit in the mobilization of ICT resources. *The Information Society*, *20*, 39–51. doi:10.1080/01972240490269997

Coe, D. T., & Helpman, E. (1995). International R&D Spillovers. *European Economic Review*, *39*, 859–887. doi:10.1016/0014-2921(94)00100-E

Coe, D. T., Helpman, H., & Hoffmaister, A. W. (1997). North-South R&D Spillovers. *The Economic Journal*, *107*, 134–149. doi:10.1111/1468-0297.00146

Cogburn, D. L., & Adeya, C. N. (2001). Prospects for the Digital Economy in South Africa: Technology, Policy, People, and Strategies. *WIDER Discussion Paper* No. 2001/77. Helsinki: UNU/WIDER.

Cogo, D., Gutiérrez, M., & Huertas, A. (Eds.). (2008). *Migraciones transnacionales y medios de comunicación: relatos desde Barcelona e Porto Alegre*. Madrid: Los Libros de la Catarata.

Cogo, D. & Brignol, L.D. (2009). Latinoamericanos en el sur de Brasil: recepción mediática y ciudadanía de las migraciones internacionales. *Comunicación y Sociedad*. Guadalajara, 11, Nueva Época, January-June 2009, 135-162.

Cohen, R. S., & Wartofsky, M. W. (1986). Editorial preface. In Mitcham, C., & Huning, A. (Eds.), *Philosophy and Technology II: Information Technology and Computers in Theory and Practice*. Boston: D. Redel.

Coleman, J. (1988). *Social Capital in the Creation of Human Capital*. Chicago: American Journal of Sociology, University of Chicago Press.

Colle, R. D. (2008). Threads of Development Communication. In Servaes, J. (Ed.), *Communication for Development and Social Change*. New Delhi: UNESCO & Sage.

Commonwealth of Learning. (n.d.). *An international scan of the use of ICT in education*. Washington, DC. *Commonwealth of Learning*.

Compaine, B. M. (2001). *The Digital Divide: Facing a Crisis or Creating a Myth?* MIT Press.

Compeau, D., Higgins, C. A., & Huff, S. (1999). Social Cognitive Theory and individual reactions to computing technology: a longitudinal study. *Management Information Systems Quarterly*, *23*(2), 145–158. doi:10.2307/249749

Compeau, D. R., & Higgins, C. A. (1995). Application of Social Cognitive Theory to training for computer skills. *Information Systems Research*, *6*(2), 118–143. doi:10.1287/isre.6.2.118

ComScore. *Comscore Brasil Report*, (2007). Retrieved January 25, 2008, from http://www.comscore.com/metrix

Cooke, B., & Kothari, U. (1998). *Participation: The new tyranny?*London: Zed Books.

Cooper, F., & Packard, R. (Eds.). (1997). *International development and the social sciences: essays on the history and politics of knowledge*. Berkeley, CA: University of California Press.

Cordes, C., & Miller, E. (2000). Fool's Gold: A Critical Look at Computers in Childhood. In *Alliance for Childhood 2000*. Retrieved December 24, 2008, from http://www.allianceforchildhood.org/projects/computers/computers_reports.htm

Cornwall, A. (2002). Making spaces, changing places: Situating participation in development. *IDS Working Paper, vol. 170*. Sussex, UK: Institute for Development Studies.

Courtright, C. (2004). Which lessons are learned? Best practices and World Bank Rural Telecommunications Policy. *The Information Society*, *20*(5), 345–356. doi:10.1080/01972240490507983

Coward, C. (2009). Second Recess, RSS feed, 11 March. Retrieved September 2009 from http://chriscoward.wordpress.com/2009/03/11/ict4d-ictd-or-what/.

Crump, B., & McIlroy, A. (2003). The digital divide: Why the "don't–want–tos" won't compute: Lessons from a New Zealand ICT Project. *First Monday, 8*(12). Retrieved December 12, 2008, from http://firstmonday.org/issues/issue8_12/crump/index.html

da Costa, A. B. (1998). *Exclusões sociais*. Lisbon: Gradiva.

Dacey, A. (2008). *The secular conscience. Why beliefs belong in public life*. New York: Promethus Books.

D'Agostino, F. (1985). Ontology and explanation in historical Linguistics. *Philosophy of the Social Sciences*, *15*, 147–165. doi:10.1177/004839318501500202

Daly, J. A. (1999). *Measuring the impact of the Internet in the developing world*. Information Impact Magazine.

Daly, H. (1996). *Beyond Growth*. Boston: Beacon Press.

Damasio, A. R. (1994). *Descartes' error. Emotion, reason and the human brain*. New York: Avon Books.

Damasio, A. R. (2004). *Looking for Spinoza. Joy, sorrow and the feeling brain*. London: Vintage.

Dart, J., & Davies, R. (2003). A dialogical, story-based evaluation tool: The most significant change technique. *The American Journal of Evaluation*, *24*(2), 137–155.

Dart, J. (2006). *Key questions for evaluation*. In Evaluation in Practice Website. Retrieved October 9, 2008 from http://www.evaluationinpractice.wordpress.com/2008/01/questionschoices.pdf

Darwin, C. (1968 / 1859). *The origin of species*. Harmondsworth, Middlesex: Penguin.

Davenport, T., & Prusak, L. (1997). *Information ecology: Mastering the information knowledge environment*. New York, Oxford: Oxford University Press.

Davidson, E., & Cotten, S. R. (2003). Connection discrepancies: Unmasking further layers of the digital divide. *First Monday*, 8(3). Retrieved September 18, 2008, from http://firstmonday.org/issues/issue8_3/davison/index.html

Davies, W. (2004). Proxicommunication - ICT and the Local Public Realm. iSociety report. London: The Work Foundation. Retrieved July 3, 2009, from http://www.theworkfoundation.com/ research/isociety/proxi_main.jsp

Davis, F. D. (1989). Perceived usefulness, perceived ease of use, and user acceptance of information technology. *Management Information Systems Quarterly*, *13*(3), 319–340. doi:10.2307/249008

Day, C., Sammons, P., & Gu, Q. (2008). Combining qualitative and quantitative methodologies in research on teachers' lives, work, and effectiveness: From integration to synergy. *Educational Researcher*, *37*(6), 330–342. doi:10.3102/0013189X08324091

De Haan, A. (2001). Social exclusion: enriching the understanding of deprivation. In World development report 2001 forum on inclusion, justice and poverty reduction. Retrieved October 2008 from http://www.sussex.ac.uk/cspt/documents/issue2-2.pdf

de Miranda, A. (2005, November) *Technological determinism and ideology: questioning the 'Information Society' and the 'digital divide'*. Paper presented in 'The Future of Research in the Information Society' event, parallel to the World Summit on Information Society, Tunis, Nov 16-18, 2005. Retrieved November 29, 2008, from http://www.ces.uc.pt/bss/documentos/2006_11_13_alvaro_de_miranda.pdf

de Moor, A. (2009). Moving Community Informatics research forward. *Journal of Community Informatics, 5*(1). Retrieved September 2009 from http://ci-journal.net/index.php/ciej/article/viewArticle/546/434

De Paula, S., & Dymski, G. (2005). *Reimagining Growth*. London: Zed Books.

de Santa Ana, J., Burity, J., Goudzwaard, B., Gurney, R., Hadsell, H., & Koshy, N. (2006). *Beyond idealism: a way ahead for ecumenical social ethics* (Gurney, R., Hadsell, H., & Mudge, L., Eds.). Grand Rapids, MI: William B. Eerdmans Publishing Co.

De Soto, H. (1989). *The other path: the economic answer to terrorism*. New York: Basic Books.

DeCerteau, M. (1994). *A Invenção do Cotidiano: Artes do fazer*. São Paulo: Vozes.

Deleuze, G., & Guatari, F. (1987). *A thousand plateaus: capitalism and schizophrenia*. Minneapolis: University of Minnesota Press.

Demo, P. (1998). *Charme da exclusão social*. Campinas: Autores Associados.

Derbyshire, H. (2003). *Gender issues in the use of computers in education in Africa*. London: Department for International Development (Imfundo).

DeRenzi, B. (2008). Thoughts on Challenges in ICTD for Young Researchers. University of Washington. Retrieved October 2009 from www.cs.washington.edu/homes/bderenzi/.../derenzi_yrictd09.pdf

Development Informatics Research Group, Council for Scientific and Industrial Research. (2009). Knowledge generation for an efficient and competitive built environment. Retrieved September 2009 from http:www.csir.co.za/Built-environment/Planning_support_system/devinfo.html

Dewan, S., & Riggins, F. (2005). The Digital Divide: Current and Future Research Directions. *Journal of the Association for Information Systems, 6*(12), 298–337.

Diamond, J. (1987). The Worst mistake in the history of the human race. *Discover*, 64–66.

Diamond, J. (2005). *Collapse. How societies choose to fail or survive*. London: Penguin.

Dicken, P. (2007). *Global Shift: mapping the changing contours of the World Economy*. New York: Guilford Press.

DiMaggio, P. J., & Hargittai, E. (2001), *From the Digital Divide to Digital Inequality': Studying Internet Use as Penetration Increases*, Working Paper 15. Princeton: Princeton University Center for Arts and Cultural Policy Studies.

DiMaggio, P., Hargittai, E., Celeste, C., & Shafer, S. (2004). Digital Inequality: From Unequal Access to Differentiated Use. In K. Neckerman. (Ed.), *Social Inequality* (pp. 355-400). New York: Russell Sage Foundation. Retrieved December 26, 2008, from http://www.webuse.org/papers?id= digitalinequality

Dirlik, A. (2000). Place-based imagination: Globalism and the politics of place. In Dirlik, A. (Ed.), *Places and politics in the age of globalization*. New York: Rowman and Littlefield.

Dishaw, M. T., & Strong, D. M. (1999). Extending the technology acceptance model with task-technology fit constructs. *Information & Management, 36*(1), 9–21. doi:10.1016/S0378-7206(98)00101-3

Distaso, W., Lupi, P., & Manenti, F. M. (2006). Platform competition and broadband uptake: Theory and empirical evidence from the European Union. *Information Economics and Policy*, *18*(1), 87–106. doi:10.1016/j.infoecopol.2005.07.002

Distler, C., & Bressand, A. (1995). *Le planete relationelle*. Paris: Flammarion.

Dockery, G. (1998). Questioning participation and equity in health. Paper presented at the symposium "Participation: The New Tyranny?" Institute for Development Policy and Management, University of Manchester.

Doheny-Farina, S. (1996). *The Wired Neighborhood*. New Haven: Yale University Press.

Dollar, D., & Kraay, A. (2004). Trade, Growth, and Poverty. *The Economic Journal*, *114*(493), F22–F49. doi:10.1111/j.0013-0133.2004.00186.x

Dos Santos, T. (1970). The Structure of Dependency. *The American Economic Review*, *60*(21).

Dosa, M. (Ed.). (1997). *Across all borders; international information flows and applications; collected papers*. London: Scarecrow Press.

Drucker, P. F. (1992). *The Age of Discontinuity: Guidelines to Our Changing Society*. Transaction Publishers.

Drucker, P. (1993). *Post-capitalist Society*. Oxford: Butterworth-Heinemann.

Dzidonu, C. K. (2001). *The Socio-Economic Development Implications of the Digital Divide within the Context of African Countries*. Paper presented at the Joint African Finance and Economic Planning Ministers Meeting of the ECA, Algiers, Algeria, 8-10 May 2001.

Earl, S., Carden, F., & Smutylo, T. (2003). *Outcome mapping*. Ottawa: IDRC.

Easterly, W. (2006). *The white man's burden. Why the west's efforts to aid the rest have done so much ill and so little good*. Oxford: Oxford Univ Press.

Easterly, W. (20020, *The Illusive Quest for Growth*. Cambridge, MA: MIT Press.

Eastin, M. S., & LaRose, R. (2000). Internet self-efficacy and the psychology of the digital divide. *Journal of Computer-Mediated Communication, 6*(1). Retrieved October 2009 from http://jcmc.indiana.edu/vol6/issue1/eastin.html

Ebo, B. (2001). 'Cyberglobalisation: Superhighway Or Superhypeway?' In B. Ebo (Ed) (2001), *Cyberimperialism? Global relations in the new electronic frontier* (pp.1-6). Westport, Ct: Praeger.

Edwards, M., (2008). Philanthrocapitalism: after the goldrush. *OpenDemocracy* 20-03-2008.

Ehrlich, P. J. (2000). *Human natures. Genes, cultures and the human prospect*. New York: Penguin.

Elder, L. (2009). Elder Musings, Monday, April 13, 2009. Retrieved September 2009 from http://eldermusings.blogs pot.com/2009/04/words-words-more-words-and-acronyms.html

Escobar, A. (1995). *Encountering development: The making and unmaking of the Third World*. Princeton, NJ: Princeton University Press.

Escobar, A. (1994). Welcome to Cyberia: Notes on the anthropology of cyberculture. *Current Journal of Anthropology*, *35*(33), 211–231. doi:10.1086/204266

Escobar, A. (2001). Culture sits in places: reflections on globalism and subaltern strategies of localization. *Political Geography*, *20*, 139–174. doi:10.1016/S0962-6298(00)00064-0

Escobar, A. (1999). Gender, place and networks. A political ecology of cyberculture. In Harcourt, W. (Ed.), *Women@Internet. Creating New Cultures in Cyberspace* (pp. 31–54). London: Zed Books.

Eubanks, V. E. (2007). Trapped in the Digital Divide: The Distributive Paradigm in Community Informatics. *The Journal of Community Informatics, 3*(2). Retrieved September 19, 2008 from http://www.ci-journal.net/

European Commission (2001, December 14). *Subject: information and communication technologies in development: The role of ICTs in EC development policy.* Communication from the Commission to the Council and the European Parliament. Brussels, 770, 3.

European Journal of Development Research. (2009). Retrieved September 2009 from http://www.plagrave-journals.com/ejdr/about.html

Evans, P. (1979). *Dependency development: the Alliance of Multinationals, State and Local Capital in Brazil.* Princeton, NJ: Princeton Univ. Press.

FAO. (1994). *Communication: A key to human development.* Rome: Communication for Development Service, FAO.

FAO. ODI, & DFID. (2002). A Livelihoods Approach to Information and Communication: A proposal. Retrieved February 9, 2007 from http://www.livelihoods.org/info/linksevents_sub/linksevents_FAO_ICTs.html

Faye, M., & John, L., Mcarthur, Jeffrey W., Sachs D., & Snow, T. (2004). The Challenges Facing Landlocked Developing Countries. *Journal of Human Development, 5*(1). doi:10.1080/14649880310001660201

FCC. (2008). FCC Consumer Facts: getting broadband. Retrieved from http://www.fcc.gov/cgb/consumerfacts/highspeedinternet.html

Feather, F. (2004). *The information society; a study of continuity and change.* London: Facet.

Federal Communications Commission (FCC). (2007). *Local Telephone Competition and Broadband Deployment.* Washington, D.C.: United States Department of Commerce. Retrieved October 2009 from http://www.fcc.gov/wcb/iatd/comp.html

Feenberg, A (2004). Modernity theory and technology studies: reflections on bridging the gap. In Misa, T. J., Brey, P., & Feenberg, A. (Eds.), *Modernity and technology* (pp. 73–104). Cambridge, MA: MIT Press.

Feeny, S., & Clarke, M. (2009). *The Millennium Development Goals and Beyond: International Assistance to the Asia-Pacific.* London: Palgrave-MacMillian. doi:10.1057/9780230234161

Ferguson, N. (2009). *The ascent of money. A financial history of the world.* London: Penguin Books.

Ferguson, J. (1990). *The anti-politics machine. 'Development,' depoliticization, and bureaucratic power in Lesotho.* Oxford: Oxford University Press.

Ferlander, S., & Timms, D. (2006). Bridging the Dual Digital Divide: A Local Net and an IT-Café in Sweden. *Information Communication and Society, 9*(2), 137–159. doi:10.1080/13691180600630732

Ferlander, S. (2003). The Internet, Social Capital and Local Community. Doctoral Dissertation. Stirling: University of Stirling. Retrieved July 3, 2009 from http://www.crdlt.stir.ac.uk/publi-cations.htm

Ferlander, S., & Timms, D. (2007). Social Capital and Community Building through the Internet: a Swedish Case Study in a Disadvantaged Suburban Area. Sociological Research Online, 12(5). Retrieved July 3, 2009 from http://www.socresonline.org.uk/12/5/8.html

Field, J. (2003). *Social Capital.* London: Routledge.

Figueroa, M., Kincaid, L., Rani, M., & Lewis, G. (2002). *Communication for social change: An integrated model for measuring the process and its outcomes.* New York: The Rockefeller Foundation.

Fink, C., & Kenny, C. (2003). W(h)ither the Digital Divide? *Info, 5*(6), 15–24. doi:10.1108/14636690310507180

Fink, C., & Kenny, C. (2002). *Whither the Digital Divide? Mimeo.* Washington, DC: World Bank.

Firth, L., & Mellor, D. (2005). Broadband: benefits and problems. *Telecommunications Policy, 29*(2), 223–236. doi:10.1016/j.telpol.2004.11.004

Fishbein, M., & Cappella, J. N. (2006). The role of theory in developing effective health communications. *The Journal of Communication, 56*(Supp. 1), S1–S17. doi:10.1111/j.1460-2466.2006.00280.x

Flora, C. B., & Flora, J. L. (1993). Entrepreneurial Social Infrastructure: A Necessary Ingredient. *The Annals of the American Academy of Social and Political Sciences, 529*, 48–58. doi:10.1177/0002716293529001005

Flora, C. B., Sharp, J., Flora, C. B., & Newlon, B. (1997b). Entrepreneurial Social Infrastructure and Locally-Initiated Economic Development. *The Sociological Quarterly, 38*(4), 623–645. doi:10.1111/j.1533-8525.1997.tb00757.x

Flora, J. L. (1998). Social Capital and Communities of Place. *Rural Sociology, 63*(4), 481–506.

Flora, C. B., Hales, B., Petrzelka, P., Trca-Black, S., & Zacharakis-Jutz, J. (1997a). *Social Reconnaissance: Field Research by a Student Team to Inform a Participatory Community Development Project*. Paper presented at the Annual Meeting of the Community Development Society, Athens, GA, July 27-30.

Foley, P., Codagnone, C., & Osimo, D. (2008). *An analysis of international digital strategies: Why develop a digital inclusion strategy and what should be the focus?* Research report. Tech4i2/Department for Communities and Local Government, London, UK Crown. Retrieved November 29, 2008, from http://www.communities.gov.uk/documents/communities/pdf/1000425.pdf.

Foucault, M. (1972). *The archaeology of knowledge*. London: Tavistock Publications.

Foucault, M. (1980). Two lectures. In Gordon, C. (Ed.), *Power/Knowledge: Selected Interviews and Other Writings 1972-1977* (pp. 78–108). New York: Harvester Wheatsheaf.

Fragoso, S., & Maldonado, A. E. (2009). (forthcoming). *A Internet na América Latina*. Porto Alegre, Sulina. *Unisinos*.

Fragoso, S. (2004). As Múltiplas Faces da Exclusão Digital. *Ícone, 6*(7), 110-122.

Frank, A. G. (1972). *Lumpen bourgeoisie: Lumpen development: Dependence, Class and Politics in Latin America* (M. D. Berdecio, Trans.). New York: Monthly Review Press. (Original published 1970).

Fredland, R. A. (2000). Technology transfer to the public sector in developing states: three phases. *The Journal of Technology Transfer, 25*, 265–275. doi:10.1023/A:1007870520985

Freire, P. (1970). The adult literacy process as cultural action for freedom. *Harvard Educational Review, 40*, 205–212.

Friberg, M., & Hettne, B. (1985) The Greening of the World. Towards a Non-Deterministic Model of Global Processes. In Hodder & Stoughton (Eds.), *Development as Social Transformation. Reflections on the Global Problematic* (pp. 204-270). Sevenoaks: The United Nations University.

Friedman, T. L. (2004). *The world is flat: a brief history of the twenty-first century*. New York: Farrar, Straus and Giroux.

Friedman, J. (1997). Simplifying complexity. In Olwig, K. F., & Hastrup, K. (Eds.), *Siting culture* (pp. 268–291). London: Routledge.

Fuchs, T., & Woessman, L. (2004). Computers and student learning: Bivariate and multivariate evidence on the availability and use of computers at home and at school. CESifo Working Paper Series (no. 1321). Munich, Germany: CESIfo.

Fukuyama, F. (1992). *The end of history and the last man*. New York: Free Press.

G8 Information Center. (2000). *Okinawa Charter on Global Information Society*. Retrieved September 6, 2006 from www.g8.utoronto.ca/summit/2000okinawa/gis.htm

Gates, B. (2008). How to fix capitalism. *Time, 172*(6), 24–29.

Gaved, M., & Anderson, B. (2006). The Impact of Local ICT Initiatives on Social Capital and Quality of Life. Chimera Working Paper 2006-6, Colchester: University of Essex.

Gefen, D., & Straub, D. W. (1997). Gender differences in the perception and use of e-mail: an extension to the technology acceptance model. *Management Information Systems Quarterly, 21*(4), 389–400. doi:10.2307/249720

Geldof, M. (2008). *The interaction between low-literate youth and ICT in Ethiopia and Malawi: Fieldwork report.* Retrieved November 1, 2008, from http://www.gg.rhul.ac.uk/ict4d/workingpapers/GeldofNGOs.pdf

Gerefi, G. (1983). *Rethinking Development Theory: Insights from East Asia and Latin America.* Princeton, NJ: Princeton Univ. Press.

Ghana News Agency (2008, July 25). ISODEC urges Parliament to pull brake on GT sale.

Giarchi, G. G. (2001). Caught in the nets: a critical examination of the use of the concept of networks in community development studies. *Community Development Journal, 16*, 63–71. doi:10.1093/cdj/36.1.63

Gibson-Graham, J. K. (1996). *The End of Capitalism as We Knew It? A Feminist Critique of Political Economy.* Blackwell.

Gibson-Graham, J. K. (2003). An ethics of the local. *Rethinking Marxism, 15*(1), 49–74. doi:10.1080/0893569032000063583

Gibson-Graham, J. K. (2005). Surplus possibilities: Post development and Community Economies. *Singapore Journal of Tropical Geography, 26*(1), 4–26. doi:10.1111/j.0129-7619.2005.00198.x

Giddens, A. (1991). *Modernity and self-identity; self and society in the late modern age.* Stanford: University Press.

Giddens, A. (1984). *The Constitution of Society.* Los Angeles: Univ California Press.

Gilchrist, A. (2004). *The well-connected community: a networking approach to community development.* London: The Policy Press.

Gillett, S., Lehr, W. H., Osario, C., & Sirbu, M. A. (2006). *Measuring the Economic Impact of Broadband Deployment.* Washington, D.C., U.S. Dept. of Commerce/EDA: 53.

Gilling, J., Jones, S., & Duncan, A. (2001). Sector approaches, sustainable livelihoods and rural poverty reduction. *Development Policy Review, 19*(4), 303–319. doi:10.1111/1467-7679.00136

Glasbergen, P. (1996). Learning to manage the environment. In W. Lafferty & J. Meadowcraft (Eds.), *Democracy and environment: Problems and prospects.* (pp. 175-193). Chettenham: Edward Elgan.

Godin, B. (2008). The information economy: the history of a concept through its measurement, 1949–2005. *History and Technology, 24*(3), 255–287. doi:10.1080/07341510801900334

Gopal, R. D., & Sanders, G. L. (2000). Global software piracy: you can't get blood out of a turnip. *Communications of the ACM, 43*(9), 82–89. doi:10.1145/348941.349002

Gordon, R. (2000). Does The New Economy Measure Up to The Great Inventions of The Past? *NBER Working Paper*, 7833.

Gore, A. (2000). Putting people first in the information age. In Leer, A. (Ed.), *Masters of the wired world: Cyberspace speaks out* (pp. 7–17). London: Financial Times.

Gould, S. J. (1996). *The mismeasure of man.* New York: WW Norton.

Gould, S. J. (2006). The richness of life. (P. McGarr & S. Rose, eds.) S. London: Vintage Books.

Govindan, P. (2005). The Digital Divide and Increasing Returns: Contradictions of Informational Capitalism. *The Information Society, 21*, 41–51. doi:10.1080/01972240590895900

Grameen Bank. (n.d.). Retrieved September 2009 from http://www.grameen-info.org/

Grameenphone (n.d.). Retrieved September 2009 from http://www.grameenphone.com/

Granovetter, M. S. (1973). The Strength of Weak Ties. *American Journal of Sociology, 78*, 1360–1380. doi:10.1086/225469

Granovetter, M. S. (1982). The Strength of Weak Ties: A network theory revisited. In Marsden, P., & Lin, N. (Eds.), *Social Structure and Network Analysis.* Beverly Hills, CA: Sage.

Grayling, A. C. (2003). *What is good? The search for the best way to live*. London: Phoenix.

Grayling, A. C. (2006). *The heart of things. Applying phisolophy to the 21st century*. London: Phoenix.

Green, L. (2001). *Communication, Technology and Society*. London: Sage Publications.

Greenwichmeantime (2008) Retrieved December 08, 2008, from http://www.greenwichmeantime.co.uk/time-zone/africa/gambia/map.htm

Grossberg, A., Struwig, J., & Tlabela, K. (1999). Contextualising the global information revolution in a development arena: A case study. *Communicare*, *18*(2), 81–103.

Grunfeld, H. (2007). Framework for evaluating contributions of ICT to capabilities, empowerment and sustainability in disadvantaged communities. Paper presented at the *CPRsouth2 Conference*. IDRC. Chennai, India, Dec. 15-17.

GSM World for Uganda. (n.d.). Retrieved September 2009 from http://www.gsm.org/cgi-bin/ni_map.pl?cc=ug&net=mt

Guba, E. (1990). *The paradigm dialogue*. Newbury Park, CA: Sage Publications.

Guijt, I., & Sidersky, P. (1996, December). Agreeing on indicators. *ILEIA Newsletter*, *12*(3), 9–11.

Gujit, I., & Shah, M. (1998). Waking up to power, conflict and process. In Gujit, I., & Shah, M. (Eds.), *The Myth of Community: Gender Issues in Participatory Development* (pp. 1–23). London: Intermediate Technology Publications.

Gunder Frank, A. (1967). *Capitalism and Underdevelopment in Latin America*. New York: Monthly Review.

Gurstein, M. (2000). *Community Informatics: Enabling Communities with Information and Communications Technologies*. Hershey, PA: Idea Group.

Gurstein, M. (2003). Effective use: A community informatics strategy beyond the digital divide. *First Monday*, *8*(12). Retrieved from http://firstmonday.org/issues/issue8_12/gurstein/index.html.

Gurstein, M. (2001). Community Informatics, Community Networks and Strategies for Flexible Networking. In Keeble, L., & Loader, B. (Eds.), *Community Informatics: Shaping Computer-mediated Social Relations*. London: Routledge.

Gurstein, M. (2005). Editorial: Sustainability of Community ICTs and its Future. Journal of Community Informatics, 1(2). Retrieved July 3, 2009 from http://ci-journal.net/index.php/ ciej/artcile/view/230/186

Habermas, J. (1987). *The philosophical discourse of modernity*. Cambridge, MA: MIT Press.

Haddad, L. (2006). *Reinventing Development Research: listening To the IDS40 Roundtables*, Conference background paper for IDS40 Conference. Brighton, UK: Institute for Development Studies, University of Sussex.

Haertel, H. (2008). *Low-cost devices in educational systems: The use of the "XO-Laptop" in the Ethiopian Educational System. Deutsche Gesellschaft fuer Technische Zusammenarbeit*. GTZ.

Haidt, J. (2006). *The happiness hypothesis. Putting ancient wisdom and philosophy to the test of modern science*. London: Arrow Books.

Hall, P. A., & Soskice, D. (Eds.). (2001). *Varieties of capitalism: The institutional foundations of comparative advantage*. Oxford: Oxford University Press.

Hall, M. (1999). Virtual Colonization. *Journal of Material Culture*, *4*(1), 39–55. doi:10.1177/135918359900400103

Hall, P. (1998). *Cities in Civilization: Culture, Innovation and Urban Order*. London: Weidenfield and Nicholson.

Halpern, D. (2005). *Social Capital*. Cambridge: Polity Press.

Hammond, A. L., Kramer, W. J., Katz, R. S., Tran, J. T., & Walker, C. (2007). *The Next 4 Billion: Market Size and Business Strategy at the Base of the Pyramid*. Washington, DC: World Resource Institute & International Finance Corporation.

Hampton, K. (2003). Grieving for a Lost Network: Collective action in a wired suburb. The Information Society, 19(5), 1-13. Retrieved July 3, 2009 from http://www.mysocialnetwork.net/downloads/mobilization-final.pdf

Hampton, K., & Wellman, B. (2003). Neighboring in Netville: How the Internet supports Community and Social Capital in a Wired Suburb. City and Community, 2, 277-311. Retrieved July 3, 2009 from http://www.chass.utoronto.ca/~wellman/publications/.../neighboring/neighboring_netville.pdf

Hargittai, E. (2002). Second-Level Digital Divide: Differences in People's Online Skills. *First Monday, 7*(4). Retrieved December 26, 2008, from http://firstmonday.org/issues/issue7_4/hargittai

Harris, P., Rettie, R., & Kwan, C. C. (2005). Adoption and use of m-commerce: a cross cultural comparison of Hong Kong and the United Kingdom. *Journal of Electronic Commerce Research*, *6*(3), 210–224.

Harrison, L. E., & Huntington, S. P. (2000). *Culture Matters: How Values Shape Human Progress*. New York: Basic Books.

Harvey, D. (1989). *The condition of post-modernity: an enquiry into the origins of cultural change*. Oxford: Basil Blackwell.

Hasan, H., & Ditsa, G. (1999). The impact of culture on the adoption of IT: an interpretive study. *Journal of Global Information Management*, *7*(1), 5–15.

He, F., Linden, L. L., & MacLeod, M. (2007). *Teaching what teachers don't know: an assessment of the Pratham English language program.* Columbia University Department of Economics. Retrieved July 28, 2008, from http://www.columbia.edu/~ll2240/PicTalk_Working_Paper_2007-03-26.pdf

Heeks, R. (2006). Theorizing ICT4D Research. *Information Technologies and International Development*, *3*(3), 1–4. doi:10.1162/itid.2007.3.3.1

Heeks, R. (Ed.). (1999). *Reinventing government in the information age. International practice in IT-enabled public sector reform*. London: Routledge. doi:10.4324/9780203204962

Heeks, R. (2002). I-development not e-development: Special issue on ICTs and development. *Journal of International Development*, *14*(1), 1–11. doi:10.1002/jid.861

Heeks, R. (2007a). Theorizing ICT4D research. *Information Technologies and International Development*, *3*(3), 1–4. doi:10.1162/itid.2007.3.3.1

Heeks, R. (2002). Information systems and developing countries: failure, success, and local improvisations. *The Information Society*, *18*, 101–112. doi:10.1080/01972240290075039

Heeks, R. (1999). Information and Communication Technologies, Poverty and Development. *Development Informatics.* Working Paper Series, Number 5. Manchester, UK: Institute for Development Policy and Management. Retrieved October 6, 2005 from: http://www.undp.org/info21/

Heeks, R. (1999). The tyranny of participation in information systems: Learning from development projects (Development Informatics Working Paper Series No. 4), University of Manchester. Retrieved May 24, 2009 from http://www.man.ac.uk/idpm_dp.htm#devinf_wp

Heeks, R. (2007b). *Impact assessment of ICT4D projects: A partial review of frameworks.* Unpublished paper prepared for the Investigating the Social & Economic Impact of Public Access to Information & Communication Technologies (IPAI) project.

Heeks, R. (2008). ICT4D 2.0: The next phase of applying ICT for international development. *IEEE Computer,* 26-33.

Heeks, R. B., & Jagun, A. (2007). Mobile Phones and Development: The Future in New Hands? *ID21 Insights Special Issue*, no.69. Retrieved May 6, 2008 from http://www.id21.org/insights/insights69/insights69.pdf

Heller, P. (1999). *The labor of development: workers and the transformation of capitalism in Kerala, India*. Ithaca, NY: Cornell University Press.

Henton, D., Melville, J., & Walesh, K. (1997). *Grassroots Leaders for a New Economy: How Civic Entrepreneurs are Building Prosperous Communities*. San Francisco: Jossey-Bass.

Hepp, P., Hinostroza, E., Laval, E., & Rehbein, L. (2004). *Technology in schools: Education, ICT and the knowledge society*. Washington: The World Bank.

Hilbert, M. R. (2001). *Latin America on Its Path Into the Digital Age: Where are We?* United Nations Publications.

Hill, C. E., Loch, K. D., Straub, D. W., & El-Sheshai, K. (1998). A qualitative assessment of Arab culture and information technology transfer. *Journal of Global Information Management*, *6*(3), 29–38.

Hillier, J. (2000). Going round the back? Complex networks and informal action in local planning processes. *Enivornment and Planning A*, *32*, 33–54. doi:10.1068/a321

Hinostroza, J. E., Guzmán, A., & Isaacs, S. (2002). Innovative uses of ICT in Chilean schools. *Journal of Computer Assisted Learning*, *18*(4), 459–469.doi:10.1046/j.0266-4909.2002.00257.doc.x

Hoffman, D. L., Novak, T. P., & Schlosser, A. E. (2000). The evolution of the digital divide: how gaps in Internet access may impact electronic commerce. *Journal of Computer-Mediated Communication, 5*(3). Retrieved October 2009 from http://jcmc.indiana.edu/vol5/issue3/hoffman.html

Hofstede, G. (1980). *Culture's Consequences: International Differences in Work-Related Values*. London: Sage Publications.

Hofstede, G. (1983). The cultural relativity of organizational practices and theories. *Journal of International Business Studies*, *14*(2), 75–89. doi:10.1057/palgrave.jibs.8490867

Hofstede, G. (1994). The business of international business is culture. *International Business Review*, *3*(1), 1–14. doi:10.1016/0969-5931(94)90011-6

Hollow, D. (2006). *The Lifeline Radio in Zambia: Findings and recommendations*. Working paper series. ICT4D Collective. Retrieved September 10, 2008, from www.gg.rhul.ac.uk/ict4d/HollowFreeplay.doc

Hollow, D. (2008). *Low-cost laptops for education in Ethiopia: Summary of Addis Ababa implementation report*. Internal report, ICT4D Collective. Retrieved December 2, 2008, from http://www.gg.rhul.ac.uk/ict4d/workingpapers/Hollowlaptops.pdf

Hongladarom, S. (2003). Exploring the philosophical terrain of the digital divide. *ACM CRPIT '03: Selected papers from conference on Computers and philosophy* (Vol. 37, pp. 85-89).

Honneth, A. (2007). *Disrespect: The Normative Foundations of Critical Theory*. Polity Press.

Hooker, M. (2008). 1:1 Technologies/computing in the developing world: Challenging the digital divide. Working paper. Global e-Schools and Communities Initiative (GeSCI).

Horrigan, J. B. (2008). Home Broadband Adoption 2008. *Pew Internet & American Life Project.* Retrieved December 28, from http://www.pewinternet.org/

Horton, D., & Mackay, R. (1999). Evaluation in developing countries: An introduction. *Knowledge. Technology & Policy*, *11*(4), 5–12.doi:10.1007/s12130-999-1001-9

Howard, P. N. (2007). Testing the Leap-frog hypothesis: the impact of existing infrastructure and telecommunications policy on the global digital divide. *Information Communication and Society*, *10*(2), 133–157. doi:10.1080/13691180701307354

Hu, P. J., Chau, P. Y. K., Liu Sheng, O. R., & Tam, K. Y. (1999). Examining the technology acceptance model using physician acceptance of telemedicine technology. *Journal of Management Information Systems*, *16*(2), 91–112.

Hudson, H. E. (2001). *The potential of ICT for development: Opportunities and Obstacles.* Background paper for the World Employment Report 2001. Retrieved April 6, 2006 from http://www.bib.ulb.ac.be/cdrom/wer_lawitie/back/hud_toc.htm

Huntington, S. P. (1996). *The clash of civilizations and the remaking of world order*. New York: Simon and Schuster.

Huntington, S. (2007). The Change to Change: Modernization, Development and Political Order in Changing Societies. In Roberts, T. R., & Hite, A. B. (Eds.), *The Globalization and Development Reader; perspectives on Development and Global Change*. Malden, MA: Blackwell Publishers.

Ianni, O. (1992). *A sociedade global*. Rio de Janeiro: Civilização Brasileira. IT for Change. (2007, January). *Development in the Information Society: exploring a social policy framework*. Workshop Report, Bangalore. Retrived May 17, 2009, from http://www.itforchange.net/media/wksp_reports/workshop_report_development_in_the_information_society_2007.pdf

IBICT. Estatítiscas. (2008). *Inclusão Digital*. Retrieved December 29, 2008, from http://inclusao.ibict.br

Igbaria, M., & Parasuraman, S. (1989). A path analytic study of individual characteristics, computer anxiety and attitudes toward microcomputers. *Journal of Management*, *15*(3), 373–388. doi:10.1177/014920638901500302

Ihde, D. (1990). *Technology and the lifeworld*. Bloomington, IN: Indiana University Press.

Information Technologies Group. (n.d.). *Readiness for the networked world: A guide for developing countries*. Cambridge, MA: Center for International Development. *Harvard University*.

Ingwersen, P. (1992). *Information retrieval interaction*. London: Taylor Graham.

Instituto Paulo Montenegro. (2007). *Indicador de Alfabetismo Funcional INAF / Brasil 2007*. Retrieved February 29, 2009, from http://www.ipm.org.br/

International Energy Agency (IEA). (2002). *World Energy Investment Outlook*. Paris: IEA.

International Tel ecommunication Union (ITU) (2007). *Measuring the Information Society*. ITU, Geneva. February 2007. Retrieved December 5, 2008 from http://www.itu.int/ITU-D/ict/publications/ict-oi/2007/index.html

International Telecommunication Union United Nations Conference on Trade and Development. World Summit on the Information Society, *Geneva 2003, Tunis2005*.

International Telecommunications Union (ITU). (2006). Telecommunication Development Report: ICT Statistics. Retrieved July 2006 from http://www.itu.org/statistics

International Telecommunications Union. (2001). *Digital divide. Overview*. Retrieved October 2009 from http://www.itu.int/ITU-D/digitaldivide

International Telecommunication Union (ITU). (2008a). *ICT in Africa: Digital Divide to Digital Opportunity*. ITU, Geneva. Retrieved on December 5, 2008 from http://www.itu.int/newsroom/features/ict_africa.html

International Telecommunication Union (ITU). (2008b). *Report on the World Summit on the Information Society Stocktaking: World Conference on the Information Society Geneva 2003 - Tunis 2007*. Place des Nations 1211 Geneva 20, Switzerland. Retrieved March 20, 2009 from http://www.itu.int/wsis/stocktaking/docs/2008/WSIS-Stocktaking2008-e.pdf

International Telecommunications Union (ITU). (2006a). World Telecommunication/ICT Development Report 2006: Measuring ICT for social and economic development. Retrieved December 5, 2008 from http://www.itu.int/ITU-D/ict/publications/wtdr_06/index.html

International Telecommunications Union (ITU). (2006b). Trends in the Information Society, ITU, Geneva. Executive Summary, Retrieved December 20, 2008 from http://www.itu.int/dms_pub/itu-s/opb/pol/S-POL-WSIS.RPT-2006-R1-SUM-PDF-E.pdf

Isserman, A. M., Feser, E., & Warren, D. (2007). Why Some Rural Communities Prosper While Others Do Not. Report prepared for the Office of the Under Secretary for Rural Development, USDA. Urbana-Champaign: University of Illinois. Retrieved November 2007 from http://www.ace.uiuc.edu/Reap/IssermanFeserWarren_070523_RuralProsperity.pdf

ITPS, Swedish Institute for Growth Policy Studies. (2003). *A Learning ICT Policy for Growth and Welfare, ITPS's final report on its assignment of evaluating the Swedish ICT policy*. Stockholm: Elanders Gotab.

ITU Internet Reports - Birth of Broadband Executive Summary. (2003). Retrieved from http://www.itu.int/osg/spu/publications/sales/birthofbroadband/exec_summary.html

ITU/UNCTAD. (2007). *World Information Society Report: Beyond WSIS*. Combined Reports for UN Summit held in Geneva December 2003 and Tunis in November 2005. ITU/UNCTAD, Geneva. Retrieved December 5, 2008 from www.itu.int/wisr and on the UNCTAD website at www.unctad.org/wisr

Ivarsson, J.-I. (1990). Medborgarinflytande i Stockholm. Levnadsförhållanden och medborgaraktiviteter i sex stadsdelar. [Stockholm: USK.]. *Utredningsrapport, 1990*, 4.

Ivarsson, J.-I. (1993). Stadsdelsnämndsförsöken i Stockholm – invånarnas reaktioner och synpunkter. [Stockholm: USK.]. *Utredningsrapport, 1993*, 3.

Ivarsson, J.-I. (1997). Så tycker brukarna om servicen i stadsdelen. [Stockholm: USK.]. *Utredningsrapport, 1997*, 3.

Ivarsson, J.-I. (2000). Servicen i stadsdelen 1999 – så tycker brukarna, jen ämförelse med 1996. [Stockholm: USK.]. *Utredningsrapport, 2000*, 1.

J. Popkin & Company. (1999). *Labor shortages, needs and related issues in small and large businesses*. Report prepared for the Office of Advocacy of the U.S. Small Business Administration. Washington, D.C. December. Retrieved Octonber 2009 from www.sba.gov/advo/research/#wages

Jackson, M.O. & Wolinsky, A. (1996) *A Strategic Model of Social and Economic Networks*. Oxford, UK: Journal of Economic Theory.

James, J. (2000). Pro-poor modes of integration into the global economy. *Development and Change, 31*(4), 765–783. doi:10.1111/1467-7660.00176

James, J. (2005). The global digital divide in the Internet: Developed countries constructs and Third World realities. *Journal of Information Science, 31*(2), 114–123. doi:10.1177/0165551505050788

James, J. (2007). From origins to implications: key aspects in the debate over the digital divide. *Journal of Information Technology, 22*(3), 284–296. doi:10.1057/palgrave.jit.2000097

James, J. (2006). Bridging the source of the digital divide. In *Digital Opportunity Forum, Background Papers and Presentations* (Seoul, Republic of Korea). Retrieved September 5, 2008 from http://www.dof.or.kr/htm/2/ann_001.asp

Janneck, M. (2009). Recontextualising Technology in Appropriation Processes. In Whitworth, B., & De Moor, A. (Eds.), *Handbook of Research on Socio-Technical Design and Social Networking Systems* (pp. 153–166). Hershey, PA: IGI.

Japanese Ministry of Foreign Affairs. (2000) *Okinawa Charter on Global Information Society*. Retrieved January 2009 from www.mofa.go.jp

Jasanoff, S. (2002). New modernities: reimagining science, technology and development. *Environmental Values, 11*(3), 253–276. doi:10.3197/096327102129341082

Javala, J., & Pohjola, M. (2001). Economic Growth in the New Economy. *Information Economics and Policy, 14*(2), 189–210. doi:10.1016/S0167-6245(01)00066-X

Jennett, P., Yeo, M., Pauls, M., & Graham, J. (2003). Organizational readiness for telemedicine: Implications for success and failure. *Journal of Telehealth and Telecare, 9*(Suppl. 2), S2: 27-39.

Jensen, R. (2007). The Digital Provide: Information (Technology), Market Performance, and Welfare in the South Indian Fisheries Sector. *The Quarterly Journal of Economics, 122*(3). doi:10.1162/qjec.122.3.879

Jensen, M. (2003). The current status of information and communication technologies in Africa. In Okpaku, J. O. Sr., (Ed.), *Information and communication technologies for African development: An assessment of progress and challenges ahead* (pp. 55–78). New York: United Nations ICT Task Force.

Jensen, M. (2002) *The African Internet – A status report.* Retrieved July 13, 2007, from http://demiurge.wn.apc.org/africa/afstat.htm.

Jiggins, J., & Röling, N. (2000). Adaptive management: Potential and limitations for ecological governance. *International Journal of Agricultural Resources. Governance and Ecology*, *1*(1), 28–43.

Jiggins, J., & Röling, N. (1997). Action research in natural resource management: Marginal in the first paradigm, core in the second. In C. Albadalejo & F. Casabianca (Eds.), *Pour une méthodologie de la recherche action.* (pp. 151-169). Versailles: INRA/SAD.

Johanson, G. (2008). Flicking the Switch: social networks and the role of Information and Communications Technologies in social cohesion among Chinese and Italians in Melbourne, Australia. In *Proceedings, The Role of New Technologies in Global Societies: Theoretical Reflections, Practical Concerns, and Its Implications for China, Conference at the Department of Applied Social Sciences.* The Hong Kong Polytechnic University, HKSAR, China, 30th-31st July 2008.

Johnson, J. (1996). *The information highway from hell: a worst-case scenario.* Paper presented at the Association for Computing Machinery's 1995 Conference on Computer-Human Interaction, 1996. Retrieved March 20, 2000 from Gopher://gopher. etext.org/h0/CPSR/nii/hell

Jorgenson, D. W. (2001). IT and The US Economy. *The American Economic Review*, *91*, 1–32. doi:10.1257/aer.91.1.1

Joseph, K. (2001). IT Boom, the Economy and Labour. *Labour and Development*, *7*(2), 1–36.

Jovanovic, B., & Rousseau, P. (2005). General Purpose Technologies. NBER Working Paper Series, No. W11093. Retrieved August 2, 2007 from http://.com/abstract=657607

Jubert, A. (1999). Developing an infrastructure for communities of practice. In B. McKenna, (Ed.), *Proceedings of the 19th International Online Meeting* (pp. 165-168). Hinksey Hill, U.K.: Learned Information.

Juhasz, A. (2006). *The Bush agenda: invading the world, one economy at a time*. New York: HarperCollins.

July 3, 2009, from http://www.isoc.org/oti/articles/0201/rao2.html

Kamppinen, M. (1998). Technology as a cultural system: the impacts of ICT upon the primary and secondary theories of the world. *Computers & Society*, 19–21. doi:10.1145/308364.308368

Kane, L. (2008). The World Bank, community development and education for social justice. *Community Development Journal*, *43*(2), 194–209. doi:10.1093/cdj/bsl043

Kanji, N., & Greenwood, L. (2001). *Participatory approaches to research and development in IIED: Learning from experience*. London: IIED.

Kapur, S. (2002). Developing Countries in the New economy: The Role of Demand-side Initiatives. *WIDER Discussion Paper 75*. Helsinki: WIDER/UNU.

Karahanna, E., Straub, D. W., & Chervany, N. L. (1999). Information technology adoption across time: a cross-sectional comparison of pre-adoption and post-adoption beliefs. *Management Information Systems Quarterly*, *23*(2), 183–213. doi:10.2307/249751

Katz, J. E., & Rice, R. E. (2002). *Social Consequences of Internet Use*. Cambridge, MA: The MIT Press.

Kauffman, R., & Techatassanasoontorn, A. A. (2005). International Diffusion of Digital Mobile Technology: A Coupled-Hazard State-Based Approach. *Information Technology Management*, *6*(2-3), 253–292. doi:10.1007/s10799-005-5882-3

Kavanaugh, A., Cohill, A., & Patterson, S. (2000). The use and impact of the Blacksburg Electronic Village. In Cohill, A., & Kavanaugh, A. (Eds.), *Community Networks: Lessons from Blacksburg*. Norwood, MA: Artech House.

Kavanaugh, A., Carroll, J. M., Rosson, M. B., Zin, T. T., & Reese, D. D. (2005). Community Networks: Where Offline Communities Meet Online. Journal of Computer-Mediated Communication, 10(4). Retrieved July 3, 2009, from http://jcmc.indiana.edu/ vol10/issue4/kavanaugh.html

Kay, J. (2008). An introduction to systems thinking. In Waltner-Toews, D., Kay, J., & Lister, N.-M. (Eds.), *The ecosystem approach: Complexity, uncertainty and managing for sustainability*. New York, Chichester: Columbia University Press.

Keeble, L. (2007). Community Informatics: building civil society in the Information Age? In Clay, C. J., Madden, M., & Potts, L. (Eds.), *People and places: an introduction towards understanding community*. New York: Palgrave Macmillan.

Kenny, C. (2002). Information and Communications Technologies for Direct Poverty Relief: Costs and Benefits. *Development Policy Review*, *20*(2), 141–157. doi:10.1111/1467-7679.00162

Kenny, C. (2002). The Internet and Economic Growth in Least Developed Countries, A Case of Managing Expectations? Discussion Paper No. 2002/75, United Nation University.

King, K. (2005). Knowledge-based Aid: A New Way of Networking or a New North-South Divide? In Stone, D., & Maxwell, S. (Eds.), *Global Knowledge networks and International Development; Bridges across Boundaries*. Oxford: Routledge. doi:10.4324/9780203340387_chapter_5

Kingsbury, D. (2004). *International Development: Issues and Challenges*. New York: Palgrave Macmillan.

Kingsbury, D., McKay, J., Hunt, J., McGillivray, M., & Clarke, M. (2008). *International Development*. London: Palgrave-MacMillian.

Kirkman, G., Osorio, C., & Sachs, J. (2002). The networked readiness index: Measuring the preparedness of nations for the networked world. In Kirkman, G., Sachs, J., Schwab, K., & Cornellius, P. (Eds.), *The Global Information Technology Report 2001-2002: Readiness for the networked world* (pp. 10–29). Oxford, UK: Oxford University Press.

Knowles, C. L. (2008). Toward a new web genre: Islamist Neorealism. *Journal of War and Culture Studies*, *1*(3), 357–380. doi:10.1386/jwcs.1.3.357_1

Ko, K. W. (2007). Internet Externalities and Location of Foreign Direct Investment: A Comparison between Developed and Developing Countries. *Information Economics and Policy Journal*, *19*, 1–23. doi:10.1016/j.infoecopol.2006.08.001

Kok, D. (2005). SAP invests in South African Technology Research Skills. Retrieved September 2009 from http://www.sap.com/southafrica/about/press/press.epx?pressid=4914

Kole, E. S. (1998). Myths and Realities in Internet Discourse: Using Computer Networks for Data Collection and the Beijing World Conference on Women. *The Gazette. The International Journal for Communication Studies*, *60*(4), 343–360.

Kole, E. S. (2002). An Alternative Conceptualization of ICT and Development, position statement for the conference. *Global E-Quality. Rethinking ICT in Africa, Asia and Latin America*. Organized by the International Institute of Infonomics and the A3 Network, 24-27 March 2002, Maastricht, the Netherlands. Retrieved October 5 2008 from: http://www.infonomics.nl/globalequality/ideas/kole.htm

Kozma, R., McGhee, R., Quellmalz, E., & Zalles, D. (2004). Closing the digital divide: Evaluation of the World Links Program. *International Journal of Educational Development*, *24*(4), 361–381.doi:10.1016/j.ijedudev.2003.11.014

Kraemer, K., Dedrick, J., & Shih, E. (2000). *Determinants of IT Investment at Country Level. CRITO*. Graduate School of Management, University of California.

Kraemer, K., & Dedrick, J. (1999). *Information Technology and Productivity: Results and Policy Implication of Cross Country Studies*. Prepared for the UNU/WIDER Study on IT and Economic Development, Working Paper No. PAC-144, 2-99.

Krannich, R. D., & Humphrey, C. R. (1986). Using Key Informant Data in Comparative Community Research: An Empirical Assessment. *Sociological Methods & Research, 14*(4), 473–493. doi:10.1177/0049124186014004006

Kranzberg, M. (1985). The Information Age: Evolution or revolution? In Guile, B. (Ed.), *Information Technologies and Societal Transformation*. Washington, DC: National Academy of Engineering.

Kuhn, T. S. (1990). *A estrutura das revoluções científicas* (Boeira, B. V., & Boeira, N., Trans.). São Paulo: Perspectiva. (Original work published 1962)

Kumar, R. (2006). Social Impact and Diffusion of Telecenter Use: A Study from the Sustainable Access in Rural India Project. *The Journal of Community Informatics, 2*(3). Retrieved December 21, 2008 from http://www.ci-journal.net/

Kusek, J. Z., & Rist, R. C. (2004). *Ten steps to a results-based monitoring and evaluation system: A handbook for development practitioners*. Washington, DC: The World Bank.

Kuttan, A., & Peters, L. (2003). *From digital divide to digital opportunity*. Lanham, MD: Scarecrow Press.

Kvasny, L., & Keil, M. (2006). The Challenges of Redressing the Digital Divide: A Tale of Two U.S. Cities. *Information Systems Journal, 16*(1), 23–53. doi:10.1111/j.1365-2575.2006.00207.x

Kyem, P. A. K., & Kyem, O. (2006). Africa's Participation in the Revolution in Information and Communications Technology: an Opportunity for Economic Development or Deprivation. In K. Konadu- Agyemang & K. Panford (Eds.), *Africa's Development in the Twenty First Century: Pertinent Socio-Economic and Development Issues* (pp. 358-390). Ashgate: UK.

Kyem, P. A. K., & LeMaire, P. K. (2006). Transforming Recent Gains in the Digital Divide into Digital Opportunities: Africa and the Boom in Mobile Phone Subscription. *Electronic Journal of Information Systems in Developing Countries, 28*. Retrieved December 5, 2008 from http://www.ejisdc.org/ojs2/index.php/ejisdc/article/view/343

Labelle, R. (2005). *ICT Policy Formulation and E-Strategy Development: A Comprehensive Guidebook*. Elsevier: UNDP-APDIP.

Lacerda, J. S. (2008). *Ambiências Comunicacionais e Vivências Midiáticas Digitais*. Unpublished doctoral thesis presented to the PostGraduate Programme in Communications of the Universidade do Vale do Rio do Sinos, Unisinos, as partial fullfilment of the requirements for the degree of Doctor in Communications.

LaRose, R., Gregg, J. L., Strover, S., Straubhaar, J., & Carperter, S. (2007). Closing the rural broadband gap: Promoting adoption of the Internet in rural America. *Telecommunications Policy, 31*(6-7), 359–373. doi:10.1016/j.telpol.2007.04.004

Latouche, S. (2007). De-growth: an electoral stake? *The International Journal Of Inclusive Democracy, 3*,1. Retrieved September 2009 from http://www.inclusivedemocracy.org/journal/vol3/vol3_no1_Latouche_degrowth.htm

Laviolette, A. (2007). Ceramic traditions, identities and population in Senegal and Gambia: Compared ethnography and historical reconstruction test. *Journal of African Archaeology, 5*(1), 149–150.

Law, P. L., Fortunati, L., & Yang, S. (Eds.). (2006). *New technologies in global societies*. Singapore: World Scientific. doi:10.1142/9789812773555

Lawn, P., & Clarke, M. (2008). *Sustainable Welfare in the Asia-Pacific: Studies Using the Genuine Progress Indicator*. London: Edward Elgar.

Leal, G. F. (2004). *A noção de exclusão social em debate: aplicabilidade e implicações para a intervenção prática*. Paper presented at the 14th Brazilian Association of Population Studies (ABEP) meeting, Caxambu, MG. Retrieved November 29, 2008, from http://www.abep.nepo.unicamp.br/site_eventos_abep/PDF/ABEP2004_42.pdf

Lederer, A. L., Maupin, D. J., Sena, M. P., & Zhuang, Y. (2000). The technology acceptance model and the World Wide Web. *Decision Support Systems*, *29*(3), 269–282. doi:10.1016/S0167-9236(00)00076-2

Lee, K. (1993). *Compass and gyroscope: Integrating science and politics for the environment*. Washington, D.C.: Island Press.

Lee, M. (2001). A refusal to define HRD. *Human Resource Development International*, *4*(3), 327–341. doi:10.1080/13678860110059348

Lee, R. M. (1994). Modernisation, post-modernism and the Third World. *Current Sociology*, *42*(2), 1–63.

Lee, K. (1995). Deliberately seeking sustainability in the Columbia River Basin. In Gunderson, L., Holling, C., & Light, S. (Eds.), *Barriers and bridges to the renewal of ecosystems and institutions* (pp. 214–238). New York: Columbia Press.

Lee, K. (1998, September 16-19.). *Appraising adaptive management.* Adaptive Collaborative Management of Protected Areas: Advancing the Potential. Cornell University, Center for International Forestry Research.

Leeuwen, G. V., & Vander Weil, H. P. (2003). Spillover Effects of ICT. *CBP Report* 2003/3.

Legris, P., Ingham, J., & Collerette, P. (2003). Why do people use information technology? A critical review of the technology acceptance model. *Information & Management*, *40*(3), 191–204. doi:10.1016/S0378-7206(01)00143-4

Lehr, W. Osorio, C.A., Gillett, S. & Sirbu, M.A. (2006). *Measuring Broadband's Economic Impact.* Cambridge, MA: MIT, Engineering Systems Division, ESD-WP-2006-02. http://esd.mit.edu/WPS/esd-wp-2006-02.pdf Last accessed October 2009.

Lengyel, G., Eranusz, E., Füleki, D., Lörincz, L., & Siklós, V. (2006). The Cserénfa experiment – on the attempt to deploy computers and Internet in a small Hungarian village. *The Journal of Community Informatics, 2*(3). Retrieved December 21, 2008 from http://www.ci-journal.net/.

Lenhart, A., Horrigan, J., Rainie, L., Allen, K., Boyce, A., Madden, M., et al. (2003). The ever-shifting internet population: a new look at internet access and the digital divide. The Pew internet and American life project. Retrieved October 2009 from http://www.pewinternet.org/Reports/2003/The-EverShifting-Internet-Population-A-new-look-at-Internet-access-and-the-digital-divide.aspx

Lerner, D. (1958). *The Passing of Traditional Society: Modernizing the Middle East*. New York: Free Press.

Lerner, D. (1967). International cooperation and communication in national development. In Lerner, D., & Schramm, W. (Eds.), *Communication and Change in the Development Countries* (pp. 103–128). Honolulu: East-West Center Press.

Liao, Z., & Wong, W. (2004). Key success factors of smartcard-based electronic payment: an empirical analysis. In *Proceedings of The Eighth Pacific-Asia Conference on Information Systems 2004 (PACIS 2004)*, Shanghai, China, 8-11 July, 2004 (pp. 2065-2071).

Liff, S., & Laegran, A. S. (2003). Cybercafés: Debating the Meaning and Significance of Internet Access in a Café Environment. *New Media & Society*, *5*, 307–312. doi:10.1177/14614448030053001

Liff, S., & Steward, F. (2001). Communities and community e-gateways: Networking for social inclusion. In Keeble, L., & Loader, B. D. (Eds.), *Community Informatics: Shaping Computer-Mediated Social Relations*. London: Routledge.

Liff, S., Fred, S., & Watts, P. (1998). Cybercafés and Telecottages: Increasing public access to computers and the Internet. Survey report, Virtual Society? Programme, Economic and Social Research Council, United Kingdom. Retrieved July 3, 2009 from http://virtualsociety.sbs.ox.ac.uk/ text/reports/access.htm

Light, D., & Manson, M. (2007). *An educational revolution to support change in the classroom: Colombia and the educational challenges of the twenty-first century*. Washington, DC: Education Development Center, Inc, Center for Children and Technology.

Light, D. Method, F., Rockman, C., Cressman, G.M., & Daly, J. (2008). *Synthesis report: Overview and recommendations to the Jordan Education Initiative.* Washington, DC: Education Development Center, Inc.

Lincoln, Y. (1990). The making of a constructivist. A remembrance of transformations past. In Guba, E. (Ed.), *The paradigm dialogue* (pp. 67–87). Newbury Park: Sage Publications.

Linden, L. L. (2008). *Complement or substitute? The effect of technology on student achievement in India.* Infodev Working Paper No. 17. Retrieved November 11, 2008, from http://www.infodev.org/en/Publication.505.html

Linden, L., Banerjee, A., & Duflo, E. (2003). *Computer-Assisted Learning: Evidence from a Randomized Experiment.* Poverty Action Lab Paper No. 5. Cambridge, MA: Poverty Action Lab. Retrieved October 4, 2008, from http://www.povertyactionlab.org/papers/5_Duflo_Computer-Assisted_Learning.pdf

Lipsey, R. G. Carlaw, K. I., & Bekar, C. (2005). *Economic Transformations: General Purpose Technologies and Long-run Economic Growth.* Oxford, UK: Oxford University Press.

Lizardo, O. (2008). *Beyond the Antinomies of Structure: Recovering the Insights of Methodological Structuralism.* Paper presented at the annual meeting of the American Sociological Association Annual Meeting, Sheraton Boston and the Boston Marriott Copley Place, Boston, MA, Jul 31, 2008. Retrieved September 2009 from http://www.allacademic.com/meta/p239582_index.html

Loader, B. D., Hague, B. N., & Eagle, D. (2000). Embedding the 'Net: community development in the age of information. In Gurstein, M. (Ed.), *Community Informatics: Enabling Communities with Information and Communications Technologies.* Hershey, PA: Idea Group.

Loader, B. D., & Keeble, L. (2004). A Literature Review of Community Informatics Initiatives. York: Joseph Rowntree Foundation. Retrieved July 3, 2009, from http://www.jrf.org.uk/publications/ literature-review-community-informatics-initiatives

Lora, E. (2004). *A Decade of Development Thinking.* Washington, DC: Inter-American Development Bank.

Lucas, H. C., & Spitler, V. K. (1999). Technology use and performance: a field study of broker workstations. *Decision Sciences, 30*(2), 291–311. doi:10.1111/j.1540-5915.1999.tb01611.x

Luyt, B. (2004). Who benefits from the digital divide? *First Monday, 9*(8). Retrieved December 24, 2008 from http://firstmonday.org/issues/issue9_8/luyt/index.html

Lynch, K. (1960). *The image of the city.* Cambridge, MA: MIT Press.

Mabogunje, A. L. (2000). Institutional radicalization, the state, and the development process in Africa. *Proceedings of the National Academy of Sciences of the United States of America, 97*(25), 14007–14014. doi:10.1073/pnas.200298097

Maclay, C., Kirkman, G., & Hawkins, R. (2005). *Global networked readiness for education: Preliminary findings from a pilot project to evaluate the impact of computers and the Internet on learning in eleven developing countries.* Washington, DC: The World Bank. Retrieved July 13, 2008, from http://siteresources.worldbank.org/EDUCATION/Resources/278200-1126210664195/1636971-1126210694253/Global_Network_Readiness.pdf

Mahiri, I. O. (1998). The environmental knowledge frontier: transects with experts and villagers. *Journal of International Development, 10,* 527–537. doi:10.1002/(SICI)1099-1328(199806)10:4<527::AID-JID543>3.0.CO;2-S

Mankiw, N. G., Romer, D., & Weil, D. N. (1992). A Contribution to the Empirics of Economic Growth. *Journal of Economics, 107*(2), 407–437.

Mansell, R. (2000). Information and communication technologies for development: Assessing the potential and the risks. *Telecommunications Policy, 23,* 35–50. doi:10.1016/S0308-5961(98)00074-3

Mansell, R. (2001). Digital Opportunities and the Missing Link for Developing Countries. *Oxford Review of Economic Policy, 17*(2), 282–295. doi:10.1093/oxrep/17.2.282

Mansell, R. (2002). Constructing the Knowledge Base for Knowledge-driven Development. *Journal of Knowledge Management*, *6*(4), 317–329. doi:10.1108/13673270210440839

Mansell, R., Steinmueller, W., & Montalvo, U. (1999). Opportunities for Knowledge-based Development: Capabilities, Infrastructure, Investment and Policy. *Science & Public Policy*, *26*(2), 91–100. doi:10.3152/147154399781782572

Mansell, R., & Wehn, U. (1998). *Knowledge societies: Information technology for sustainable development.* Oxford: Published for and on behalf of the United Nations by Oxford University Press.

Marcel, L. (1998). *The Role Of The Gambian Electronic Media In The Implementation Of The Gambian Foreign Policy*. Unpublished Masters dissertation, Dept of Mass Communication, University Of Nigeria.

Marcelle, G. (2000). Getting gender into African ICT policy: A strategic view. In Rathgeber, E. M., & Adera, E. O. (Eds.), *Gender and the information revolution in Africa* (pp. 35–84). Ottawa: IDRC.

Marker, P., McNamara, K., & Wallace, L. (2002). *The significance of information and communication technologies for reducing poverty*. Department of International Development UK, London. Retrieved April 25, 2009 from http://www.dfid.gov.uk/Pubs/files/ictpoverty.pdf

Martin, S. (2003). Is the Digital Divide Really Closing? A critique of inequality measurement in a nation online. *Information Technology and Society*, *1*(4), 1–13. doi:10.1080/0268396031000077413

Martín Barbero, J. (1995). Modernidad, posmodernidad, modernidades: discursos sobre la crisis y la diferencia. *Intercom - Revista Brasileira de Comunicação, 23*(2).

Mason, S., & Hacker, K. (2003). Applying Communication Theory to Digital Divide Research, *Information. Technology and Society*, *1*(5), 40–55.

Masuda, J. (1981). *The Information Society as Post-Industrial Society*. Washington: World Future Society.

Mathieson, K. (1991). Predicting User Intentions: Comparing the Technology Acceptance Model with the Theory of Planned Behavior. *Information Systems Research*, *2*(3), 173–191. doi:10.1287/isre.2.3.173

Mathieson, K., Peacock, E., & Chin, W. W. (2001). Extending the technology acceptance model: the influence of perceived user resources. *The Data Base for Advances in Information Systems*, *32*(3), 86–112.

Mattelart, A. (2002). *História da sociedade da informação* (Campanário, N. N., Trans.). São Paulo: Loyola. (Original work published 2001)

Maturana, H. R., & Varela, F. J. (1980). *Autopoiesis and cognition*. Dordrecht, Holland: D Reidel.

Max-Neef, M. A. (1982). *From the outside looking in. Experiences in 'barefoot economics*. London: Zed Books.

Maxwell, S., & Conway, T. (2000). *Perspectives on partnership*. Washington, DC: World Bank.

Mayer, C. (2002). Financing the New economy: Financial Institutions and Corporate Governance. *Information Economics and Policy*, *14*(2), 311–326. doi:10.1016/S0167-6245(01)00072-5

Mayur, R., & Daviss, B. (1998). *The Technology of Hope: Tools to Empower the World's Poorest Peoples. The Futurist*. January-February.

Mbeki, M., Vale, P. (2009). Community is the new state. *Mail & Guardian*, 27.

McAlpine, P., & Birnie, A. (2006). Establishing sustainability indicators as an evolving process: Experience from the Island of Guernsey. *Sustainable Development*, *14*(2), 81–92. doi:10.1002/sd.301

McAnany, E. (1983). From Modernization and Diffusion to Dependency and Beyond: Theory and Practice in Communication for Social Change in the 1980s. *Development Communications in the Third World, Proceedings of a Midwest Symposium.* University of Illinois.

McCaston, M. K. (2005). *Documenting our methods and major findings*. CARE Newsletter. Retrieved August 10, 2008, from http://pqdl.care.org/Practice/Documenting%20Methods%20and%20Findings.pdf

McCloskey, D. (2003). Evaluating electronic commerce acceptance with the technology acceptance model. *Journal of Computer Information Systems*, *44*(2), 49–57.

McConnaughey, J. W., & Lader, W. (1998). Falling through the Net II. New Data on the Digital Divide. NTIA Report. Retrieved October 2008 from http://www.ntia.doc.gov/ntiahome/net2/falling.html

McGowan, P. J., Cornelissen, S., & Nel, P. (Eds.). (2006). *Power, wealth and global equity. An international relations textbook for Africa*. Cape Town: UCT Press.

McIlvenny, P. (1999). Avatars R Us? Discourses of Community and Embodiment in Intercultural Cyberspace. *Journal of Intercultural Communication, 1*(2). Retrieved 24 December, 2008 from http://www.immi.se/

McIver, W. (2003). Community Informatics for the Information Society. In Girard, B., & Siochru, S. O. (Eds.), *Communicating in the Information Society*. Geneva: United Nations Research Institute for Social Development.

McLuhan, M. (1965). *Understanding media: The extensions of man*. New York: McGraw-Hill Book Company.

McLuhan, M., & Fiore, Q. (1967). *The medium is the massage*. New York, London, Toronto: Bantam Books.

McNamara, K. S. (2000). *'Why be Wired? The Importance of Access to Information and Communication Technologies'. TechKnowLogia, March/April 2000*. Knowledge Enterprise, Inc.

McNamara, K. S. (2003, December). Information and communication technologies, poverty and development: learning from experience. Background paper for the infoDev Annual Symposium, Geneva, Switzerland.

Meichenbaum, D. (1990). Review: paying homage: providing challenges. *Psychological Inquiry*, *1*(1), 96–100. doi:10.1207/s15327965pli0101_25

Melucci, A. (2001). *A invenção do presente: movimentos sociais na sociedade complexa*. (M.C.A. Bomfim, Trans.) Petrópolis: Vozes. (Original work published 1982).

Menou, M. (2008, April). *ImperialICTism: the highest stage of capitalism?* Keynote address [La Habana, Cuba, Foro Sociedad del Conocimiento: nuevos espacios para su construcción.]. *Info*, ▪▪▪, 2008.

Menou, M., & Taylor, R. (2006, November-December). A "Grand Challenge": Measuring information societies. *The Information Society, 22*(5), 261–267. doi:10.1080/01972240600903904

Menou, M. J., Poepsel, K. D., & Stoll, K. (2004). Latin American Community Telecenters: "It's a long way to TICperary". *The Journal of Community Informatics, 1*(1), 39-57. Retrieved September 18, 2008 from http://www.ci-journal.net

Mercer, C. (2004). Engineering Civil Society: ICT in Tanzania. *Review of African Political Economy*, *31*(99), 49–64. doi:10.1080/0305624042000258414

Mercer, C. (2006). Telecentres and transformations: Modernizing Tanzania through the Internet. *African Affairs*, *105*(419), 243–264. doi:10.1093/afraf/adi087

Meredith, M. (2006). *The state of Africa. A history of fifty years of independence*. Johannesburg: Jonathan Ball.

Mertons, D. (2005). *Research and evaluation in education and psychology: Integrating diversity with quantitative, qualitative, and mixed methods*. Thousand Oaks, CA: Sage Publications, Inc.

Michel, L. (2006). *Compendium on development cooperation strategies*. Brussels, Belgium: European Commission.

Milanovic, B. (2005). *Worlds Apart: Measuring International and Global Inequality*. Princeton, NJ: Princeton University Press.

Minges, M. (2002). *Counting the Net: Internet Access Indicators*. Geneva: ITU.

Minsky, M. (1986). *The society of mind*. New York: Simon & Schuster.

Mitchell, W. J. (1995). *City of Bits: Space, Place and the Infobahn*. Cambridge, MA: MIT Press.

Mitchell, M. M. 2003. Possible, Probable and Preferable Futures of the Digital Divide. *Informing Science InSITE - Where Parallels Intersect.* Retrieved May 8, 2007 from http://www.informingscience.org/proceedings/IS2003Proceedings/docs/083Mitch.pdf

Mitra, S., Dangwal, R., Chatterjee, S., Jha, S., Bisht, R. S., & Kapur, P. (2005). Acquisition of computing literacy on shared public computers: Children and the "hole in the wall". *Australasian Journal of Educational Technology, 21*(3), 407–426.

Moore, S. K., & Gardner, A. (2007, June). Megacities by the numbers. *IEEE Spectrum*, ▪▪▪, 16–17. doi:10.1109/MSPEC.2007.295503

Moore, G. C., & Benbasat, I. (1991). Development of an instrument to measure the perceptions of adopting an information technology innovation. *Information Systems Research, 2*(3), 193–222. doi:10.1287/isre.2.3.192

Moore, D. (Ed.). (2007). *The World Bank development, poverty, hegemony*. Scottsville: University of Kwazulu-Natal Press.

Morris, M. G., & Venkatesh, V. (2000). Age differences in technology adoption decisions: implications for a changing workforce. *Personnel Psychology, 53*(2), 375–403. doi:10.1111/j.1744-6570.2000.tb00206.x

Moshiri, S. (2010). *Information Technology and Changing the Workplace; Firm Level Evidence*. St. Thomas More College, University of Saskatchewan.

Mossberger, K., Tolbert, C. J., & Stansbury, M. (2003). *Virtual Inequality: Beyond the Digital Divide*. Washington, DC: Georgetown University Press.

Mosse, D. (1993). *Authority, gender and knowledge: Theoretical reflections on the practice of participatory rural appraisal. Network Paper* (*Vol. 44*). London: ODI.

Moyo, D. (2009). *Dead Aid. Why aid is not working and how there is another way for Africa*. London: Allen Lane.

Mulder, J. (2008). *Knowledge dissemination in Sub-Saharan Africa: What role for Open Educational Resources (OER)?* Master's thesis, International School for Humanities and Social Sciences, University of Amsterdam. Retrieved October 4, 2008, from http://www.gg.rhul.ac.uk/ict4d/workingpapers/mulderOER.pdf

Muller, P. (2002). Internet Use in Transition Economies. *WIDER Discussion Paper No. 95*. Helsinki: WIDER/UNU.

Mumford, L. (1934). *Technics and civilization*. New York: Harcourt, Brace and Company.

Munir, L. Z. (2002) "He is your garment and you are his...": Religious precepts, interpretations, and power relations in marital sexuality among Javanese Muslim women. *Sojourn: Journal of Social Issues in Southeast Asia, 17.*

Murdock, G. (2002). Review Article: Debating Digital Divides. *European Journal of Communication,* 17, 385. Retrieved December 7, 2008 from http://ejc.sagepub.com/cgi/content/abstract/17/3/385

Murphy, J. T. (2008). Economic Geographies of the Global South: Missed Opportunities and promoting intersections with development studies. *Geography Compass, 2*(3), 851–873. doi:10.1111/j.1749-8198.2008.00119.x

Musa, P. F. (2006). Making a case for modifying the technology acceptance model to account for limited accessibility in developing countries. *Information Technology for Development, 12*(3), 213–224. doi:10.1002/itdj.20043

Mutula, S. M. (2003). Cyber café industry in Africa. *Journal of Information Science, 29*(6), 489–497. doi:10.1177/0165551503296006

Myers, M. (2004). *Evaluation methodologies for information and communication for development (ICD) programmes*. London: DFID.

Narayan, D., Chambers, R., Shah, M. K., & Petesch, P. (2000). Crying Out for Change. *Voices of the Poor: Vol. 2. World Bank*. Oxford: University Press.

Narayan, D., Patel, R., Schafft, K., Rademacher, A., & Koch-Schulte, S. (1999). Can Anyone Hear Us? *Voices of the Poor*: *Vol. 1*. *World Bank*. Oxford: University Press.

Narayan, D., & Petesch, P. (2002). From Many Lands. *Voices of the Poor*: *Vol. 3*. *World Bank*. Oxford: University Press.

Nardi, B. A., & O'day, V. L. (1999). *Information ecologies: Using technology with heart*. Cambridge/Massachusetts & London: The MIT Press.

Nasar, J. (1998). *The evaluative image of the city*. Thousand Oaks, CA: Sage.

National Agriculture Statistics Service. (2007). *Farm Computer Usage and Ownership*. Washington, DC: USDA, NASS.

National Center for Educational Statistics (NCES). (2001). *The nation's report card: Mathematics 2000*. Washington, D.C.: National Center for Educational Statistics.

National Research Council. (1998). *Internet counts: Measuring the impacts of the internet*. Retrieved February 9, 2005 from www.bsos.umd.edu/cidcm/wilson/xnasrep2.htm

National Social Inclusion Programme of the UK. (n.d.). Retrieved March 2008 from http://www.socialinclusion.org.uk/

National Telecommunciations Cooperative Assocation (NTCA). (2003). *NTCA 2003 Internet/Broadband Availability Survey Report*. Washington, DC. NTCA. Retrieved October 2009 from http://www.ntca.org/content_documents/2003broadband.pdf

National Telecommunications and Information Administration (NTIA). (1995). *Falling through the Net: a survey of the 'have-nots' in rural and urban America*. Washington, DC, U.S. Department of Commerce. Retrieved December 23, 2008, from http://www.ntia.doc.gov/ntiahome/fallingthru.html

National Telecommunications and Information Administration (NTIA). (1999). *Falling through the Net: defining the digital divide*, Washington, DC, U.S. Department of Commerce. Retrieved December 23, 2008, from http://www.ntia.doc.gov/ntiahome/fttn99/contents.html

National Telecommunications and Information Administration (NTIA). (2000). *Falling through the Net: toward digital inclusion*, Washington, DC, U.S. Department of Commerce. Retrieved December 23, 2008, from http://www.ntia.doc.gov/ntiahome/fttn00/contents00.html

Naudé, W. (2009). The Financial Crisis of 2008 and the Developing Countries. *Discussion Paper No. 2009/01*. World Institute for Development Economics Research, UNU, Helsinki.

NBTF. (2006).*National Broadband Task Force Report - The National Dream: networking the nation for broadband access*. Retrieved December 20, 2008 from http://www.collectionscanada.gc.ca/.

Neef, D. (Ed.). (1998). *The Knowledge Economy*. Oxford: Butterworth-Heinemann.

Negroponte, N. (1995). *Being digital*. New York: Routledge.

Neves, B. C., & Gomes, H. F. (2008). A convergência dos aspectos de inclusão digital: experiência nos domínios de uma universidade. *Encontros Bibli: Revista Eletrônica de Biblioteconomia e Ciência da Informação, 13*(26). Retrieved December 23, from http://www.periodicos.ufsc.br/

Nguyen, T. T. (2008). *Seeking Thachsanh's Rice Bowl: an Exploration of Knowledge, ICTs and Sustainable Economic Development in Vietnam*. Unpublished Doctor of Philosophy dissertation. Melbourne, Australia: Faculty of Information Technology, Monash University.

Niitamo, V. (2000). *IT, Economic Growth and Development*. Presented at UN ECOSOC 2000 Preparatory Process. New York: 5 May.

Nnadi, N., & Gurstein, M. (2007). Towards Supporting Community Information Seeking and Use. *Journal of Community Informatics, 3*(1). Retrieved October 2009 from http://ci-journal.net/index.php/ciej/article/view/325

Nora, S., & Minc, A. (1978). *A informatização da sociedade* (de Vasconcelos, P., Trans.). Lisbon: Sociedade Astória.

NORDICOM. (Nordiskt Informationscenter för Medie- och Kommunicationsforskning) (2002). Internetbarometern 2002. Retrieved July 3, 2009, from http://www.nordicom.gu.se/mt/filer/Inetbar%202002-3.pdf

Norris, P. (2003). *Digital divide. Civic engagement, information poverty, and the internet worldwide*. Cambridge: Cambridge University Press.

Norris, P. (2001). *Digital Divide: Civic Engagement, Information Poverty, and the Internet Worldwide*. Cambridge: Cambridge University Press.

Norris, K. (2001). *Dakota: A Spiritual Geography*. New York: Houghton-Mifflin Co.

Notley, T., & Foth, M. (2008). *Extending Australia's digital divide policy: an examination of the value of social inclusion and social capital policy frameworks*. Retrieved October 2008 from http://eprints.qut.edu.au/

NTIA. (1995). Falling Through the Net: A Survey of the 'Haves' and 'Have Nots' in Rural and Urban America. Retrieved October 2008 from http://www.ntia.doc.gov/ntiahome/fallingthru.html

NTIA. *Falling through the net: defining the digital divide* - a *report on the telecommunications and information technology gap in America*. (1999). National Telecommunications & Information Administration, U.S. Department of Commerce. Retrieved June 3, 2008, from http://www.ntia.doc.gov

NTIA. *Falling through the net: toward digital inclusion – a report on American's access to technology tools*, October. (2000). National Telecommunications & Information Administration, Economic and Statistics Administration. Retrieved June 3, 2008, from http://www.ntia.doc.gov

NTIA. *Falling Through the Net II: New Data on the Digital Divide*. (1998). National Telecommunications & Information Administration, U.S. Department of Commerce. Retrieved June 3, 2008, from http://www.ntia.doc.gov

O'Donnell, d., & Henriksen, L.B. (2002). Philosophical foundations for a critical evaluation of the social impact of ICT. *Journal of Information Technology*, *17*(2), 88–99.

O'Hara, K., & Stevens, D. (2006). *Inequality.Com: Power, Poverty and the Digital Divide*. New York: OneWorld Publications.

OECD (2004). *ICT and Economic Growth in Developing Countries*. Development and Co-Operate Directorate, Development Assistant Committee Network on Poverty Reduction, 10-Dec.

OECD. (2008). *2007). Development Cooperation Report*. Paris, France: OECD.

OECD. (2003). *ICT and Economic Growth: Evidence from OECD Countries, Industries and Firms*. Paris, France. Retrieved October 2009 from http://www.labs-associados.org/docs/OCDE_TIC.PDF

Ojo, T. (2004). Old paradigm and Information & Communication Technologies for Development Agenda in Africa: Modernization as Context. *Journal of Information Technology Impact, 4*(3), 139-150. Retrieved May 7, 2009 from http://www.allacademic.com/meta/p14108_index.html.

Oldenburg, R. (1989/1997). *The Great Good Place*. New York: Paragon House.

Oldenburg, R. (2001). *Celebrating the Third Place: Inspiring Stories about the "Great Good Places" at the Heart of Our Communities*. New York: Marlowe and Company.

Oliner, S. D., & Sichel, D. E. (2000). The Resurgence of Growth in The Late 1990s: Is The Story? *The Journal of Economic Perspectives*, *14*, 3–22. doi:10.1257/jep.14.4.3

Oliner, S. D., & Sichel, D. E. (2002). IT and Productivity: Where Are We Now and Where Are We Going? *Economic Review (Federal Reserve Bank of Atlanta)*, *87*(3), 15–44.

Omar Dengo Foundation. (2007). *Multistakeholder Partnerships and Digital Technology for Development in Latin America and the Caribbean.* San Jose, Costa Rica: Edicciones innova.

O'Neil, D. V., & Baker, P. M. A. (2003). The role of institutional motivations in technological adoption: Implementation of DeKalb County's Family Technology Resource Centers. *The Information Society, 19*(4), 305–314. doi:10.1080/01972240309485

Ong, A. (2007). Neoliberalism as a mobile technology. *Transactions of the Institute of British Geographers, 32*, 3–8. doi:10.1111/j.1475-5661.2007.00234.x

Ongley, E. D. (1996). Control of water pollution from agriculture - FAO irrigation and drainage paper 55. FAO (Food and Agriculture Organization of the United Nations). Retrieved September 2009 from http://www.fao.org/docrep/W2598e/w2598e00.htm

Onis, Z., & Senses, F. (2005). Rethinking the Emerging Post-Washington Consensus. *Development and Change, 36*(2), 263–290. doi:10.1111/j.0012-155X.2005.00411.x

Organization for Economic Cooperation and Development (OECD). (1996a). *Technology, Productivity and Job Creation*. Paris: OECD.

Organization for Economic Cooperation and Development (OECD). (1996b). *Employment and Growth in the Knowledge-Based Economy*. Paris: OECD.

Orozco, M. (2004). Remesas económicas y migración – cuestiones y perspectivas sobre el desarollo. *Vanguardia Dossier: Los hispanos en Estados Unidos*. Barcelona, *13*, 75-81.

Pacey, A. (1999). *Meaning in technology*. Cambridge, MA: The MIT Press.

Palmer, B. (1990). *Descent into discourse: the reification of language and the writing of social history*. Philadelphia, PA: Temple University Press.

Parkinson, S. (2005). *Telecentres, access and development: Experience and lessons from Uganda and South Africa*. Warwickshire, UK; Kampala; Ottawa: ITDG, Fountain Publishing, IDRC.

Parkinson, S., & Ramírez, R. (2006). Using a sustainable livelihoods approach to assessing the impact of ICTs in development. *Journal of Community Informatics, 2*(3). Retrieved October 7, 2008 from http://ci-journal.net/index.php/ciej/issue/view/15

Parks, W. with Felder-Gray, D., Hunt, J., & Byrne, A. (2005). *Who measures change? An introduction to participatory monitoring and evaluation of communication for social change.* New Jersey, USA: Communication for Social Change Consortium.

Partnership on measuring ICT for development. (2005). *Core ICT Indicators*. Beirut, Lebanon: United Nations - ESCWA.

Pasquali, A. (2003). A brief descriptive glossary of communication and information (aimed at providing clarification and improving mutual understanding). In Girard, B., & Siochru, S. O. (Eds.), *Communicating in the Information Society*. Geneva: United Nations Research Institute for Social Development.

Patton, M. (2008). *Utilization-focused evaluation, 4th*. Thousand Oaks, CA: Sage Publications.

Peñaranda Cólera, M. C. (2008). Tecnologías que acercan distancias? Sobre los "claroscuros" del estudío de las tecnologías en los procesos migratorios transnacionales. In Santamaría, E. (Ed.), *Retos epistemológicos de las migraciones transnacionales* (pp. 133–164). Barcelona: Anthropos.

Pereira, P. A. (2007). *Necessidades humanas. Subsídios à crítica dos mínimos sociais*. São Paulo: Cortez. (Original work published 2000)

Persaud, A. (2001). The knowledge gap. *Foreign Affairs (Council on Foreign Relations), 80*(2), 107–117.

Phahlamohlaka, J., Braun, M., Romijn, H., & Roode, D. (Eds.). (2008). *Community-driven projects: reflections on a success story. A case study of science education and information technology in South Africa*. Pretoria: Van Schaik Publishers.

Phelps, E. S. (1997). *Rewarding Work. How to Restore Participation and Self-Support to Free Enterprise*. Cambridge, Massachusetts: Harvard Univ Press.

Picot, A., & Wernick, C. (2007). The role of governments in broadband access. *Telecommunications Policy, 31*(10/11), 660–674. doi:10.1016/j.telpol.2007.08.002

Pigg, K. E., & Crank, L. D. (2005). Do Information Communication Technologies Promote Rural Economic Development. *Community Development: Journal of the Community Development Society, 36*(1), 65–76. doi:10.1080/15575330509489872

Pigg, K. E., & Bradshaw, T. K. (2003). Catalytic Community Development: A Theory of Practice for Changing Rural Society. In Brown, D. L., & Swanson, L. (Eds.), *Challenges for Rural America in the 21st Century. State College*. PA: State University Press.

Pigg, K. E., & Crank, L. D. (2004) Building Community Social Capital: The Potential and Promise of Information and Communication Technologies. Journal of Community Informatics, 1(1). Retrieved July 3, 2009, from http://ci-journal.net/index.php/ciej/article/view/184/132

Pilat, L., & Ark, V. (2002). Production and Use of ICT: A Sectoral Perspective on Productivity Growth in OECD Area. *OECD Economic Studies*, 35.

Pimienta, D. (2005). *Measuring Linguistic Diversity on the Internet*. Paris, France: UNESCO.

Pisón, J. M. (1998). *Políticas de bienestar. Un estudio sobre los derechos sociales*. Madrid: Universidad de la Rioja.

Pitula, K., & Radhakrishnan, T. (2007). A Framework and Process for Designing Inclusive Technology. In *International Conference on Software Engineering Advances (ICSEA 2007)*. Retrieved October 2009 from http://ieeexplore.ieee.org/stamp/stamp.jsp?tp=&arnumber=4299944&isnumber=4299877

Pochmann, M. (2002). *e-Trabalho*. São Paulo: Publisher Brasil.

Pohjola, M. (Ed.). (2001). *Information Technology, Productivity and Economic Growth*. Oxford: Oxford University Press.

Pohjola, M. (2002a). The New economy: Facts, Impacts and Policies. *Information Economics and Policy, 14*(2), 133–144. doi:10.1016/S0167-6245(01)00063-4

Pohjola, M. (2001b). Information Technology and Economic Development: A Cross-Country Analysis. In Pohjola, M. (Ed.), *Information Technology, Productivity and Economic Growth*. Oxford: Oxford University Press.

Pohjola, M. (1998). Information Technology and Economic Development: An Introduction to the Research Issue, UNU/WIDER Working Paper, No:153.

Pohjola, M. (2002). New Economy in Growth and Development, Presented in UNU/WIDER Conference on New Economy in Development, 10-11 May, Helsinki.

Pohjola, M. (2002b). New Economy in Growth and Development. *WIDER Discussion Paper No. 67*. Helsinki: WIDER/UNU.

Polikanov, D., & Abramova, I. (2003). Africa and ICT: a chance for breakthrough? *Information, Communication & Society, 6*(1), 42:56.

Porter, C. E., & Donthu, N. (2006). Using the technology acceptance model to explain how attitudes determine Internet usage: The role of perceived access barriers and demographics. *Journal of Business Research, 59*(9), 999–1007. doi:10.1016/j.jbusres.2006.06.003

Portes, A. (1997). Globalization from below: the rise of transnational communities. *Transnational Communities Programme Working paper series*, WPTC-98-01. Retrieved May, 12, 2009, from http://www.transcomm.ox.ac.uk/working%20papers/portes.pdf

Postman, N. (1993). *Technopoly. The surrender of culture to technology*. New York: Vintage Books. (Original work published 1992)

Postman, N. (1990). *Informing ourselves to death*. Speech at the German Informatics Society (Gesellschaft für Informatik). In Stuttgart, October 11, 1990. Retrieved January 1, 2009, from http://world.std.com/~jimf/informing.html

Potashnik, M. and Douglas Adkins (1996). Cost Analysis of Information Technology Projects in Education: Experiences from Developing Countries. *Education and Technology Series, 1*(3). Washington, DC: The World Bank.

Powell, M. (2006). *Rethinking education management information systems: Lessons from and options for less-developed countries.* Infodev Working Paper No. 6. Retrieved November 8, 2008, from http://www.infodev.org/en/Publication.504.html

Prell, C. (2003). Community Networking and Social Capital: Early investigations. Journal of Computer-Mediated Communication, 8(3). Retrieved July 3, 2009, from http://jcmc.indiana.edu/vol8/issue3/prell.html

Pretty, J. (1994). Alternative systems of inquiry for a sustainable agriculture. *IDS Bulletin, 25*(2), 37–48. doi:10.1111/j.1759-5436.1994.mp25002004.x

Pringle, I., & Subramanian, S. (Eds.). (2004). *Profiles and experiences in ICT innovation for poverty reduction.* Paris: UNESCO.

Pritchett, L. (1997). Divergence, Big Time. *The Journal of Economic Perspectives, 11*(3), 3–17.

Procter, D. (2005). *Civic Communion: The Rhetoric of Community Building*. Manhattan, KS: Kansas State University Press.

Pryor, J., & Ampiah, J. G. (2003). *Understandings of education in an African village – the impact of ICTs. Report on DFID Research Project Ed2000-88*. London: DFID. Retrieved December 2, 2008, from http://www.dfid.gov.uk/Pubs/files/understandedafricaedpaper52.pdf

Putnam, R. D. (1993). *Making Democracy Work: Civic traditions in modern Italy*. Princeton, NJ: Princeton University Press.

Putnam, R. D. (1995). Tuning in, Tuning out: The strange disappearance of social capital in America. *Political Science and Politics, 28*, 664–683. doi:10.2307/420517

Putnam, R. (2000). *Bowling Alone: The Collapse and Revival of American Community*. New York: Simon & Schuster.

Qiang, C. Z.-W., & Pitt, A. (2003). *Contribution of Information and Communication Technologies to Growth.* Washington: The World Bank. doi:10.1596/0-8213-5722-0

Quaddus, M., & Achjari, D. (2005). A model for electronic commerce success. *Telecommunications Policy, 29*(2-3), 127–152. doi:10.1016/j.telpol.2004.11.009

Quaghebeur, K., & Masschelin, J. (2003, 27-28 February). *Participation making a difference? Critical analysis of the participatory claims of change, reversal and empowerment.* Paper presented at the workshop on Participation: From tyranny to transformation? Exploring new approaches to participation in development. University of Manchester.

Quah, D. (2001). The Weightless Economy in Economic Development. In Pohjola, M. (Ed.), *Information Technology Productivity, and Economic Growth*. Oxford: Oxford University Press.

Quarry, W., & Ramírez, R. (2009). *Communication for another development: Listening before telling*. London: Zed Books.

Quaynor, N. (1997). *Computer networking and accessing the Internet in Ghana: Problems and prospects* (pp. 28–40). Interlibrary Lending and Document Delivery in Developing Countries.

Ramamritham, K., & Bahuman, A. (2005). Developmental Informatics: Potential and Challenges. *Media Asia, 32*(3), 165–167.

Ramdas, K. N. (2008). Philanthrocapitalism in denial. *OpenDemocracy*. Retrieved October 2008 from http://www.opendemocracy.net/article/globalization/philanthrocapitalism_in_denial

Ramírez, R. (2001). Understanding the approaches for accommodating multiple stakeholders' interests. *International Journal of Agricultural Resources. Governance and Ecology, 1*(3/4), 264–285.

Ramírez, R. (2003). Bridging disciplines: The natural resource management kaleidoscope for understanding ICTs. *The Journal of Development Communication*, *14*(1), 51–64.

Ramírez, R. (2007, March-April). Appreciating the contribution of broadband ICT with rural and remote communities: Steppingstones towards an alternative paradigm. *The Information Society*, *23*(2), 85–94. doi:10.1080/01972240701224044

Ramírez, R., & Lee, R. (2008). Service delivery systems for natural stakeholders: Targeting information and communication functions and policy considerations. *Agronomía Colombiana*, *XXV*(2), 357–366.

Ramírez, R., & Richardson, D. (2005). Measuring the impact of telecommunication services on rural and remote communities. *Telecommunications Policy*, *29*, 297–319. doi:10.1016/j.telpol.2004.05.015

Ramírez, R. 2008. A 'meditation' on meaningful participation and engagement. *The Journal of Community Informatics 4*(3). Retrieved 1 June 2009 from http://ci-journal.net/index.php/ciej/article/view/390/424

Ramírez, R., Murray, D., Kora, G., & Richardson, D. (2000). *Evaluation Report: Rural Resources Partnership for Oxford County Library and HRDC*. Guelph, Canada: University of Guelph.

Ransom, J. S. (1997). *Foucault's Discipline*. Durham, NC: Duke University Press.

Rao, M. (2001). *Local Community Networks: The Human Face of the Internet Economy*. Retrieved.

Rathgeber, E. M. (2000). Women, Men, And ICTs in Africa: Why gender is an issue. In E. M. Rathgeber & E. O. Adera (Eds.) (2000) *Gender and the information revolution in Africa* (pp.17-34). Ottawa: IDRC.

Reed, M., Fraser, E., & Dougill, A. (2006). An adaptive learning process for developing and applying sustainability indicators with local communities. *Ecological Economics*, *59*(4), 406–418. doi:10.1016/j.ecolecon.2005.11.008

Reinert, E. S. (2004). *Globalisation, Economic Development and Inequality*. Northampton, UK: Edward Elgar Publishing.

Renfrew, C., & Bahn, P. (2004). *Archaeology: theories, methods and practice*. London: Thames and Hudson.

Research, I. C. T. Africa (2007) Success and failures of ICT projects in Africa. *Research ICT Africa*. Retrieved on November 22, 2007 from www.researchictafric.net

Retrieved October 2008 from http://www.opendemocracy.net/article/globalization/visions_reflections/philanthrocapitalism_after_the_goldrush

Rheingold, H. (2000). *The Virtual Community: Homesteading on the Electronic Frontier* (Revised ed.). Cambridge, MA: The Harvard University Press.

Richardson, D., & Paisley, L. (1998). *The first mile of connectivity: Advancing rural telecommunications through a communication for development approach*. Rome: FAO.

Richardson, D., Ramírez, R., Aitkin, H., & Kora, G. (2002). Sustaining ICTs for rural development. *INASP Newsletter*, *20*, 5.

Roberts, T. R., & Hite, A. B. (Eds.). (2007). *The Globalization and Development Reader; perspectives on Development and Global Change*. Malden, MA: Blackwell Publishers.

Robinson, J. P., Barth, K., & Kohut, A. (1997). Social Impact Research – Personal Computers, Mass Media, and Use of Time. *Social Science Computer Review*, *15*(1), 65–82. doi:10.1177/089443939701500107

Rodino-Colocino, M. (2006). Laboring under the digital divide. *New Media Society, 8 (487).* Sage Publications, London, Thousand Oaks; CA and New Delhi. Retrieved November 29, 2008, from http://nms.sagepub.com/cgi/content/abstract/8/3/487

Roduner, D., & Schläppi, W. (2008). Logical framework approach and outcome mapping: A constructive attempt of synthesis. *Rural Development News*, *2*, 9–19.

Rogers, P. (2008). Using programme theory to evaluate complicated and complex aspects of interventions. *Evaluation, 14*(1), 29–48. doi:10.1177/1356389007084674

Rogers, E. M. (2003). *The diffusion of innovations* (5th ed.). New York: Free Press.

Röling, N. (2003). From causes to reasons: The human dimension of agricultural sustainability. *International Journal of Agricultural Sustainability, 1*(1), 73–88. doi:10.3763/ijas.2003.0108

Röling, N., & Jiggins, J. (1998, September 16-19, 1998). *The soft side of land: An incomplete exploration of the implications of seeing ecological sustainability as emerging from human learning and inter-action*. Paper presented at the Symposium on Adaptive Collaborative Management of Protected Areas. Cornell University and CIFOR.

Room, G. (1995). Poverty and social exclusion: the new European agenda for policy and social research. In Room, G. (Ed.), *Beyond the Threshold: The Measurement and Analysis of Social Exclusion* (pp. 1–9). Bristol: The Policy Press.

Rosanvallon, P. (1998). *A nova questão social* (Bath, S., Trans.). Brasília: Instituto Teotônio Vilela. (Original work published 1995)

Rose, S. (2005). *Lifelines. Life beyond the gene*. London: Vintage.

Rose, G. (2001). *Visual Methodologies: An introduction to the interpretation of visual materials*. Thousand Oaks, CA: Sage Publications.

Rose, J. (1998). Evaluating the contribution of structuration theory to the information sciences discipline. Paper presented at the 6th European Conference on Information Systems, Aix-en-Provence, France.

Rosenau, P. M. (1992). *Post-modernism and the social sciences: insights, inroads, and intrusions*. Princeton, NJ: Princeton University Press.

Rostow, W. (1960). *The Stages of Economic Growth*. London: Cambridge University Press.

Rousseau, J. (1984). *A discourse on inequality*. London: Penguin. (Original work published 1754)

Sachs, J. (2005). *The end of poverty. How we can make it in our lifetime*. London: Penguin Books.

Sachs, J. (2008). *Common wealth. Economics for a crowded planet*. London: Penguin Books.

Sachs, J. D. (2005). *The end of poverty: economic possibilities for our time*. New York: Penguin Press.

Sairosse, T. M., & Mutula, S. M. (2004). Use of cyber cafes: Study of Gaborone city, Botswana. *Program-Electronic Library and Information Systems, 38*(1), 60–66. doi:10.1108/00330330410519206

São Paulo. (2008). *Decreto nº 49.914, de 14 de agosto de 2008*. Diário Oficial da Cidade de São Paulo.

Sardar, Z. (1996). Alt.civilizations.faq: cyberspace as the darker side of the West. In Sardar, Z., & Ravetz, J. R. (Eds.), *Cyberfutures: Culture and Politics on the Information Superhighway* (pp. 14–41). New York: New York University Press.

Saul, J. R. (2009). *The collapse of globalism and the reinvention of the world*. London: Atlantic Books. (Original work published 2005)

Scan-ICT Report. (2007) Status of ICT, usage and exploitation in the Gambia. *Government of the Gambia and UNECA*. Retrieved July 05, 2008 from http://www.scanict-gbos.gov.gm

Schech, S. (2002). Wired for change: The links between ICTs and development discourses. *Journal of International Development, 14*, 13–23. doi:10.1002/jid.870

Schiller, D. (2000). *Digital Capitalism: Networking the Global Market System*. MIT Press.

Schmidt, V. H. (2006). Multiple modernities or varieties of modernity? *Current Sociology, 54*(1), 77–97. doi:10.1177/0011392106058835

Scholfield, J., Eurich-Fulcer, R., & Britt, C. (1994). Teachers, computer tutors, and teaching: The artificially intelligent tutor as an agent for classroom change. *American Educational Research Journal, 31*(3), 579–607.

School of Informatics, University of Edinburgh. (2002). What does Informatics mean? Retrieved September 2009 from http://www.dai.ed.ac.uk/homes/cam/informatics.shtml.

Schramm, W. (1979). *Mass Media and National Development*. Paris: UNESCO.

Schuler, D. (1996). *New Community Networks: Wired for change*. New York: Addison-Wesley.

Schumacher, E. F. (1973). *Small is Beautiful. Economics as people mattered*. New York: Simon and Schuster.

Schuurman, F. (1993). Development Theory in the 1990s, introduction. In Schuurman, F. (Ed.), *Beyond the Impasse. New Directions in Development Theory*. London, New Jersey: Zed Books.

Sciadas, G. (2005a). Infostates across Countries and Over Time: Conceptualization, Modeling, and Measurements of the Digital Divide: View from Practice. *Information Technology for Development, 11*(3), 299–304. doi:10.1002/itdj.20018

Sciadas, G. (Ed.). (2005b). *From digital divide to digital opportunities; measuring infostates for development*. Montreal, Quebec: Claude-Yves Charron. Retrieved May 20, 2009 from http://www.itu.int/ITU-D/ict/publications/dd/material/index_ict_opp.pdf

Scoones, I. (1998). Sustainable rural livelihoods: A framework for analysis. *IDS Working Paper, vol. 72*. Brighton, UK: IDS.

Sein, M. K., & Harindranath, G. (2004). Conceptualizing the ICT Artefact: Toward understanding the role of ICT in national development. *The Information Society, 20*, 15–24. doi:10.1080/01972240490269942

Sein, M. K., & Harindranath, G. (2004). Conceptualising the ICT artifact: towards understanding the role of ICT in national development. *The Information Society, 20*(1), 15–24. doi:10.1080/01972240490269942

Selener, D. (1997). *Participatory action research and social change*. Quito, Ecuador: Global Action Publications.

Selwyn, N., Gorard, S., & Furlong, J. (2005). Whose Internet is it Anyway? Exploring Adults' (Non)Use of the Internet in Everyday Life. *European Journal of Communication, 20*(1), 5–26. doi:10.1177/0267323105049631

Sen, A. (2001). *Development as freedom*. Oxford: University Press.

Sen, A. (2006). *Identity and violence: the illusion of destiny*. New York: W.W. Norton.

Sen, A. (2005). Human Rights and Capabilities. *Journal of Human Development, 6*(2), 151–166. doi:10.1080/14649880500120491

Sen, A. (1992). *Inequality Reexamined*. Oxford: Oxford University Press.

Sen, A. (1997a). *Development and Thinking at the Beginning of the 21st Century. STICERD – Development Economics Papers 02. Suntory and Toyota International Centres for Economics and Related Disciplines*. LSE.

Sen, A. (1997b). *What's the Point of a Development Strategy. STICERD – Development Economics Papers 03. Suntory and Toyota International Centres for Economics and Related Disciplines*. LSE.

Sen, A. (1999). *Development as Freedom*. New York: Anchor Books.

Sen, A. (1985). *Commodities and Capabilities*. Amsterdam: North Holland.

Sen, A. (2002). *Rationality and Freedom*. Harvard: Harvard Belknap Press.

Sengupta, A. (2000). Realizing the Right to Development. [Blackwell Publishers.]. *Institute of Social Studies, 31*, 553–578.

Servaes, J. (Ed.). (2002). *Approaches to Development Communication, Part 1 Development Theory*. Paris: UNESCO.

Servaes, J., & Malikhao, P. (2002). Development communication approaches in international perspective. In Servaes, J. (Ed.), *Approaches to development communication*. Paris: UNESCO.

Servaes, J. (1989). *One World, Multiple Cultures. A new paradigm on Communication for Development*. Leuven: Acco.

Servon, L. J. (2002). *Bridging the digital divide: Technology, community, and public policy*. Oxford: Blackwell Publishing. doi:10.1002/9780470773529

Shaffer, R., Deller, S., & Marcouiller, D. (2006). Rethinking Community Economic Development. *Economic Development Quarterly*, *20*(1), 59–74. doi:10.1177/0891242405283106

Shanin, T. (1997). The idea of progress. In Rahnema, M., & Bawtree, V. (Eds.), *The post-development reader* (pp. 65–72). London: Zed Books.

Sheehan, P., Pappas, N., Tikhomirova, G., & Sinclair, P. (1995). *Australia and the Knowledge Economy: An Assessment of Enhanced Economic Growth through Science and Technology*. Melbourne: Centre for Strategic Economic Studies, Victoria University.

Shulha, L. M., & Cousins, J. B. (1997). Evalution use: theory, research, and practice since 1986. *Evaluation Practice*, *18*(3), 195–208. doi:10.1016/S0886-1633(97)90027-1

Shulman, L. S. (1986). Those who understand: Knowledge growth in teaching. *Educational Researcher*, *15*(2), 4–14.

Simpson, L. (2005). Community Informatics and Sustainability: Why Social Capital Matters. Journal of Community Informatics, 1(2). Retrieved July 3, 2009, from http://ci-journal.net/ index.php/ciej/articles/view/184/132

Singer, P. (1972). Famine, affluence, and morality. *Philosophy & Public Affairs*, *1*(3), 229–243.

Singer, H. (1970). Dualism Revisited: A new approach to problems of the dual society. *The Journal of Development Studies*, *7*(1), 60–75. doi:10.1080/00220387008421348

Smillie, I. (1999). Narrowing the digital divide: Notes on a global Netscape. Retrieved December 28, 2008 from http://www.unites.org/html/resource/smillie/smillie0.htm

Smith, A. (2003). *The wealth of nations* (Classic, B., Ed.). Kruger, AB. (Original work published 1776)

Snowden, D. (2005). *Multi-ontology sense making: A new simplicity in decision making*. Singapore: The Cynefin Centre.

Snyman, M., & Snyman, M. M. M. (2003). Getting information to disadvantaged rural communities: the centre approach. *South African Journal of Library and Information Science*, *69*(2), 95–107.

Social Inclusion Board, Government of South Australia. (n.d.). Retrieved March 2008 from http://www.socialinclusion.sa.gov.au/

Solé, C., & Parella, S. (2005). *Negocios étnicos: Los comercios de los inmigrantes no comunitarios en Cataluña*. Barcelona: Fundación CIDOB.

Solow, R. (1987). We'd better watch out. *New York Times Book Review*. New York: 12 July.

Sorj, B. (2005). *Civil Societies North-South Relations: NGOs and Dependency*. Sao Paulo, Brazil: Edelstein Center for Social Research.

Souter, D. (2004). ICT and Economic Growth in Developing Countries, Part I. *The DAC Journal*, *5*(4).

Spirakis, P., Manolopoulos, C., & Efstathiadou, R. (2008, May). *The socioeconomic aspects of digital divide*. Proceedings of the Bridging the digital divide in rural communities: practical solutions and policies Conference. Athens, Greece.

Srinivasan, R. (2006). Where information society and community voice intersect. *The Information Society*, *22*, 355–365. doi:10.1080/01972240600904324

Statistics Sweden. (2004). Use of Computers and the Internet by Private Persons in 2004. Stockholm: Statistiska centralbyrån.

Stenbacka, R. (2001). Microeconomic Policies in the New Economy. *WIDER Discussion Paper No. 73*. Helsinki: WIDER/UNU.

Stewart, J. (2007). Local Experts in the Domestication of Information and Communication Technologies. *Information Communication and Society*, *10*(4), 547–569. doi:10.1080/13691180701560093

Steyn, J. (2006, October). *Community memory and ICT in a developing economy*. Paper presented at the Constructing and Sharing Memory: Community Informatics conference, Community Informatics Research Network, Prato, Italy.

Steyn, J. (2007, December). *e-Post: networking remote areas*. Paper presented at the 2007 Community Informatics Research Network conference, Prato, Italy.

Steyn, J. (2008). Book series, in IDIA International Development Informatics Association. Retrieved September 2009 from http://www.developmentinformatics.org/projects/book/index.html.

Steyn, J. (2009). ePost: Networking Remote Areas. In L. Stillman, G. Johanson, & R. French (Eds.), *Communities in Action: Papers in Community Informatics* (pp. 60-67). Cambridge Scholars Publishing.

Stiglitz, J. E. (2002). *Globalization and its discontents*. London: Penguin.

Stiglitz, J. E. (2006). *Making globalization work*. London: Penguin Books.

Stiglitz, J. E. (2003). *Globalization and Its Discontents*. New York: W.W. Norton.

Stiglitz, J. (2007). *Making Globalization Work*. New York: Norten.

Stillman, L., Johanson, G., & French, R. (Eds.). (2009). *Communities in Action: Papers in Community Informatics*. Cambridge Scholars Publishing.

Stiroh, K. J., & Jorgenson, D. (2000a). Raising the Speed Limit: U. S. Economic Growth in the Information Age. *BPEA*, *1*, 125–211.

Stiroh, K. J. (2002). ICT Spillovers Driving The New Economy? Federal Reserve Bank of New York. *Review of Income, Wealth Series*, *48*(1).

Stivers, R. (1999). *Technology as magic. The triumph of the irrational*. New York: Continuum Publishing.

Stoecker, R. (2005). *Research Methods for Community Change: A Project-based Approach*. Thousand Oaks: Sage.

Stoll, K., Menou, M., Camacho, K., & Khellady, Y. (2002). *Learning about ICTs' role in development: A framework towards a participatory, transparent and continuous process*. Ottawa: IDRC.

Stowasser, B. (1998). Gender issues and contemporary Qur'an interpretation. In Haddad, Y. Y., & Esposito, J. L. (Eds.), *New Islam, gender and social change* (pp. 30–44). New York: Oxford University Press.

Straton, J. (1997). Cyberspace and the globalization of culture. In Porter, D. (Ed.), *Internet Culture* (pp. 253–276). Routledge.

Straub, D. W. (1994). The effect of culture on IT diffusion: email and fax in Japan and the U.S. *Information Systems Research*, *5*(1), 23–27. doi:10.1287/isre.5.1.23

Streeck, W. & Yamamura, Kozo (Eds.). (2001). *The origins of nonliberal capitalism: Germany and Japan in comparison*. Ithaca, NY: Cornell University Press.

Streeter, T. (1999). "That deep romantic chasm": libertarianism, neoliberalism, and the computer culture. In Calabrese, A., & Burgelman, J. C. (Eds.), *Communication, Citizenship, and Social Policy: Re-Thinking the Limits of the Welfare State* (pp. 49–64). Rowman & Littlefield.

Strigel, C., Ariunaa, L., & Enkhjargal, S. (2007). *Where desert meets technology: Findings from ICT in education initiatives in rural schools in Mongolia*. Washington, DC: RTI International.

Strover, S., Chapman, G., & Waters, J. (2004). Beyond community networking and CTCs: access, development and public policy. *Telecommunications Policy*, *28*(7/8), 465–485. doi:10.1016/j.telpol.2004.05.008

Sunkel, O. (1972, April). "Big Business" and "Oependencia.". *ForeignAffairs (Council on Foreign Relations)*, 517–531.

Sweeney, S. (2007). Globalization, Multiple Threats and the Weakness of International Institutions: a Community-centred Response. In Clay, C. J., Madden, M., & Potts, L. (Eds.), *People and Places: an Introduction towards Understanding Community*. New York: Palgrave Macmillan.

Szajna, B. (1996). Empirical evaluation of the revised technology acceptance model. *Management Science*, *42*(1), 85–92. doi:10.1287/mnsc.42.1.85

Tait, R. (2006). Iran bans fast Internet to cut West's influence. *The Guardian*, October 18th, 2006. Retrieved May 25, 2008 from http://www.guardian.co.uk/technology/2006/oct/18/news.iran

Tallon, P. P., & Kraemer, K. L. (1999). IT and Economic Development: Ireland's Coming of With Lessons for Developing Countries. In *Proceedings of Hawaii International Conference on System Science*, Jan, 5-8.

Tapscott, D. (1996). *The Digital Economy: Promise and Peril in the Age of Networked Intelligence*. New York: McGraw-Hill.

Taylor, S., & Todd, P. (1995). Understanding information technology usage: a test of competing models. *Information Systems Research*, *6*(2), 144–176. doi:10.1287/isre.6.2.144

Telecentre.org. (2009). From the Ground Up: The Evolution of the Telecentre Movement. Retrieved July 3, 2009, from http://ebook.telecentre.org/

Thatcher, A., & Ndabeni, M. (2005). HCI accessibility guidelines and illiteracy: developing a model of illiteracy and engagement with technology. *Ergonomics SA*, *17*(1), 13–24.

The Gambia Population Secretariat (2005). Unpublished Draft Population Policy obtained from the Population Secretariat.

Thompson, R. L., Higgins, C. A., & Howell, J. M. (1991). Personal computing: toward a conceptual model of utilization. *Management Information Systems Quarterly*, *15*(1), 124–143. doi:10.2307/249443

Thompson, J., & Guijt, I. (1999). Sustainability indicators for analysing the impacts of participatory watershed management programmes. In Hinchcliffe, F., Thompson, J., Pretty, J., Guijt, I., & Parmesh, S. (Eds.), *Fertile ground: The impacts of participatory watershed management* (pp. 13–31). London: IT Publications.

Tinio, V. L. (2003). *ICT in education*. Bangkok: UNDP-Asia Pacific Development Information Programme (APDIP). Retrieved September 1, 2008, from http://www.apdip.net/publications/iespprimers/eprimer-edu.pdf

Tong, D. (2001). Cybercolonialism: Speeding along the superhighway or stalling on a beaten track. In Ebo, B. (Ed.), *Cyberimperialism? Global Relations In The New Electronic Frontier* (pp. 1–6). Westport, CT: Praeger.

Tongia, R., & Wilson, E. J. (2007). Turning Metcalfe on his head: the multiple costs of network exclusion. *Telecommunications Policy Research Conference* (TPRC). September 2007. Retrieved October 2008 from http://web.si.umich.edu/tprc/papers/2007/772/TPRC-07-Exclusion-Tongia&Wilson.pdf

Tongia, R., Subrahmanian, E., & Arunachalam, V. (2005). *Information and communications technology for sustainable development*. Bangalore: Allied Publishers Pvt. Ltd. Retrieved on November 02, 2008 from http://www.cstep.in/docs/ict4sd.pdf

IFPRI. (2006). *Information and Communication Technologies for Development and Poverty Reduction: The Potential of Telecommunications* (Torero, M., & von Braun, J., Eds.). Washington, DC: Johns Hopkins University Press and IFPRI.

Toyama, K., Dias, M.B. (2008). Guest Editors' Introduction: Information and Communication Technologies for Development. *Computer*.

Trucano, M. (2005). *Knowledge maps: ICT in education*. Washington, DC: infoDev / World Bank. Retrieved November 22, 2008, from http://www.infodev.org/en/Publication.8.html

Turkle, S. (1984). *The second self: Computers and the human spirit*. New York: Simon & Schuster.

Turkle, S. (1995). *Life on the screen: Identity in the age of the Internet*. New York: Simon & Schuster.

Underwood, J. (2004). Research into information and communication technologies: Where now? *Technology, Pedagogy and Education, 13*, 135–145. doi:10.1080/14759390400200176

UNDP. (1998). Information and Communication Technologies for Development. UNDP, New York. Retrieved June 20, 2005 from http://www.undp.org/info21/

UNDP. (2005) *UN Millennium Project: Investing in Development: A practical Plan to Achieve the Millennium Development, Goals. Overview*. Washington, D.C. UNDP. Retrieved June 20, 2008 from http://www.unmillennium-project.org/reports/index_overview.htm van Dijk, Jan (2005). Pitfalls of a Metaphor. Retrieved November 10, 2008 from http://www.gw.utwente.nl/vandijk/research/digital_divide/Digital_Divide_overigen/a_framework_for_digital_divide.doc/

UNESCO Institute of Statistics. (2006). *ICTs and education indicators: Suggested core indicators based on meta-analysis of selected International School Surveys*. Montreal: UNESCO Institute of Statistics. Retrieved October 15, 2008, from http://www.itu.int/ITU-D/ict/partnership/material/ICT_Education_Paper_Nov_2006.pdf

United Nation, Millennium Development Goals. (n.d.). Retrieved January 2009 from www.un.org/millenniumgoals

United Nations Conference on Trade and Development (UNCTAD), (2006). *The Digital Divide Report: ICT Development Indices.* United Nations, New York and Geneva. Retrieved June 20, 2008 from http://www.unctad.org/en/docs/sdteecb20061_en.pdf

United Nations Conference on Trade and Development (UNCTAD) (2008). *Information Economy Report; Science and Technology for Development; the new paradigm for ICT.* UNITED NATIONS New York and Geneva. Retrieved June 28, 2008 from: http://www.unctad.org/en/docs/sdteecb20071_en.pdf

United Nations Development Programme. (2000). Millennium Development Goals (MDG). Retrieved September 2009 from http://www.undp.org/mdg

United Nations Development Programme. (2001). *Evaluation Unit. Essentials: Information and Communications Technology for Development. Synthesis of Lessons Learned, no 5. September*. New York: UNDP.

United Nations. (UN) (2000), Millennium Development Goals (MDGs). Retrieved October 2009 from http://www.un.org/millenniumgoals/

Unwin, T. (Ed.). (2009). *ICT4D*. Cambridge: Cambridge University Press.

Unwin, T. (2009a). Blog. Retrieved October 2009 from http://unwin.wordpress.com/2009/04/27/reflections-on-ict4d-the-british-council-manchester

USK. (2008). Statistics by time. Retrieved July 3, 2009, from http://www.usk.stockholm.se/ internet/omrfakta/tabellappl.asp?omrade=0&appl=Tidserier&resultat=Andel&sprak=eng

Van der Vyver, A. G. (2006). Personal communication.

Van Dijk, J. A. G. M. (2006). *The network society. Social aspects of new media*. London: Sage.

Van Dijk, J. A. G. M. (2005). *The Deepening Divide: Inequality in the Information Society*. Thousand Oaks, CA: SAGE Publications.

van Dijk, J. (2001, 15-17 November). *The ideology behind "closing digital divides": Applying static analysis to dynamic gaps.* Paper presented at the IAMCR/ICA Symposium on the Digital Divide. Austin, Texas: University of Texas.

Van Leeuwen, G., & Vander Weil, H. (2003). Spillover Effects of ICT, *CPB Report*, 2003/3.

Van Rossem, R. (1996). The world-system paradigm as general theory of development: A cross-national test. *American Sociological Review*, *61*, 508–527. doi:10.2307/2096362

Van Wart, M., Rahm, D., & Sanders, S. (2000). Economic Development and Public Enterprise: The Case of Rural Iowa's Telecommunications Utilities. *Economic Development Quarterly*, *14*(2), 131–145. doi:10.1177/089124240001400201

Varian, H. (2006, January 12). American Companies Show an Edge in Putting Information Technology to Work. *New York Times*. Retrieved August 10, 2007 from http://www.topics.mytimes.com/top/reference/timestopics/people/v/hal_r_varian/index.html/?offset¼10& Vehovar, V., Sicherl, P., H¨using, T., & Dolnicar, V. (2006). Methodological Challenges of Digital Divide Measurements. *The Information Society*, 22, 279–290.

Vehovar, V., Sicherl, P., Hüsing, T., & Dolnicar, V. (2006, November-December). Methodological challenges of digital divide measurements. *The Information Society*, *22*(5), 279–290. doi:10.1080/01972240600904076

Veiga, J. F., Floyd, S., & Dechant, K. (2001). Towards modelling the effects of national culture on IT implementation and acceptance. *Journal of Information Technology*, *16*(3), 145–158. doi:10.1080/02683960110063654

Venkatesh, V. (1999). Creation of favorable user perceptions: exploring the role of intrinsic motivation. *Management Information Systems Quarterly*, *23*(2), 239–260. doi:10.2307/249753

Venkatesh, V., & Brown, S. A. (2001). A longitudinal investigation of personal computers in homes: adoption determinants and emerging challenges. *Management Information Systems Quarterly*, *25*(1), 71–102. doi:10.2307/3250959

Venkatesh, V., & Davis, F. D. (2000). A theoretical extension of the technology acceptance model: four longitudinal field studies. *Management Science*, *46*(2), 186–204. doi:10.1287/mnsc.46.2.186.11926

Venkatesh, V., & Morris, M. G. (2000). Why don't men ever stop to ask for directions? Gender, social influence, and their role in technology acceptance and usage behavior. *Management Information Systems Quarterly*, *24*(1), 115–139. doi:10.2307/3250981

Venkatesh, V., Morris, M. G., Davis, G. B., & Davis, F. D. (2003). User acceptance of information technology: toward a unified view. *Management Information Systems Quarterly*, *27*(3), 424–478.

Vishwanath, A. (2003). Comparing online information effects. *Communication Research*, *30*(6), 579–598. doi:10.1177/0093650203257838

Vodafone, (2005). Africa: Impact of Mobile Phones: Moving the debate forward. *Mobile Phone Paper Series* No. 3. Retrieved July 2, 2008 from http://www.vodafone.com/etc/medialib/attachments/cr_downloads.Par.78351.File.tmp/GPP_SIM_paper_3.pdf

Wagner, D. A., & Daswani, C. J. (2006). *Bridges to the Future Initiative: Project update and research report.* Philadelphia, PA: International Literacy Institute, University of Pennsylvania.

Wagner, D. A., Day, B., & Sun, J. (2004). *Information technologies and education for the poor in Africa (ITEPA): Recommendations for a pro-poor ICT4D non-formal education policy. Technical Report*. Philadelphia, PA: International Literacy Institute, University of Pennsylvania.

Wagner, D. A., & Day, B. James, T., Kozma, R.B., Miller, J. & Unwin, T. (2005). *Monitoring and evaluation of ICT in education projects: A handbook for developing countries*. Washington, DC: infoDev/World Bank. Retrieved July 24, 2008, from http://www.infodev.org/en/Publication.9.html

Waisbord, S. (2001). *Family Tree of Theories, Methods and Strategies in Development Communication.* Prepared for The Rockefeller Foundation, May 2001. Retrieved on July 2nd 2008 from: http://www.comminit.com/pdf/familytree.pdf

Waitayangkoon, P. (2004). *Thai project on the uses of low-cost technology in science and mathematics*. Paper presented at the International Conference on Educational Technology, Singapore.

Wakeford, N. (2003). The Embedding of Local Culture in Global Communication: Independent Internet Cafés in London. *New Media & Society*, *5*, 379–399. doi:10.1177/14614448030053005

Wallerstein, I. (2005). After developmentalism and globalization, what? *Social Forces*, *83*(3), 1263–1278. doi:10.1353/sof.2005.0049

Wallerstein, I. (1974). *The Modern World System*. Academic Press.

Wallerstein, I. (1979). *The Capitalist World Economy*. Cambridge: Cambridge University Press.

Walsham, G., & Sahay, S. (2006). Research on Information Systems in Developing Countries: Current Landscape and Future Prospects. *Information Technology for Development*, *12*(1), 7–24. doi:10.1002/itdj.20020

Warscahuer, M. (2003). Dissecting the "Digital Divide": A Case Study in Egypt. *The Information Society*, *19*, 297–304. doi:10.1080/01972240309490

Warschauer, M. (2004). *Technology and social inclusion*. Cambridge, Mass: MIT Press.

Warschauer, M., Knobel, M., & Stone, L. (2004). Technology and Equity in Schooling: Deconstructing the Digital Divide. *Educational Policy*, *18*(4), 562–588. doi:10.1177/0895904804266469

Warschauer, M. (2006). *Tecnologia e inclusão social: a exclusão digital em debate* (Szlak, C., Trans.). São Paulo: Senac. (Original work published 2003)

Warschauer, M. (2002). Reconceptualizing the digital divide. *First Monday*, *7*(7). Retrieved October 2009 from http://firstmonday.org/htbin/cgiwrap/bin/ojs/index.php/fm/article/view/967/888

Waverman, L., Meschi, M., & Fuss, M. (2005). The Impact of Telecoms on Economic Growth in Developing Countries. *The Vodafone Policy Paper Series*, 3, 10-23. Retrieved July 2, 2008 from http://www.vodafone.com/etc/medialib/attachments/cr_downloads.Par.78351.File.tmp/GPP_SIM_paper_3.pdf

Weick, K. (1995). *Sense making in organizations*. Thousand Oaks, CA: Sage.

Weil, M. M., & Rosen, L. D. (1998). *TechnoStress: Coping With Technology @work, @home, @play*. New York: John Wiley & Sons.

Wellman, B. (1986). Are Personal Communities Local? A Dumpterian Reconsideration. *Social Networks*, *18*(3), 347–354.

Wenger, E. (1998). *Communities of Practice: Learning, Meaning and Identity*. Cambridge: Cambridge University Press.

Wenger, E. (2000). Communities of practice: The key to knowledge strategy. In Lesser, E., Fontaine, M., & Slusher, J. (Eds.), *Knowledge and communities* (pp. 3–51). Boston: Butterworth Heinemann. doi:10.1016/B978-0-7506-7293-1.50004-4

Werle, R. (2005). *The Dynamics of the digital divide*. Retrieved June 15,2008 from http://www.mpi-fg-koeln.mpg.de/people/we/Links/Digital%20Divide%20Dynamics.pdf

West, A. R. (2006). Related Dangers: The issue of Development and Security for Marginalized Groups in South Africa. *The Journal of Community Informatics*, *2*(3). Retrieved December 21, from http://www.ci-journal.net/

Wheeler, D. (2001). New technologies, old culture: A look at women, gender and the Internet in Kuwait. In Ess, C., Sudweeks, F., & Herring, S. C. (Eds.), *Culture, technology, communication: Towards an intercultural global village* (pp. 187–212). Albany, New York: Sunny Press.

White African. (2009). Spray against caterpillar. Retrieved September 2009 from http://whiteafrican.com/wp-content/uploads/2009/06/Caterpillar.jpg

Widrow, B., Hartenstein, R., & Hecht-Nielsen, R. (2005). 1917 Karl Steinbuch 2005. In *IEEE Computational Intelligence Society Newsletter, August* (p. 5). Eulogy.

Wilhelm, A. G. (2000). *Democracy in the Digital Age: Challenges to Political Life in Cyberspace*. New York: Routledge.

Williams, J., Sligo, F., & Wallace, C. (2004). What a difference it makes? The internet in the Everyday lives of new user families. In *Anzca04: making a difference annual conference of the Australian and New Zealand Communication Association*. University of Sydney. Retrieved September 18, 2008, from http://conferences.arts.usyd.edu.au/

Willis, K. (2005). *Theories and practices of development*. London: Routledge.

Wilson, G., & Heeks, R. (2000). Technology, Poverty and Development. In Allen, T., & Thomas, A. (Eds.), *Poverty and Development into the 21st Century*. Oxford: Oxford University Press.

Wilson-Grau, R. (2006). *Complexity and Evaluation in International Networks*. Toronto, Canada: International Development Research Centre.

Wolf, S. (2001). *Determinants and Impact of ICT Use for African SMEs: Implications for Rural South Africa*. Paper prepared for TIPS Forum.

Wolff, L. (1999). Costa Rica: Are computers in school effective? *TechKnowLogia*. Retrieved December 2, 2008, from http://www.techknowlogia.com/TKL_active_pages2/CurrentArticles/main.asp?IssueNumber=2&FileType=PDF&ArticleID=42

Wong, P.-K. (2002). ICT Production and Diffusion in Asia: Digital Dividends or Digital Divides? *Information Economics and Policy*, *14*(2), 167–188. doi:10.1016/S0167-6245(01)00065-8

Wong, P.-K. (2001). The Contribution of Information Technology to the Rapid Economic Growth of Singapore. In Pohjola, M. (Ed.), *Information Technology, Productivity and Economic Growth*. Oxford: Oxford University Press.

Woodhill, J., & Röling, N. (1998). The second wing of the eagle: The human dimension in learning our way to more sustainable futures. In Röling, N., & Wagemakers, M. (Eds.), *Facilitating sustainable agriculture: Participatory learning and adaptive management in times of environmental uncertainty* (pp. 46–71). Cambridge: Cambridge University Press.

World Bank. (1999). *World Development Report: Knowledge for Development*. New York: Oxford University Press.

World Bank. (2003). *World Development Indicators*. New York: World Bank.

World Bank. (2004). *Monitoring and evaluation: Some tools, methods and approaches*. Washington, DC: The World Bank.

World Bank. (1998). *World Development Report*. World Bank, Washington, DC. Retrieved September 6, 2006 from www.g8.utoronto.ca/summit/2000okinawa/gis.htm

World Bank. (2000). Global development gateway project proposal draft. World Bank: Washington, DC. http://www.worldbank.org/gateway/ (accessed September 2000).

World Bank. (2005). Financing information and communication infrastructure in the Developing World: public and private roles. Draft for discussion. *Global Information and Communication Technologies Department (Gict)*. The World Bank.International Development Agencies: World Bank

World Bank. (2006). World development indicators. Retrieved December 2006 from http://www.worldbank.org/data

World Bank. (2007) World Development Indicators. Retrieved on October 13, 2007 from http://go.worldbank.org/k2ckm78cc0

World Bank. (2009a). Data & Statistics. Retrieved September 2009 from http://web.worldbank.org/WBSITE/EXTERNAL/DATASTATISTICS/0,contentMDK:20535285~menuPK:1192694~pagePK:64133150~piPK:64133175~theSitePK:239419,00.html. Accessed in September 2009.

World Bank. (2009b). Information and Communications for Development 2009: Extending Reach and Increasing Impact. Retrieved September 2009 from http://web.worldbank.org/WBSITE/EXTERNAL/TOPICS/EXTINFORMATIONANDCOMMUNICATIONANDTECHNOLOGIES/EXTIC4D/0,contentMDK:22229759~menuPK:5870649~pagePK:64168445~piPK:64168309~theSitePK:5870636,00.html.

World Development Report2001Forum on Inclusion, Justice and Poverty Reduction, prepared for the World Development Report 2001 Forum on 'Inclusion, Justice and Poverty Reduction'. Retrieved October 2008 from http://www.dfid.gov.uk/pubs/files/sdd9socex.pdf

World Factbook, C. I. A. (2009). Retrieved September 2009 from https://www.cia.gov/library/publications/the-world-factbook/

World Summit on the Information Society. (2005). Declaration of Principles; Building the Information Society: a global challenge in the new Millennium. Geneva, United Nations/International Telecommunications Union. Retrieved September 2009 from http://www.itu.int/wsis/docs/Geneva/official/dop.html

Wright, K. (2008). *What role for evaluation in the context of performance-based management? INTRAC Briefing Paper No. 22*. Oxford, UK: INTRACT.

Wu, J., & Wang, S. (2005). What drives mobile commerce? An empirical evaluation of the revised technology acceptance model. *Information & Management*, *42*(5), 719–729. doi:10.1016/j.im.2004.07.001

Wynn, B. O., Dutta, A., & Nelson, M. I. (2005). *Challenges in program evaluation of health interventions in developing countries*. Santa Monica, CA: RAND.

Xiberras, M. (1996). *As teorias da exclusão: para uma construção do imaginário do desvio* (Rego, J. G., Trans.). Lisbon: Instituto Piaget. (Original work published 1993)

Xue, L., & Sheehan, P. (2002). China's Development Strategy. In Grewal, B. (Eds.), *China's Future in the Knowledge Economy: Engaging the New World*. Beijing: Tsinghua University Press.

Yamamura, K., & Streeck, W. (Eds.). (2003). *The end of diversity? Prospects for German and Japanese capitalism*. Ithaca, NY: Cornell University Press.

Yan, Y. (2002). Managed globalization: State power and cultural transition in China. In P. L. Berger & S. P. Huntington (Eds.), *Many Globalizations: Cultural diversity in the contemporary world* (pp. 19-47). Oxford: Oxford University Press.

Yoon, C., & Kim, S. (2007). Convenience and TAM in a ubiquitous computing environment: the case of wireless LAN. *Electronic Commerce Research and Applications*, *6*(1), 102–112. doi:10.1016/j.elerap.2006.06.009

Young, J. R. (2001). Does 'Digital Divide' Rhetoric Do More Harm Than Good? *The Chronicle of Higher Education - Information Technology*. Retrieved December 20, 2008 from http://chronicle.com/free/v48/i11/11a05101.htm

Yu, L. (2006). Understanding information inequality: Making sense of the literature of the information and digital divides. *Journal of Librarianship and Information Science*, 38 (229). London, Thousand Oaks, CA and New Delhi: Sage Publications. Retrieved November 29, 2008, from http://lis.sagepub.com/cgi/content/abstract/38/4/229

About the Contributors

Giuliana Dettori has been working as researcher for the Italian National Research Council since 1978. She is currently with the Institute for Educational Technology. After an initial involvement in Applied Mathematics, her research interests have been in Educational Technology for most of her professional life. She has been working, in particular, on self-regulated learning, narrative learning and the mediating role of ICT in teaching and learning, in relation to school settings, distance education and teacher training. She is teaching in the PhD program of the University of Genoa "Languages, cultures and ICT", has authored numerous scientific papers, is carrying out editorial collaboration with many international journals and conferences and has been responsible for her institute in several international and national projects.

Donatella Persico is senior researcher at the Institute for Educational Technology of the Italian National Research Council (CNR). She has been active in the field of educational technology, theory and applications, since 1981. Her major interests include instructional design, e-learning, self-regulated learning and teacher training. She is author of educational material and scientific publications of various kinds, including books, educational software, multimedia material and research papers concerning various aspects of educational technology. She was lecturer of educational technology at the post-graduate teacher training school of the University of Genoa from 2000 to 2006. She is co-editor of the Italian Journal Tecnologie Didattiche, collaborates with international and national journals on Educational Technology and has been in charge of several national and international projects, mostly funded by the European Community. She sits on the IFIP (International Federation for Information Processing) WG3.8 Life Long Learning committee as well as several other professional boards at national and international level.

* * *

Kamariah Abu Bakar has dedicated about 30 years of her life pursuing Science Education as a teacher and a teacher educator. Her contributions in teaching and research revolve around Science Education and the utilisation of Information, Communication and Technology in teaching-learning strategies. To date she has been involved in more than 30 research projects in her area of interest some of which are projects funded by the Ministry of Education, Ministry of Higher Education (MOE), Ministry of Science, Technology and Innovation (MOSTI), and Multimedia Development Council. One of the research projects was conducted at the University of Iowa, Iowa City, U.S.A. while she was a Fulbright Scholar during her sabbatical in 1996. She had also been involved for about 20 years as resource person to the

Science Unit, Curriculum Development Centre, MOE; was the Head of the Education and Training Sector for Prioritised Research under MOSTI until the end of the 8th Malaysia Plan. Prof. Dr. Abu Bakar has also held the posts of Deputy Dean of Academic Affairs, and the Dean of the Faculty of Educational Studies while at the same time being the Director of the Multimedia Instructional Design Centre, UPM.

Anita Aguilar received her Ph.D. in the Psychological Study of Education at Temple University in 2008. In 2004 she earned her Master's Degree in Special Education from Temple University. Her research focuses on using self-reflective processes to develop and transfer self-regulated learning behaviors in elementary school students. From 2001 to the present, she has been a teacher in the Philadelphia School District where she has primarily taught 5th grade.

Ottavia Albanese is full professor of Developmental and Educational Psychology at the University of Milan Bicocca, Italy. Her scientific research activity concerns cognitive and metacognitive processes on learning and vocational guidance at the university. Her interest in the metacognitive approach broadened to meta-emotion and emotional competence. She is also involved in research on teacher training. She is author of national and international articles and books.

Maureen S. Andrade is the associate dean of University College at Utah Valley University. She has extensive experience teaching ESL and TESOL courses, and has served as an ESL program administrator and department chair. She is also a former editor of the TESL Reporter. Her professional interests include teaching English for academic purposes, content-based language instruction, program assessment, adjustment and retention issues for international students, and the scholarship of teaching and learning. She is a regular presenter at national and international conferences related to her scholarship interests, and has published on related topics in academic journals and books. Dr. Andrade can be contacted at Utah Valley University, 800 W. University Parkway MS #177, Orem, UT 84058, USA.

Alessandro Antonietti is full professor of Cognitive Psychology and head of the Department of Psychology at the Catholic University of the Sacred Heart in Milano (Italy). He is also the head of the Service of Learning and Educational Psychology, the coordinator of the Cognitive Psychology Lab and of the master in Cognitive Assessment and Training. He is member of the international advisory board of Learning and Instruction and of the Open Education Journal, academic editor of PlosONe and reviewer of various scientific journals. He carried out experimental studies about creativity, problem-solving, decision-making, mental imagery, and analogy. He investigated the role played by media in cognition. He is interested in the applications of cognitive issues in the field of instruction and rehabilitation. He devised tests to assess thinking skills and programs to train cognitive abilities.

Oscar Ardaiz Villanueva is a professor of computer technologies at the Public University of Navarre (Pamplona, Spain). He has taken part in projects of research on application of the computer technologies to the learning. He has organized two International Workshop on Application of Grid Technologies to Collaborative Learning in CCGrid Conference. He has created software tools to support learning UlabGrid and Wikideas. His research has been published in various journals indexed in the ISI of Knowledge, e.g. Future Generation Computer Systems, Applied Intelligence, Lecture Notes in Computer Science or IEEE Distributed Systems Online. He has been a reviewer for conferences, e.g. IEEE Social Computing and International Conference Computer Supported Cooperative Work in

Design. His scientific interest is focused on the application of information technology in educational settings to improve group creativity.

Lucy Barnard-Brak is an assistant professor in the department of Educational Psychology at Baylor University. She received her doctoral degree in educational psychology from Texas Tech University. Her research interests pivot on examining the educational experiences and outcomes of persons with disabilities along with individuals with other special learning needs. This research interest drew her to the area of self-regulated learning as a tenable catalyst to improving the educational experiences and outcomes with respect to both learners with and without disabilities. As a result, dr. Barnard-Brak has a body of scholarly work examining self-regulated learning in a variety of learning contexts (e.g. online learning environment, blended learning environment, across time or longitudinally) across a diverse body of learners (e.g. individuals with ADHD and autism).

Per Bergamin is involved since the early nineties in research in Distance Education and Technology Enhanced Learning. He participated to different European Education & Training and Research projects and also to projects on a national and regional level. He has done several publications in this area. He is founder and Co-Director of the IFeL-Institute (2006). The focus of his research lies on the two topics: Self-Regulated Learning (and Learning success) and quality of Virtual Learning Environments. Another area where he has done some publications is Open Educational Resources. He is member of the advisory committee for eLearning (FeL) of the Rector's Conference of the Swiss Universities of Applied Sciences, member of the advisory board of the Foundation SSAB and director of the program committee of the Swiss Forum for Educational Media. In 2000 he founded an enterprise (brain-tec ltd.) which is active in the area of content management and eLearning services. Today he holds the chair of the executive board of this company.

Matthew Bernacki is a doctoral candidate in the Psychological Study of Education at Temple University in Philadelphia, Pennsylvania. He holds Master's degrees in Experimental Psychology from Saint Joseph's University and Social Work from Temple University. His current research focuses on the facilitation of self-regulated learning behaviors in computer based learning environments, and how such behaviors influence knowledge acquisition. His previous research includes investigating the impact of service-learning coursework on moral development, prosocial attitudes and behaviors and academic achievement. He has also worked as a researcher on the National Institute for Health and Human Development's Study of Early Child Care and the Pathways to Desistance Project which investigates factors related to recidivism amongst adjudicated youth.

Marco Bettoni is Director of Research & Consulting at the Swiss Distance University of Applied Sciences (FFHS). From 1977 to 2005 researcher, engineer and lecturer with industrial and academic organizations in the domains of machine design, engineering education, IT development, knowledge engineering and knowledge management. My main research interest is e-collaboration, especially online community development. I like to learn and share knowledge with everyone. Besides that I am interested in pretty much everything and my areas of knowledge are knowledge technologies (especially knowledge engineering), knowledge management (focusing on human aspects) and knowledge cooperation (focusing on communities of practice), distance cooperation, distance- and e-learning. Finally since 1981 I do research in knowledge theory, especially Radical Constructivism, Operational Methodology and Kantian Criticism. I enjoy reading ancient Greek philosophy, sailing, walking, cooking (Italian, Greek) and dinner with friends, especially in Greek restaurants. Motto: We need to be the change we wish to see in the world.

Canan Blake is a researcher at the Institute of Educational Technology and the co-convener of Computers and Learning Research Group (CALRG) at the Open University, UK. Dr. Blake has a background in Science Education and her recent area of research is Technology Enhanced Learning, focusing on mobile and Computer Supported Collaborative Learning using collaborative discussion tools.

Ellen L. Bunker is the distance learning coordinator in the English as an International Language program at Brigham Young University Hawaii. Dr. Bunker is leading the development and implementation of multi-skill, multi-level distance ESL courses. She has extensive teaching experience in secondary, higher education, and adult education and has consulted on distance education projects with international, military, and corporate organizations. Her educational experiences and consulting has taken her around the world to places such as Tonga, Brunei, Hong Kong, Thailand, Nigeria, and Israel. Dr. Bunker can be contacted at Brigham Young University Hawaii, Box 1940, 55-220 Kulanui Street, Laie, HI 96762, USA.

Nicoletta Businaro is graduate in Psychology at the University of Turin and she is now PhD student at the University of Milan Bicocca (Department of Educational Science). She has conducted interventions with university students to promote self regulated skills. Her studies deal with emotional and metacognitive processes on study learning. She also carried out studies on cognitive aspects and emotions which could promote children's wellbeing.

James Byrnes received his PhD in Developmental Psychology from Temple University in 1985. Prior to his return to Temple as Professor in the Psychological Studies in Education department in 2004, he held academic appointments at the City University of New York (Postdoctoral Fellow, 1985-1986), the University of Michigan, Ann Arbor (Visiting Assistant Professor, 1986-1988) and University of Maryland (Assistant through full Professor; 1988-2004). In 2006, he assumed the position of Associate Dean for Academic Affairs in the College of Education at Temple. He has served as Vice President of the Jean Piaget Society and currently serves as Associate Editor of the Journal of Cognition and Development. He is a Fellow of Division 15 (Educational Psychology) of the American Psychological Association and has received grant funding from the National Science Foundation, National Institutes of Health, and U. S. Department of Education. He has received awards for his teaching and mentoring of undergraduate and graduate students. He has published over 70 books, chapters, or articles on several different areas of cognitive development (e.g., logical reasoning, math learning), but his most recent work has primarily focused on developing two comprehensive theoretical models (one on adolescent decision-making and the other on academic achievement). The model of academic achievement has been specifically designed to provide insight into ways to eliminate or substantially reduce gender, ethnic, and racial gaps in achievement.

Rita Calabrese teaches English Language and Second Language Pedagogy at the University of Salerno. She researches and publishes in the areas of Second Language Acquisition, Interlanguage Analysis, and distance learning. She is the author of a number of publications on the use of English as the medium of instruction in non-language subjects (Content and Language Integrated Learning) and the use of corpora technology to study interlanguage (Insights into the Lexicon-Syntax Interface in Italian Learners' English. A Generative Framework for a Corpus-based Analysis, Aracne, 2008).

Kevin Chin earned his PhD in Educational Psychology from McGill University. His doctoral research focused on the field of human rights education (HRE), with a specific emphasis on educators' personal beliefs and professional practices in both formal and non-formal educational contexts. He is currently a Medical Council of Canada Postdoctoral Fellow with the Centre for Medical Education at McGill University. As principal investigator, he has been awarded a grant from the Royal College of Physicians and Surgeons of Canada to explore the use of video as an evaluation tool for physicians interested in incorporating narrative skills in their clinical teaching and practice. In 2005-2006 he was a Canada-U.S. Fulbright Fellow based at the University of Minnesota Human Rights Center.

Barbara Colombo is researcher of cognitive psychology at the Catholic University of Sacred Heart. Her main research areas are related to multimedia, use of new technologies, creative thinking, naive conceptions, psychology of music. She has participated in numerous research projects, both national and European presenting her works at international conferences and on international scientific journals. She is reviewer of scientific journals. She built several tools and training, both to investigate the cognitive aspects related to multimedia and to enhance the use of specific thinking strategies (visual thinking, creativity, motivation, attentional, etc.).

Jesús De la Fuente is Associate Professor of Educational Psychology, in the Department of Developmental & Educational Psychology, University of Almeria (Spain); Editor-in-Chief of the Electronic Journal of Research in Educational Psychology; Director of the HUM-746 Research Group on Educational & School Psychology; Coordinator for the Master's Degree in Educational Psychology; Science & Technology Manager of the University of Almeria spin-off company, Education & Psychology I+D+i. His research interests include personal self-regulation (motivational and cognitive) when learning in formal contexts, and use of ICTs for diverse educational issues. Dr. De la Fuente is presently working in the area of regulated teaching and self-regulated learning and their assessment, in the university context, in an e-learning format.

Manuela Delfino is a researcher for the Institute for Educational Technology of the Italian National Research Council (ITD-CNR). She earned her PhD in "Languages, Cultures, and Information and Communication Technology" from the University of Genoa (Italy), and a BA in Humanities from the University of Pisa (Italy); she also qualified as a secondary school teacher of humanities. Her major research interests include distance education, teacher training and information literacy. Her current activity is focused on: Computer-Supported Collaborative Learning (CSCL), especially related to the social and emotional dynamics that occur in learning processes mediated by computers; Self-Regulated Learning; approaches to teaching and learning in secondary school, with a special attention to the uses of ICT in education.

Barbara De Marco is research fellow in Developmental and Educational Psychology at the University of Milan Bicocca (Department of Educational Science). Her interests in study and research regards the development of study method and self-regulation in younger and elder students as well as the impact of ICT and collaborative learning technologies on SRL. In January, 2009 she discussed her PhD Thesis about "Promoting a self-regulated approach to study: a research with university students in blended learning situation". Her research fellowship project is about "Self-regulation and motivation at University".

Sisira Edirippulige is the coordinator of e-healthcare programs at the Centre for Online Health (COH), University of Queensland (UQ), Australia. Dr Edirippulige obtained his PhD from the University of Auckland, New Zealand. His main responsibilities involve teaching and coordinating of undergraduate and graduate courses in e-healthcare and the Continuing Professional Development (CPD) courses in telehealth offered at the COH. His research interests include the development, promotion and integration of e-health education and telemedicine applications into the healthcare sector. Before joining the UQ, dr. Edirippulige taught at Kobe Gakuin University in Japan and at the University of Auckland in New Zealand. He has extensive experience in development studies working in a number of countries including Russia, Sri Lanka, South Africa, Japan and New Zealand.

Rylan Egan is an Educational Psychology doctoral candidate at Simon Fraser University. Rylan completed his Master's of Education at Queen's University, and his Bachelor of Business Administration at Brock University. His interests focus on factors influencing self-regulated learning through metacognitive judgment accuracy. Specifically, the role of prejudgment study priming and assessment expectations on heuristic decision-making and metacomprehenison accuracy. He is also a member of the Learning Kit Lab headed by his supervisor Dr. Philip Winne designing learning software to record fine-grained in situ instances of self-regulated learning behavior. He is currently investigating the impact of text vs. situation model study priming on metacomprehension judgment accuracy. He is a sessional instructor in Assessment in Education and a seminar leader in Educational and Instructional Psychology.

Cindy Eggs studied Contemporary History, Political Sciences and Public Law at the University of Fribourg (Switzerland) and at the Universidad Complutense in Madrid. After her studies she carried out a lessons learned study in an international cooperation project in Nicaragua. Next she worked for the Swiss Red Cross in the Strategy Section, especially in the field of Knowledge Management. Now she works in the Research Management Team at the Swiss Distance University of Applied Sciences in Brig. Her main research interests are trust aspects in knowledge management and eCollaboration.

Cath Ellis, BA(Hons), PhD, has been a Principal Lecturer in Humanities (English Literature) at the University of Huddersfield, West Yorkshire, United Kingdom Since 2005. Prior to this she was a Lecturer in English Studies in the Faculty of Arts at the University of Wollongong, New South Wales, Australia. In both of these roles she has been actively involved in the development of innovative curriculum design using flexible and distributed learning. She has led two large teaching and learning projects: one to do with the adoption of distributed learning strategies and one to do with the development of flexibly delivered Continued Professional Development offerings. She has published in the area of postcolonial studies and in the scholarship of teaching and learning. Her main pedagogical research interest is in student agency, critical pedagogy and teaching technology.

Filomena Faiella is a researcher at the University of Salerno and the coordinator of the eLearning_Lab (www.eformazione.unisa.it). She teaches Learning and Instructional Technologies at the Faculty of Education and in master and doctoral courses. She is currently participating, as coordinator or partner, in projects on e-learning models for higher education, e-tutoring and online cooperative learning, collaborative knowledge building, semantic web. She has published a number of books (Progettare la didattica costruttivista – Learning Design, Pensa Editore, 2009) and papers in refereed journals (Teaching in virtual worlds: educational experiences in Second Life, Je-lks, 5(2), 2009) and proceedings (Innovare i processi educativi. Costruzione Collaborativa di Conoscenza multimediale - Innovative educational processes. Collaborative knowledge building, 1st National Conference of Collaborative Knowledge Building Group, ScriptaWeb, 2007).

Eleonora Farina is researcher of Developmental and Educational Psychology at the University of Milan Bicocca. Her scientific research activity concerns cognitive and metacognitive processes involved in self-regulated learning at university. Her interests recently broadened to emotional and motivational aspects in their relationships with typical and atypical development and in learning processes.

Sue Folley, BSc (Hons), PGCE MSc, is an Academic Development Advisor working with academic staff across the University of Huddersfield, to encourage the take up of learning technologies and advise on the effective use of learning technologies. Her previous role was on a project on Distributed Learning working with staff to develop materials in a pedagogically sound way for use with distributed cohorts of students. She has an MSc in Multimedia and E-learning and has previously taught in the FE and HE sector. Sue also works with academic staff on various teaching and learning projects involving learning technologies, assisting with funding applications, and project management. She is currently working towards an EdD researching tutors experience of teaching online.

Bruce Harris is a professor in the Department of the Instructional Design and Technology at Western Illinois University. He has been directly involved in (both in teaching and researching) online learning for over 15 years. He has published refereed articles and presented papers at major conferences on the topic of self-regulated learning, with a particular emphasis on integrating self-regulated learning strategies in online learning environments. Dr. Harris has worked as an external evaluator and consultant for distance learning initiatives, grants involving advanced learning technologies, and large corporations and school districts. He is on the advisory board of a professional organization and on the editorial board of a refereed journal.

Thieme Hennis is a PhD researcher in the Systems Engineering Group of the Faculty of Technology, Policy, and Management of Delft University of Technology. Before joining the university, he was involved in the development of social software to support professionals in their day-to-day tasks at Peers Communities and Filtering (http://aboutpeers.com). In his research, he focuses on reputation as a social currency in online professional communities. More specifically, he researches the influence of reputation on interaction and contributions in communities, and focuses on building reputation standards that support the exchange and creation of knowledge in learning communities, improve relevancy of recommendations, and sustain self-organization.

Yoko Hirata is currently teaching English at Faculty of Engineering, Hokkai-Gakuen University in Sapporo, Japan. She has been teaching EFL for more than 10 years at secondary and tertiary levels. She formerly worked at Otaru University of Commerce as a language advisor, and was involved in the development of a self-access center. Her current research interests include computer-assisted language learning, concordance-based material development, as well as the development of learner autonomy by means of self-access facilities and technologies in language education.

Hermann Koerndle is professor of Psychology of Learning and Instruction at Dresden University of Technology. He has an extensive background in both cognitive psychology and man-machine interaction, specifically in applications of psychological and ergonomic principles for designing technology enhanced tools for fostering the active construction and communication of knowledge. Koerndle received his PhD at Oldenburg University, worked at Regensburg University in the field of applied psychology, then worked at the Technical University of Aachen in the field of man-machine interaction. Since October 1993, he has been at Dresden University where he is currently engaged in (a) research on the factors in and effects of technology-enhanced interactive learning tasks, and (b) research on open-ended authoring tools in various instructional contexts.

Bracha Kramarski, PhD, is a senior researcher in school of education, University of Bar-Ilan, Israel. Her interests include the study of mathematics teaching and learning, metacognition and self-regulation in technology advanced learning environments, and teacher professional development.

William Y. Lan is a professor of Educational Psychology and the Chair of the Department of Educational Psychology and Leadership in the College of Education at Texas Tech University. Dr. Lan received his master's and doctoral degrees in Educational Psychology from the University of Iowa and has been teaching and researching at Texas Tech University since 1990. His research interests include online teaching and learning, self-regulated learning, student motivation and achievement and secondary analysis of national databases on dropout.

Reinhard Lindner is Professor in, and Chair of, the Department of Educational and Interdisciplinary Studies at Western Illinois University. Trained as an applied cognitive scientist, he has published refereed articles and presented papers at major conferences on the topic of self-regulated learning for over a decade. He has taught at the University level for over 20 years. Dr. Lindner currently serves on the editorial board of the Journal of Teacher Education, and serves on the Committee for Research and Dissemination of the American Association of Colleges for Teacher Education (AACTE).

Antonia Lozano earned her undergraduate degrees in Education and in School Psychology, and a PhD in Educational Psychology in 2008 from the University of Almeria, with the thesis "Design and validation of software for assessing self-regulated learning strategies in Early Childhood Education". Dr. Lozano began professional practice as a teacher of early childhood education, and is presently a school psychologist. She is a member of the HUM-746 Research Group on Educational & School Psychology, at the University of Almeria. Her interests include self-regulated learning processes in young children, computer-assisted assessment, and e-learning. She is presently an adjunct professor in the Department of Didactics and School Organization at the University of Almeria.

Laura Maffei graduated from the Department of Mathematics of the University of Pisa, Italy, in 2004, and she obtained her PhD in Telematics and Information Society at the University of Florence, Italy, in 2009. Her research concerns Mathematics Education and in particular issues related to the introduction of ICTs tools in the school practice, difficulties encountered by weak students in Algebra, and the use of ICTs tools in remedial activities in Algebra. She collaborated in some Italian and European Projects aimed at investigating how new technologies could be exploited to enhance teaching and learning processes.

Rohana B. Marasinghe is a Senior Lecturer at the Department of Medical Education and Health Sciences, Faculty of Medical Sciences, University of Sri Jayewardenepura (USJP), Sri Lanka. He obtained his M.B.B.S. (Colombo-Sri Lanka) and M. Phil degree in Medical Education (USJP-Sri Lanka) and is currently undertaking his PhD at the Centre for Online Health, University of Queensland, Australia. His PhD research investigates the potential of telemedicine applications in preventing suicide/self-harm. He is also interested in Medical Education as well as ICT/Innovative technology use in education and teaching.

Maria Alessandra Mariotti graduated in 1974 from the Department of Mathematics at the University of Pisa - Italy, she obtained her PhD in Mathematics Education at the University of Tel Aviv – Israel. Currently she is Professor at the Department of Mathematics and Computer Science of the University of Siena, where she teaches Mathematics and Mathematics Education. Her research concerns Mathematics Education and the main fields of her study have been "Argumentation and Proof" and "Integration of New Technologies in School Practice". She has been the principal investigator of several National and International research projects and coordinator and developer of a research group involving secondary school teachers. Her current interest resides the role of artifacts in the teaching and learning of mathematics. Within this perspective she investigates the role of specific computational environments that enhance teaching and learning processes.

Mark McMahon is Program Director for Creative Industries and Contemporary Arts at Edith Cowan University, Australia, where he co-ordinates the university's courses in Digital Media and Game Design & Culture. He has published research in the application of educational psychology within the field of instructional technology with a focus on the use of ICT to promote cognitive self-regulation. More recently he has conducted research into the design and development of Serious Games as tools to motivate learners and engage them in deep learning approaches such as problem solving and situated cognition, particularly through the use of 3D immersive technologies. He consults widely to industry and is the lead mentor for the Flexible Learning Toolbox project – a multimillion dollar initiative that has overseen the development of national elearning resources for the Vocational Education and Training sector over the last 13 years.

Susanne Narciss is a lecturer and researcher at the Department of Psychology of Learning and Instruction at Dresden University of Technology. She received her PhD from Heidelberg University, and then moved to Dresden University. Her current interests include (a) research on the role of motivation and metacognition for technology-enhanced learning and instruction, and (b) research on the factors in and effects of informative tutoring feedback (ITF). Her work on ITF was considered cutting-edge research by the American Association on Educational Communication and Technology (AECT). Her AECT handbook chapter "Feedback strategies for interactive learning tasks" received the AECT Distinguished Development Award 2007.

Eunice Olakanmi is a holder of MA in ICT and Education from the University of Leeds in the UK. For several years, Eunice worked as a science teacher in Nigeria before she left for further studies in the UK. She completed a Master in Research Method (MRes) from the Open University in 2008. Presently, she is a PhD student at the Open University, Milton Keynes where she is investigating "Self - and Co-Regulation in a Computer Supported Collaborative Learning Environment among Key Stage Three Students". Within the general field of technology enhanced learning, her area of interest is computer supported collaborative learning especially supporting learner with metacognitive behaviours when learning science in computer-based learning environment.

Valerie Osland Paton reports to the Provost as Vice Provost for Planning and Assessment, leading institutional effectiveness in administrative and academic affairs at Texas Tech University (TTU). As a faculty member in the higher education program in TTU's College of Education, her research interests include higher education policy and practice, engagement, planning and assessment. Dr. Paton received a B.A. in American Studies from San Jose State University, a M.A. Counseling Psychology from Santa Clara University, and a PhD in Education from the University of Southern California.

Jane Pilling-Cormick is a researcher with the Hamilton-Wentworth District School Board in Hamilton, Ontario, Canada. She is a prominent researcher in the field of self-directed learning with a proven track record of developing practical research projects and achieving significant, easy to use results. She has published and presented internationally in the area of self-directed learning, online learning, literacy development, and career, co-operative and alternative education. She enthusiastically continues to share realistic and positive messages about how to build opportunities for self-directed learning in educational settings through her role as consultant with Professional Learning & Training in Burlington, Ontario, Canada. Her research interests include developing supportive self-directed learning environments for elementary, secondary and post-secondary education. For more details about her work, visit www.prolt.com or contact Jane at info@prolt.com.

Anthony Piña is Dean of Online Studies for the Sullivan University System. He has both academic and industry experience and has been a professional consultant to Fortune 500 corporations, small businesses, government agencies, university consortia and the U.S. Department of Defense. He has taught at the K-12, college and university levels. Dr. Piña is on the advisory board of two higher education institutions and on the editorial board of two refereed journals. He is the author of the book "Distance Learning and the Institution" and has published several peer-reviewed journal articles and book chapters on instructional technology and distance learning. Dr. Piña is a Past President of the Division of Distance Learning of the Association for Educational Communications & Technology (AECT) and serves currently on the AECT Executive Board.

Antje Proske is a research assistant at the Department of Psychology of Learning and Instruction at Dresden University of Technology. She received her PhD in Psychology (2006) on the development and evaluation of interactive training tasks in academic writing. She was actively involved in several joint projects of the German funding program "New Media in Education" dealing with the question of how to support efficient web-based learning in various instructional contexts (http://studierplatz2000.tu-dresden.de). Her current research interests include the development and experimental investigation of computer-based scaffolding for academic writing and self-regulated learning, as well as the construction of interactive learning tasks.

María Luisa Sanz de Acedo Lizarraga is a professor of cognitive psychology at the Public University of Navarre (Pamplona, Spain). She has directed research projects on metacognition, critical thinking skills and factors that influence decision making. She has published several books such as Psychology: Mind and Behaviour, Individual and Group Creativity, Making Appropriate Decisions and the most recent Cognitive Skills in Higher Education (2010). She has also published the results of her research in journals indexed in the ISI of Knowledge and she is a reviewer of several journals, e.g. Cognition and Instruction, Thinking Skills and Creativity, and Spanish Journal of Psychology. Her scientific interest is focused on implementation of cognitive psychology to improve self-regulation of learning, thinking skills and individual and social creativity.

María Teresa Sanz de Acedo Baquedano is a professor of psychology and education at the Public University of Navarre (Pamplona, Spain) and a teaching assistant at the National University of Distance Education (Tudela, Spain). She has participated in research projects on creativity and problem solving and she is co-author of some books, such as Development of Creativity and Decision Making and Problem Solving. Her research has been published in journals indexed in the ISI of Knowledge, e.g. Learning and Instruction, Educational Studies and School Psychology International. She belongs to the research group Cognitive Models and their Applications, directed by the professor María Luisa Sanz de Acedo Lizarraga. Her scientific interest is focused intervention in the cognitive processes of learning and stimulating creativity.

Eileen Scanlon is Professor of Educational Technology and Associate Director (Research and Scholarship) in the Institute of Educational Technology at the Open University in the UK. Her research interests in the area of Information and Communication Technologies are wide-ranging. Prof. Scanlon is currently directing projects on science learning in formal and informal settings concentrating on the development of an inquiry learning pedagogy and innovative approaches to evaluation.

Jan-Paul van Staalduinen is a PhD researcher in the Systems Engineering Group of the Faculty of Technology, Policy and Management of Delft University of Technology. His research interests are education, social software and serious games. His PhD thesis focuses on the integration of educational theory and game design methods. In 2004 he got his Masters in Systems Engineering & Policy Analysis at the TU Delft, with a thesis on scenarios for education support infrastructures. After graduating he worked as an e-learning consultant for the TU Delft. After that he worked at Unisys Netherlands, as a process analyst and trainer. From 2006 to 2008 he worked for the consultancy firm Verdonck, Klooster & Associates, where he helped governmental bodies with projects on ICT policy and strategy, information management and quality management. He is a fulltime PhD researcher since July 2008. In his spare time he teaches adults presentation and debating skills, for an educational foundation.

I-Pei Tung is a PhD Candidate in McGill University's Educational Psychology program, majoring in Applied Cognitive Science. As an educational psychologist and educational consultant, she has been promoting student-centered learning environments by using technology to better facilitate instruction and assessment to enhance student success and achievement. Her research interests are situated in digital learning environments and include self-regulated learning, video research, and student assessment for learning. She has taught courses at the National University of Tainan and McGill University, and presented at national and international conferences to audiences in the fields of education and psychology.

Wim Veen is full professor of Education & Technology in the Systems Engineering Department of the Faculty of Technology, Policy and Management of Delft University of Technology. He has been involved in institutional strategies for educational innovations at his university. He researches the development of new concepts and models for ICT enabled learning in both the private and public sector. He introduced the concept of Homo Zappiens, a generation of learners with new learning strategies demanding flexible and participative education. Together with a multidisciplinary team of researchers he contributes to e-learning developments in organizations where new learning cultures are emerging for a knowledge intensive and creative economy. Key ideas in his view are: Game developers are excellent educationalists; Learning and working are intimate lovers; Knowledge is communication about meaning.

Vighnarajah graduated from the Universiti Teknologi Malaysia (UTM) in 2004 with a Bachelor of Science in Computer Science and Education (Physics). In 2005, he pursued his Masters in Science (Multimedia-based teaching and learning) at the Faculty of Educational Studies, Universiti Putra Malaysia (UPM) and graduated in 2008. In the same year, he pursued his PhD (Pedagogy) at the same institution. His areas of interest and expertise include self-regulated learning; role of pedagogy in the development of soft skills; application of constructivist pedagogy in technology-enhanced learning environments; instrument development; research methods and qualitative research. He has also held positions as Teaching Assistant for subjects such as Computer Application and Research Methods.

Su Luan Wong graduated from Universiti Pertanian Malaysia (now known as Universiti Putra Malaysia) in 1996 with a Bachelor of Education majoring in Agricultural Science. She joined UPM as a tutor in 1997 and pursued her Masters in Science (Information Technology) at Loughbourough University, UK. In 1999, she pursued her PhD at UPM in Educational Technology and graduated in 2002. Dr. Wong is current an Associate Professor at the Faculty of Educational Studies, Universiti Putra Malaysia. Her areas of interest and expertise include the integration of IT into pre-service teacher & in-service teacher education; teaching & learning with IT; constructivism in IT; instrument development.

Mingming Zhou is now an assistant professor at National Institute of Education in Nanyang Technological University, Singapore. She received her Ph.D. in 2008 from Educational Psychology at Simon Fraser University, and her master degree from Educational Studies in University of Leuven in 2004. Her research focuses on developing innovative research methods for researching the development of cognition, motivation, and self-regulated learning. She is also interested in integrating advanced computer data mining algorithms into educational research and investigating modern technology to enhance teaching and learning experience. She has been teaching courses in educational psychology at various levels.

Simone Ziska studied Media-Psychology at the University of Bern. During her practical course, she participated in the development and realization of a linear web-based learning system. In her master thesis, she evaluated the learning activities of the students in the VLC and demonstrated empirically that learning efficiency was significantly enhanced by the use of the VLC. Now she's working at the Institute for Research in Open-, Distance- and eLearning (IFeL). Her main research interests are self-regulated learning, learning strategies and transfer.

Index

A

B

C

D

E

F

G

H

U

V

W

Y

www.ingramcontent.com/pod-product-compliance
Lightning Source LLC
LaVergne TN
LVHW080924210926
848485LV00004B/723
9781616929015